ACS SYMPOSIUM SERIES **653**

Transition Metal Sulfur Chemistry

Biological and Industrial Significance

Edward I. Stiefel, EDITOR
Exxon Research and Engineering Company

Kazuko Matsumoto, EDITOR
Waseda University

Developed from a symposium sponsored by the
1995 International Chemical Congress of Pacific Basin Societies

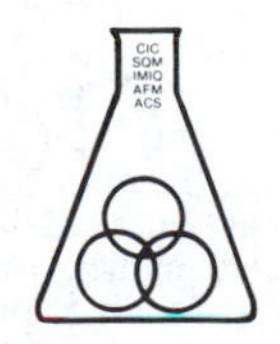

American Chemical Society, Washington, DC

Library of Congress Cataloging-in-Publication Data

Transition metal sulfur chemistry: biological and industrial significance / Edward I. Stiefel, editor, Kazuko Matsumoto, editor.

p. cm.—(ACS symposium series, ISSN 0097–6156; 653)

"Developed from a symposium sponsored by the 1995 International Chemical Congress of Pacific Basin Societies, Honolulu, Hawaii, December 17–22, 1995."

Includes bibliographical references and indexes.

ISBN 0–8412–3476–0

1. Transiton metal sulphur compounds—Congresses.

I. Stiefel, Edward I., 1942– . II. Matsumoto, Kazuko, 1949– III. International Chemical Congress of Pacific Basin Societies (1995: Honolulu, Hawaii) IV. Series.

QD411.8.T73T7 1996
546′.6—dc20

96–45738
CIP

This book is printed on acid-free, recycled paper.

PRINTED IN THE UNITED STATES OF AMERICA

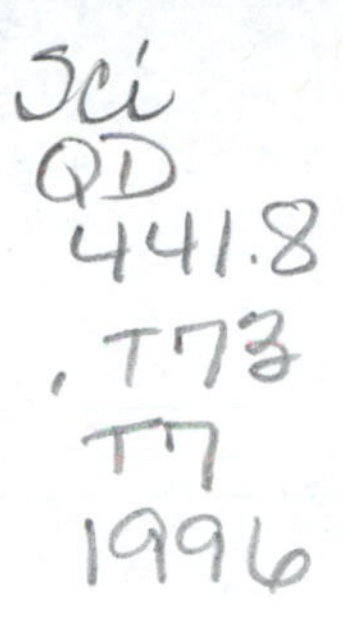

Advisory Board

ACS Symposium Series

Foreword

THE ACS SYMPOSIUM SERIES was first published in 1974 to provide a mechanism for publishing symposia quickly in book form. The purpose of this series is to publish comprehensive books developed from symposia, which are usually "snapshots in time" of the current research being done on a topic, plus some review material on the topic. For this reason, it is necessary that the papers be published as quickly as possible.

Before a symposium-based book is put under contract, the proposed table of contents is reviewed for appropriateness to the topic and for comprehensiveness of the collection. Some papers are excluded at this point, and others are added to round out the scope of the volume. In addition, a draft of each paper is peer-reviewed prior to final acceptance or rejection. This anonymous review process is supervised by the organizer(s) of the symposium, who become the editor(s) of the book. The authors then revise their papers according to the recommendations of both the reviewers and the editors, prepare camera-ready copy, and submit the final papers to the editors, who check that all necessary revisions have been made.

As a rule, only original research papers and original review papers are included in the volumes. Verbatim reproductions of previously published papers are not accepted.

ACS BOOKS DEPARTMENT

Contents

HYDRODESULFURIZATION AND RELATED SYSTEMS

NOVEL STRUCTURES

SMALL-MOLECULE ACTIVATION

Preface

ELEMENTAL SULFUR, "BRIMSTONE", and its basic chemistry were known to the ancients. The simple sulfur anion, S_3^-, is likely responsible for the blue color of the ancient gemstone *Lapis lazuli* (lazurite). These long-known features of sulfur chemistry illustrate some of the variety of redox and aggregation states available to sulfur, even in the absence of a transition metal ion.

Transition metal sulfur compounds also have a lengthy history. Pyrite, the familiar fool's gold, is iron disulfide, and cinnabar, mercuric sulfide, has been used for more than 2,000 years as a red pigment and as a source of mercury. Molybdenite, molybdenum disulfide, now known to have a graphite-like layered structure, has long been appreciated for its soft and flaky texture, which makes it useful in lubrication and in writing implements (*molybdos* is the Greek word for pencil!). In more recent times, sulfide ores, often formed hydrothermally (as occurs presently at deep-sea hydrothermal vents), are used as a source of metal raw materials. For example, molybdenite is the source of metallic molybdenum, which has major uses in steels, catalysts, lubricants, and other applications.

The ubiquity of transition metal sulfur compounds in nature has been augmented by the synthesis of thousands of new transition metal coordination and cluster compounds. The redox and reactive character of the sulfur, combined with that of the transition metal, leads to versatile chemistry that has been exploited both industrially and biologically. We now have in our hands a dazzling array of structurally and electronically interesting compounds whose reactivity is just beginning to be appreciated. The exposition of this newly discovered chemistry, in juxtaposition with industrial uses and biological manifestations, constitutes the main theme of this volume.

The industrial uses of transition metal sulfur compounds are of great current interest and economic importance. Various compounds are important in lubrication, semiconductor applications, and catalysis. Hydrotreating catalysis, that is, hydrodesulfurization, hydrodenitrogenation, and hydrodemetallation, plays a major role in the processing of petroleum. The commercial hydrotreating catalysts contain molybdenum or tungsten sulfides promoted by cobalt or nickel, and much work has been done on the catalysts themselves and on their putative model systems.

Of more recent apprehension is the presence of transition metal sulfur sites at many metalloprotein active centers. These sites are involved with simple electron-transfer reactions and with such noteworthy chemical reactions as oxygen atom transfer (on nitrogen and sulfur oxyanions and heterocyclic molecules); activation of small molecules such as dinitrogen, dihydrogen, and dioxygen; structural recognition of DNA; and important biological metal-sensing, -processing, and -detoxification systems.

Seven of the 10 biologically essential transition metals use sulfur coordination in some or all of their biological manifestations. Iron–sulfur sites are biologically ubiquitous. All molybdenum and tungsten enzymes use sulfur coordination in their respective Mo and W cofactors. Many Cu, Ni, and Zn proteins have sulfur in their metal-coordination spheres. Perhaps the most spectacular of the transition metal sulfur sites found in nature are the unusual clusters of the iron–molybdenum protein of the nitrogen-fixation enzyme (nitrogenase) and the iron–vanadium protein of the alternative nitrogenase. X-ray crystallography has revealed structures that were not fully expected. Model systems have played and will continue to play a key role in the development of our understanding of all of the biological systems.

This volume presents a broad exposition of molecular transition metal sulfur systems in the context of their biological and industrial importance. These systems have inherently interesting and potentially important chemistry and are also of great value for the insights they provide concerning industrial and biological systems.

The first chapter provides an overview of biological and industrial aspects of transition metal sulfur chemistry and highlights some of the key trends that are emerging in the study of these systems. The chapters in the next section deal with biological systems and their models. Here it is seen that the interplay of biological and model system study represents a powerful juxtaposition, leading both to new chemistry and to increased understanding of the biological systems.

Chapters 8 through 11 involve hydrodesulfurization systems and, especially, model molecular systems that react with the thiophenes and benzothiophenes. The reactions studied in the molecular systems provide food for thought concerning the functioning and, possibly, the improvement of industrial hydrotreating catalysts.

In chapters 12 through 18, the emphasis is on novel structures that, to an increasing extent, are being prepared in high yields by systematic approaches. The resultant clusters and complexes reveal intriguing structural, electronic–structural, and reactivity properties and interesting analogies to solid-state and enzymatic structures.

Chapters 19 through 21 show aspects of the reactivity of mononuclear and polynuclear transition metal sulfur compounds. The chemistry is diverse, involving metal-based as well as ligand-based reactions, which have implications for both industrial and biological systems.

By placing the molecular studies in the context of the enzymatic and industrial catalytic systems, we hope that the collected contributions will have a stimulatory effect on all of the areas discussed. This book should be useful to those entering any of these exciting fields and to those seeking an overview of some of the fascinating work that is in progress worldwide.

This volume was developed from a symposium presented at the International Chemical Congress of Pacific Basin Societies in Honolulu, Hawaii, in December 1995. Researchers at the frontiers of transition metal sulfur chemistry from the Pacific Basin and from Europe have contributed to this book. The worldwide representation allows broad and up-to-date coverage of this rapidly expanding area of chemistry.

Acknowledgments

We are grateful to Joyce Stoneking for her assistance in handling many of the administrative details of the symposium. Support for the symposium was provided by the Donors of the Petroleum Research Fund (administered by the American Chemical Society), the ACS Division of Inorganic Chemistry, Inc., the Esso Company (Japan), the Exxon Research and Engineering Company, and the Suzuki Motor Company. We are grateful to Michelle Althuis and Marc Fitzgerald of ACS Books for their assistance during the assembly and production of this volume.

EDWARD I. STIEFEL
Exxon Research and Engineering Company
Clinton Township
Route 22 East
Annandale, NJ 08801

KAZUKO MATSUMOTO
Department of Chemistry
Waseda University
Tokyo 169
Japan

September 3, 1996

OVERVIEW

Chapter 1

Transition Metal Sulfur Chemistry: Biological and Industrial Significance and Key Trends

Edward I. Stiefel

Exxon Research and Engineering Company, Clinton Township, Route 22 East, Annandale, NJ 08801

Transition metal sulfur (TMS) sites in biology comprise mononuclear, and homo- and heteropolynuclear centers in metalloproteins. In industry, the use of TMS systems in lubrication and in hydrotreating catalysis is of great technological significance. Trends in the structure and reactivity of molecular TMS systems include: increasing nuclearity with higher *d*-electronic configuration; structural overlap of molecular and solid state systems; redox reactivity of ligand as well as metal sites; internal redox reactivity; diversification of synthetic strategies; and versatile small molecule activation. Potential relationships between biological, technological, and molecular systems are emphasized.

The chemistry of the transition metals is exquisitely exploited in both biology and industry. In many of these applications, the transition metal is coordinated by sulfur, either in the form of a sulfur-containing organic ligand or in the form of a variety of inorganic sulfur-donor groups. This book deals with the chemistry of the biological and industrial systems, and, in large part, with related molecular systems. This introductory chapter sets the background for the collected papers in this volume.

First, the scope of transition metal sulfur systems in biology is discussed. From ferredoxins, plastocyanins, and zinc fingers to cytochrome P450, hydrogenase, nitrogenase, and cytochrome oxidase, sulfur coordination is necessary for the functioning of numerous biological transition metal centers. These centers encompass seven different transition metals and with nuclearity (number of metal centers) up to eight. In some cases, the role of sulfur involves the modulation of the activity of the transition metal, but often the sulfur ligand itself is involved in substrate binding, acid-base activity, or redox processes crucial to active-site turnover. Much work in biomimetic chemistry is directed at duplicating, or at least imitating, features of the metalloenzyme active-site structure, spectra, magnetism, and reactivity.

The broad scope of transition metal sulfur species in industry also encompasses a number of different metals. The industrial interest includes catalysis, corrosion, lubrication, antioxidancy, and battery technology. Most, but not all, of the activity is associated with solid state systems and heterogeneous catalysis. Molecular species serve as precursors to the solid-state systems and, moreover, may have

0097–6156/96/0653–0002$19.25/0

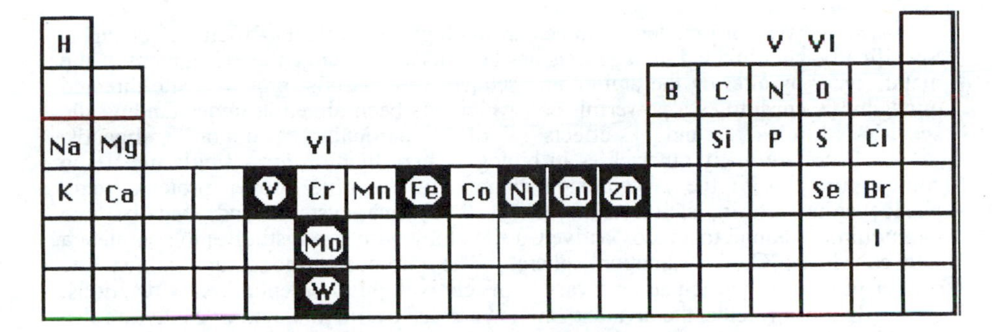

Figure 1. Bioinorganic Periodic Table highlighting those essential elements that use sulfur coordination for some of their biological functions.

catalytic activity of their own or display chemical reactivity relevant to the study of surface active sites.

In addition to the intrinsic interest in the novel chemistry of transition metal sulfur systems, the molecular systems that are the principal subject of this volume, clearly have relevance to our apprehension of the functioning of both the biological and industrial systems. This opening chapter briefly introduces the biological and industrial contexts of transition metal sulfur systems. Then, various trends that come into play (and interplay) in transition metal sulfur chemistry are discussed. We illustrate these trends by referring to work from our own laboratory, to review articles, and to particular chapters in this volume.

Transition Metal Sulfur Systems in Biology

The explosive growth of the field of bioinorganic chemistry is replete with the publication of five new "text" books (*1-5*). A significant component of the interest in transition metal sulfur chemistry stems from the use of transition metal sulfur species as reagents in biochemical, physiological, and pharmacological contexts and, more so, from the presence of a variety of transition metal sulfur sites in proteins and enzymes. The metal-sulfur component imparts critically important reactivity to the biological macromolecule by serving as a (or the) key part of its active site.

In Figure 1, a periodic table template shows the biologically relevant elements that are known to have essential roles in at least one organism. Of the ten transition metals so involved, seven of them, highlighted in the figure, are found to have sulfur coordination in many of their biological occurrences. In some cases, such as molybdenum and tungsten enzymes, *all* known systems involve coordination with sulfur ligands.

The transition metal sulfur sites that occur in metalloenzymes can be classified into two types. The first are mononuclear sites with specific protein or cofactor ligation. The second are polynuclear sites in which sulfur bridges bind together two or more metals.

Mononuclear Sites: The simplest of the mononuclear systems are the rubredoxin proteins (*6-9*). These small proteins (M ≈ 6,000) are involved in electron transfer and contain a single tetrahedrally coordinated iron site (Figure 2a). Four cysteine thiolates from the protein side chains provide the ligation to the iron, which can be either in the ferrous or ferric state. Despite the simplicity of the system, much control is possible over the redox properties of the site. The tetrahedral coordination of Fe in rubredoxin is illustrated in Figure 2a. In Chapter 2, Wedd and co-workers reveal the immense power of modern molecular biology to effect structural changes. Specifically, site-directed mutagenesis has been used to change the evironment of the metal center by altering the amino acid sequence of the host protein. Site-directed mutagenesis constitutes a powerful tool, which has been aimed at understanding the specific roles and effects of particular donor ligands.

There are many other cases in biology where ligation for a single metal site comes purely from the protein side chain. In the blue copper proteins (e.g., plastocyanins, azurin, stellacyanin, etc.) (*10-12*) one cysteine and, usually, one methionine, is bound to a redox-active Cu site along with two histidines to complete a four-coordinate Cu coordination sphere. The cuprous/cupric couple allows the copper proteins to participate in a variety of relatively high-potential redox reactions.

In zinc proteins, the divalent zinc ion can have structural and/or catalytic roles (*13-15*). Since zinc has no redox ability, its catalytic role involves its acid-base or polarizing properties. In zinc finger proteins (*16*), the Zn^{2+} ion plays a structural role. In these crucial DNA-binding and recognizing systems, tetrahedral coordination of zinc is usually provided by two thiolates and two imidazole ligands from, respectively, cysteine and histidine protein side chains. In alcohol dehydrogenase (*17*), four thiolate ligands from cysteine side chains coordinate to tetrahedral zinc, which serves as an organizing center for the protein.

Proteins clearly constitute very versatile 'multidentate ligands.' An interesting and adaptable class of multidentate ligands, which similarly illustrates the propensity of sulfur donors to bind to heavy metals, is given in Chapter 18 by Lindoy et al.

In contrast to the relative simplicity of the rubredoxin, copper, and zinc finger systems, are the mononuclear systems represented by the molybdenum and tungsten cofactors (Moco and Wco, respectively). Enzymes that use Moco play a wide variety of redox roles in plants, animals, and bacteria (*18-24*). Included in this group are aldehyde (including retinal) oxidoreductase, xanthine oxidase and dehydrogenase, sulfite oxidase, DMSO reductase, nitrate reductase, and sulfite oxidase (*18-24*). In the first two of these enzyme types, a terminal sulfido ligand is present in the Mo coordination sphere in the active form of the enzyme in addition to the cofactor ligand. To date, tungsten enzymes have been found mainly in thermophilic bacteria (*25,26*). The cofactors of these enzymes represent an example of a special non-protein ligand, elaborated by nature to bind to a particular transition metal or metals.

The common structure of the molybdenum and tungsten cofactors shown in Figure 3 reveals a pterin-dithiolene unit wherein the dithiolene is the direct ligand to Mo or W. This mode of coordination was inferred by the chemical work of Rajagopalan and co-workers (*27,28*) and recently confirmed crystallographically in the structures of two Mo (*29,30*) and one W enzyme (*31*). The dithiolene coordination of molybdenum and tungsten have been extensively studied by coordination chemists (*32-34*), for whom it is gratifying to see that this interesting ligand has also been selected by Nature. Dithiolene complexes are known for their reversible redox reactivity, undoubtedly important for their biological function. While significant progress has been made in the synthesis of analogs of pterin-dithiolene structures (*35,36*), the complete cofactor has not yet been chemically synthesized. A very common non-protein ligand involves the porphyrin family, with heme iron used extensively in electron-transfer and oxygen-binding proteins, and in oxygen-activating enzyme systems. Of course, the porphyrin itself is a tetranitrogen donor ligand, but, a major determinant of the specific functionality of hemoproteins is the ligation of iron in the non-porphyrin fifth and/or sixth coordination position. Here the ligands often come from one of the sulfur-containing amino acids, cysteine and methionine. In cytochrome *c* (*37*), which undergoes simple electron transfer involving ferrous and ferric states, there is thioether ligation from a single methionine, whereas the heme (of yet unknown function) in bacterioferritin has bis(methionine) coordination (*38*). In contrast to the methionine coordination in these proteins, the heme in cytochrome P450 contains a cysteine thiolate ligand in the position trans to the dioxygen activation site of the

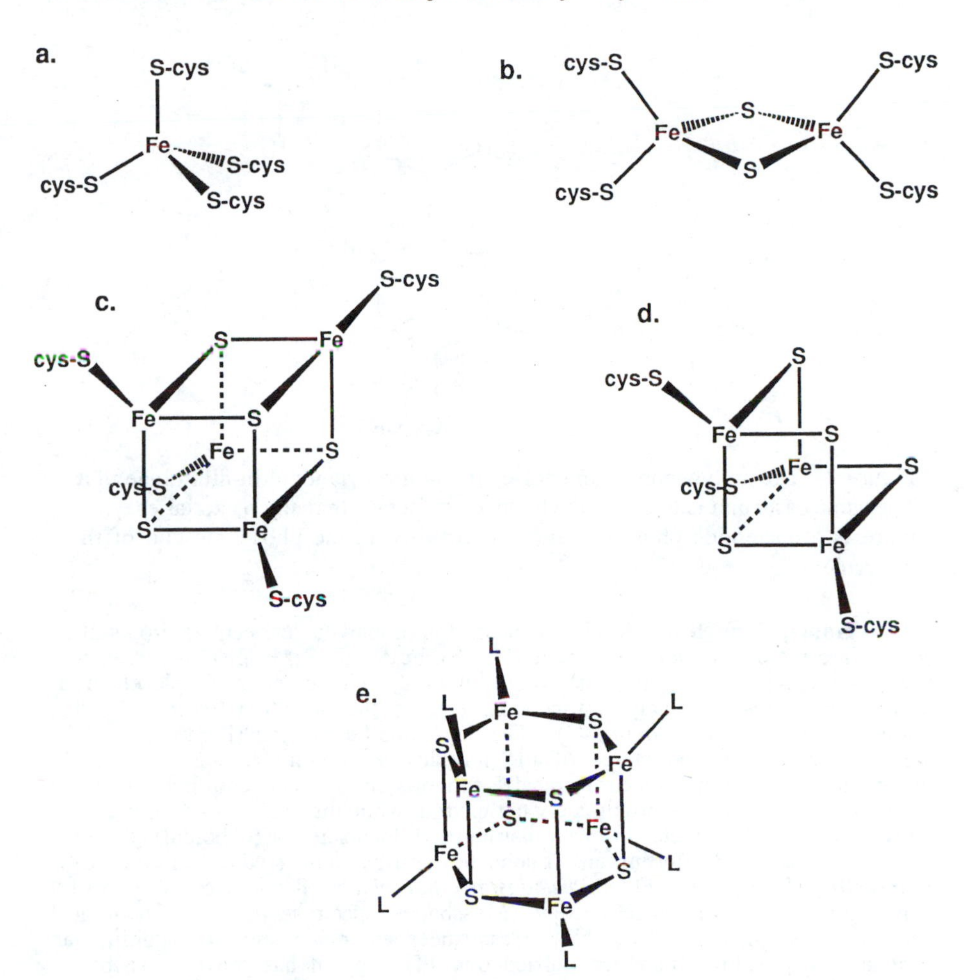

Figure 2. The Fe-S sites found in biological systems. a. The tetrahedral tetrathiolate site of rubredoxins. b. The dinuclear Fe_2S_2 site of ferredoxins. c. The tetranuclear thiocubane structure of ferredoxins (including HiPIPs). d. The trinuclear Fe_3S_4 site typical of enzymes such as aconitase. e. The hexanuclear prismane structure implicated spectroscopically in a number of proteins.

enzyme (*39,40*). This thiolate coordination contributes to the ability of P450 site to 'stabilize' forms of oxygen capable of reacting with substrate C-H bonds (*41*).

Polynuclear Sites. The mononuclear metalloprotein sites discussed above owe their versatility to the specific protein and prosthetic group ligands that bind the metal ions. In multinuclear situations, *added* to this is the presence of varied states of aggregation and geometric arrangements of the metal ions. Both homopolynuclear and heteropolynuclear sites are known.

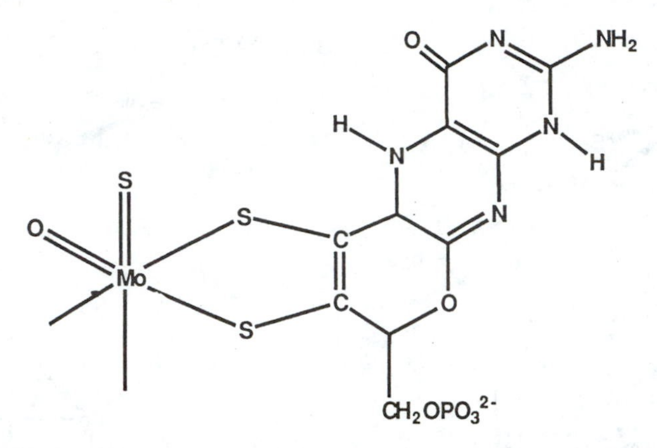

Figure 3. The molybdenum cofactor containing a pyranopterin-dithiolene unit. The same basic unit is found in the tungsten cofactor. In many bacterial enzymes there is a nucleotide phosphate linkage attached to the phosphate end of the molecule.

Homopolynuclear Sites. In the class of homopolynuclear centers, iron sulfide centers are prominent and these include Fe_2S_2, Fe_3S_4, Fe_4S_4, Fe_8S_8, and, probably, Fe_6S_6 cores (*6-8*). These centers, shown in Figure 2, are involved in 'simple' electron-transfer processes, in catalytic functions, and in the detection of iron for the regulation of the metabolism (*42,43*). The Fe_2S_2 and Fe_4S_4 cores (Figure 2b and 2c) are the redox active structures of the ferredoxins, which are relatively small, metabolically ubiquitous electron-transfer proteins (*6-8*). The coordination about individual iron atoms is approximately tetrahedral, while the overall structures can be viewed as complete or partial thiocubane (cuboidal) units.

In Chapter 3, Bertini and Luchinat describe their studies on the Fe_4S_4 ferredoxin (HiPIP) from *Ectothiorhodospira halophila.* They use the powerful combination of NMR spectroscopy, Mössbauer spectroscopy, and theoretical analysis. Assignments of the NMR resonances are made and full Hamiltonian analysis of the spin systems are carried out. Thereby, it has proven possible to establish precisely the identity of the iron atom(s) involved in redox activity and the extent of electron delocalization in the oxidized and reduced systems.

The Fe_3S_4 core structure (Figure 2c) can be viewed as a thiocubane missing a single iron atom. In some of the proteins that have Fe_3S_4 cores, it is possible to add the missing iron atom. This iron-binding ability appears to provide a control mechanism for aconitase (*44*), the fumarate nitrate reduction protein of *Escherichia coli* (*45*), and for the biosyntheses of ferritin and the transferrin receptor, which are regulated by an Fe_3S_4/Fe_4S_4-containing iron-binding protein (iron regulatory element) (*46,47*). Recently, synthetic efforts (*48,49*) have succeeded in duplicating the structure of the Fe_3S_4 core by using multidentate ligands that engender site-specific reactivity in Fe_4S_4 systems, thereby allowing extraction of a single iron atom and formation of the desired Fe_3S_4 core.

Biological occurrences of both six-iron and eight-iron clusters are known. Although at present there is no protein crystal structure for the six-iron cluster, analytical data and comparison with model compounds strongly support its existence and implicate a prismane structure (*50,51*) first reported in synthetic systems (*52*) (Figure 2d). The eight-iron site is crystallographically established in the P-clusters of the iron-molybdenum protein of nitrogenase (*53-56*). As shown in Figure 4, the P-cluster consists of two Fe_4S_4 units fused with a disulfide group (or possibly a single

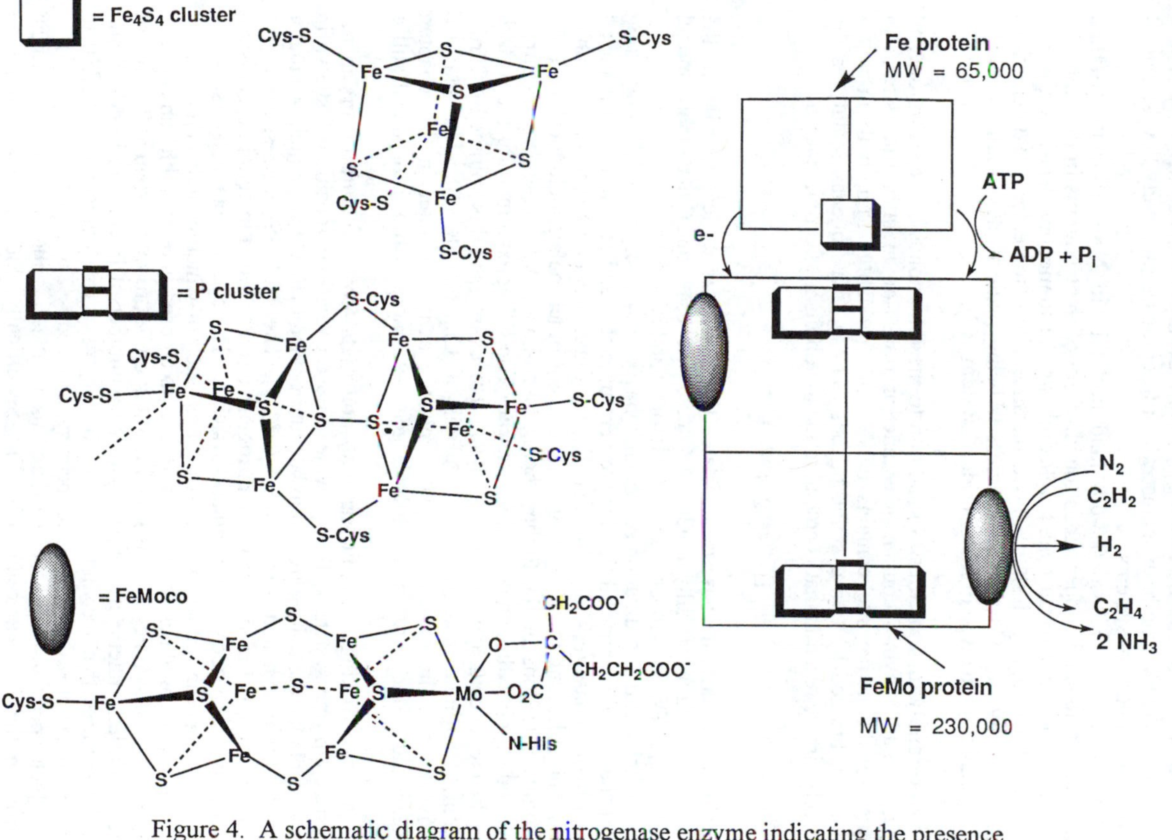

Figure 4. A schematic diagram of the nitrogenase enzyme indicating the presence of an Fe_4S_4 center in the iron protein and two P clusters and two FeMoco clusters in the iron-molybdenum protein.

sulfide ligand), and additionally bridged by two protein cysteine thiolate ligands. In addition to those clusters found in biological systems, a large array of polynuclear Fe-S sites have been synthesized and structurally characterized (*6*). It remains to be seen if more of these will be found in biology. In any event, inorganic chemistry clearly provides an invaluable data base against which to compare putatively new structural types found in biological systems.

In cytochrome oxidase (*57-60*) [and probably in N_2O reductase (*61*)], a bis(cysteinate) bridged dicopper site is found. Both of the Cu atoms in the dinuclear Cu_A unit of cytochrome oxidase are tetrahedral (similar to mononuclear blue Cu) and the bis(thiolate) bridge and short Cu-Cu distance allow for extensive interactions between the individual Cu atoms. Progress is being made in developing synthetic analogs containing this type of metal-bridged system (*62*).

Heteropolynuclear Sites. In addition to these homonuclear systems, several important metalloenzymes contain heteronuclear transition metal sulfide sites. In hydrogenase, the active site contains nickel and (probably) iron, both in sulfur coordination, bridged by two cysteine ligands (*63,64*). Hydrogenase catalyzes the deceptively simple reaction that is crucial to the metabolism of certain bacteria:

$$H_2 \rightleftharpoons 2\,H^+ + 2\,e^-$$

The hydrogenase enzyme has several states that are more or less reactive in the uptake or evolution of dihydrogen (*6,64,65*). These states and their model compounds are described in Chapter 4 by Maroney and co-workers.

Nickel-sulfur coordination compounds have proven very useful in suggesting some of the chemical possibilities for the various states and reactions of the nickel center of hydrogenase (*66-70*). For example, the thiolate-bound Ni sites are capable, through their sulfur atoms, of serving as ligands for other metals including iron (*70*). Interestingly, questions similar to those being asked about hydrogenase, for example, the location of the hydrogen activation sites, are also being examined for the surface sites on heterogeneous hydrotreating catalysts. The chapters in this volume by Curtis, Bianchini, Boorman, and Rakowski-DuBois (Chapters 8, 10, 11, and 16, respectively), and the section on small molecule activation later in this chapter, consider some of the ways in which hydrogen can bind at a transition metal sulfur site.

Another striking heteronuclear transition metal core structure occurs in the nitrogenase enzyme system. This enzyme, whose overall composition is shown in Figure 4, contains, the unique iron-molybdenum cofactor (FeMoco), which is a major part of the dinitrogen activation system (*53-56*). The $MoFe_7S_8$ core structure [or VFe_7S_8 core structure in an alternative nitrogenase (*6,71*) found in certain organisms] is chemically unprecedented in synthetic systems. The overall octanuclear structure consists of two partial thiocubane substructures, each of which contains an open Fe_3 face. The manner in which these Fe_3 faces are juxtaposed is remarkable: an *eclipsed* arrangement of the two Fe_3 faces is found forming an Fe_6 trigonal prism at the core of FeMoco. While synthetic efforts have yielded a number of interesting Fe-Mo-S and V-Mo-S structures that resemble portions of the respective cofactors (*6,72,73*), to date, no complete chemical analog has been synthesized.

Much speculation has been offered as to the possible sites for dinitrogen activation on the FeMoco center (*74-77*). Specifically, the six low-coordination number iron atoms at the core of the structure have been suggested as possible sites wherein multiple binding and hence activation of dinitrogen could occur. The work presented by Dance in Chapter 7 adds the weight of computational chemistry to the discussion and suggests a four-iron binding site for the dinitrogen. It must be stressed, however, that to date there is no hard evidence as to the manner in which dinitrogen binds to the active site or on how it is activated for reduction.

In Chapter 6, Coucouvanis et al. describe some of the clusters that have been used as models for the iron-molybdenum cofactor of nitrogenase. While none of these cluster systems has the stoichiometry or the reactivity of the nitrogenase cofactor, they do show significant (partial) structural overlap with the nitrogenase FeMoco site (*6,73,74*). Moreover, as Coucouvanis et al. demonstrate, reactions with acetylene and hydrazine, both nitrogenase substrates, are catalyzed by the synthetic cluster systems (*78*). Interestingly, in sharp contrast to most of the speculation on the functioning of the enzyme, the model systems clearly suggest the ability of the molybdenum site in thiocubane analogs to bind and activate nitrogenase substrates. The work of Sellmann et al. in Chapter 5 and of Matsumoto et al. in Chapter 15 reveal how hydrazine and diazene can bind to transition metal sulfur sites and give us additional food for thought about the mode in which the FeMoco of nitrogenase may behave.

In addition to nitrogenase, many other bioinorganic systems contain two different subsites that are closely held by sulfur bridges, providing a biologically functional unit. For example, in sulfite reductase a siroheme is bridged to an Fe_4S_4 thiocubane cluster by a thiolate ligand (*79*). Synthetic analogs have been reported for the bridged system (*80*).

A great deal of activity is currently involved in learning how multinuclear biological centers are synthesized in various organisms (*81*). In addition, the work of Dance in Chapter 7 discusses cluster formation and others (*82,83*) are involved in the application of physical theoretical tools to the understanding of molecular and electronic structures and reactivity of the individual redox centers.

In many proteins that catalyze redox reactions, there are multiple redox-active sites that are not simply or directly bridged by coordinated ligands (*84*). In these enzyme systems, long-distance electron transfer (>10 Å) is an important part of the catalytic cycle. The mode of reactivity is clearly analogous to electrochemical systems, with the 'anode' reaction, where oxidation occurs, clearly separated from the 'cathode' reaction, where reduction occurs. This organizational strategy avoids (undesirable) direct contact between the oxidant and the reductant (*85*). Many such enzymes use sulfur-bound metals in one or more of their redox centers. In addition, metal centers coordinated by sulfur ligands may be primed for the activation of small molecules and/or for the facilitation of an electron-transfer pathway to regenerate the active site. Chapter 5 by Sellmann et al. and Chapter 16 by Matsumoto et al. show that insight can be obtained into the chemistry of the active sites in such systems by using combinations of metals and ligands that, while not themselves found in the biological system, have structural and/or electronic features that resemble biological systems.

Clearly, biological systems have been able to utilize a large variety of sulfur ligand types, both organic and inorganic, as well as varied states of metal aggregation and coordination geometries. Among the tasks ahead is to appreciate the *raison d'etre* of these varied constructs in the context of the functional behavior and evolutionary origins of the enzyme systems. The emulation of these active sites constitutes a major thrust of bioinorganic chemistry.

TMS Sites in Industry

Metal-sulfide sites are well known in industrial and commercial contexts. For the purposes of this chapter and this book we focus mostly on the transition metal sulfur based systems that are important in catalysis. Nevertheless, it must be noted that transition metal sulfur systems play important roles in: lubrication (see below); electro- and photocatalysis (Mo-S, Re-S, Ru-S, and Cd-S systems) (*86,87*); corrosion (Fe-S systems) (*88*); battery technology (electrointercalation batteries containing MoS_2) (*89*); photovoltaic materials (Cd-S and Mo-S systems) (*90*); and magnetic resonance imaging (MRI) contrast enhancement agents (chromium sulfide clusters) (*91*).

Lubrication. Lubrication is one of the earliest known uses of transition metal sulfide materials. Solid molybdenum disulfide, which occurs naturally, is a lubricant comparable to graphite in many of its properties (*92,93*). The layered structure of MoS_2 (*94*) is shown in Figure 5. The molybdenum is found in trigonal-prismatic six-coordination between sheets of eclipsed close-packed sulfur atoms. The individual layers of MoS_2 stack in a variety of ways to give the different polytypes of MoS_2. The van der Waals gap between layers is indicative of weak binding that is not strongly directional. Therefore, single layers of MoS_2 slide laterally with respect to one another with minimal resistance. This sliding is considered responsible for the lubrication activity. The soft flaky structure of MoS_2 is obvious upon visual inspection or physical probing of natural MoS_2 crystals.

In motor oils and greases it is possible to use molecular molybdenum sulfur complexes (*95,96*) as precursors for MoS_2. Presumably, the molecular complexes decompose thermally or under shear to produce coatings of MoS_2 on the rubbing surfaces. The coated surfaces have significantly reduced friction coefficients.

Catalysis. The commercial use of transition metal sulfur catalysis in industry is confined at present to heterogeneous systems. Numerous reactions are catalyzed by transition metal sulfur systems and some of these are summarized in Table 1 (*97,98*). While significant research attention has been given to many of these reactions, the major commercial use involves the set of reactions known as hydrotreating.

Hydrotreating is a mainstay of the petroleum and petrochemical industries. Relatively high pressures and high temperatures are used in the hydrogenation of unsaturated molecules including aromatics, and, more importantly, in the removal of sulfur, nitrogen, oxygen, and metal atoms from the petroleum feedstocks (*99-102*). These processes are called, respectively, hydrodesulfurization (HDS), hydrodenitrogenation (HDN), hydrodeoxygenation (HDO) and hydrodemetallation (HDM). The sulfur, nitrogen, oxygen, and metals to be removed are found in organic molecules that are present in the crude oil. Examples of some of these components of crude oil are given in Figure 6.

Greater than 50% of all refinery streams undergo catalytic hydrotreating. The volume of catalyst is very high (*103*), with estimated use for 'western' refineries at 50 x 10^3 ton of catalyst/year valued at roughly \$500 x 10^6 (*103*). In hydrotreating reactions, relatively high pressure of dihydrogen (from 10 to 150 atm) and relatively high temperature (from 320-440°C) are used to assure that applicable kinetic and thermodynamic limitations are overcome (*100*). Typical equations for hydroprocessing reactions are given in Table 1.

The industrial catalysts are generally supported and 'promoted.' The support is usually alumina, although other supports, including carbon, titania, and magnesia, have also been investigated (*99*). The catalysts ordinarily contain combinations of the metals molybdenum or tungsten with nickel or cobalt (*99-102*). Nickel and cobalt are said to promote the activity of molybdenum or tungsten. Curtis et al., in Chapter 8, comment on aspects of the reactivity of the so-called 'CoMoS phase,' postulated as the active site in Co-Mo hydrotreating catalysts.

Despite the fact that combinations of Group VI and Group VIII metals are used in the industrially favored catalyst, in bulk (unsupported) catalysts, analysis of periodic trends reveals that noble metal sulfides, such as those of ruthenium, rhodium, and iridium, are the most reactive binary metal-sulfide systems (*100,104,105*). This result has spawned considerable experimental work (*106-108*) and theoretical studies (*109-111*) attempting to understand the underlying electronic structural reason(s) for the observed periodic trend.

Over the last fifteen years several groups have studied molecular transition metal sulfur systems of relevance to understanding reactions that may occur on surfaces of the heterogeneous catalysts (Chapters 8-11) (*112-120*). Some of these studies, such as those reported by Rauchfuss and co-workers in Chapter 9 and by

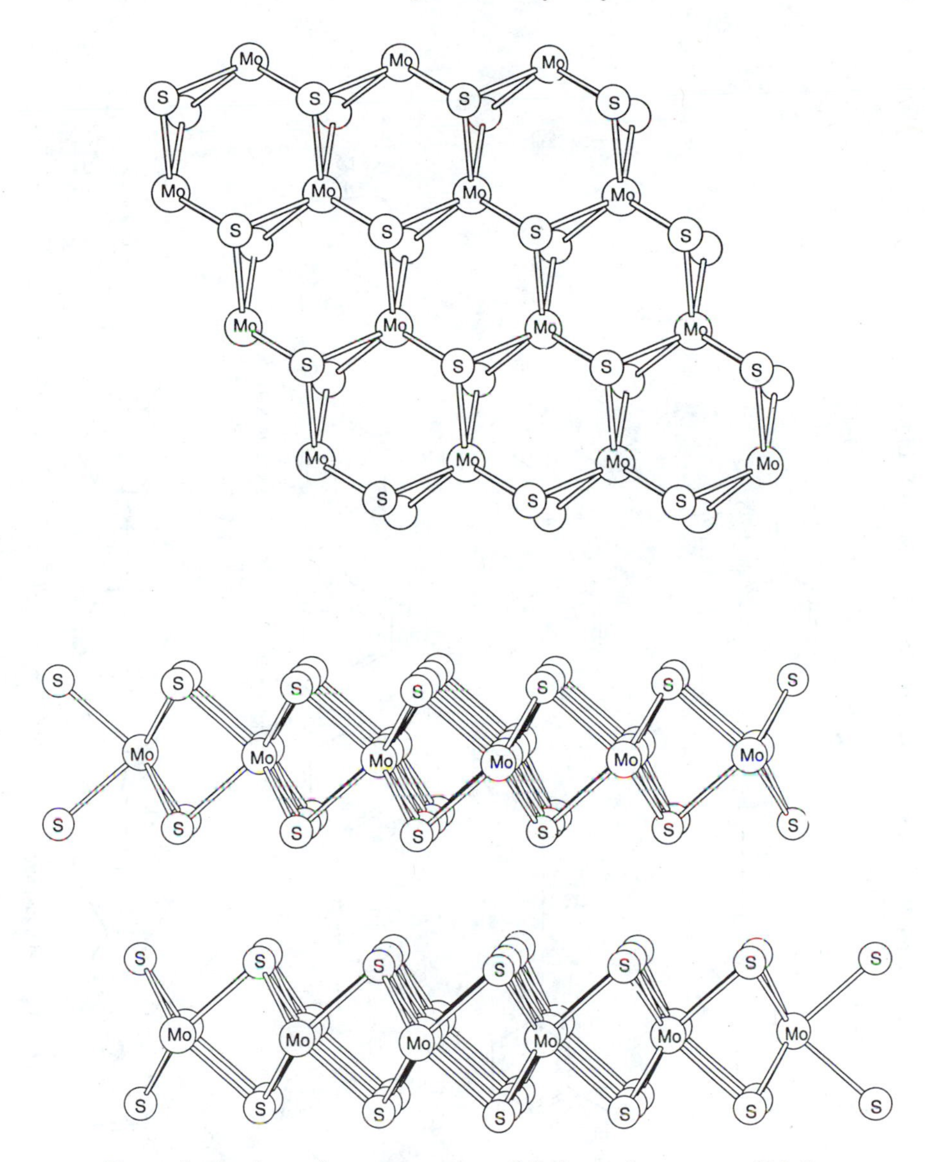

Figure 5. A schematic representation of the layered structure of MoS_2.

Bianchini and co-workers in Chapter 10, have utilized organometallic noble metal systems to investigate the reactivity of model feed molecules such as thiophenes, benzothiophenes, and dibenzothiophenes.

A parallel line of investigation involves reactions of dihydrogen with transition metal sulfur sites. These highlight the ability of the sulfur ligands to be the main site of dihydrogen activation and binding (*121-123*). (See the section in this chapter on small molecule activation.) Metal-bound S-H groups in the molecular

Table I. Some Reactions Catalyzed by Transition Metal Sulfides

Hydrogenation of Aromatics

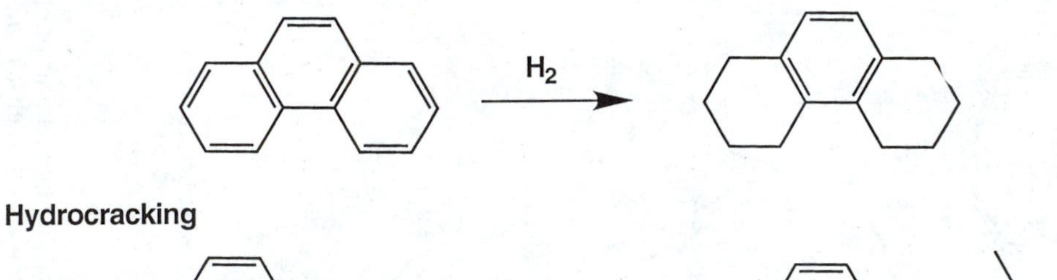

Hydrocracking

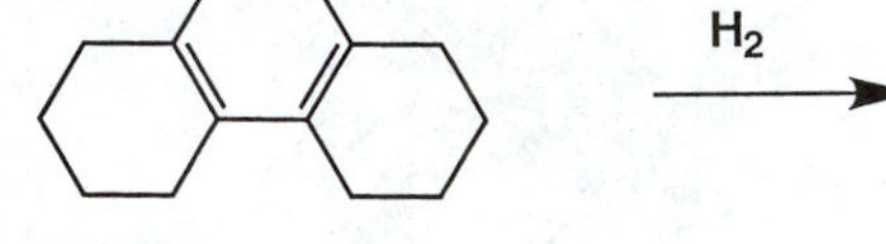

Hydrodesulfurization

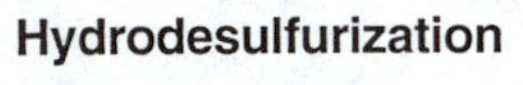

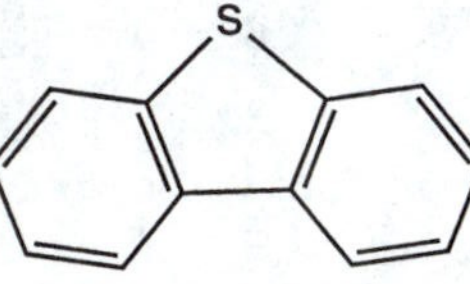

$\xrightarrow{H_2}$ Biphenyl + Cyclohexylbenzene + H_2S

Hydrodenitrogenation

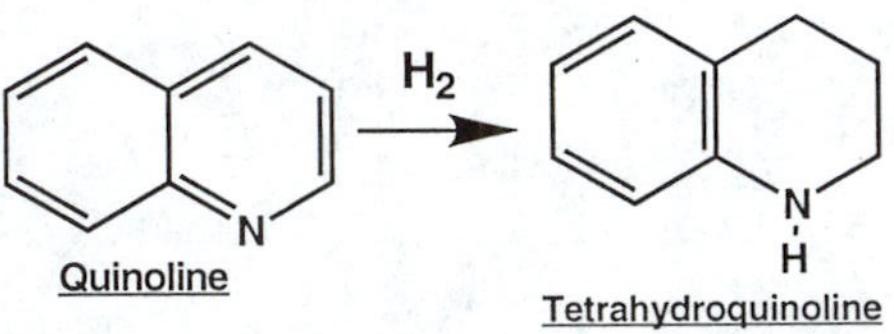

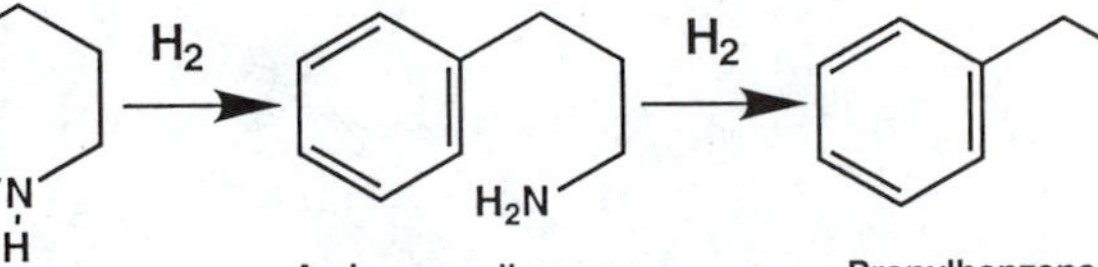

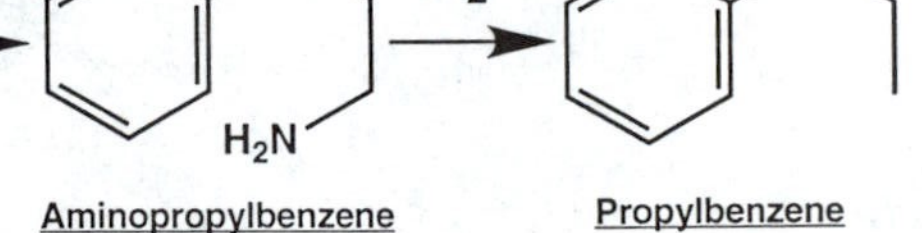

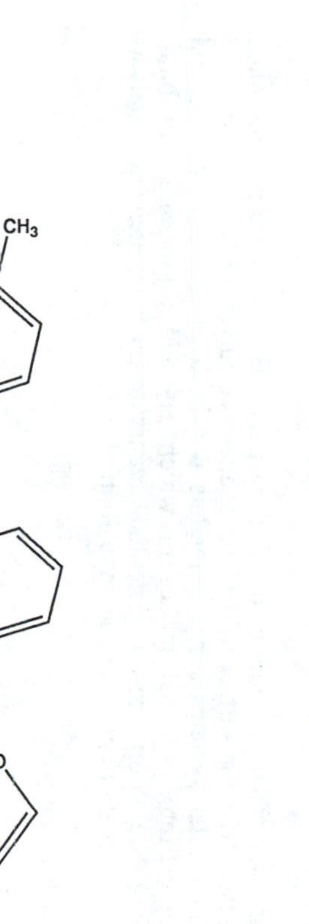

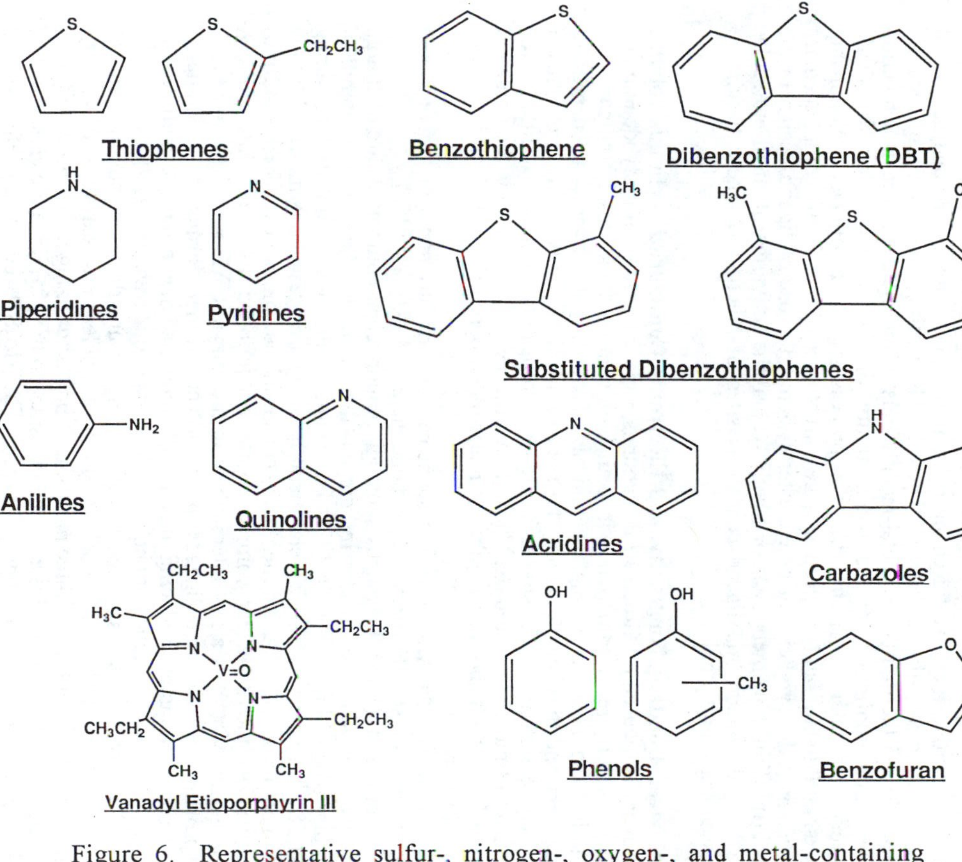

Figure 6. Representative sulfur-, nitrogen-, oxygen-, and metal-containing compounds in petroleum that are subject to heteroatom removal as, respectively, H_2S, NH_3, H_2O, and metal sulfide.

complexes are significant in view of the finding that S-H groups are present when dihydrogen reacts with the surface of the heterogeneous catalysts (*124*).

The molecular systems serve as models for the hydrotreating catalysts and give us insights into potential modes in which H_2 and thiophenes can bind. Moreover, the model complexes are potential precursors for heterogeneous catalysts formed by the thermal or chemical decomposition of the molecular systems.

Trends in TMS Chemistry

Work on transition metal sulfur chemistry is now of sufficient scope and depth that we can begin to identify key trends. These trends involve: the state of aggregation (nuclearity) and the number of metal-metal bonds of the metal centers as a function on *d*-electron configuration; the redox ability of ligand and metal; the facile occurrence of internal electron-transfer processes; the recognized relationship between solid-state and molecular core structures; the use of complementary approaches in the synthesis of new cluster systems; the use of ligand design in the control of affinity, geometry, and/or reactivity; and the activation of small molecules using both the metal and ligand centers as sites of reaction. These trends are discussed sequentially below.

Metal-Metal Bonding, Nuclearity, and Electronic Configuration In this volume, much work is focused on Group VI compounds, especially those of molybdenum and tungsten, which are of importance in both biological and industrial contexts. A correlation can be seen among the number of metal-metal bonds, the degree of aggregation, and the *d*-electron count for the individual Mo or W atoms in the cluster (*98*). The correlation is illustrated in Figure 7, mostly with simple sulfide-ligated complexes of molybdenum in oxidation states II to IV. We can follow the argument by considering the maximum number of metal-metal bonds that can form in each of these oxidation states.

Obviously, the Mo(VI), $4d^0$, systems cannot form any metal-metal bonds and the most common sulfido species is the mononuclear MoS_4^{2-} ion (*125*).

For the Mo(V), $4d^1$, systems, although mononuclear compounds are common, and some complexes of nuclearity greater than two are known, the dominant molecular type is dinuclear (*126,127*). For example, dinuclear complexes containing the $Mo_2S_2^{2+}$ core form a single metal-metal bond whose presence is indicated by the short metal-metal distance and the diamagnetism of the complexes. The bis(sulfido) bridge makes the dinuclear complex quite stable once formed. Interestingly, *mononuclear* Mo(V) is a key intermediate in the enzymic Mo systems (*18-24,128*) and protein and cofactor ligands must prevent dimer formation. Similarly, biomimetic attempts at producing analogs of Moco enzymes often use multidentate ligands, specifically designed or chosen to discourage dimerization (*129*).

For Mo(IV), $4d^2$, systems, trinuclear centers are the most common metal-metal bonded unit (*130-132*). Each Mo in the trinuclear unit can form two metal-metal bonds, giving a total of three metal-metal bonds, utilizing all six 4*d* electrons in the triangular $Mo_3S_4^{4+}$ core cluster. The open sites on the $Mo_3S_4^{4+}$ core cluster can be filled by a wide variety of ligands (*130-132*). The $Mo_3S_4^{4+}$ core is a useful synthon (see below) for forming thiocubane complexes (Chapters 12, 13, and 19), raft structures (Chapter 14), and molecular Chevrel (hexanuclear) clusters (*133*).

For Mo(III), $4d^3$, systems, tetranuclear centers are common (*133,134*). In particular, a tetrahedral arrangement of four Mo atoms allows each Mo to form three metal-metal bonds giving six metal-metal bonds in the $Mo_4S_4^{4+}$ core cluster. This thiocubane core is also maintained in oxidized forms that have $Mo(III)_3Mo(IV)$ or $Mo(III)_2Mo(IV)_2$ oxidation states (*135*). [Alternatively, to form three metal-metal bonds, two Mo(III) units can form a triple bond such as those found in alkoxide and related complexes studied by Chisholm and co-workers (136).]

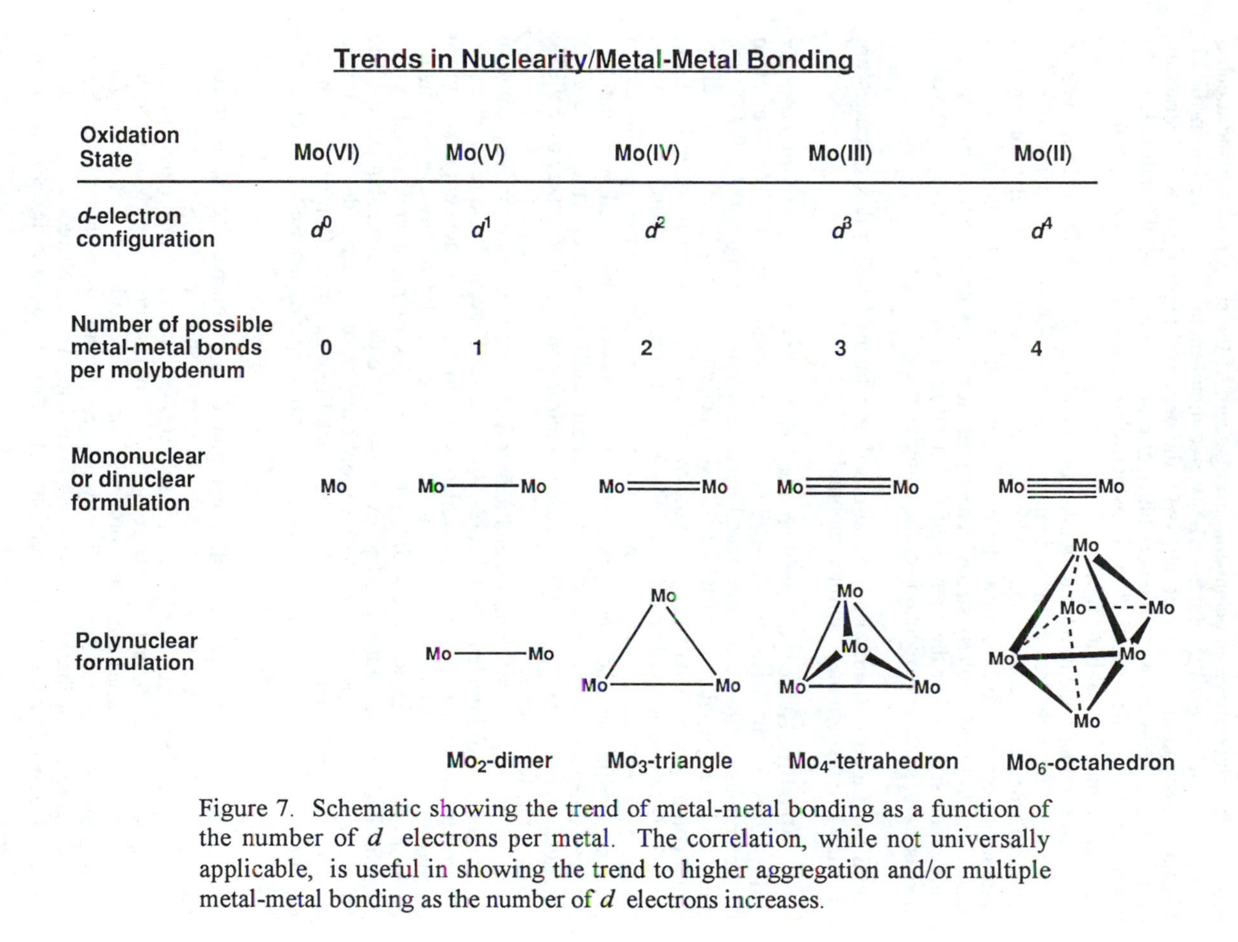

Figure 7. Schematic showing the trend of metal-metal bonding as a function of the number of *d* electrons per metal. The correlation, while not universally applicable, is useful in showing the trend to higher aggregation and/or multiple metal-metal bonding as the number of *d* electrons increases.

Finally, for Mo(II), $4d^4$, hexanuclear structure in which each Mo atom can form four metal-metal bonds is well known in solid state Chevrel phases (*137*), and, of late, has been prepared in hexanuclear molecular clusters (*138,139*). These $Mo_6S_8^{4+}$ structures have an octahedral arrangement of the six Mo atoms and a cubic arrangement of the eight S atoms. The twelve metal-metal bonds present in the all Mo(II) compound are only achievable with the observed regular octahedral structure of the metals. [Alternatively, the possibility of multiple metal-metal bonding leads to the quadruply bonded $Mo(II)_2^{4+}$ unit, which also allows full utilization of the 4*d* electrons in metal-metal bonding.]

In summary, the maximum number of metal-metal bonds formed per metal is the same as the *d*-electron count and ranges from none for the d^0 complexes to four per metal for the d^4 compounds. While there are exceptions, the electron precise systems shown in Figure 7 form the maximum number of metal-metal bonds and hence tend to by particularly stable. The correlation helps us to understand why the aggregation state and/or the strength of the metal-metal bonds are often observed to increase with successive reduction of the complexes.

Metal vs. Ligand Redox. Both metal-based and ligand-based redox reactions are common in transition metal sulfur compounds.

Metal-Based Redox. Transition metal systems are recognized for their redox ability due to the viability of a given metal center in multiple *d*-electron configurations (*140*). Transition metal systems with sulfur ligation (*140-143*) are found to have considerable metal-based redox activity. Redox reactions of sulfur-donor complexes can entail changes in the metal coordination sphere, often involving atom-transfer reactions. For example, the Mo(IV) complex MoOL(dmf) shown in Figure 8 accepts an oxo group (oxygen atom) from, for example, a sulfoxide to produce a sulfide and the Mo(VI) complex MoO_2L (*144*). This reaction is related to the oxo transfer reactions carried out by certain molybdenum enzymes (*145-146*). The sulfur-donor ligands, perhaps because of their π-donor ability, appear to facilitate the oxo transfer process.

Perhaps the most dramatic and simplest examples of redox activity come in the complexes of dithiolene ligands (*32-34,147-150*). Here, multiple reversible one-electron transfer reactions are common. However, in the dithiolene complexes, much discussion has focused on the question of the extent of ligand involvement in the electron-transfer processes (*147-150*). The closeness of metal and ligand orbital energies and their favorable overlap gives rise to the extensive delocalization. The HOMOs and LUMOs relevant to electron transfer are clearly delocalized and, therefore, the redox reactions are not solely metal in character. Consonant with this situation is the extensive redox ability of the sulfur ligands themselves, even in the absence of the transition metal.

Ligand-Based Redox. Sulfur compounds are conspicuous in their redox reactivity. The simple inorganic molecules or ions of sulfur (*151-154*), which include S^{2-}, S_2^{2-} S_4^{2-} S_8, SO_2, SO_3, $S_2O_3^{2-}$, SO_3^{2-}, and SO_4^{2-}, contain sulfur in oxidation states ranging from -II to VI, and are interconverted by redox processes. Moreover, these inorganic sulfur species and most organosulfur compounds, are potential ligands, binding to transition metals by sulfur, by oxygen, or through both atoms (*155,156*). The redox ability of these ligands makes it critical that ligand redox be considered along with metal redox in the chemistry of sulfur-donor complexes of transition metals.

An example of ligand redox occurs in the reaction of $Mo_2(S_2)_6^{2-}$ with thiolate ligands (*157,158*), which gives rise to complexes containing the $Mo_2S_4^{2+}$ core.

$Mo_2(S_2)_6^{2-}$ + 24 RS^- --> $Mo_2S_4(S_2)_2^{2-}$ --> $Mo_2S_4(SAr)_4^{2-}$

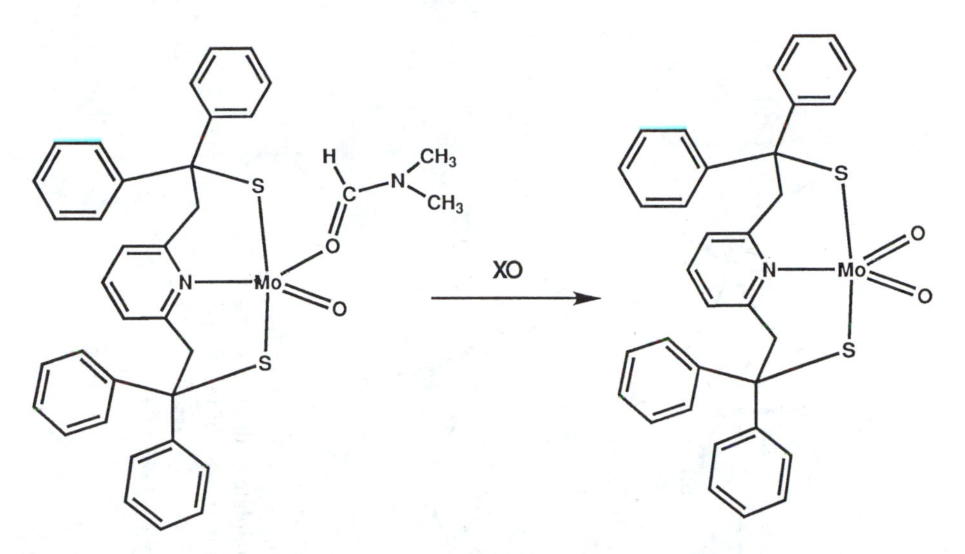

Figure 8 . Metal redox accompanied by oxide transfer as seen in the Mo(VI)/Mo(IV) couple. The oxygen atom (oxo) transfer shown here for a model system (144,145) is potentially related to oxo transfer reactions seen in Mo enzymes.

Here, as illustrated in Figure 9, the disulfide linkages are sequentially reduced; first the bridging S_2^{2-} ligands, and then the terminal S_2^{2-} ligands. Note that the initial reactant is a Mo(V)-Mo(V) dimer and the intermediate and final products are also binuclear complexes of pentavalent molybdenum. Therefore, the overall reaction involves the reduction of six disulfide ligands to twelve sulfide level (S^{2-}, HS^-, or H_2S) species. This is a twelve-electron redox process, in which, remarkably, the metal oxidation state remains unchanged. Clearly, there is a great deal of redox ability present in the sulfur ligands of transition metal complexes.

Another example of ligand redox occurs in the chemistry of coordinated thiolate or sulfide, which can be oxidized to coordinated sulfinate or sulfonate groups by the addition of dioxygen, peroxide, or other oxidants. This type of reactivity is well established in Co(III) (*159*), Ni(II) (*68*), and Mo(V) (*160*) complexes and generally occurs without any change in the metal oxidation state. Sulfur ligand oxygenation is illustrated in the contribution of Maroney and co-workers in Chapter 4. Similar sulfur- donor ligand modification is important for understanding the deactivation of the Ni enzyme hydrogenase, which has an inactive form that may have an oxygen-bound sulfur ligand (*65*).

Internal Redox Reactions. Because of the closeness of the redox potentials of transition metals and sulfur ligands, the intriguing possibility arises that definable internal redox processes can occur between the metal and ligand. Such processes are now well established and add significantly to the richness of the redox chemistry of sulfur-coordinated transition metal complexes.

We illustrate the idea through the chemistry of the tetrathiomolybdate ion, MoS_4^{2-}. This ion contains molybdenum in its highest oxidation state, VI, and sulfur in its lowest oxidation state, -II. Yet, MoS_4^{2-} is a stable entity in solution, known for over 150 years (*161*), and has been isolated in stable salts with a variety of cations (*125*). However, in the presence of oxidants this ion readily undergoes internal redox

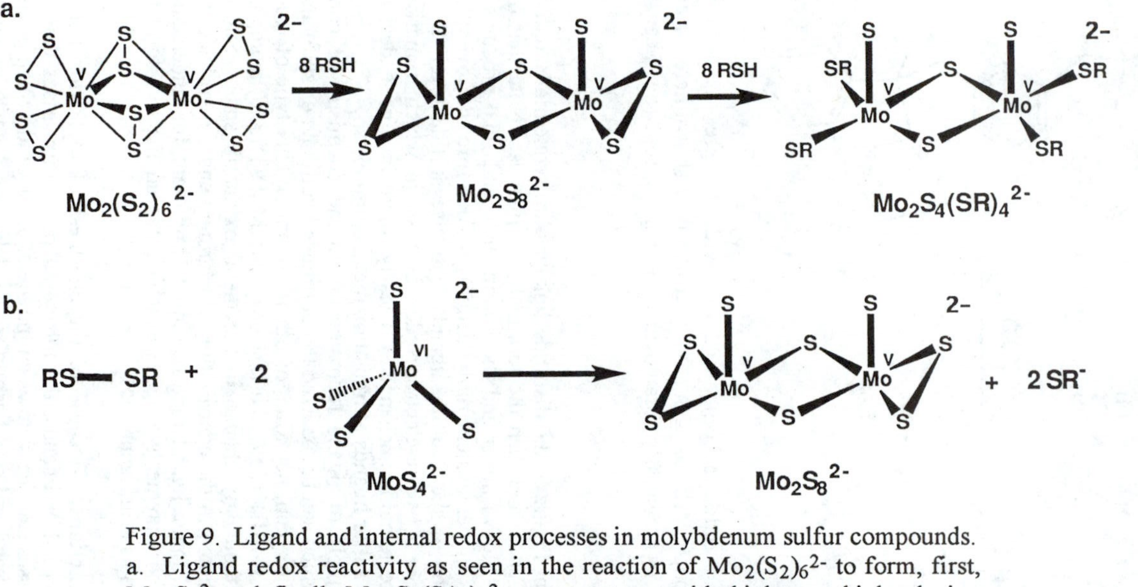

Figure 9. Ligand and internal redox processes in molybdenum sulfur compounds. a. Ligand redox reactivity as seen in the reaction of $Mo_2(S_2)_6^{2-}$ to form, first, $Mo_2S_8^{2-}$ and, finally $Mo_2S_4(SAr)_4^{2-}$ upon treatment with thiolate or thiol reducing agents. The metal oxidation state (V) stays the same during this set of reactions. b. Induced internal electron transfer to form the Mo(V) dimer $Mo_2S_8^{2-}$ from the Mo(VI) monomer MoS_4^{2-} upon treatment with an organic disulfide oxidant.

reactions. An example involves the reaction of MoS_4^{2-} with an organic disulfide (*158,162,163*):

$$2\,MoS_4^{2-} + RSSR \rightarrow Mo_2S_8^{2-} + 2\,RS^-$$

This reaction involves *reduction* of hexavalent Mo in the MoS_4^{2-} ion to pentavalent Mo in the $Mo_2S_8^{2-}$ dimer, which is identical to that shown in Figure 9 (produced by ligand redox). The ability of an *oxidant* (RSSR) to effect *reduction* of the metal involves the participation of the coordinated ligand. In this case, the coordinated sulfide ligands (four on two Mo centers) are oxidized to two disulfido ligands, which remain in the coordination spheres of the two Mo atoms. Of the four electrons that are made available by the oxidation of sulfide [2 (2 $S^{2-} \rightarrow S_2^{2-} + 2\,e^-$)], two electrons reduce the organic disulfide (RSSR) oxidant to thiolate (2 RS^-), while the other two electrons reduce two Mo(VI) ions, each by one electron. This internal electron transfer is said to be induced by the external oxidant. The reaction is designated an induced internal electron transfer process (*164,165*).

Induced internal electron-transfer reactions have been demonstrated in a number of tetrathiometallate ions including VS_4^{3-} (*166*), MoS_4^{2-} (*167*), WS_4^{2-} (*168*), and ReS_4^- (*169,170*). The latter species undergoes striking induced internal electron transfer reactions involving the reduction of Re(VII) to Re(IV) and Re(III) (*169,170*). In contrast, the WS_4^{2-} ion seldom undergoes internal redox (*167*). Why is internal redox more facile in some tetrathiometallates than in others?

In seeking a correlation between internal redox ability and a physical parameter, we were drawn to the first ligand-to-metal charge transfer (LMCT) band of the tetrathiometallate ions (*171*). The LMCT process, caused by the absorption of a photon, is closely related to the chemical process of moving an electron from a ligand-based orbital to a metal-based orbital. Indeed, the ease of internal redox correlates well with the position of the lowest LMCT. For example, the lowest LMCT for the red MoS_4^{2-} ion occurs at 21,300 cm^{-1}, while that for the yellow WS_4^{2-} ion occurs at 25,300 cm^{-1}. Obviously, in these isostructural compounds, it takes less energy to move an electron from S to Mo than from S to W. This trend correlates with the more facile and common induced internal electron transfer processes observed in MoS_4^{2-} versus WS_4^{2-}. An interesting verification of this correlation comes from the reactivity of WSe_4^{2-}, which resemble that of MoS_4^{2-} rather that of WS_4^{2-} (*172*). Interestingly, the substitution of sulfur by selenium lowers energy of the lowest LMCT of the red WSe_4^{2-} ion to 21,600 cm^{-1}, a position very close to that of the MoS_4^{2-} ion. Moreover, the ReS_4^- ion has a very low first LMCT at 19,800 cm^{-1}, correlating nicely with the extreme reactivity of ReS_4^- toward internal redox. Clearly, the position of the LMCT band is a *spectroscopic indicator* of the facility with which internal electron transfer can occur in *chemical reactions*.

The versatile redox activity of sulfur-coordinated transition metal compounds is clearly important in their behavior in a wide variety of circumstances of relevance to the action of both enzymatic and industrial catalyst systems.

Relationship of Solid-State and Molecular Systems. Solid-state transition metal sulfide materials are well known and their technological importance has been described above. Many of these materials have historically been prepared by high-temperature thermal techniques, although of late CVD and hydrothermal (or solvatothermal) processes have been added to the synthetic repertoire (*173,174*). Despite the vastly different modes of preparation of the solid-state compared to molecular materials, increasingly, it is being recognized that there many structural resemblances between the two (*175,176*). The similarities involve core structures as well as overall organizational patterns.

The relationship of the core structure in extended lattices and core structures in molecular species is illustrated in vanadium-sulfur chemistry. The reaction product of VS_4^{3-} and thiuram disulfide $[(R_2NCS_2)_2]$ is $V_2(S_2)(S_2CNR_2)_4$ formed by the internal redox reaction described above (*166*). The tetravalent $V_2(S_2)^{4+}$ core of this molecular structure is virtually identical in dimensions to the structure of the $V_2(S_2)^{4+}$ units found in the mineral patronite (*177*), which consists of a linear chain structure of $V_2(S_2)$ units bridged by additional bis(disulfide) linkages. Another example of the congruence of core structures comes in the Chevrel phases where the $Mo_6S_8^{4-}$ cores are found in the solid state and, more recently, have also been synthesized in molecular complexes (138,139). Many such overlaps have been identified between solid-state and molecular core structures (*175,176*).

Some solid-state materials can be characterized as two dimensional. Here again there is an interesting structural relationship with molecular materials. For example, in ReS_2 the layered structure typical of early transition metal chalcogenides, is present (*178*). Interestingly, several Re-S complexes with high-sulfur coordination about Re can clearly be viewed (*170*) as 'layered' with quasi-close-packed planes of sulfur sandwiching a 'plane' of Re atoms, as in the solid state system. Interestingly, similar Re-S distances are found in the molecular and solid state materials.

Other solid-state structures are three dimensional in nature. Once again a similarity is seen with related molecular clusters. For example, fragments of Zn or Cd sulfide/thiolate structures have adamantane-like arrangements resembling the zinc-blende structures present in ZnS and CdS (*175*). Indeed, some of the Cd structures formed are sufficiently large that quantum effects are found that are characteristic of the borderline region between what is considered a molecular species and what is considered an extended lattice (quantum size effects) (*179*). Chapter 7 by Dance discusses some of these large clusters.

The relationship between solid-state and molecular systems is not simply of academic interest. It can be exploited in new syntheses. Recognition of a structural unit in one of these media often stimulates attempts to produce that structural unit in the other. Ideas of bonding in solids are helped by understanding the electronic structures of the core units in molecular species. Finally, in recent years synthesis of solids using molecular precursors, and synthesis of molecular systems by cluster excision reactions from solids, have both become useful and coveted methods. Consequently, the recognized relationship between solid-state and molecular materials has proven a stimulus to new synthetic strategies.

Synthetic Strategies. For the synthesis of molecular transition metal sulfide systems three distinct strategies have evolved. These are often called 'spontaneous self-assembly', 'designed (building block) synthesis,' and 'cluster excision reactions,' respectively.

Spontaneous Self-Assembly. In this approach (*179*), the desired ingredients are mixed in solution, heating may be used, and the molecular material is usually crystallized directly from the reaction mixture. This methodology has been particularly fruitful in producing Fe_2S_2- and Fe_4S_4-core clusters (*6-8*) as well as heteronuclear clusters containing the $MoFe_3S_4$ unit (*6,180*), which resemble, in part, some of the features of the nitrogenase FeMoco. A key feature of the spontaneous self-assembly approach, in contrast to the other two approaches (below), is that the structures of the reactants are not necessarily preserved in the structure of the product.

A classic example of the spontaneous self-assembly approach is the reaction of MoS_4^{2-} or WS_4^{2-} with $FeCl_3$ and thiolate reagents to produce double thiocubane structures (*180*). To illustrate some points about spontaneous self-assembly we discuss the analogous reaction of $MoSe_4^{2-}$ or WSe_4^{2-} (*181*). In the case of the selenium analogs, an isostructural double selenocubane structure is formed. The

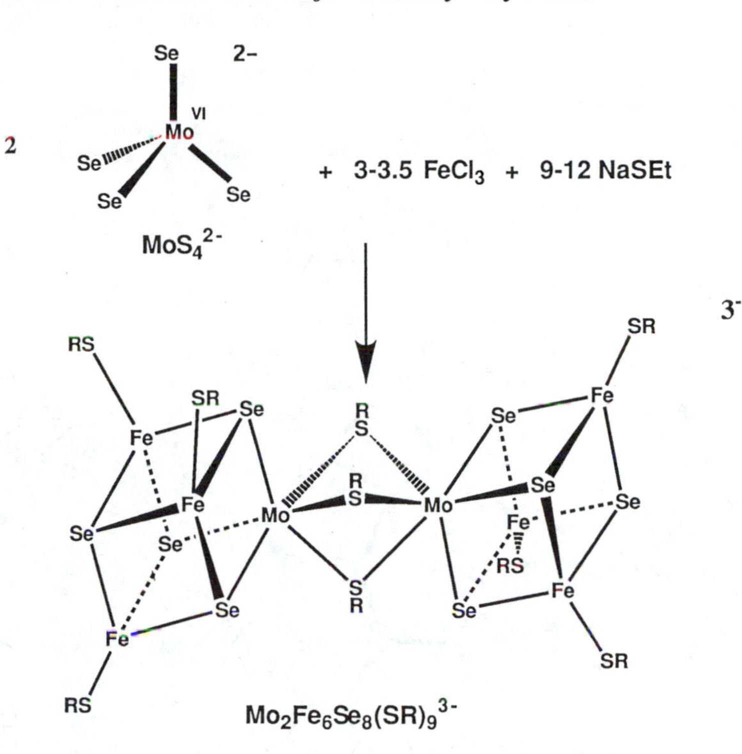

Figure 10. The reaction on $MoSe_4^{2-}$ (or WSe_4^{2-}) with thiol and $FeCl_3$ to yield the double selenocubane clusters. Note: Mo coordination changes from tetrahedral to distorted octahedral and only three of the four selenium atoms are directly coordinated to Mo in the final product.

reaction is illustrated in Figure 10, where it is quite clear that the tetrahedral four-coordination of the tetraselenometallate ion is not preserved in the final structure. The Mo (or W) is cleanly six coordinate with a distorted octahedral structure. Moreover, one of the selenium atoms of the starting material is no longer coordinated to the Mo or W, being found on the diagonally opposite corner of the cube. Clearly, while the $MoSe_4^{2-}$ (or WSe_4^{2-}) stoichiometry (1:4) is maintained in the product double cubane cluster, the *structure* of the Mo (or W) center is not preserved.

Spontaneous self-assembly has proven exceeding fruitful for certain classes of compounds and it remains a useful tool for exploratory syntheses. However, it has the distinct disadvantage that there is no structural predictability in the outcome of the reaction. As synthetic inorganic chemistry seeks to approach the legendary prowess of synthetic organic chemistry, more predictively systematic approaches, such as those described below, are required.

Building-Block Syntheses The building block strategy has often been labeled as 'designed synthesis' in which the individual units, sometimes referred to as 'synthons', are used in the planned aggregation of particular structures. The individual building blocks can be mononuclear, dinuclear, trinuclear and of even higher nuclearity. Examples are given in Figure 11.

Mononuclear Synthons. Figure 11a shows the simple reaction of a tetrathiomolybdate ion with cyclopentadienyl cobalt dicarbonyl. The tetrahedral

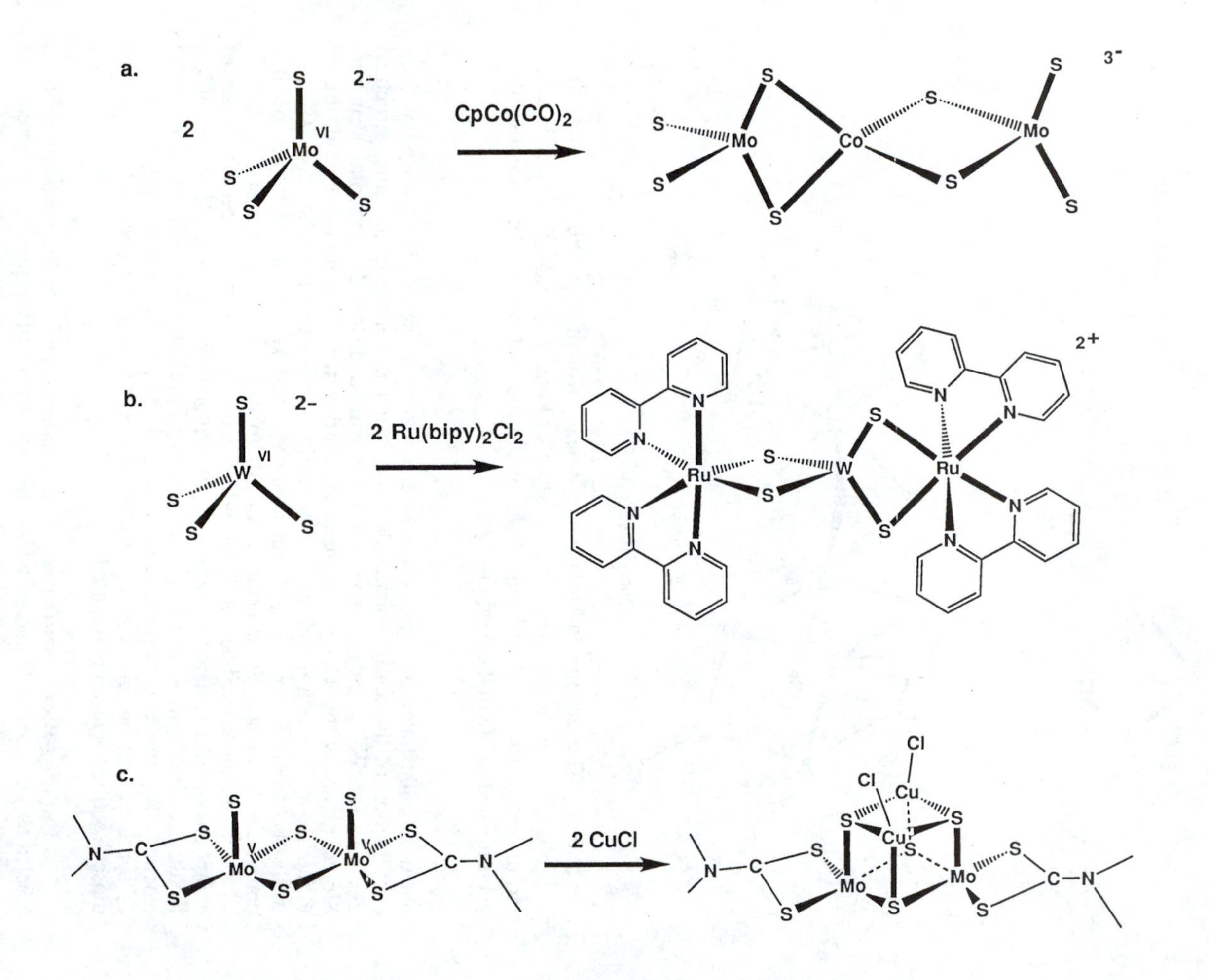
a.
2
CpCo(CO)2
b.
2 Ru(bipy)2Cl2
c.
2 CuCl

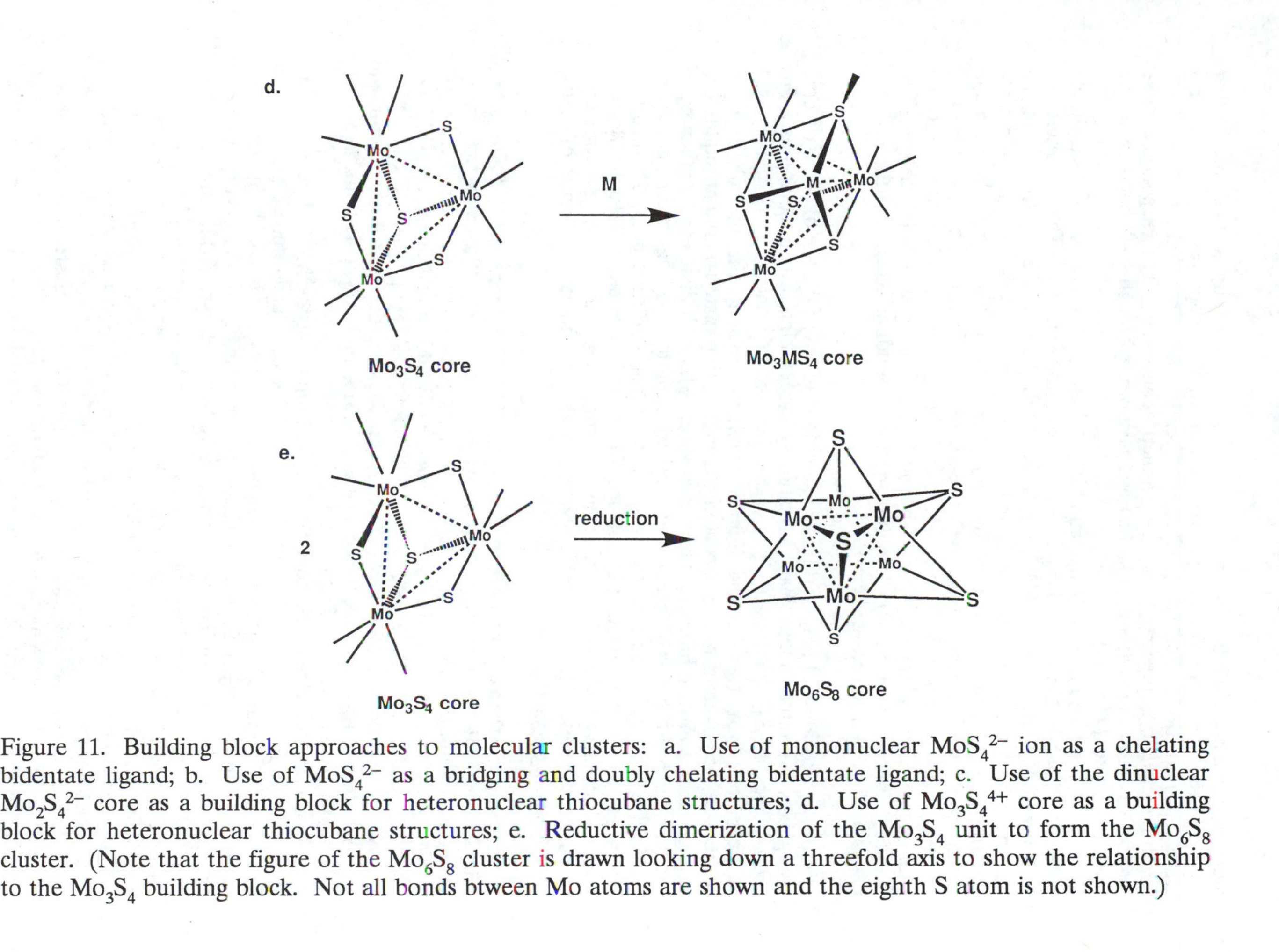

Figure 11. Building block approaches to molecular clusters: a. Use of mononuclear MoS_4^{2-} ion as a chelating bidentate ligand; b. Use of MoS_4^{2-} as a bridging and doubly chelating bidentate ligand; c. Use of the dinuclear $Mo_2S_4^{2-}$ core as a building block for heteronuclear thiocubane structures; d. Use of $Mo_3S_4^{4+}$ core as a building block for heteronuclear thiocubane structures; e. Reductive dimerization of the Mo_3S_4 unit to form the Mo_6S_8 cluster. (Note that the figure of the Mo_6S_8 cluster is drawn looking down a threefold axis to show the relationship to the Mo_3S_4 building block. Not all bonds btween Mo atoms are shown and the eighth S atom is not shown.)

anion acts in a bidentate fashion and clearly is a good enough ligand to displace other strong ligands such as CO and Cp. Tetrathiometallate complexes are known for a variety of transition metal ions (*182-184*). Clearly, the tetrahedral structure of the tetrathiometallate is preserved in these synthetic constructs. The tetrathiometallate 'ligand' has both donor (through sulfur) and acceptor (through back-bonding to the metal) character (*185,186*).

As shown in Figure 11b, the tetrathiometallate ion not only serves as a simple bidentate ligand, but can also act in a bridging mode. The structure of $Ru(bipy)_2WS_4Ru(bipy)_2^{2+}$ (*187*) reveals that both the octahedral geometry of the starting Ru(II) center and the tetrahedral geometry of the WS_4^{2-} doubly bidentate bridging ligand are preserved in the final complex. At the limit, a tetrathiometallate ligand could chelate six metals, one on each of the edges of the S_4 tetrahedron. Indeed, in the work of Secheresse, Jeannin, and co-workers (*188,189*) there are examples of five and six Cu(I) atoms bound to a single tetrathiometallate ion. The use of tetrathiomolybdate and tetrathiotungstate building blocks is discussed in Chapter 17 by Wu and co-workers .

Binuclear Synthons. In Figure 11c, an example is given of the use of $Mo_2S_4^{2+}$ core complexes to produce a thiocubane structure by the addition of two metal ions (*190-192*). The $Mo_2S_4^{2+}$ unit clearly possesses six (two Mo and four S) of the eight vertices (two Mo, two M, and four S), which constitute the thiocubane core. Moreover, the positions of these six atoms in the starting material are not required to change significantly to form the cubane structure. Although there is a lengthening of the terminal Mo-S bonds of the starting complex upon forming the cubane, otherwise only relatively small structural changes occur. Electronic structural considerations [from simple electron counting to detailed theoretical calculations (*193,194*)] allow the understanding and prediction of viable electronic configurations for the formation of stable cubanes. These favorable electron counts maximize the degree of metal-metal bonding present in the clusters.

Trinuclear Synthons. The work of Shibahara (*131,132*), Sykes (*195*), Saito (*133*), and Hidai (*196*) and their respective co-workers beautifully illustrates the use of trinuclear building blocks. Aspect of this work are presented In Chapters 12, 13, 14, and 19. Units such as the $Mo_3S_4^{4+}$ core can cleanly add an additional metal (Figure 11d) to produce thiocubane species of stoichiometry MMo_3S_4, where M can be Fe, Co, Ni Cu, Zn, Mo, Pd, Sn, and W (*131,132*). In addition, the trinuclear synthons can formally fuse to give raft structures containing Mo_4S_6 or Mo_6S_8 core units, as shown in Chapter 14 in the work of Saito et al. (*133*).

Under highly reducing conditions, it is found that the $Mo_3S_4^{4+}$ core can be dimerized to form the $Mo_6S_8^{4-}$ core characteristic of the Chevrel phase (*197*). This last reaction (Figure 11e) aptly illustrates two of the previous trends discussed in this chapter. First, aggregation upon reduction is predicted from the general trend of increased nuclearity with increasing numbers of 4*d* electrons. While, the trinuclear $Mo_3S_4^{4+}$ center is favored by the Mo(IV), $4d^2$ configuration, the $Mo_6S_8^{4-}$ core is favored by the Mo(II) $4d^4$ configuration. Clearly, a powerful reductant is required to effect the conversion. Second, both the $Mo_3S_4^{4+}$ and $Mo_6S_8^{4-}$ cores are found in solid-state structures and the former has been prepared by cluster excision reactions from such structures (see below). Finally, the Mo_3S_4 structural unit, somewhat distorted, is seen in the final Mo_6S_8 structure. The 'Chevrel' cluster may therefore be viewed as a reductive dimerization product of the $Mo_3S_4^{4+}$ core, with reduction leading to the formation of the additional metal-metal bonds characteristic of the hexanuclear structure.

Cluster-Excision Reactions. The relationship between solid-state and molecular systems suggests the intriguing possibility that molecular species could be prepared by extraction (excision) of a cluster from a solid. In some cases, this has indeed been accomplished (*198,199*). These so-called cluster-excision reactions represent a powerful approach, especially when combined with modifications of solid- state syntheses to make the solid-state structures more amenable to cluster extrusion. For example, dimensional reduction approaches (205, 206) may 'soften' solids toward extraction by reducing three-dimensional and two-dimensional structures to "one-dimensional' structures that contain isolated cluster units.

An example of successful cluster excision comes from the chemistry of molybdenum. The structure of $Mo_3S_7Cl_4$ (*200*), prepared by conventional solid state approaches, contains trinuclear $Mo_3S_7^{4+}$ units bridged by chloride ions. The trinuclear clusters can be "extracted" from the lattice with aqueous polysulfide to produce $Mo_3S_{13}^{2-}$ (*130,198*). In related work, reaction of $W_3S_7Br_4$ with tetraphenylphosphonium bromide yields the molecular cluster $[(C_6H_5)_4P]_2[W_3S_7Br_6]$ (*199*). In each of these cases, the $M_3S_7^{4+}$ (M = Mo, W) cluster unit in the molecular species is virtually identical to that found in the solids.

In rhenium chemistry, a variety of structures containing the 'Chevrel type' $Re_6S_8^{2+}$ cluster are known (*201-203*). For example, the three-dimensional structure of $Re_6S_8Cl_2$ has bridging chlorides connecting the octahedral Re_6S_8 cores (*204*). Dimensional reduction using added CsCl in the preparations (*205*) should lead successively to two-dimensional structures, linear chain structures, and isolated clusters. The latter two should be more amenable to extrusion, and indeed treatment of $Cs_5Re_6S_8Cl_7$ (*205*) using 1 M HCl leads to molecular $Re_6S_8Cl_6^{4-}$ clusters (*206*). The hexanuclear unit extruded has the same stoichiometry, shape, and dimensions as the cluster found in the solid (*206*).

Clearly, successful cluster excision is dependent on the relatively loose binding of the cluster core in the solid state lattice. Solid-state structures in which the desired cluster is strongly covalently bound to other clusters of the same type are considerably more difficult to extrude. Nevertheless, cluster-excision reactions, combined with solid- state syntheses tailored to make core clusters more accessible, should prove a powerful technique that may allow preparation of molecular cluster types that are difficult to obtain by other methods.

Design of Sulfur-Donor Ligands

Ligand design has become a powerful tool in the control of affinity, geometry, and reactivity of transition metal ions. New sulfur ligands tailored for specific uses have been synthesized. For example, a bulky tridentate pyridine bis(thiolate) ligand has been used to provide steric hinderance, thereby assuring that oxo molybdenum compounds do not dimerize (*144*) (Figure 8). The resultant complexes are reactivity models for some of the oxo molybdenum enzymes that catalyzed oxo transfer reactions (*145*). Chapter 18 by Lindoy et al. shows how ligand design can affect the affinity of heavy metals for specific ligand systems.

Small Molecule Activation

The enzymes and heterogeneous catalysts that utilize transition metal sulfide centers as key parts of their active sites are capable of reacting with a large number of small molecules. These include, H_2, N_2, N_2H_4, C_2H_2, C_2H_4 (*207*), O_2 (*208*), CO (*209*), NO (*210*), SO_2 (*211, 212*), and H_2S (*213*). The binding and activation of three of these species, H_2, C_2H_2, and N_2H_4, are briefly discussed. The reaction of transition metal sulfur compounds with each of these molecules reveals that, as with the redox reactivity discussed above (and for similar reasons), both the metal and ligand are potentially reactive sites.

Dihydrogen Binding. Dihydrogen activation is an important part of the reactivity of hydrogenase, nitrogenase, and industrial hydrotreating catalysts. The literature reveals

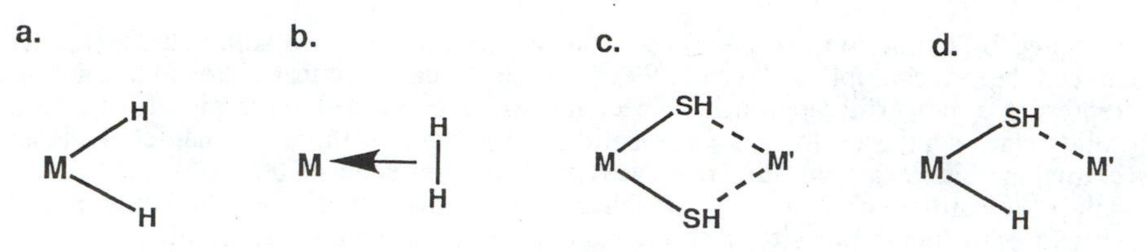

Figure 12. Possible modes for the binding of dihydrogen at a transition metal sulfur site: a. Dihydride; b. Dihydrogen; c. Bis(sulfhydryl); d. Hydride sulfhydryl.

that dihydrogen can react with transition metal centers in several different ways. These are illustrated in Figure 12.

The classical way in which H_2 reacts with a metal center involves oxidative addition to a low-valent metal to form a dihydride complex (*214*). Alternatively, a simple dihydrogen adduct must also be considered as possible (*215*), although it is not clear that any such entity has yet been prepared in a complex containing a predominantly sulfur ligand donor set. At the other extreme, H_2 can react with a transition metal sulfide center to form coordinated SH groups (*216*) as described in Chapter 16 by Rakowski DuBois et al. Such a reaction generally involves reduction of the metal center coupled with protonation of the sulfur site, that is, a sort of coupled proton-electron transfer of the type proposed for Mo enzymes (*217,218*). Alternatively, if a disulfide bond is present, the metal oxidation state need not change and the H_2 can simply cleave the S-S bond to yield the bis(SH) complex.

An intermediate situation is also possible as indicated by the work of Bianchini and co-workers (*122*) shown in Figure 13. Here, addition of two equivalents of H_2 to a dinuclear rhodium complex leads to the formation of hydrido-thiolo complex wherein H is bound both at the metal and at the sulfur. Clearly, both ligand- and metal-centered reactivity is possible in the reaction of H_2 with transition metal sulfur systems (*219*).

What type of dihydrogen binding and activation is important for hydrogenase, nitrogenase, and for hydrotreating catalysis? There seems to be strong evidence for SH groups in hydrotreating catalysts, but in the enzymes there is at present no direct evidence to implicate either bound SH groups or metal hydrides as present during the catalytic cycles. The significant variety of possibilities for hydrogen activation and binding indicates that much spectroscopic and mechanistic work will be required to understand, at the atomic level, the reactivity of this simplest of reagents, the dihydrogen molecule.

Acetylene Binding. Some of the possible modes of binding of acetylene to sulfur-coordinated transition metal complexes are summarized in Figure 14 (*219*). The binding of acetylene classically occurs through its π bond(s) acting as donor(s) and with its π^* orbital(s) acting as acceptor(s), thereby avoiding buildup of excess electron density on the metal (*220-222*). In the extreme, a metallacyclopropene construction is one manner of describing the metal acetylene binding. Such direct binding of acetylene to the metal in transition metal sulfur complexes has been established for some time (*223,224*). This mode of binding and activation of acetylene is likely occurring in the work of Hidai and Mizobe in Chapter 19, which shows the Mo_3PdS_4 cluster-catalyzed addition of alcohols and carboxylic acids to alkynes (*225*). Similarly, the work of Coucouvanis et al., described in Chapter 6, implicates the Mo on $[MoFe_3S_4]^{n+}$ clusters as the site of acetylene reduction by protons and reducing equivalents.

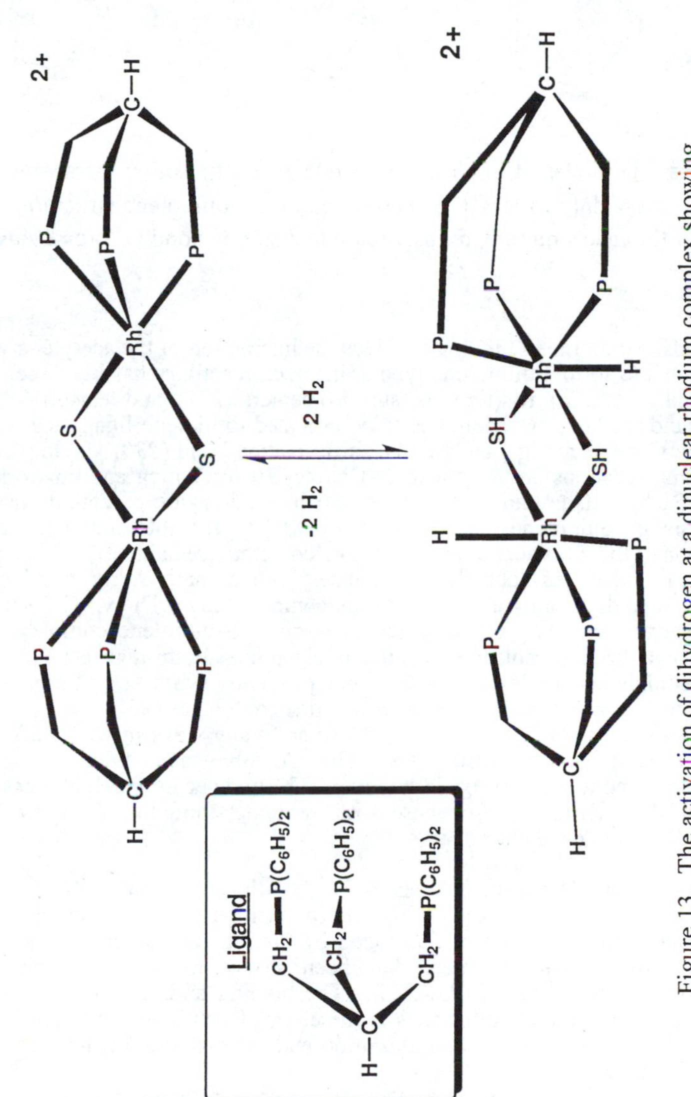

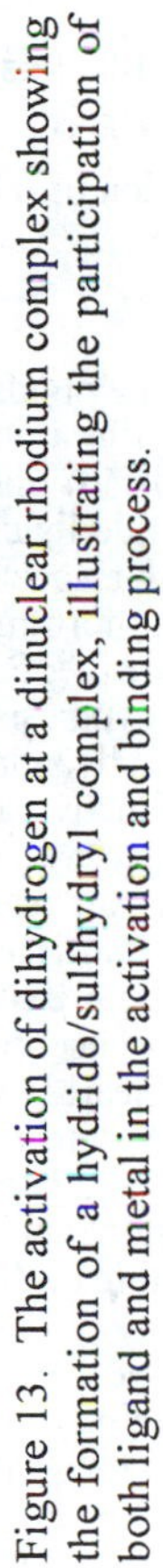

Figure 13. The activation of dihydrogen at a dinuclear rhodium complex showing the formation of a hydrido/sulfhydryl complex, illustrating the participation of both ligand and metal in the activation and binding process.

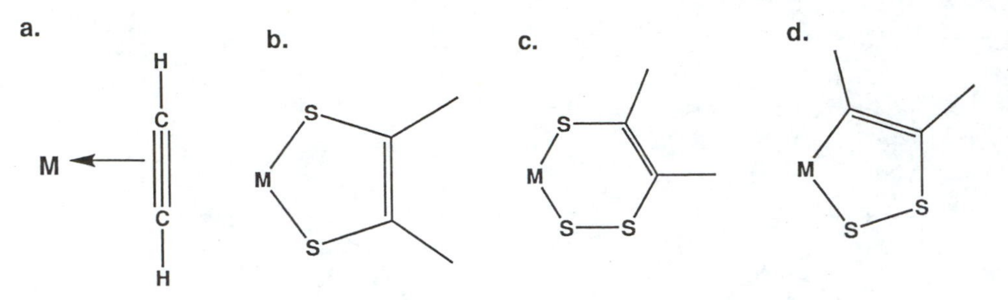

Figure 14. Possible modes of binding for acetylene at a transition metal sulfur site: a. Classic π bonding to metal; b. Formation of 1,2-dithiolene structure; c. Formation of trithiolene structure; d. Insertion into disulfide bond to form a vinyl disulfide.

The opposite extreme in binding involves the interaction of the acetylene with ligand sulfur atoms to form a dithiolene type unit. Such reactivity has been seen for bis(μ-sulfido) linkages (*226*), tris(terminal sulfido) centers (*227*), and tetasulfide and pentasulfide ligands (*228,229*). Moreover, a coordinated dithiolene ligand can itself react with an acetylene, forming an additional sulfur-carbon bond (*230*). Examples of dithiolene-forming reactions are described in Chapter 20 by Young and co-workers and in Chapter 21 by Tatsumi and co-workers. Such reactions are potentially useful for the synthesis of dithiolenes (*229*), some of which resemble the pterin-ene-dithiolene ligands of the molybdenum and tungsten cofactors (see above).

Finally, an (activated) acetylene can insert into a metal-sulfur bond of a coordinated S_2^{2-} ligand, forming a vinyl disulfide complex (*231*). The vinyl disulfide complex can rearrange to the corresponding isomeric 1,2-dithiolene complex in a reaction that is catalyzed by sulfur (*232*) and likely proceeds through a trithiolene intermediate. Trithiolene complexes have been independently synthesized through the reaction of acetylenes with transition metal tetrasulfide complexes (*229*).

Clearly, as with the reactivity for H_2, there are many ways in which alkynes can react with transition metal sulfur sites. These involve metal bonding, ligand bonding, and combined metal and ligand bonding. The manner in which alkynes are bound and reduced to olefins by nitrogenase is not yet understood, but likely involves one or more of these modes of coordination.

Hydrazine and Diazene Binding Hydrazine (N_2H_4), diazene (diimine, N_2H_2), and other 'dinitrogen hydride' intermediates have been postulated as intermediates on route to ammonia in the reduction of dinitrogen by the nitrogenase enzyme system (*233-236*). Moreover, hydrazine itself has been shown to be a substrate for nitrogenase, reducible to ammonia (*237,238*). The binding and reactivity of N_2H_4 and N_2H_2 with transition metal sulfur sites is therefore of continuing interest (*239*). Much work on the formation of related diazenido and hydrazido(2-) complexes has been reported and reviewed (*240,241*).

In Chapter 15, Matsumoto and co-workers describe ruthenium sulfur species that show great structural variety and interesting reactivity. Some of the compounds form hydrazine complexes and one of these has been oxidized to a *cis*-diazene complex, which is the first structurally characterized example of a coordination complex containing the cis ligand.

The *trans*-diazene complex, discussed in Chapter 5 by Sellmann et al. was also prepared by the oxidation of the corresponding hydrazine complex. The structure reveals stabilization of the *trans*-diazene by two transition metal complexes

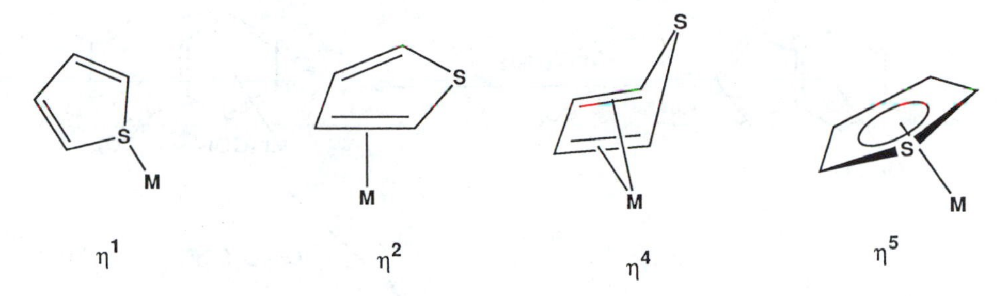

Figure 15. Modes of binding of thiophene at a single transition metal center (*112,113*).

involving both coordinate-covalent bonds of nitrogen to iron and N-H--S hydrogen bonds (*242*). In view of the presence of sulfur-bound iron in nitrogenase, modes of binding related to that found by Sellmann et al. must receive serious consideration as specific models for diazene formation during nitrogenase turnover.

Coucouvanis and co-workers report in Chapter 6 the catalytic reduction of hydrazine to ammonia by heteronuclear thiocubane clusters of $[MoFe_3S_4]^{n+}$, which have been considered as partial structural models for the FeMoco of nitrogenase. Evidence points to the reaction occurring at the molybdenum site of the cluster.

Binding and reaction of hydrazine are of significant interest with respect to nitrogenase function. Of great interest would be the binding and reduction of dinitrogen itself at transition metal sulfur cluster sites. The lack of success in this endeavor to date is a telling comment on the state of our ignorance of the nature of the reactivity of the nitrogenase active site and its putative model systems.

Thiophene Binding and Activation

Substituted thiophenes (including benzo- and dibenzothiophenes) are among the more recalcitrant sulfur compounds found in petroleum and must be desulfurized during the refining of fuels and petrochemicals. In view of the difficulty of getting direct information from the surfaces of heterogeneous catalysts, high vacuum experiments on surfaces and solution chemistry of molecular complexes have been carried out to ascertain the possible modes in which thiophenes may bind to a metal center. The structures of the thiophene complexes and the reactivity of the bound thiophenes give potential insights into the way in which thiophenic molecules are activated in the heterogeneous catalysts (see Chapters 8-11).

Figure 15, after Angelici (*112*), is a compilation of the modes of thiophene binding at transition metal centers. Binding can occur at the sulfur or with the unsaturated carbon framework of the molecule. The resultant activation that occurs is discussed by Curtis (Chapter 8), Rauchfuss (Chapter 9), and Bianchini (Chapter 10), and their respective co-workers. Figure 16 shows examples from the work of Sweigart (*115*) and co-workers revealing that activation of benzothiophenes can occur on single or on bimetallic sites. Chapter 11 by Boorman et al. addresses C-S bond breaking, which is a necessary part of hydrodesulfurization reactions.

The thiophene activation observed by Curtis (Chapter 8) is of particular interest since a cluster of cobalt, molybdenum and sulfur (related to the CoMoS catalyst of choice) is shown to be capable of thiophene desulfurization. Moreover, related clusters serve as precursors for supported catalysts that display good hydrodesulfurization activity (*243-245*). The work of Rauchfuss (Chapter 11) shows that the binding of benzothiophene leads to altered acid-base properties of the coordinated ligand, suggesting that acid base behavior on catalyst surfaces may be important in the hydrodesulfurization reaction. Bianchini and co-workers (Chapter

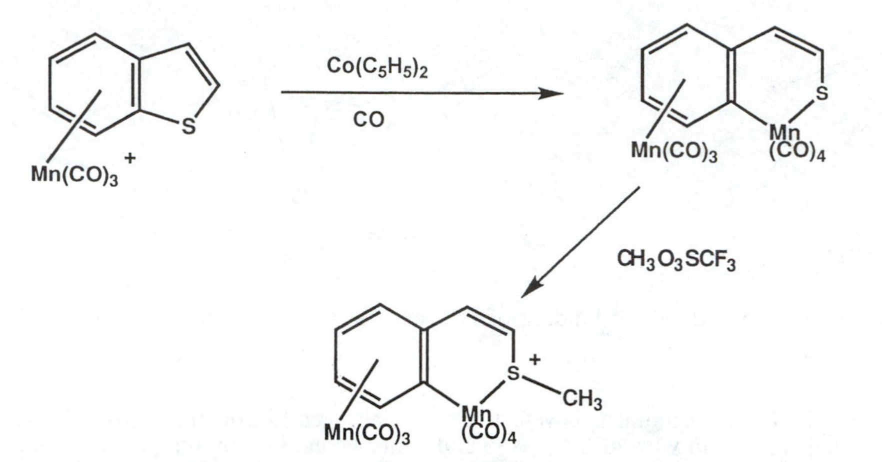

Figure 16. Reaction of thiophene complexes showing increased nucleophilicity of S in thiophene bound to one or two metal centers (*120*).

12) have developed homogeneous catalytic hydrodesulfurization systems, indicating that, at least in principle, a heterogeneous catalyst is not a *sine qua non* for hydrodesulfurization.

The binding and activation of thiophene will continue to provide input into the understanding of the mechanism of substituted thiophene activation and reaction. Combined with surface science data, theoretical treatments, and information on the catalysts themselves the mechanism of hydrodesulfurization should increasingly come into focus.

Conclusion

In this chapter, the biological and industrial contexts in which transition metal sulfur chemistry plays an important role have been introduced. The chemistry of transition metal sulfur systems has attracted increasing attention in part due to the biological and industrial relevance and in part due to the inherently interesting behavior of the molecular systems, for which our explorations have barely scratched the surface. We are now beginning to see certain key trends discernible in the chemistry of the transition metal sulfur species. These include: the relationship of electronic configuration and state of aggregation (nuclearity); the ligand-based, metal-based, and internal redox ability of the transition metal sulfur system; the interesting congruences between the core structures of the molecular and solid-state systems; the synthetic schemes involving spontaneous self-assembly, building-block approaches, and cluster excision reactions; the versatile activation of small molecules including dihydrogen and various nitrogenase substrates; and the binding and activation of substituted thiophene ligands, a key step in understanding the industrially significant hydrodesulfurization catalysis reaction. The papers in this volume amplify these trends and reveal the connections that exist between the molecular, solid-state, and biological systems. There is much to be gained from the cross-fertilization of research in these teleologically independent but chemically related areas of research.

Acknowledgments: I am grateful to Drs. J. A. McConnachie, H. H. Murray, and M. A. Walters for helpful comments on this manuscript.

Literature Cited

1. Bertini, I.; Gray, H. B.; Lippard, S. J.; Valentine, J. S. Eds., *Bioinorganic Chemistry*, University Science Press, Mill Valley, CA, **1994**.
2. Lippard, S. J.; J. M. Berg, *Principles of Bioinorganic Chemistry*, University Science Press, Mill Valley, CA, **1994**.
3. Kaim, W.; Schwederski, B. *Bioinorganic Chemistry: Inorganic Elements in the Chemistry of Life* J. Wiley & Sons, New York, NY, **1994**.
4. Cowan, J. A. *Inorganic Biochemistry: An Introduction*, VCH, New York, NY, **1993**.
5. Williams, R. J. P.; Frausto da Silva, J. J. R. *The Biological Chemistry of the Elements,* Clarendon Press, Oxford, UK, **1991**.
6. Stiefel, E. I.; George, G. N. in *Bioinorganic Chemistry*, Bertini,I.; Gray, H. B.; Lippard, S. J.; Valentine, J. S. Eds., University Science Press, Mill Valley, CA, **1994**, pp. 365-453.
7. Bertini, I.; Ciurli, S.; Luchinat, C. *Struct. Bonding (Berlin)* **1995**, *83*, 1-53.
8. Johnson, M. K. in *Encyclopedia of Inorganic Chemistry*, King, R. B., Ed. John Wiley and Sons, NY, **1994**, pp. 1896-1915.
9. Cammack, R. *Adv. Inorg. Chem.* **1992**, ***38***, 281-322.
10. Kyritsis, P.; Dennison, C.; Sykes, A. G. *NATO ASI Ser., Ser. C* **1995**, *459 (Bioinorganic Chemistry)*, 67-76.
11. Chapman, S. K. *Perspect. Bioinorg. Chem.* **1991**, *1*, 95-140.
12. Solomon, E. I.; Lowery, M. D.; Root, D. E.; Hemming, B. L. in *ACS Advances in Chemistry Series. No. 225*, **1995**, pp.121-164.
13. Bertini, I.; Briganti, F.; Scozzafava, A. *Handb. Met.-Ligand Interact. Biol. Fluids: Bioinorg. Chem.* , Vol.1, Berthon, G., Ed., Marcel Dekker, New York, N. Y., **1995**, pp.175-191.
14. Vallee, B. L.; Auld, D. S. *Acc. Chem. Res.* **1993**, *26*, 543-551.
15. Berg, J. M.; Shi, Y. *Science* **1996**, *271*, 1081-1085.
16. Klug, A. Schwabe, J. W. R. *FASEB J.* **1995**, *9*, 597-604.
17. Christianson, D. W.; *Adv. Prot. Chem.* **1991**, *42*, 281-355.
18. Coughlan, M. P., Ed. *Molybdenum and Molybdenum-Containing Enzymes*, Pergamon, NY, **1980**.
19. Hille, R. *Biochim. Biophys. Acta*, **1994**,*1184*, 143-169.
20. Stiefel, E. I. in *Molybdenum Enzymes, Cofactors, and Model Systems*, E. I. Stiefel, D. Coucouvanis, W. E. Newton, Eds., ACS Symposium Series. No. 535, American Chemical Society, Washington, D. C., **1993**, pp.1-19.
21. Enemark, J. H.; C. G. Young, *Adv. Inorg. Chem.*, **1993**, *40*, 1-88.
22. Pilato, R. S.; Stiefel, E. I. in *Bioinorganic Catalysis*, J. Reedijk, Ed., Marcel Dekker, NY, **1993**, pp.131-188.
23. Wootton, J. C.; et al., *Biochim. Biophys. Acta*, **1991**, *1057*, 157-185.
24. Young, C. G.; Wedd, A. G. in *Encyclopedia of Inorganic Chemistry*, King, R. B., Ed. John Wiley and Sons, NY, **1994**, pp. 2330-2346.
25. Adams, M. W. W. *Ann. Rev. Microbiol.* **1993**, *47*, 627-658.
26. Johnson, J. L.; Rajagopalan, K. V.; Makund, W.; Adams, M. W. W. *J. Biol. Chem.* **1993**, *268*, 4848-4852.
27. Johnson, J.; Bastian, N. R.; Rajagopalan, K. V. *Proc. Nat. Acad. Sci. USA,* **1990**, *87*, 3090-3044.
28. Rajagopalan, K. V. in *Molybdenum Enzymes, Cofactors, and Model Systems*, E. I. Stiefel, D. Coucouvanis, W. E. Newton, Eds., ACS Symposium Series. **No. 535**, American Chemical Society, Washington, D. C., **1993**, pp. 38-49.
29. Ramao,M. ; Archer, M.; Moura, I.; Moura, J. J. G.; LeGall, J.; Engh, R.; Schneider, M.; Hof, P.; Huber, R. *Science*, **1995**, *270*, 1170-1176.
30. Schindelin, H.; Kisker, C.; Hilton, J.; Rajagopalan, K. V.; Rees, D. C. *Science*, **1996**, *272*, 1615-1621.

31. Chan, M. K.; Makund, W.; Kletzin, A.; Adams, M. W. W.; Rees, D. C. *Science*, **1995**, *267*, 1463-1469.
32. Mahadevan, C. *J. Crystallogr. Spectrosc. Res.* **1986**, *16*, 347-416.
33. Burns, R. P.; Mcauliffe, C. A. *Adv. Inorg. Chem. Radiochem.* **1979**, *22*, 303-348.
34. McCleverty, J. A. *Progr. Inorg. Chem.* **1968**, *10*, 49-221.
35. Pilato, R. S.; Eriksen, K.; Greaney, M. A.; Gea, Y.; Taylor, E. C.; Goswami, S.: Kilpatrick, L.: Spiro, T. G.; Rheingold, A. L.; Stiefel, E. I. in *Molybdenum Enzymes, Cofactors, and Model Systems*, E. I. Stiefel, D. Coucouvanis, W. E. Newton, Eds., ACS Symposium Series. No. 535, American Chemical Society, Washington, D. C., **1993**, pp. 83-97.
36. Garner, C. D.; Armstrong, E. M.; Ashcroft, M. J.; Austerberry, M. S.; Birks, J. H.; Collison, D.; Goodwin, A. J., Larsen, L. Rowe, D. J.: Russell, J. R. in *Molybdenum Enzymes, Cofactors, and Model Systems*, E. I. Stiefel, D. Coucouvanis, W. E. Newton, Eds., ACS Symposium Series. No. 535, American Chemical Society, Washington, D. C., **1993**, pp. 98-113.
37. Scott, R. A. ; Mauk, G. *Cytochrome c: A Multidisciplinary Approach,* University Science Books, Mill Valley, CA **1996**.
38. George, G. N.; Richards, T.; Bare, R. E.; Gea, Y.; Prince, R. C.; Stiefel, E. I.; Watt, G. D. *J. Amer. Chem. Soc.*, **1993**, *115*, 7716-7718.
39. Ortiz de Montellano, Ed., Cytochrome P450, 2nd Ed., **1995**, Plenum, NY.
40. Poulos, T. L. *Curr. Opin. Struct. Biol.* **1995**, *5*, 767-774.
41. Watanabe, Yoshihito; Groves, John T. in *Enzymes,* 3rd Ed., Sigman, D. S. Ed., *Vol. 20*, Academic, San Diego, CA, **1992**, pp. 405-452.
42. Theil, E. C. *Adv. Enz. Rel. Areas Mol. Biol.* **1991**, 67, 421-449.
43. Mascotti, D. P. ; Rup, D.; Thach, R. E. *Annu. Rev. Nutr.* **1995**, *15*, 239-261.
44. Beinert, H.; Kiley, P. *FEBS Letters*, **1996**, *382*, 218-219.
45. Rouault, T. A.; Klausner, R. D. *Trends Biochem. Sci.* **1996**, *21*, 174-177.
46. Eisenstein, R. S.; Munro, H. N. *Enzyme* **1990**, *44*, 42-58.
47. Leipold, E. A.; Guo, B. *Annu. Rev. Nutr.* **1992**, *12*, 345-368.
48. Zhou, J.; Holm, R. H. *J. Am. Chem. Soc.,* **1995**, *117*, 11,353-11,354.
49. Zhou, J.; Hu, Z; Münck, E.; Holm, R. H. *J. Am. Chem. Soc.* **1995**, *117*, 1966-1980.
50. Hagen, W. R.; Pierik, A. J.; Veeger, C. *J. Chem Soc. Faraday Trans. I* **1989**, *85*, 4083-4090.
51. Stokkermans, J. P. W. G.; Pierik, A. J.; Wolbert, R. B. G.; Hagen, W. R.; van Dongen, W. M. A. M. *Eur. J. Biochem.* **1992**, *208*, 435-442.
52. Kanitzidis, M. G.; Hagen, W. R.; Dunham, W. R.; Lester, R. K.; Coucouvanis, D. *J. Am. Chem. Soc.* **1985**, ***107***, 953-961.
53. Rees, D. C.; Kim, J.; Georgiadis, M. M.; Komiya, H.; Chirino, A. J.; Woo, D.; Schlessman, J.; Chan, M. K.; Joshua-Tor, L.; Santillan, G.; Chakrabarti, P. in *Molybdenum Enzymes, Cofactors, and Model Systems*, E. I. Stiefel, D. Coucouvanis, W. E. Newton, Eds., ACS Symposium Series. No. 535, American Chemical Society, Washington, D. C., **1993**, pp. 170-185.
54. Bolin, J. T.; Campobasso, N.; Muchmore, S. W.; Morgan, T. V.; Mortenson, L. E. in *Molybdenum Enzymes, Cofactors, and Model Systems*, E. I. Stiefel, D. Coucouvanis, W. E. Newton, Eds., ACS Symposium Series. No. 535, American Chemical Society, Washington, D. C., **1993**, pp. 186-195.
55. Kim, J.; Woo, D.; Rees, D. C. *Biochemistry* **1993**, *32*, 7104-7115.
56. Chan, M. K.; Kim, J.; Rees, D. C. *Science* **1993**, *260*, 792-794.
57. Musser, S. M.; Stowell, M. H. B.; Chan, S. I. in *Adv.in Enzym. and Rel. Areas Mol Biol.* Meister, A. Ed., Wiley, NY, **1995**, pp.79-208.
58. Iwata, S.; Ostermeier, B.; Ludwig, B.; Michel, H. *Nature*, **1995**, *376*, 660-669.
59. Tsukihara, T.; Aoyama, H.; Yamashita, E.; Tomizaki, T.; Yamaguchi, H.;

Shinzowa-Itoh, K.; Nakashima, R.; Yaono, R.; Yoshikawa, S.; *Science*, **1995**, *269*, 1069-; *272*, 1136-1144.
60. Kadenbach, B. *Angew. Chem. Int. Ed. Engl.* **1995**, *34*, 2635-2637.
61. Antholine, W. E.; Kastrau, D. H. W.; Steffens, G. C. M.; Buse, G.; Zumft, W. G.; Kroneck, P. M. H. *Eur. J. Biochem.* **1992**, *209*, 875-881.
62. Houser, R. P.; Young Jr., V. G.; Tolman, W. B. *J. Am. Chem. Soc.* **1996**, *118*, 2101-2102.
63. Volbeda, A.; Charon, M.-H.; Piras, C.; Hatchikian, E. C.; Frey, M.; Fontecilla-Camps, J. C. *Nature* **1995**, *373*, 580-587.
64. Halcrow, M. A. *Angew. Chem. Int. Edit. Eng.* **1995**, *34*, 1193-1195.
65. Albracht, S. P. J. *Biochim. Biophys. Acta* **1994**, *1188*, 167-204.
66. Franolic, J. D.; Wang, Y. W.; Millar, M. *J. Am. Chem. Soc.* **1992**, *114*, 6587-6588.
67. Cha, M.; Gatlin, C. L. ; Critchlow, S. C.; Kovacs, J. A. *Inorg. Chem.* **1993**, *32*, 5868-5877.
68. Morganian, C. A., Vazir, H.; Baidya, N.; Olmstead, M. M.; Mascharak, P. K. *J. Am. Chem. Soc.* **1995**, *117*, 1584-1594.
69. James, T. L.; Smith, D. M.; Holm, R. H. *Inorg. Chem.* **1994**, *33*, 4869-4877.
70. Musie, G.; Farmer, P. J.; Tuntulani, T.; Reibenspies, J. H.; Darensbourg, M. Y. *Inorg. Chem.* **1996**, *35*, 2176-2183.
71. Eady, R. R. *Met. Ions Biol. Syst.* **1995**, *31*, 363-405.
72. Coucouvanis, D. in *Molybdenum Enzymes, Cofactors, and Model Systems*, E. I. Stiefel, D. Coucouvanis, W. E. Newton, Eds., ACS Symposium Series. No. 535, American Chemical Society, Washington, D. C., **1993**, pp. 302-331.
73. Bazhenova, T. A.; Shilov, A. E. *Coord. Chem. Rev.* **1995**, *144*, 69-145.
74. Deng, H.; Hoffmann, R. *Angew. Chem. Int. Edit. Eng.* **1993**, *32*, 1062-1064.
75. Dance, I. G. *Aust. J. Chem.* **1994**, *47*, 979-990.
76. Leigh, G. J. *Eur. J. Biochem.* **1995**, *229*, 14-20.
77. Stavrev, K. K.; Zerner, M. C. *Chem. Eur. J.* **1996**, *2*, 83-87.
78. Coucouvanis, D. *Adv. Inorg. Biochem.* **1994**, *9*, 75-122.
79. McRee, D. E.; Richardson, D. C.; Richardson, J. S.; Siegel, W. M. *J. Biol. Chem.* **1986**, *261*, 10277-10281.
80. Zhou, E.; Cai, L.; Holm, R. H.; *Inorg. Chem.* **1996**, *35*, 2767-2772.
81. Hausinger, R. P.; Eichhorn, G. L.; Marzilli, L. G., eds., *Mechanisms of Metallocenter Assembly*, VCH Publishers, NY **1996**.
82. Solomon, E. I.; Lowery, M. D. *Science* **1993**, *259*, 1575-1581.
83. Welch, G. R.; Somogyi, B.; Damjanovich, S. *Prog. Biophys. Molec. Biol.* **1982**, *39*, 109-146.
84. J. Reedijk, Ed., *Bioinorganic Catalysis*, Marcel Dekker, NY, **1993**.
85. Stiefel, E. I. in *Bioinorganic Catalysis*, J. Reedijk, Ed., Marcel Dekker, NY **1993**, pp. 21-27.
86. Alonso-Vante, N; Tributsch, H. *Electrochem. Novel Mater.* Lipkowski, J.; Ross, P. N. Eds., VCH, New York, NY, **1994**, pp. 1-63.
87. Roundhill, D. M. *Photochemistry and Photophysics of Transition Metal Complexes,* Plenum, NY, **1994**, pp. 103-105.
88. Bryant, R. D.; Kloeke, F. V. O.; Laishley, E. J. *Appl. Env. Microbiol.* **1993**, 59, 491-495.
89. Miki, Y.; Nakazato, D.; Ikuta, H.; Uchida, T; Wakihara, M. *J. Power Sources* **1995**, *54*, 508-510.
90. Jager-Wasdau, A.; Lux-Steiner, M. C.; Bucher, E. *Solid State Phenomena*, **1994**, *37-38*, 479-484.
91. Bino, A.; Johnston, D. C.; Goshorn, D. P.; Halbert, T. R.; Stiefel, E. I. *Science*, **1988** *241*, 1479-1481.
92. Winer, W. O. *Wear* **1967**, *10*, 422-452.

93. Farr, J. P. G. *Wear* **1975**, *35*, 1-22.
94. Dickenson, R.; Pauling, L. *J. Am. Chem. Soc.* **1923**, *45*, 1466-1471.
95. Mitchell, P. C. H. *Wear*, **1984**, *100*, 281-300.
96. Yamamoto, Y.; Gondo, S. *Tribology Trans.* **1989** *32*, 251-257.
97. Weisser, O.; Landa, S. *Sulfide Catalysts, Their Properties and Applications* Pergamon Press, NY, **1973**.
98. Stiefel, E. I. in *Kirk-Othmer Encyclopedia of Chemical Technology*, Fourth Edition, Vol.16 **1995**, pp. 940-962.
99. H. Topsøe, B. S. Clausen, F. E. Massoth, *Hydrotreating Catalysis: Science and Technology*, Springer Verlag, Berlin **1996**.
100. Chianelli, R. R.; Daage, M.; Ledoux, M. J. "Fundamental Studies of Transition Metal Sulfide Catalytic Materials" *Advances in Catalysis*, **1994**, *40*, 177-232.
101. Delmon, B. *Bull. Soc. Chim. Belg.* **1995,** *104*, 173-187.
102. Ho, T. C. *Catal. Rev. - Sci. Eng.* **1988**, *30*, 117-160.
103. Pratt, K. C. *Chem. Austr.* **1989**, *56*, 192-195.
104. Chianelli, R. R. *Catal. Rev. - Sci. Eng.* **1984**, *26*, 361-393.
105. Pecoraro, T. A.; Chianelli, R. R. *J. Catal.* **1981**, *67*, 430-445.
106. Lacroix, M.; Boutarfa, N.; Guillard, C.; Vrinat, M.; Breysse, M. *J. Catal.* **1989**, *120*, 473-477.
107. Hillerova, E.; Zdrazil, M. *Collect. Czech. Chem. Commun.* **1989**, *54*, 2648-2656.
108. Harvey, T. G.; Matheson, T. W. *J. Catal.* **1986**, *101*, 253-261.
109. Harris, S.; Chianelli, R. R. *J. Catal.* **1984**, *86*, 400-412.
110. Burdett, J. KL.; Chung, J. T. *Surf. Sci. Lett*. **1990**, *236*, L353-L357.
111. Nørskov, J. K.; Clausen, B. S.; Topsøe, H. *Catal. Lett.* **1992**, *13*, 1-8.
112. Angelici, R. J. *Coord. Chem. Rev.* **1990**, *105*, 61-76.
113. Angelici, R. J. *Bull. Chem. Soc. Belg.* **1995**, *104*, 265-282.
114. Bianchini, C.; Frediani, P.; Herrera, V.; Jimenez, M. V.; Meli, A., Rincon, L.; Sanchez-Delgado, R. Vizza, F. *J. Am. Chem. Soc.* **1995**, *117*, 4333-4346.
115. Boorman, P. M.; Gao, X.; Fait, J. F.; Parvez, M. *Inorg. Chem.* **1991**, *30*, 3886-3893.
116. Druker, S. C.; Curtis, M. D. *J. Am. Chem. Soc.* **1995**, *117*, 6366-6367.
117. Myers, A. W.; Jones, W. D.; McClements, S. M. *J. Am. Chem. Soc.* **1995**, *117*, 11704-11709.
118. Koczaja Dailey, K. M.; Rauchfuss, T. B.; Rheingold, A. L.; Yap, G. P. A. *J. Am. Chem. Soc.* **1995**, *117*, 6396-6397.
119. Garcia, J. J.; Mann, B. E.; Adams, H.; Bailey, N. A.; Maitlis, P. M. *J. Am. Chem. Soc.* **1995**, *117*, 2179-2186.
120. Dullaghan, C. A.; Sun, S.; Carpenter, G. B.; Weldon, B.; Sweigart, D. A. *Angew. Chem. Int. Ed. Engl.* **1995**, *22*, 212-214.
121. Rakowski DuBois, M.; Van Derveer, M. C.; DuBois, D. L.; Haltiwanger, R. C.; Miller, W. K. *J. Am. Chem. Soc.* **1980**, *102*, 7456-7461.
122. Bianchini, C.; Mealli, C.; Meli, A.; Sabat, M. *Inorg. Chem.* **1991**, *30*, 3886-3893.
123. Lough, A. J.; Park, S. Ramachandran, R.; Morris, R. H. *J. Am. Chem. Soc.* **1994**, *116*, 8356-8357.
124. Mitchell, P. C. H.; Green, D. A.; Payon, E.; Evans, A. E. *J. Chem. Soc. Faraday Trans.*, **1995**, *91*, 4467-4469.
125. Stiefel, E. I. in *Comprehensive Coordination Chemistry*, Wilkinson, G.; Gillard, R. D.; McCleverty, J. A., Eds., *Vol. 3,* Pergamon, NY, **1987**, pp. 559-577.
126. Stiefel, E. I. *Prog. Inorg.Chem.* **1977**, *22*, 1-223.
127. Garner, C. D.; Charnock, J. M. in *Comprehensive Coordination Chemistry*, Wilkinson, G.; Gillard, R. D.; McCleverty, J. A., Eds., Pergamon, NY, **1987**, pp. 1329-1374.
128. Hille, R. in *Molybdenum Enzymes, Cofactors, and Model Systems*, E. I. Stiefel, D. Coucouvanis, W. E. Newton, Eds., ACS Symposium Series. No. 535, American Chemical Society, Washington, D. C., **1993**, pp. 22-37.

129. Holm, R. H.; Berg, J. M. *Acc. Chem. Res.* **1986**, *19*, 363-370.
130. Müller, A.; Jostes, R.; Cotton, F. A. *Angew. Chem. Int. Edit. Eng.* **1980**, *19*, 875-882.
131. Shibahara, T. *Adv. Inorg. Chem.*, **1991**, *37*, 143-173.
132. Shibahara, T. *Coord Chem. Rev.*, **1993**, *123*, 73-147.
133. Saito, T., in *Chalcogenide Clusters Complexes of the Early Transition Metals*, Chisholm, M.; VCH, New York, **1995**, pp. 63-164.
134. Ooi, B.-L.; Sharp, C.; Sykes, A. G. *J. Am. Chem. Soc.* **1989**, *111*, 125-130.
135. Coyle, C. L.; Eriksen, K. A.; Farina, S.; Francis J.; Gea, Y.; Greaney, M. A.; Guzi, P. J.; Halbert, T. R.; Murray, H. H.; Stiefel, E. I. *Inorg. Chim. Acta* **1992**, *198-200*, 565-575.
136. Chrisholm, M. H. *in Early Transition Metal Clusters with π-Donor Ligands,* Chisholm, M. H., Ed., VCH, NY, **1995**, 165-215.
137. Chevrel, R.; Sergent, M.; Prigent, J. *J. Solid State Chem.* **1971,** 3, 515-519; Chevrel, R.; Hirrien, M.; Sergent, M. *Polyhedron*, **1986**, *5*, 87-94.
138. Zhang, X.; McCarley, R. E. *Inorg. Chem.* **1995**, *34*, 2678-2683.
139. Ebihara, M.; Kawamura, T.; Isobe, K.; Saito, K. *Bull. Chem. Soc. Jpn.* **1994**, *67*, 3119-3121.
140. Jørgensen, C. K. *Oxidation Numbers and Oxidation States*, Springer-Verlag, NY, **1969**.
141. Müller, A.; Diemann, E. in *Comprehensive Coordination Chemistry*, Wilkinson, G.; Gillard, R. D.; McCleverty, J. A., Eds., Pergamon, NY, **1987**. pp. 515-550.
142. Garner, C. D.; Charnock, J. M. in *Comprehensive Coordination Chemistry*, Wilkinson, G.; Gillard, R. D.; McCleverty, J. A., Eds., *Vol. 3*, Pergamon, NY, **1987**. pp. 1329-1374.
143. Livingstone, S. E. in *Comprehensive Coordination Chemistry*, Wilkinson, G.; Gillard, R. D.; McCleverty, J. A., Eds., *Vol. 2*, Pergamon, NY, **1987**. pp. 633-659.
144. Holm, R. H. *Chem. Rev.* **1987**, *87*, 1401-1449.
145. B. E. Schultz, R. Hille, R. H. Holm, *J. Am. Chem. Soc.* **1995**, *117*, 827-828.
146. Hille, R.; Sprecher, H. *J. Biol. Chem.* **1987**, *262*, 10,914-10,917.
147. Stiefel, E. I.; Eisenberg, R.; Rosenberg, R. C.; Gray, H. B. *J. Am. Chem. Soc.*, **1965**, *88*, 2956-2966.
148. Eisenberg, R. *Progr. Inorg. Chem.* **1970**, *12*, 295-369.
149. Tian, Z.-Q.; Donahue, J. P.; Holm. R. H. *Inorg. Chem.* **1995**, *34*, 5567-5572.
150. Greenwood, N. N.; Earnshaw, A. *Chemistry of the Elements*, Pergamon, NY **1984**, 757-881.
151. Cotton, F. A.; Wilkinson, G. *Advanced Inorganic Chemistry*, 5th Edit., Wiley Interscience, **1988**, pp 491-543.
152. Fergusson, J. E. *Inorganic Chemistry and the Earth*, Pergamon, NY, **1982**, pp. 221-228.
153. Müller, A.; Krebs, B. *Sulfur: Its Significance for Chemistry, for the Geo-, Bio- and Cosmosphere and Techology,* Elsevier, NY, **1984**.
154. Nickless, G., Ed., *Inorganic Sulphur Chemistry*, Elsevier, NY, **1968**.
155. Goggin, P. L. in *Comprehensive Coordination Chemistry*, Wilkinson, G.; Gillard, R. D.; McCleverty, J. A., Eds., *Vol. 2*, Pergamon, NY, **1987**. pp. 487-503.
156. Tomita, A.; Mitsuru, S. *Inorg. Chem.* **1994**, *33*, 5825-5830.
157. Miller, K. F.; Wherland, S.; Bruce, A. E.; Corbin, J. L.; Stiefel, E. I. *J. Am. Chem. Soc.* **1980**, *102*, 5102-5104.
158. Coyle Lee, C. L.; Halbert, T. R.; Pan, W.-H.; Harmer, M. A.; Wei, L.; McKenna, S.; Dim, C. O. B.; Miller, K. F.; Bruce, A. E.; Pariyadath, N.; Corbin, J. L.; Wherland, S.; Stiefel, E. I. *Inorg. Chim. Acta*, **1996**,*70*, 147-160.
159. Lange, B. A.; Libson, K.; Lydon, J. D.; Elder, R. C.; Deutsch, E. *Inorg. Chem.* **1979**, *18*, 303-311.
160. Wang, R.; Mashuta, M.; Richardson, J. F.; Noble, M. *Inorg. Chem.* **1996**, *35*, 3022-3030.

161. Müller, A.; Diemann, E. in *Comprehensive Coordination Chemistry*, Wilkinson, G.; Gillard, R. D.; McCleverty, J. A., Eds., Pergamon, NY, **1987**. pp. 559-577.
162. W.-H. Pan, M. A. Harmer, T. R. Halbert and E. I. Stiefel, *J. Am. Chem. Soc.* **1984**,*106*, 459-460.
163. Coyle, C. L. ; Harmer, M. A. ; George, G. N. ; Stiefel, E. I. *Inorg. Chem.* **1990** *29*, 14-19.
164. Taube H. *Electron Transfer Reaction of Complex Ions in Solution*, Academic Press, NY, **1970**.
165. Harmer, M. A.; Halbert, T. R.; Pan, W.-H.; Coyle, C. L.; Cohen, S. A.; Stiefel, E. I. *Polyhedron*, **1986**, *5*, 341-347.
166. Halbert, T. R.; Hutchings, L L.; Rhodes, R.; Stiefel, E. I. *J. Am. Chem. Soc.* **1986**,*108*, 6437-6438.
167. Pan, W.-H.; Halbert, T. R.; Hutchings, L L.; Stiefel, E. I. *J. Chem, Soc. Chem. Commun.* **1985**, 927-928.
168. Cohen, S. A.; Stiefel, E. I. *Inorg. Chem.* **1985**, 24, 4657-4662.
169. Wei, L.; Halbert, T. R.; Murray, H. H.; Stiefel, E. I. *J. Am. Chem. Soc.* **1990**,*112*, 6431-6433.
170. Murray, H. H.; Wei, L.; Sherman, S. E.; Greaney, M. A.; Eriksen, K. A.; Carstensen, B.; Halbert, T. R.; Stiefel, E. I. *Inorg. Chem.* **1995**, 34, 841-853.
171. Lever, A. B. P. Inorganic Electronic Spectroscopy, 2nd Edit., Elsevier, **1984**.
172. Gea, Y.; Greaney, M.A.; Coyle, C. L.; Stiefel, E. I. *J. Chem. Soc. Chem. Commun. 1992*, 160-161.
173. Rao, C. N. R.; Gopalakrishnan, J. *New Directions in Solid State Chemistry*, Cambridge University Press, NY, **1989**.
174. Wold, A.; Dwight, K. *Solid State Chemistry*, Chapman & Hall, NY, **1993**.
175. Lee, S. C.; Holm, R. H. *Angew. Chem. Int. Ed. Eng.* **1990**, *29*, 840-856.
176. Krebs, B.; Henkel, G. *Angew. Chem. Int. Ed. Eng.* **1991**, *30*, 769-788.
177. Allman, R.; Baumann, I.; Kutoglu, A.; Rösch, H.; Hellner, E. *Naturwissenschaften* **1964**, *51*, 263-264.
178. Murray, H. H.; Kelty, S. P.; Chianelli, R. R.; Day, C. S. *Inorg. Chem.*, **1994**, *33*, 4418-4420.
179. Berg, J. M.; Holm, R. H., in *Iron-Sulfur Proteins*, Spiro, T. G., Ed., John Wiley, NY, **1982**.
180. Holm, R. H. *Adv. Inorg. Chem.*, **1992**, *38*, 1-71.
181. M. A. Greaney, C. L. Coyle, R. S. Pilato, E. I. Stiefel, *Inorg. Chim. Acta*, **1991**, *189*, 81-96.
182. Müller, A.; Diemann, E. in *Nitrogen Fixation: Chem.-Biochem.-Genet. Interface*, Müller, A.; Newton, W.E., Eds. Plenum, New York, NY, **1981**, pp.183-210.
183. Pan, W.-H.; Johnston, D. C.; Mckenna, S. T.; Chianelli, R. R.; Halbert, T. R.; Hutchings, L. L.; Stiefel, E. I. *Inorg. Chim. Acta*, **1985**, *97*, L17-L19.
184. Müller, A.; Bogge, H.; Schimanski, U.; Penk, M.; Nieradzik, K.; Dartmann, M.; Krickemeyer, E.; Schimanski, J.; Romer, C.; Romer, M.; Dornfeld, H.; Wienboker, U.; Hellmann, W.; Zimmerman, M. *Monatsh. Chem.* **1989**, *120*, 367-391.
185. Bernholc, J.; Stiefel, E. I. *Inorg. Chem.* **1985**, 24, 1323-1330.
186. Müller, A.; Diemann, E.; Jostes, R.; Bogge, H. *Angew. Chem. Int. Ed. Eng.* 1981, **20**, 934-954.
187. Greaney, M. A.; Coyle, C. L.; Harmer, M. A.; Jordan, A.; Stiefel, E. I. *Inorg. Chem.* **1989**, 28, 912-920.
188. Sécheresse, F.; F. Bernes, S.; Robert, F.; Jeannin, Y. *Bull. Soc. Chim. Fr* **1995**, *132*, 1029-1037.
189. Bernes, S.; Sécheresse, F.; Jeannin, Y. *Inorg. Chim. Acta* **1992**, *191*, 11-13.
190. Brunner, H.; Wachter, J. J. *J. Organomet. Chem.* **1982**, *240*, c41-c44.

191. Cohen, S. A.; Halbert, T. R.; Stiefel, E. I. *Organometallics*, **1985**, *4*, 1689-1690.
192. Stiefel, E. I;. Halbert, T. R.; Coyle, C. L.; Wei, L.; Pan, W.-H.; Ho, T. C.; Chianelli, R. R.; Daage, M. *Polyhedron 1989*, *8*, 1625-1629.
193. Harris, S. *Polyhedron* **1989**, *8*, 2843-2882.
194. Harris, S. *Inorg. Chem.* **1987**, *26*, 4278-4285.
195. Rutledge, C. A.; Humanes, M.; Li, Y.-J.; Sykes, A. G. *J. Chem. Soc. Dalton Trans.* **1994**, 1274-1282.
196. Murata, T.; Mizobe, Y.; Gao, h.; Ishii, Y.; Wakabayashi, T.; Nakano, F.; Tanase, T.; Yano, S.; Hidai, M.; Echizen, I.; Nanikawa, H; Motomura, S. *J. Am. Chem. Soc.* **1995**,*116*, 3389-3398.
197. Saito, T.; Yamamoto, N.; Yamagata, T.; Imoto, H. *J. Am. Chem Soc.* **1988**, *110*, 1646-1647.
198. Fedin, V. P.; Kolesov, B. A.; Mironov, Y. V. *Polyhedron* **1989**, *8*, 2419-2423.
199. Fedin, V. P.; Sokolov, M. N.; Geras'ko, O. A.; Kolesov, B. A.; Federov, V. Y.; Mironov, A. V.; Yufit, D. S.; Slovohotov, Y. L.; Struchkov, Y. T. *Inorg. Chim. Acta,* **1990**, *175*, 217-229.
200. Federov, V. Y.; Mironov, Y. V.; Kuzmina, O. A.; Fedin, V. P. *Zh. Neorg. Khim.* **1986**, *31*, 2476-2479.
201. Spangenberg, M.; Bronger, W. *Angew. Chem. Int. Ed. Eng.* **1978**, *17*, 368-369.
202. Nemrudy, A.; Schollhorn, R. *J. Chem. Soc. Chem. Commun.* **1994**, 2617-2618.
203. Bronger, W.; Loevenich, M.; Schmitz, D. *J. Alloys. Comps.* **1994**, *216*, 25-28.
204. Fischer, C.; Fiechter, S. Tributsch, H.; Reck, G.; Schultz, B. *Ber. Bunsenges Phys. Chem.* 1992, **96**, 1652-1658.
205. Long, J. R.; Williamson, A. S.; Holm, R. H. *Angew. Chem. Int. Ed. Eng.* **1995**, *34*, 226-229.
206. Long, J. R.; McCarty, L. S.: Holm, R. H. *J. Am. Chem Soc.* **1996**, *118*, 4606-4616.
207. Birnbaum, J.; Laurie, J. C. V.; Rakowski DuBois, M. *Organometallics* **1990**, *9*, 156-164.
208. Schrauzer, G. N.; Zhang, C.; Chadha, R. *Inorg. Chem.* **1990**, *28*, 4104-4107.
209. Dilworth, J. R.; Hutchinson, J.; Zubieta, J. A. *J. Chem. Soc., Chem. Commun.* **1983**, 1034-1035.
210. Sellmann, D. *J. Organometal. Chem.* **1989**, *372*, 99-107.
211. Kubas, J. G.; Ryan, R. R. *J. Am. Chem Soc.* **1985**,*107*, 6138-6140.
212. Schenk, W. *Angew. Chem. Int. Ed. Eng.* **1987**, *26*, 98-109.
213. Rabinovich, D.; Parkin, G. *J. Am. Chem Soc.* **1991**,*113*, 5904-5905.
214. Crabtree, R. H. *Acc. Chem. Res.* **1990**, *23*, 95-101.
215. Kubas, J. G. *Acc. Chem. Res.* **1988**, *21*, 120-128.
216. Laurie, J. V. C.; Duncan, L.; Haltiwanger, R. C.; Weberg, R. T.; Rakowski DuBois, M. *J. Am. Chem Soc.* **1986**, *108*, 6234-6241.
217. Stiefel, E. I. *Proc. Natl. Acad. Sci., U.S.A.* **1973**, *70*, 988-993.
218. Stiefel, E. I.; Gardner, J. K. *J. Less-Common Metals* **1974**, *36*, 521-533.
219. Stiefel, E. I.; Thomann, H.; Jin, H.; Bare, R. E.; Morgan, T. V.; Burgmayer, S. J. N.; Coyle, C. L. In *Metal Clusters in Proteins*, Que, L., Ed. ACS Symposium Series No.372, American Chemical Society, Washington, DC, **1988**, pp. 372-389.
220. Templeton, J. L. *Adv. Organomet. Chem.* **1989**, *29*, 1-100.
221. Otsuka, S.; Nakamura, *Adv. Organomet. Chem.* **1976**, *14*, 245-283.
222. Green, M. *J. Organometal. Chem.* **1986**, *300*, 93-109.

223. Newton, W. E.; McDonald, J. W.; Corbin, J. L.; Ricard, L.; Weiss, R. *Inorg. Chem.* **1980**,*19*, 1997-2006.
224. Herrick, R. S.; Templeton, J. L. *Organometallics*, **1982**, *1*, 842-851.
225. Wakabayashi, T.; Ishii, Y.; Murata, T.; Mizobe, Y.; Hidai, M. *Tetrahedron* **1995**, *36*, 5585-5588.
226. Rakowski DuBois, M; Haltiwanger, R. C.; Miller, D. J.; Glazmaier, G. *J. Am. Chem Soc.* **1979**, *101*, 5245-5252.
227 Kawaguchi, H.; Tatsumi, K. *J. Am. Chem Soc.* **1995**, *117*, 3885-3886.
228. Bolinger, C. M.; Rauchfuss, T. B. *Inorg. Chem.* **1982**, *21*, 3947-3954.
229. Pilato, R. S.; Eriksen, K. A.; Greaney, M. A.; Stiefel, E. I.; Goswami, A.; Kilpatrick, L.; Spiro, T. G.; Taylor, E. C.; Rheingold, A. L. *J. Am. Chem Soc.* **1991**, *113*, 9372-9374.
230. Schrauzer, G. N. *Acc. Chem. Res*. **1969**, *2*, 72-80.
231. Halbert, T. R.; Pan, W.-H.; Stiefel, E. I. *J. Am. Chem Soc.* **1983**, *105*, 5476-5477.
232. Coucouvanis, D.; Toupadakis, A.; Lane, J. D.; Koo, S. M.; Kim, C. G.; Hadjikyriacou, A. *J. Am. Chem Soc.* **1991**, *113*, 5271-5282.
233. Burgess, B. K.; Wherland, S.; Newton, W. E.; Stiefel, E. I. *Biochemistry* **1981**, *20*, 5140-5146.
234. Thorneley, R. N. F.; Lowe, D. J. *Biochem. J.* **1984**, *224*, 887-894.
235. Thorneley, R. N. F.; Eady, R. R.; Lowe, D. J. *Nature* **1978**, *272*, 557-558.
236. Davis, L. C. *Arch. Biochem. Biophys.* **1980**, *204*, 270-276.
237. Hadfield, K. L.; Bulen, W. E. *Biochemistry*, **1969**, *8*, 5103-5108.
238. Stiefel, E. I.; Newton, W. E.; Watt, G. D.; Hadfield, K. L.; Bulen, W. A. in *Bioinorganic Chemistry II.* ACS Advances in Chemistry, No. 162., Raymond, K. N., Ed.,**1977**, 353-388.
239. Johnson, B. F. G.; Haymore, B. L.; Dilworth, J. R. in *Comprehensive Coordination Chemistry*, Wilkinson, G.; Gillard, R. D.; McCleverty, J. A., Eds., *Vol. 2*, Pergamon, NY, **1987**, pp. 99-159.
240. Bazhenova, T. A.; Shilov, A. E. *Coord. Chem. Rev.* **1995**, *144*, 69-145.
241. King, R. B. *J. Organomet. Chem.* **1995**, *500*, 187-194.
242. Sellmann, D.; Soglowok, W.; Knoch, F.; Moll, M. *Angew. Chem. Int. Ed. Eng.* **1989**, *28*, 1269-1270.
243. Curtis, M. D. *Applied Organomet. Chem.* **1992**, *6*, 429-436.
244. Ho, T. C. in *Catalytic Hydroprocessing of Petroleum and Distillates*, Oballa, M. C.; Shih, S. S., Eds., Marcel Dekkert, NY, **1994**.
245. Stiefel, E. I.; Pan, W. H.; Chianelli, R. R.; Ho, T. C. *U. S. Patent 4,581,125*, **1986**.

Biological and Model Systems

Chapter 2

Rubredoxin from *Clostridium pasteurianum*: Mutation of the Conserved Cysteine and Glycine Residues

Mustafa Ayhan, Zhiguang Xiao, Megan J. Lavery, Amanda M. Hammer, Sergio D. B. Scrofani, and Anthony G. Wedd[1]

School of Chemistry, University of Melbourne, Parkville, Victoria 3052, Australia

The rubredoxin from *Clostridium pasteurianum* is an electron transfer protein featuring an Fe(S-Cys)$_4$ active site with *pseudo* 2-fold symmetry. The effect upon its properties of substitution of the metal atom, the sulfur ligand atoms and conserved glycine amino acid residues is presented. Mutant proteins have been expressed in *Escherichia coli* in stable form. Substitution of cysteine ligand by serine at positions 6, 9, 39 and 42 in turn produces proteins whose properties reflect both the O for S substitution and the *pseudo* symmetry of the native site. An X-ray crystal structure of the C42S protein confirms the presence of an Fe(III)-O$^{\gamma}$-Ser link. The O$^{\gamma}$-atom is protonated in the Fe(II) and Cd(II) forms. Glycine residues at positions 10 and 43, related by the *pseudo* symmetry, have been mutated to alanine and valine. Physical properties of the four single mutant proteins, as well as two double mutants (G10V/G43A and G10V/G43V), are modified by steric interactions of the new β and γ carbon substituents with the CO functions of C9 and C42 and with other adjacent groups.

The rubredoxins (Rd) are the simplest of the iron-sulfur proteins and feature a single Fe(S-Cys)$_4$ in a protein of molar mass approximately 6,000 Da (*1-5*). Figure 1 emphasizes the *pseudo* 2-fold symmetry present in the vicinity of the active site for Rd from the dinitrogen-fixing bacterium *Clostridium pasteurianum* (*Cp*).

Rds are presumed to be electron transport proteins and this role has been demonstrated in the sulfate-reducing bacterium *Desulfovibrio gigas* where specific redox interactions occur between Rd and NADH-rubredoxin oxidoreductase and rubredoxin-dioxygen oxidoreductase enzymes (*4,6*). An electron transfer function has been identified in the aerobe *Pseudomonas aleovorans* but here the protein is more complex, with two iron sites in a molar mass of 15,000 Da (*7*). In *Desulfovibrio* bacteria, there are a number of proteins that contain at least two iron atoms (*8-11*). In addition to Rd-like sites (*10*), some of these contain other iron centers which feature less than four cysteinyl ligands (*8,9*).

Thirteen Rds have been sequenced and all conserve the four Cys ligands (6, 9, 39 and 42 in Rd*Cp*) and the three residues (G10, P40, G43) involved in the formation of

[1]Corresponding author

0097–6156/96/0653–0040$15.00/0

the tight turns of the "chelate loops" defined by residues 5-11 and 38-44 (Scheme Ia) (*4,12*). In addition, each of the centers defined by X-ray crystallography (*4*) exhibits the *pseudo* 2-fold symmetry for both the $Fe(S\text{-}Cys)_4$ site and the loops and features six NH...S interactions (Figure 2). Sulfur ligand atoms S-C9 and S-C42 are close to the molecular surface while S-C6 and S-C39 are buried (*2,4*). Using the type I-IV classification for hydrogen bonds (*13*), buried metal ligands Cys S(i) (i = 6, 39) interact with the peptide NH function i + 2 (type I) and i + 3 (type III) while surface metal ligands Cys S(i + 3) interact with i + 5 (type II) (Scheme Ib; Figure 2).

Rd*Cp* has been expressed in *Escherichia coli* (*12,14,15*), providing recombinant protein for further study (*15,16*). Generation of mutant forms has begun (*15,17,18*). A detailed understanding of the structure and function of the active site can be pursued by variation of the metal atom, the ligand atoms and the surrounding protein structure, including the NH...S interactions. Initial excursions into each of these aspects is discussed below, starting with the production of mutant proteins.

Characterization of Recombinant and Mutant Proteins.

The Rd*Cp* gene has been cloned previously and the recombinant protein expressed in *E. coli* as a mixture of forms containing Zn(II) as well as the native metal Fe(III) (*12*). In contrast to the native protein, the N terminus was not formylated in either recombinant form. The protein sequence was shown to vary from the previously reported sequence (*19*) in that asparagine is present at positions 14 and 22 and glutamine at position 48, rather than the corresponding acids.

The sequence was confirmed in the present work which also reports high yield expression (40 mg L^{-1} of starting culture) of non-formylated recombinant protein via the pKK223-3/JM109 system after reconstitution of the accompanying Zn(II) forms with Fe(III) (*15*). The reconstitution can be carried out conveniently at the crude lysate stage of the isolation. Equivalent procedures lead to yields of ca.10-20 mg L^{-1} for the mutant proteins discussed below (*15,20*). Substitution of zinc for iron has been observed for other iron proteins expressed in *E. coli*, including one of the two $Fe(S\text{-}Cys)_4$ sites in desulforedoxin from *D. gigas* (*11,21*).

The identity of these proteins was confirmed by N-terminal sequencing for the recombinant protein and those forms with mutations at positions 6, 9 and 10, by metal analysis (Table I) and by electrospray ionisation mass spectrometry (Table II; Figure 3). The native protein contains a mixture of formylated and unformylated forms. The protein ion peaks seen to higher mass numbers are due to Na^+- and/or NH_4^+-protein adducts.

Metal Substitution

Substitution of the native iron by other metals is useful for a comparison of the intrinsic properties of the tetrathiolato site as well as permitting application of a wider variety of physical techniques. For example, NMR methods are evolving for the detection of rapidly relaxing protons close to the active sites in functional paramagnetic iron-sulfur proteins (*22-24*). Meanwhile, the NH protons involved in the four type I and III NH...S interactions detailed above have been detected by $^{1}H\text{-}^{113}Cd$ and $^{1}H\text{-}^{199}Hg$ correlation experiments in metal-substituted forms of Rd from *Pyrococcus furiosis* (*25,26*). These observations point to the significant covalent character of these NH...S "hydrogen bonds" and this, plus the detection of ^{199}Hg coupling to alkyl protons of nearby residues (*26*), appears to be relevant both to the tuning of redox potential and the mechanism of electron transfer.

Metal substitution via trichloroacetic acid precipitation followed by addition of the substituting metal under reducing conditions is a straightforward procedure for

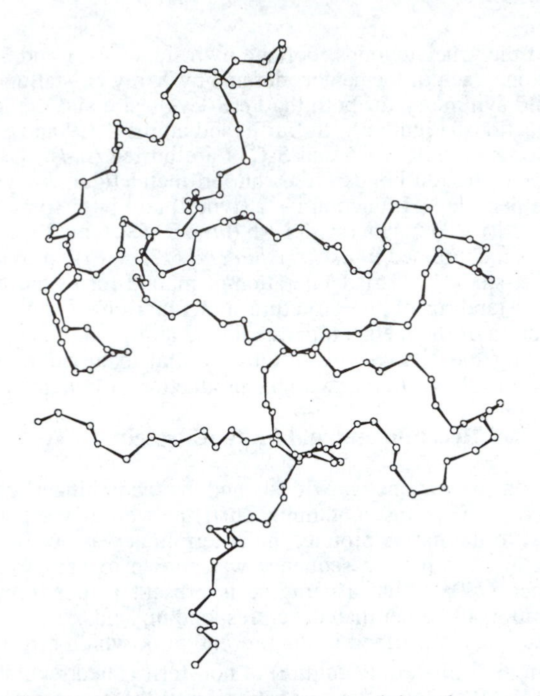

Figure 1. Backbone diagram of Rd*Cp*, emphasizing the *pseudo* two fold symmetry in the vicinity of the iron atom.

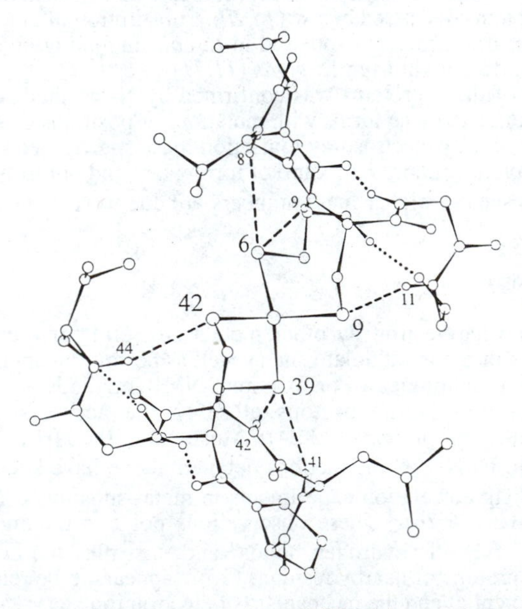

Figure 2. Ball-and-stick representation of NH...S interactions (- - - -) around the $Fe(S\text{-}Cys)_4$ center and NH...OC interactions (·· ·· ··) within the chelate loops in Rd*Cp* . The *pseudo* two fold axis (see text) is perpendicular to the page, passing though the Fe atom. (Figures generated from the coordinates of ref. 2: pdb5rxn.ent in the Brookhaven Protein Databank)

Table I. Metal Content of *Cp*Rd[a]

Protein	Fe	Zn
native	0.94(1)	0.05(1)
recombinant	1.01(1)	0.02(1)
C9S	1.03(2)	n.d[b]
C42S	1.05(2)	n.d[b]
G10A	1.01(1)	0.03(1)
G10V	1.12(2)	0.04(1)
G43A	0.99(3)	0.01(1)
G43V	1.02(1)	0.01(1)
G10VG43A[c]	n.d[b]	1.3(1)
G10VG43A[d]	0.89(3)	n.d[b]
G10,43V[d]	0.84(6)	n.d[b]

[a] metal:protein molar ratio determined by ICP-MS: single standard deviations are quoted in brackets.

[b] n.d., not detected.

[c] isolated without reconstitution with Fe(III) (see text).

[d] isolated with reconstitution with Fe(III) (see text).

Table II. Molar Masses (Da) of *Cp*Rd determined by Electrospray Mass Spectrometry[a]

Protein	Calc.	Found[b]	Protein	Calc.	Found[b]
native	6047.6[c,d]	6047(1)[f]	G10V	6142.5	6143(1)
	6075.6[c,e]	6075(1)[f]			
	6100.5[d]	6100(1)			
	6128.5[e]	6129(1)			
recombinant	6100.5	6101(1)	G43A	6114.5	6115(1)
C6S	6084.4	6084(1)	G43V	6142.5	6143(1)
C9S	6084.4	6084(1)	Zn^{II}-G10V/G43A[g]	6167.0	6167(1)
C39S	6084.4	6084(1)	G10V/G43A	6156.5	6157(1)
C42S	6084.4	6084(1)	G10,43V	6184.5	6185(1)
G10A	6114.5	6115(1)	$^{113}Cd^{II}$-recomb	6158.7	6159(1)
Zn^{II}-G10V[g]	6153.0	6152(1)	$^{113}Cd^{II}$-G10,43V	6242.7	6244(1)

[a] Determined under basic conditions in negative ion mode unless otherwise stated (see text)

[b] Single standard deviations given in brackets.

[c] Apo protein.

[d] Unformylated N terminus.

[e] Formylated N terminus.

[f] Determined under acid conditions in positive ion mode.

[g] Colorless Zn form as isolated: see text.

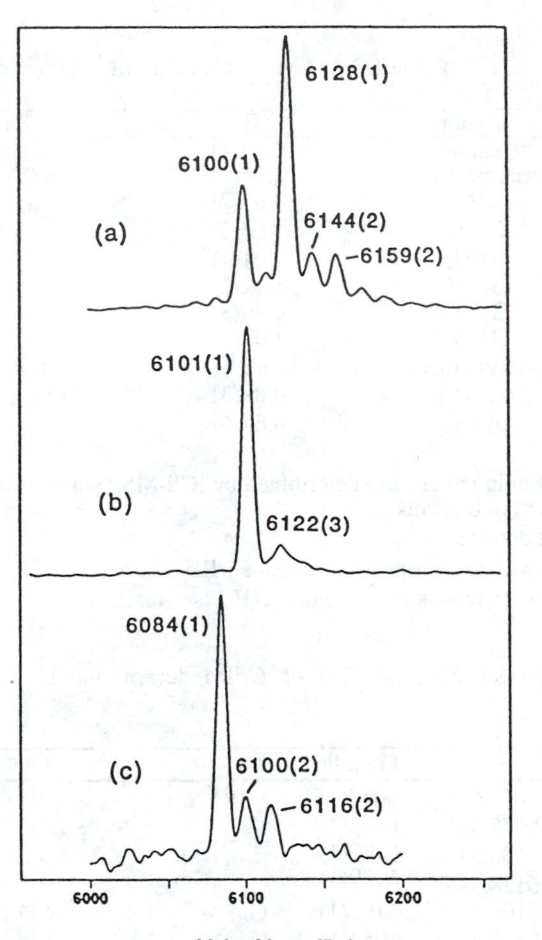

Figure 3. Negative-ion electrospray ionization mass spectra of *Cp*Rds. The proteins (0.2-1.0 mM) suspended in distilled water or 10 mM ammonium acetate were infused into a mixture of MeOH:H_2O:NH_3(aq) or Et_3N (30:70:0.1 v:v:v). (a) native; (b) recombinant; (c) C6S. Weak peaks with higher molar masses appear to be due to the formation of Na^+- and/or NH_4^+-protein adducts.

Rd*Cp* (*15,27,28*). It appears that the hydrophobic core of the protein (dominated by aromatic residues) remains intact upon precipitation, minimizing refolding problems. Intramolecular contact between the two chelate loops, 6-11 and 39-44, is weak (Figure 4a), a feature expected to favor local denaturation and easy access to the active site.

Mutation of Cysteinyl Ligands

The first mutant form of Rd*Cp*, C42S, has been reported: the surface cysteine at position 42 was substituted with serine, providing an oxygen ligand atom in place of sulfur (*17*). In the present work, each cysteine has been mutated in turn, providing C6S, C9S, C39S and C42S forms (*20*).

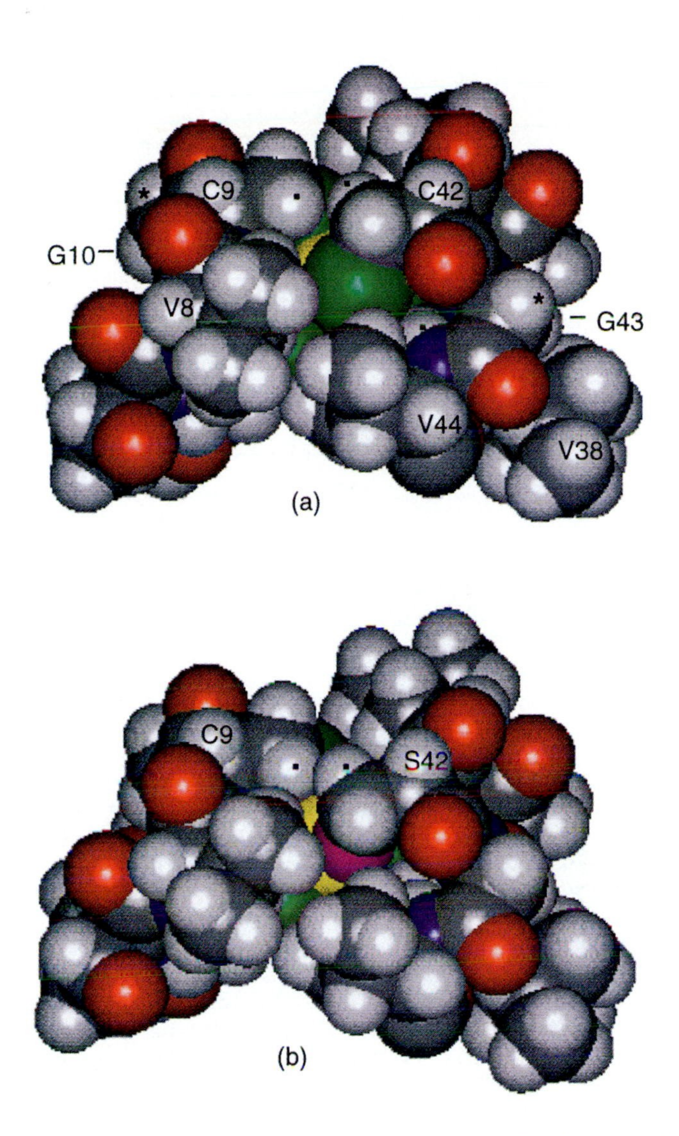

Figure 4. Partial structures of Rd*Cp*. (a) native Fe(III) form. Individual amino acid residues are identified at $H^{\alpha 1}$, except for V38 which is labelled at a H^{γ} atom. The C9 and C42 $H^{\beta 2}$ atoms and the V44 NH atom are identified by ·. The points of mutation at the G10 and G43 $H^{\alpha 2}$ atoms are indicated by *. The iron atom is yellow and the S^{γ} atoms are green. Other atoms have their usual CPK colors.
(b) Equivalent view of the C42S mutant form (generated from coordinates supplied by M. Guss (*20*)).

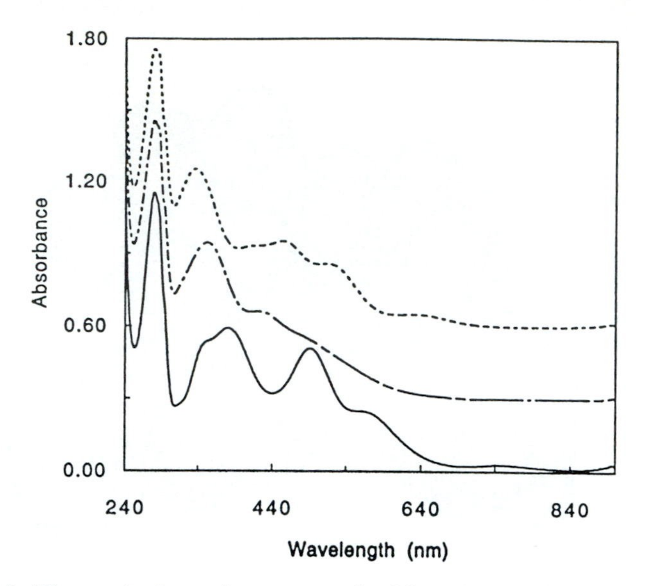

Figure 5. Electronic absorption spectra of oxidized *Cp*Rd proteins (0.15 mM in 50 mM Tris-HCl (pH, 7.4)): recombinant (——); C6S (-- - -- -); C9S (- - - -). The spectra have been offset by 0.1 absorption units for clarity.

The C42S protein has been crystallized and its structure solved by X-ray diffraction (*20*). An $FeOS_3$ core is confirmed (Figure 4b). At the present stage of refinement, the three Fe-S distances are the same, 2.36(4) Å, compared to a range of 2.23-2.34(1) Å in the native protein (*4*). The Fe-O^{γ} and C^{β}-O^{γ} distances of 1.90(2) and 1.42(2) Å, respectively, are both 0.4 Å shorter than those involving sulfur in the native protein and result in the C^{α} atom of residue 42 moving 0.43 Å towards the interior of the molecule. This "kink" perturbs the conformation of residues 41-43 and causes closer packing of atoms in this local surface region (compare Figures 4a and 4b). For example, the separation of $H^{\beta 2}$ atoms on residues 9 and 42 (each identified by a • in Figure 4) is 2.29(2) Å in the native protein but only 1.95(5) Å in C42S (assuming r(C-H), 1.09 Å; H...H van der Waals separation is about 2.4 Å).

The NH...X (X = O,S) interactions (c.f., Figure 1b; Table III) provide an interesting pattern. The V44-NH...O^{γ}-S42 distance in the C42S mutant is 3.07(5) Å (assuming r(N-H), 1.03 Å). This is well outside the range consistent with an hydrogen bonding interaction (1.7-2.1 Å) and, indeed, outside van der Waals contact of about 2.6 Å (compare Figures 4a and 4b). The ^{1}H resonance of the V44-NH group in the $^{113}Cd(II)$-substituted C42S protein shifts 1.65 ppm upfield from its position in the native protein. In fact, in each C → S mutant, there is an upfield shift in the range 1.6-1.9 ppm for every NH proton that interacts with the relevant S^{γ} atom of the native protein (Figure 2). It is apparent that substitution of O for S eliminates these hydrogen bonds in each mutant protein.

The properties of the four mutant proteins reflect the *pseudo* 2-fold symmetry of the native protein. Mutation at surface cysteines 9 and 42 provides products that are significantly more stable than does mutation at the buried 6 and 39 positions. Presumably, structural changes are more easily accommodated near the surface. Electronic spectra of the pairs (C6S, C39S) and (C9S, C42S) differ (Figure 5) but are essentially superimposible within each pair.

Table III. Hydrogen Bonds in the Chelate Loops

residue	NH donor to	CO acceptor from[a]	Sγ acceptor from
C6	OC-Y11	G10-NH	V8-NH C9-NH
T7	OC-Q48	surface	
V8	Sγ-C6	surface	
C9	Sγ-C6	surface	Y11-NH
G10	OC-C6	surface	
Y11	Sγ-C9	C6-NH	
C39	OC-V44	G43-NH	L41-NH C42-NH
P40		surface	
L41	Sγ-C39	surface	
C42	Sγ-C39	surface	V44-NH
G43	OC-C39	surface	
V44	Sγ-C42	C39-NH	

[a] "surface" indicates that this CO group is part of the exterior surface of *Cp* Rd.

Detailed electron transfer properties are under investigation. One interesting observation is that, in contrast to the native protein, the redox potentials of the mutant proteins are pH dependent. For example, those for the C9S and C42S proteins exhibit a slope of 59(3) mV/pH unit below pH7, consistent with the couple :-

$$Fe^{III}(O\text{-}Ser) + H^{+} + e^{-} = Fe^{II}\ (HO\text{-}Ser)$$

A new resonance, connected to S42 H^{α} and H^{β} protons, appears in the ^{1}H NMR spectrum of the Cd(II) form of the C42S protein. It appears that the S42 O^{γ} atom is protonated in this form.

Mutation of the Conserved Glycine 10 and 43 Residues.

Type II NH...S interactions exist between Cys S(9,42) and NH(11,44) (Scheme Ib; Figure 2). It has been suggested that a β carbon on the intervening residue (10 or 43) would destabilize such an interaction by eclipsing the carbonyl oxygen of 9 or 42 (*13*). The presence of conserved Gly residues at positions 10 and 43 following Cys 9 and 42 is consistent with this idea.

Gly 10 and 43 were mutated to alanine and valine, i.e., substitution of side chain H by CH_3 and $CH(CH_3)_2$, in an attempt to perturb the Rd site and the pattern of NH...S interactions. Figure 4a shows the relative disposition of the mutation point on G43 $H^{\alpha 2}$ and the V44-NH...S^{γ}-C42 interaction. Besides the four single mutants, double mutants G10V/G43A and G10V/G43V have been generated (*15*). Electronic and resonance Raman spectra for the mutant forms vary minimally from those of the recombinant protein. They confirm the presence of an intact FeS_4 center in each protein.

Structural Aspects. The chelate loops T5-Y11 and V38-V44 each contain 7 residues and determine the *pseudo* 2-fold symmetry of the $Fe(S\text{-}Cys)_4$ site of Rd*Cp* (Scheme I; Figure 2). Besides the three NH...S interactions characteristic of a loop, one hydrogen bond connects the ends of each loop (C6-NH...OC-Y11; C39-NH...OC-V44) and another reaches across each loop (G10-NH...OC-C6 ; C43-NH...OC-C39) (Table III; Figures 7 and 8). Note that C6 and C39 are each involved in four hydrogen bonds. Each loop defines a pocket in the surface of the molecule: the amide CO groups not involved in hydrogen bonding (7-10 ; 40-43) help define this pocket.

The crystal structure of the Cd(II)-substituted form of the recombinant protein, Cd-rRd*Cp*, has been solved (*29*). The overall structure is similar to that of the native protein (*2,4*), but there are differences in and around the active site. These are outlined briefly here to assist interpretation of ^{1}H NMR spectra. The four Cd-S distances are the same at the present stage of refinement and, at 2.58(5) Å, are 0.29(6) Å longer than the Fe-S bonds in Rd*Cp* (*4*). Table IV lists changes in hydrogen bonding distances for the chelate loops. The four NH...S distances to the buried ligands C6 and C39 decrease by 0.17-0.28(6) Å in response to the expansion of the MS_4 core. Those to surface ligands C9 and C42 contrast in their response: the distance Y11-NH...S-C9 decreases by 0.12 Å but V44-NH...S-C42 has not changed significantly. The differential behaviour can be traced to the effect of non-bonded interactions between the H_3C^{γ} groups of the Pr^{i} side chains of V44 and V8 and the S ligand atoms of C6 and C42. $H_3C^{\gamma 1}$-V44 contacts both $H_3C^{\gamma 1,2}$ -V8 as well as S-C42 while $H_3C^{\gamma 2}$ -V44 contacts $H_3C^{\gamma 2}$-V8, HC^{β}-V8 and S-C6 (note the relative positions in Figure 4a).

In contrast to the buried residues, the surface sidechains of V8 and V44 are able to adjust to the core expansion. The non-bonded interations mean that V44-NH alters

(a)	----	5	6	7	8	9	10	11	---	38	39	40	41	42	43	44	----
		T	C	T	V	C	G	Y		V	C	P	L	C	G	V	

(b) H-bond type[a]	Donor NH	Acceptor S^{γ}
I	8/41	6/39
III	9/42	6/39
II	11/44	9/42

[a]see ref 16

Scheme I *Cp*Rd (a) Partial sequence; (b) NH...S interactions

Table IV. Comparison of Hydrogen Bonding Distances in Fe^{III}-*Cp*Rd[a] and Cd^{II}-r*Cp*Rd[b]

bond	Δr, (Å)[c]
C6-NH....OC-Y11	-0.02
C39-NH...OC-V44	+0.04
G10-NH...OC-C6	+0.07
G43-NH...OC-C39	+0.14
V8-NH...S-C6	-0.28
C9-NH...S-C6	-0.24
L41-NH...S-C39	-0.28
C42V8-NH...S-C39	-0.17
Y11-NH...S-C9	-0.12
V44-NH...S-42	+0.06

[a] ref. 4. [b]ref. 29. [c]change in H...X distance (X = O,S) assuming r(NH), 1.03 Å; a positive value indicates an increase in the *Cd* protein relative to the iron protein. Standard deviation, 0.06 Å.

position and, in contrast to the other NH...S interactions, the V44-NH...S-C42 distance is not compressed.

Nuclear Magnetic Resonance The ^{1}H NMR spectrum of Cd-rRd*Cp* has been assigned (*30*). A comparison of the spectrum in the NH region with those of the Cd(II)-substituted G10V, G43V and G10V/G43V proteins is made in Table V. Structural changes induced by the mutations appear to be highly localized. The sums of the chemical shift changes for a given resonance in the single mutants correlate closely with the observed shift change in the double mutant (Table V). To first order, the covalently bound Cd(S-Cys)$_4$ center does not transmit the magnetic changes induced by the mutations from one chelate loop to the other. The localization is also consistent with the weak contacts which exist between the chelate loops (Figure 4a, but see below).

In the G10V mutant, large shift changes are observed for the NH protons of the T5-Y11 loop (Table V). This correlates with the Pr^i sidechain of V10 occupying part of the surface pocket defined by the loop. The T5-Y11 NH functions line the bottom of the pocket (Figure 7b). The significant shift changes also observed for I12 and Y13 NH protons may be associated with their proximity to the Y11 aromatic ring.

The observations for the V38-Y44 loop in G43V are very different. A single dominant shift change of -0.36 ppm is seen for the V44NH proton with small changes (< 0.06 ppm) for the other members of the loop (Table V). Examination of the Cd-rRd*Cp* structure reveals that the sidechain of V38 may block an orientation of the V43 sidechain equivalent to that discussed above for V10 in the G10V mutant (Figure 8; c.f., Figure 4a). The V43 sidechain would experience non-bonded repulsions from both C42 OC and V38 $H_3C^{\gamma1}$. An alternative orientation between the 42 and 43 OC functions would impose non-bonded repulsions with C42-CO and V44-$C^{\gamma1}H_3$ (Figure 8; c.f., Figure 4a). These are just the residues involved in the V44-NH...S-C42 interaction whose NH function exhibits the dominant chemical shift change and which, as discussed above, was structurally sensitive to the incorporation of cadmium.

This difference in preferred orientation is driven by differences in the molecular surface presented by the two chelate loops. The orientation into the loop pocket observed in the G10V mutant is favoured as T5-O^{γ}H interacts with the C6-NH...OC-Y11 hydrogen bond (Figure 7a,b). Such an interaction is not possible for the isopropyl sidechain of V38, related to T5 by the *pseudo* symmetry, and its $C^{\gamma1}H_3$ group protrudes into the V38-V44 chelate loop pocket (Figures 4a and 8a,b).

Two significant shift changes are seen in the non-mutated loops: G10V: V44, 0.29 ppm; G43V: Y11, 0.11 ppm (Table 5). Y11 and V44 are related by the *pseudo*-symmetry, as are C9 and C42. These perturbations appear to be propagated via the close contacts between C9 and C42 $H^{\beta2}$ and S^{γ} atoms. A C9 $H^{\beta2}$...$H^{\beta2}$-C42 close contact of 2.29 Å in native Rd*Cp* (van der Waals distance, 2.4-2.9 Å) has increased to 2.46 Å in Cd-rRd*Cp*, but C9 $H^{\beta2}$...S^{γ}-C42 has decreased significantly from 3.28 to 3.05 Å (van der Waals distance, 3.2-3.45 Å). In the G10V mutant, the effect of the structural change can be communicated via the C9 $H^{\beta2}$...C42 $H^{\beta2}$ and/or C9 $H^{\beta2}$...S^{γ}-C42 contacts to the V44-NH...S-C42 interaction (c.f, Figure 4a), leading to the observed chemical shift change in V44-NH. In the G43V mutant, it can be communicated similarly to the Y11-NH...S-C9 interaction.

Electrochemistry. Square wave voltammetry (SWV) was carried out at a pyrolytic graphite edge electrode, surface modified with poly-L-lysine as an electron transfer promotor. Results are given in Table VI and Figure 6. Electron transfer between the electrode and Rd*Cp* is promoted efficiently by poly-L-lysine and electrochemistry is reversible under all conditions: the peak potentials remain constant at different pulse frequencies and a plot of the peak currents versus the square root of the frequency is

Table V. ^{1}H Chemical Shifts δ (ppm) for NH protons in ^{113}Cd-substituted rubredoxins[a]

	δ	Change in δ[b]			
residue	recombinant	G10V	G43V	G10,43V	predicted[c]
C6	9.19	-0.76	-0.04	-0.79	-0.80
T7	8.39	-0.66	0.00	-0.71	-0.66
V8	9.41	0.59	0.05	0.60	0.64
C9	9.35	0.02	0.04	0.09	0.06
G10 (V10)	7.82	-0.72	0.01	-0.69	-0.71
Y11	8.97	0.37	0.11	0.41	0.48
I12	7.38	-0.56	0.00	-0.58	-0.56
Y13	9.45	0.22	0.01	0.22	0.23
V38	6.43	-0.07	-0.02	-0.10	-0.09
C39	8.84	-0.01	-0.09	-0.15	-0.10
P40					
L41	8.91	0.04	0.05	0.09	0.09
C42	8.74	0.11	-0.06	0.05	0.05
G43 (V43)	7.95	0.01	-0.04	-0.06	-0.03
V44	7.95	0.29	-0.36	-0.03	-0.07
G45	8.19	-0.06	-0.04	-0.13	-0.10
K46	8.39	-0.07	-0.02	-0.19	-0.09
E50	9.18	-0.16	0.01	-0.17	-0.15

[a] Chemical shifts measured in phosphate buffer (20 mM; pH 6.8), 0.2 M NaCl at 303 K referred to $^{1}H_2O$ at 4.73 ppm. Error ± 0.02 ppm.

[b] The difference between the recombinant and mutated proteins. Positive shifts are downfield. Error ± 0.04 ppm. Only residues which show a shift change of at least 0.04 ppm in one of the given proteins are listed.

[c] The sum of the experimental changes in δ observed for G10V and G43V.

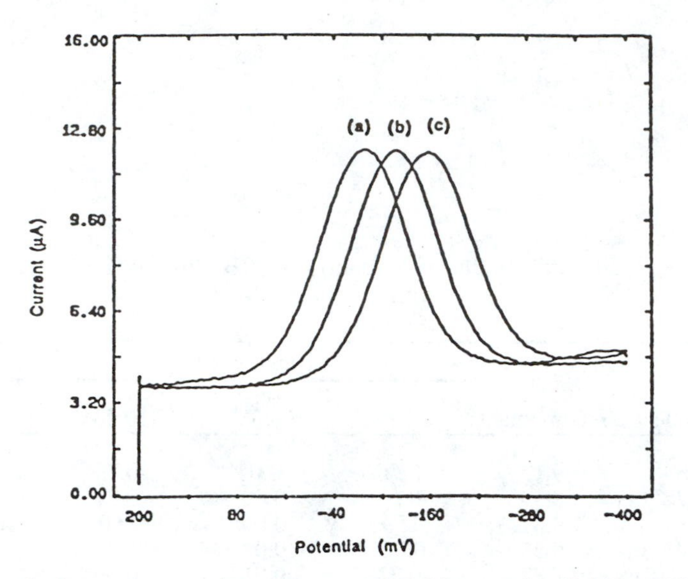

Figure 6. Square wave voltammograms of Rd*Cp* proteins (0.080 mM in 30 mM Tris-HCl (pH, 7.4) and 0.1 M NaCl). (a) recombinant; (b) G10V and (c) G10,43V. Conditions: E_p= 50 mV, E_s = 1 mV and τ^{-1} = 30 Hz.

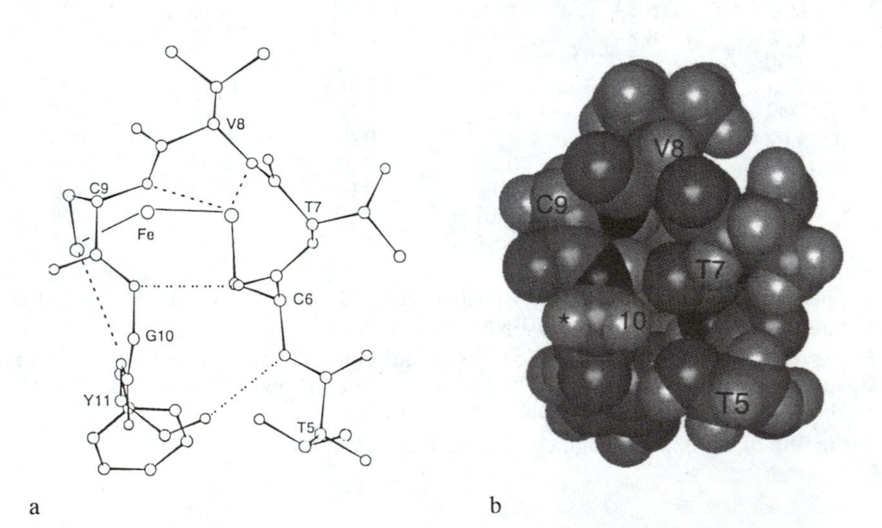

Figure 7. Structure around the chelate loop T5-Y11 of Cd-rRd*Cp* (generated from coordinates supplied by the authors of ref. 39). (a) ball-and-stick model (H atoms are absent); (b) space-filling model. Amino acid residues are identified at C^{α} in (a) and at $H^{\alpha 1}$ in (b), except for T5, which is labelled at a H^{γ} atom. The point of mutation at the G10 $H^{\alpha 2}$ atom is indicated by *

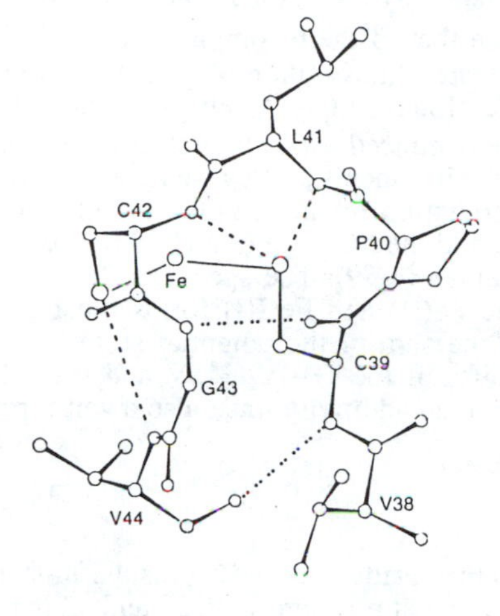

a

b

Figure 8. Structure around the chelate loop V38-V44 of Cd-rRd*Cp*. (a) ball-and-stick model; (b) space-filling model. The labelling follows that in Figure 7. V38 and V44 are labelled at a H^{γ} atom.

Table VI. Mid-point Potentials E (mV) for *Cp*Rd Proteins [a]

Protein	E	ΔE[b]	Predicted[c]
native	-76		
recombinant	-77		
G10A	-104	-27	
G10V	-119	-42	
G43A	-93	-16	
G43V	-123	-46	
G10V/G43A	-134	-57	-58
G10,43V	-163	-86	-88

[a] Determined by square wave voltammetry, *vs* SHE.

[b] E_{mutant} - $E_{recombinant}$.

[c] Sum of observed ΔE for the proteins mutated at individual single sites.

linear. This is a one-electron couple, Fe^{III}/Fe^{II}, as characterized by the observed peak width at half-height of 127 ± 1 mV (*31-32*).

The observed mid-point potentials of the native and recombinant Rd*Cp* are the same within experimental error (-76(2) mV versus SHE) and consistent with a previously reported value (-74 mV) also estimated by SWV (*32*). The values for the six G → A,V mutants are more negative than that of the recombinant form (Table VI), i.e., the mutations stabilize the oxidized state. Substitution of glycine by valine has a larger effect than substitution by alanine. Interestingly, turning on four NH...S interactions in a model $[Fe^{II}(SR)_4]^{2-}$ complex induced a positive shift of 240 mV while exposing similar centers to more polar environments also caused positive shifts (*33,34*). Overall, it is proposed that a combination of the detailed orientation of peptide amide dipoles, access to solvent water and the presence of NH...S interactions modulate redox potentals in iron-sulfur proteins (*34-39*). The specific effect of the present mutations upon these factors for the Fe(II) and Fe(III) forms must await further work. One intriguing aspect is that the sum of the potential shifts for the single mutants predicts the experimental shifts of the G10V/G43V and G10,43V double mutants (Table VI). The origin of such an additivity must also await further structural characterization.

Conclusions

Mutation of the conserved cysteine and glycine residues of Rd*Cp* results in stable proteins. Substitution of cysteinyl ligand by serine at positions 6, 9, 39 and 42 in turn provides proteins whose properties reflect both the O for S substitution and the *pseudo* 2-fold symmetry of the native site. Peptide NH...S^{γ}-Cys interactions are a feature of the native protein. An X-ray crystal structure of the C42S protein and ^{1}H NMR studies of the $^{113}Cd(II)$-substituted forms indicate that equivalent NH...O^{γ}-Ser hydrogen bonds are absent in each mutant.

Mutation of conserved residues G10 and G43, related by the *pseudo* 2-fold symmetry, to alanine and valine leads to stable mutant forms of Rd*Cp*. Physical properties are perturbed by the steric interactions between the β and γ carbon substituents of the new sidechains with the CO functions of C9 and C42 and with other adjacent groups.

Substitution of Fe(III) by Cd(II) leads to significant structural change at the metal site and in the NH...S distances, in particular (Table IV). Such differential changes must temper interesting and significant conclusions concerning electron tranfer mechanisms for native rubredoxins from study of Cd(II)- and Hg(II)-substituted forms (*26*). The present mutations produce further perturbation of the chelate loops 5-11 and 39-44 in the Cd(II) derivatives. ^{1}H NMR results (Table V) indicate that the Pr^{i} sidechain of V10 in the G10V mutant occupies the surface pocket defined by loop 5-11 and thereby modifies the environment of the 5-11 NH protons. The equivalent sidechain of V43 in G43V is denied the same access to the 39-44 pocket. This leads to a specific perturbation of the V44-NH...S-C42 interaction in this mutant. These effects are additive in the double mutants, consistent with the different structural changes being localized in each loop, to a first approximation.

Given the similar surface features of the native Fe(III) and recombinant Cd(II) molecules (*4,29*), similar differentiation of properties might be expected in the Fe(III) mutants. This is manifest in the half-wave potentials where, again, an additivity of the differential effects is seen in the double mutants (Table VI).

Interestingly, similar G → A mutations in the $2[Fe_4S_4]$-ferredoxin from *Cp* did not affect potentials significantly (*40*), nor did mutation of surface carboxylates to carboxamides (*41*). Of course, detailed effects will vary from system to system (c.f., ref 42). Overall, the present report represents one approach to a systematic exploration of the influence of the metal atom, the ligand atoms and the surrounding protein structure upon the fundamental properties of rubredoxin molecules.

Acknowledgements

The work was supported by Australian Research Council Grant A29330611 to AGW.

References

1. Eaton, W. A.; Lovenberg, W. in *Iron-Sulfur Proteins*; Academic Press: New York, 1993; Vol. II. p. 131.
2. Watenpaugh, K. D.; Sieker, L. C.; Jensen, L. H. (a) *J. Mol. Biol.* **1979,** *131*, 509; (b) *J. Mol. Biol.* **1980,** *138*, 615.
3. Day, M. W.; Hsu, B. T.; Joshua-Tor, L.; Park, J.-B.; Zhou, Z. H.; Adams, M. W. W.; Rees, D. C. *Protein Sci*, **1992,** *1*, 1494.
4. Sieker, L. C.; Stenkamp, R. E.; LeGall, J. *Methods.Enzymol.* **1994,** *243*, 203.
5. Gebhard, M. S.; Deaton, J. C.; Koch, S. A.; Millar, M.; Solomon, E. I. *J. Am. Chem. Soc.* **1990,** *112*, 2217.
6. Santos, H.; Faraleira, P.; Xavier, A. V.; Chen, L.; Liu, M.-Y.; Le Gall, J. *Biochem. Biophys. Res. Commun.* **1993,** *195*, 551
7. Eggink, G.; Engel, H.; Vriend, G.; Terpastra, P.; Witholt, B. *J. Mol. Biol.* **1990,** *212*, 135.
8. Chen, L.; Sharma, P.; Le Gall, J.; Mariano, A. M.; Teixeira, M.; Xavier, A. V. *Eur. J. Biochem.* **1994**, *226*, 613.
9. Verhagen, M. F. J. M.; Voorhorst, W. G. B.; Kolkman, J. A.; Wolbert, R. B. G.; Hagen, W. R. *FEBS Lett.* **1993**, *336*, 13.
10. Archer, M.; Huber, R.; Tavares, P.; Moura, I.; Moura, J. J. G.; Carrondo, M.; Sieker, L. C.; LeGall, J.; Romao, M. J. *J. Mol. Biol.* **1995**, *251*, 690.
11. Czaja, C.; Litwiller, R.; Tomlinson, A. J.; Naylor, S.; Tavares, P.; LeGall, J.; Moura, J. J. G.; Moura, I.; Rusnak, F. *J. Biol. Chem.* **1995**, *270*, 20273
12. Mattieu, I.; Meyer, J.; Moulis, J.-M. *Biochem. J.* **1992**, *285*, 255.
13. Adman, E.; Watenpaugh, K.; Jensen, L. H. *Proc. Natl. Acad. Sci. USA* **1975**, *72*, 4854.
14. Eidness, M. K.; O'Dell, J. E.; Kurtz, D. M., Jr.; Robson, R. L.; Scott, R. A. *Prot. Eng.* **1992**, *5*, 367.
15. Ayhan, M.; Xiao, Z.; Lavery, M. J.; Hamer, A. M.; Nugent, K. W.; Guss, M.; Scrofani, S. D. B.; Wedd, A. G. *Inorg. Chem.*, in press.
16. Petillot, Y.; Forest, E.; Mathieu, I.; Meyer, J.; Moulis, J.-M. *Biochem. J.* **1993**, *296*, 657.
17. Meyer, J.; Gaillard, J.; Lutz, M. *Biochem. Biophys. Res. Commun.* **1995**, *212*, 827.
18. Richie, K. A.; Teng, Q.; Elkin, C. J.; Kurtz, D. M., Jr. *Protein Sci*. **1996**, *5*, 883.
19. Watenpaugh, K. D.; Sieker, L.C.; Herriot, J. R.; Jensen, L. H. *Acta Crystallogr.* **1973**, *B29*, 943.
20. Ayhan, M.; Xiao, Z.; Lavery, M. J.; Scrofani, S. D. B.; Guss, M.; Wedd, A. G., manuscript in preparation.
21. Mann, G. J.; Graslund, A.; Ochai, E.-I.; Ingamarson, R.; Thelander, L. *Biochem.* **1991**, *30*, 1939.
22. Sadek, M.; Scrofani, S. D. B.; Brownlee, R. T. C.; Wedd, A. G. *J. Chem. Soc., Chem. Commun.* **1995**, 105.
23. Xia, B.; Westler, M.; Cheng, H.; Meyer, J.; Moulis, J.-M.; Markley, J. L. *J. Am. Chem. Soc.* **1995**, *117*, 5347.
24. Chen, Z.; de Ropp, J. S.; Hernandez, G.; La Mar, G. N. *J. Am. Chem. Soc.* **1994**, *116,* 8772.
25. Blake, P. R.; Park, J.-B.; Adams, M. W. W.; Summers, M. F. *J. Am. Chem. Soc.* **1992**, *114,* 4931.
26. Blake, P. R.; Lee, B.; Summers, M. F.; Park, J.-B.; Zhou, Z. H.; Adams, M. W. W. *New. J. Chem.* **1994**, *18,* 387.

27. Lovenberg, W.; Williams, W. M. *Biochem.* **1969**, *8*, 141.
28. Moura, I.; Teixeira, M.; LeGall, J.; Moura, J. J. G. *J. Inorg. Biochem.* **1991**, *44*, 127.
29. Lavery, M. L.; Guss, M. unpublished observations. The nominal resolution limit on the data is 1.5 Å. At the present stage of refinement, the R-factor is 0.167 for all data from 7-1.5 Å.
30. Scrofani, S. D. B., unpublished observations.
31. Armstrong, F. A.; Hill, H. A. O.; Oliver, B. N.; Walton, N. J. *J. Am. Chem. Soc.* **1984**, *106,* 921
32. Smith, E. T.; Feinberg, B. A. *J. Biol. Chem.* **1990**, *265*, 14371.
33. Nakata, M.; Ueyama, N.; Fuji, M.-A.; Nakamura, A.; Wada, K.; Matsubara, H. *Biochim. Biophys. Acta* **1984**, *788*, 306.
34. Ueyama, N.; Okamura, T.; Nakamura, A. *J. Chem. Soc., Chem. Commun.* **1992**, 1019.
35. Jensen, G. M.; Warshel, A.; Stephens, P. J. *Biochem.* **1994**, *33*, 10911.
36. Backes, G.; Mino,Y.; Loehr, T. M.; Meyer, T. E.; Cusanovich, M. A.; Sweeney, W. V.; Adman, E. T.; Sanders-Loehr, J. *J. Am. Chem. Soc.* **1991**, *113*, 2055.
37. Walters, M. A.; Dewan, J. C.; Min, C.; Pinto, S. *Inorg. Chem.* **1991**, *30,* 2656.
38. Shenoy, V. S.; Ichiye, T. *Proteins: Struct. Funct. Gen.* **1993**, *17*, 152.
39. Yelle, R. B.; Park, N.-S.; Ichiye, T. *Proteins: Struct. Funct. Gen.* **1995**, *22*, 154.
40. Scrofani, S. D. B.; Brereton, P. S.; Hamer, A. M.; Lavery, M. J.; McDowall, S. J.; Vincent, G. A.; Brownlee, R. T. C.; Hoogenraad, N. J.; Sadek, M.; Wedd, A. G. *Biochem.* **1994**, *33*, 14486.
41. Brereton, P. S., Ph.D. thesis, University of Melbourne, 1995.
42. Butt, J. N.; Sucheta, A.; Martin, L. L.; Shen, B.; Burgess, B. K.; Armstrong, F. A. *J. Am. Chem. Soc.* **1993**, *115*, 12587.

Chapter 3

Electronic Isomerism in Oxidized Fe_4S_4 High-Potential Iron–Sulfur Proteins

I. Bertini[1] and C. Luchinat[2]

[1]Department of Chemistry, University of Florence, Via Gino Capponi 7, 50121 Florence, Italy
[2]Institute of Agricultural Chemistry, University of Bologna, Viale Berti Pichat 10, 40127 Bologna, Italy

The experimental Mössbauer and NMR data on the oxidized HiPIP II from *Ectothiorhodospira halophila* provide the key i) to the understanding of its electronic structure and ii) to the rationalization of the data obtained on a series of other HiPIPs in terms of electronic isomerism between two electronic distributions. The data are discussed in the light of recent theoretical considerations. The different valence distributions in the series are then related to the relative magnitudes of the reduction potentials of the individual iron ions.

Oxidized high potential iron-sulfur proteins (HiPIP) contain the polymetallic center $[Fe_4S_4]^{3+}$. Formally they contain one Fe^{2+} and three Fe^{3+} (*1*). However, after the first Mössbauer reports on the oxidized HiPIP from *Chromatium vinosum* it was apparent that the electronic structure was more complex. The isomer shifts indicate two Fe^{3+} and two $Fe^{2.5+}$ (*2*). The hyperfine coupling with the ^{57}Fe nuclei was negative for the latter set of iron ions and positive for the former set. The first theoretical attempt to rationalize these data (*3*) followed the analysis presented by Girerd et al. on the $[Fe_3S_4]^0$ polymetallic center (*4*). It was assumed that the antiferromagnetic coupling between the two ferric pairs (1 and 2 in Figure 1A) was larger than for the other five iron pairs, and a double exchange parameter, B_{34}, was added in the expression for the energies (*3*). Such a term allows the iron ions 3 and 4 to share one electron. The corresponding Hamiltonian is:

$$\mathcal{H} = J\left[S_3^* \cdot S_4 + (S_3^* + S_4)\cdot S_1 + (S_3^* + S_4)\cdot S_2\right]\cdot O_3 + J\left[S_3 \cdot S_4^* + (S_3 + S_4^*)\cdot S_1 + (S_3 + S_4^*)\cdot S_2\right]\cdot O_4 + +\Delta J_{12} S_1 \cdot S_2 + B_{34} V_{34} T_{34} \quad (1)$$

0097–6156/96/0653–0057$15.00/0

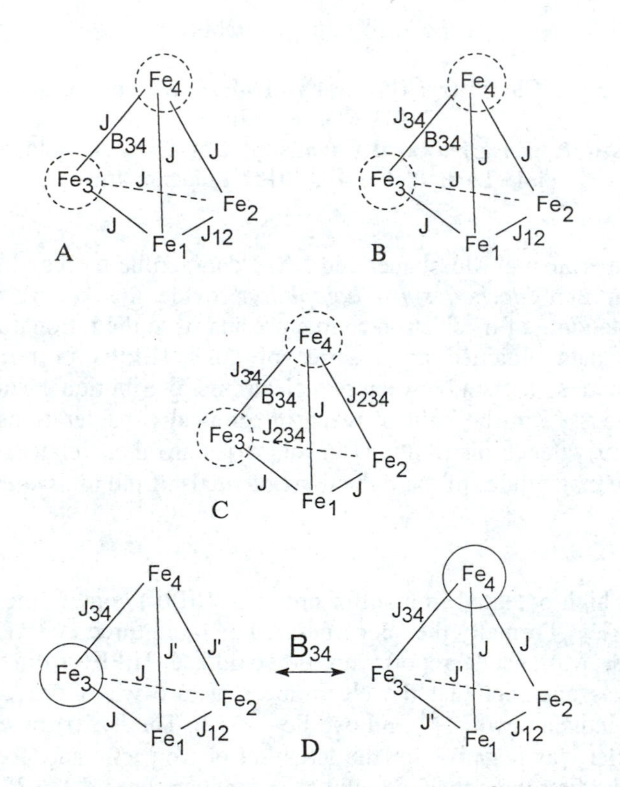

Figure 1. Magnetic coupling schemes in $[Fe_4S_4]^{3+}$ clusters. Double exchange on the 3-4 pair without (A) (*3*) or with (B) (*11*) differentiation of the J_{34} value, and in a pseudo-C_{3v} symmetry (C) (*16,17*). Resonance between two C_{2v} symmetries is shown in D (*18,19*).

where J is the Heisenberg coupling constant experienced by all iron-iron pairs except the ferric pair, ΔJ_{12} is the departure of J_{12} from J, B_{34} is the double exchange parameter operative in the mixed-valence pair, $\boldsymbol{T}_{34}$ is the transfer operator, V_{34} is an operator producing as eigenvalues $S_{34} + 1/2$ (S_{34} being the subspin quantum number of the 3-4 pair) and $\boldsymbol{O}_3$ and $\boldsymbol{O}_4$ are occupation operators.

This treatment provided a semiquantitative interpretation of the experimental data based on a mixed valence pair with a larger subspin (S_{34}) and a ferric pair with a smaller subspin (S_{12}). Now we discuss the case of the HiPIP II from *Ectothiorhodospira halophila*, which is the only example of an oxidized HiPIP containing a species with a single electronic distribution. This system has provided the key for the interpretation of the behavior of the other oxidized HiPIPs on the basis of an equilibrium between two species with different electronic distributions, which we call electronic isomerism. Electronic isomerism is discussed on the basis of a theoretical requirement for the presence of mixed valence pairs. Finally, we will discuss the different valence distributions in the series of HiPIPs and the relative magnitudes of the reduction potentials of the individual iron ions.

The Electronic Structure in the Oxidized HiPIP II from *E. halophila*

The ^{1}H NMR spectrum of the oxidized HiPIP II from *E. halophila* is reported in Figure 2, together with the assignment of the hyperfine-shifted signals (*5*). It appears that four β-cysteine protons experience negative hyperfine shifts, and four experience positive shifts. This is typical behavior when there are two spins antiferromagnetically coupled (*6,7*). In the ground state, the experimental value of $\langle S_z \rangle$ relative to the larger subspin is negative as it is in monomers, whereas the experimental value of $\langle S_z \rangle$ relative to the smaller subspin has a reverse sign owing to the requirement of antiferromagnetic coupling. Each pair of β-CH_2 groups is well spread. This is due to a Karplus type relationship, which relates the hyperfine coupling to the Fe-S-C-H dihedral angle (*8,9*). The different values of the center of gravity for each pair of CH_2 protons is presumably due to the lack of symmetry.

Therefore, the ^{1}H NMR spectra indicate the presence of two subspins antiferromagnetically coupled. The Mössbauer spectra at 4.2 K can be interpreted (*10*) in a fashion similar to the analysis of the spectra of the HiPIP from *C. vinosum* (*2*). According to the isomer shifts there are two iron ions with oxidation number 3+ and two with oxidation number 2.5+. The hyperfine coupling with ^{57}Fe has the usual negative sign for the iron ions with oxidation number 2.5+ and reversed sign for the iron(III) ions. It follows that the mixed valence pair has the larger subspin in the ground state, which according to Figure 1A will be labelled S_{34}, and the ferric pair (S_{12}) has the smaller subspin. The upfield shifted signals in the NMR spectra belong to the ferric domain and, vice versa, the downfield shifted signals belong to the mixed valence domain (*5*). The population of excited levels at room temperature, where the NMR spectra are recorded, does not

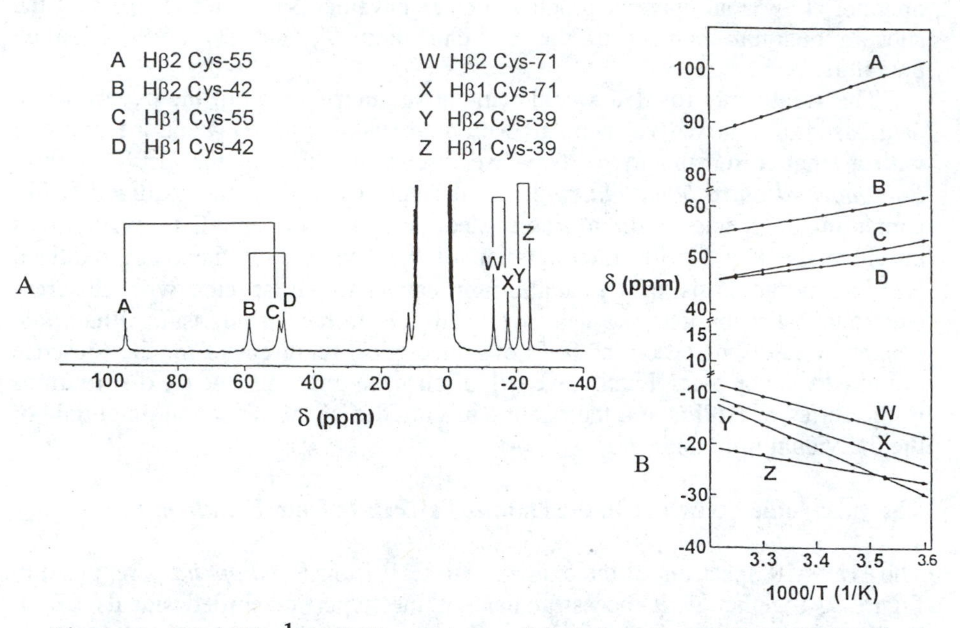

Figure 2. 600 MHz ^{1}H NMR spectrum of *E. halophila* oxidized HiPIP II and assignment of the hyperfine shifted signals (A) together with the temperature dependence of their shifts (B) (Adapted from ref. *5*).

Table I. Comparison of experimental hyperfine constants and g_{av} values for the oxidized form of *E. halophila* HiPIP II with values calculated[a] by different models (*14*)

	Exp. values	|1/2,9/2,4⟩	|1/2,7/2,3⟩	0.95|1/2,9/2,4⟩ + 0.31|1/2,7/2,4⟩
A_{12} (MHz)	21.4 ± 1.5	26.7	20.0	21.5
A_{34}[b] (MHz)	-31.5 ± 1.0	-38.3	-31.5	-33.1
g_{av}	2.07	2.054	2.042	2.066

[a] All calculations are performed using monomer hyperfine constants $A_{Fe^{3+}}$ = -20 MHz and $A_{Fe^{2+}}$ = -22 MHz (*12*).
[b] Average value of A_3 and A_4.

change the pattern dictated by the ground state. The sequence-specific assignment of the cysteine β-CH_2 signals permits the determination of the valence distribution within the protein framework. In other words, we know which cysteine is bound to iron with a given oxidation state.

At this point the problem arises of describing the magnetically coupled system in order to reproduce the hyperfine coupling of the ^{57}Fe in the ground state and the hyperfine coupling of β-CH_2 protons at room temperature, which is averaged over the populated levels according to the Boltzmann analysis. The most general Hamiltonian which can be used is:

$$\mathcal{H} = \sum_{i \neq j} J_{ij} \mathbf{S}_i \cdot \mathbf{S}_j \tag{2}$$

where each of ther six iron-iron pairs *i-j* is allowed to experience a difference Heisenberg coupling, described by the coupling constant J_{ij}. A further double exchange term can in principle be added, in analogy with equation 1, to impose delocalization between two ions with different oxidation states. The disadvantage of this approach is that the wavefunctions do not provide any obvious information on the behavior of pairs of spins. With the same symmetry of Figure 1A the following Hamiltonian can be written (Figure 1B) (*11*) which has analytical solutions

$$\begin{aligned}\mathcal{H} = {} & J\left[\mathbf{S}_3^* \cdot \mathbf{S}_4 + (\mathbf{S}_3^* + \mathbf{S}_4) \cdot \mathbf{S}_1 + (\mathbf{S}_3^* + \mathbf{S}_4) \cdot \mathbf{S}_2\right] \cdot \mathbf{O}_3 + \\ & J\left[\mathbf{S}_3 \cdot \mathbf{S}_4^* + (\mathbf{S}_3 + \mathbf{S}_4^*) \cdot \mathbf{S}_1 + (\mathbf{S}_3 + \mathbf{S}_4^*) \cdot \mathbf{S}_2\right] \cdot \mathbf{O}_4 + \\ & + \Delta J_{12} \mathbf{S}_1 \cdot \mathbf{S}_2 + \Delta J_{34} \mathbf{S}_3 \cdot \mathbf{S}_4 + B_{34} V_{34} \mathbf{T}_{34}\end{aligned} \tag{3}$$

where ΔJ_{34} is the departure of J_{34} from J and all other terms are as defined in equation 1. The hyperfine constant with any nucleus is:

$$A_i = A_0 \frac{<S_{zi}>}{<S_z'>} \tag{4}$$

In equation 4, A_0 is the hyperfine coupling constant with ^{57}Fe in the absence of magnetic coupling which can be estimated from a suitable monomeric system like, for instance, rubredoxin (*12*). On the other hand, $<S_{zi}>$ and $<S_z'>$ are the expectation values for any of the four individual spins, *i*, and for the total spin of the cluster, respectively. These can be calculated from either Hamiltonian 1 or 3 as a function of the values of S' and of the S_{12} and S_{34} subspins. The ground state would be formed by $S_{34} = 9/2$ and $S_{12} = 4$, both antiferromagnetically coupled to provide $S' = 1/2$. As already noted by Noodleman (*3*), the agreement with the experimental *A* values is approximate (Table I). It has then been noted that, if the ground state were $S_{34} = 7/2$ and $S_{12} = 3$, the agreement with the experimental *A* values would be better (Table I) (*13*). However, this ground state occurs only with a large B_{34} and in a narrow range of ΔJ_{12} and ΔJ_{34} parameters (*14*).

It is in our opinion important to stress that the same ground state and similar patterns of excited states are obtained under a large variety of B_{34} and ΔJ_{34} values, if both are allowed to vary independently (Figure 1B). This means that there is covariance between the two parameters and that fitting the experimental data hardly provides an independent evaluation of B_{34} (*15*).

If we assume a geometry with pseudo-C_{3v} symmetry (Figure 1C) described by the following Hamiltonian (*16,17*):

$$\begin{aligned} \mathcal{H} = {} & J\left[S_3^* \cdot S_4 + (S_3^* + S_4)\cdot S_2 + (S_2 + S_3^* + S_4)\cdot S_1\right]\cdot O_3 + \\ & J\left[S_3 \cdot S_4^* + (S_3 + S_4^*)\cdot S_2 + (S_2 + S_3 + S_4^*)\cdot S_1\right]\cdot O_4 + \\ & + \Delta J_{34} S_3 \cdot S_4 + \Delta J_{234} S_2 \cdot S_{34} + B_{34} V_{34} T_{34} \end{aligned} \tag{5}$$

the agreement with experimental results is even worse than in the previous case. Belinskii has recently figured out that we can lower the symmetry starting from either C_{2v} or pseudo-C_{3v} symmetry, and treat double exchange as resonance between the two configurations shown in Figure 1D (*18*). We can express the wavefunctions as combinations of the solutions of Hamiltonian 3. Now the ground state, as judged from the A and g values (*14*), as well as the excited states, as judged from the fitting of the temperature dependence of the proton NMR spectra (*19*), seem more reasonable (Table I). The ground state would be:

$$\psi = 0.95|4,9/2,1/2> - 0.30|4,7/2,1/2> \tag{6}$$

The range of parameter values which provide this ground state is relatively large. Despite the covariance between B_{34} and ΔJ_{34}, a small value of B_{34} is required (*14,19*). It is possible, however, that sets of parameters with larger B_{34} provide the same ground state, as required by independent theoretical considerations (*20*).

Electronic Isomerism in Oxidized HiPIPs

If we look at the NMR spectra of all the oxidized proteins up to now investigated we see that they are significantly different (Figure 3) (*21*). The ^1H NMR spectra of the reduced form of several HiPIPs are reported in Figure 4 (*22*). The spectra also look somehow different in the Cys β-CH_2 region but, indeed, the differences can be explained in terms of the different Fe-S-C-H dihedral angles (*8,21*). Otherwise the NMR spectra agree with the Mössbauer data, which show that all the iron ions are equivalent and at an oxidation state +2.5 (*2*).

In contrast, when analyzing the ^1H NMR spectra of the oxidized species (Figure 3) we note that the β-CH_2 proton signals of one cysteine (cysteine III) are typical of a cysteine bound to an iron at the oxidation state 2.5+, those of another cysteine (cysteine I) are typical of a cysteine bound to an iron(III), while the other two cysteines have an intermediate behavior. The earliest explanation

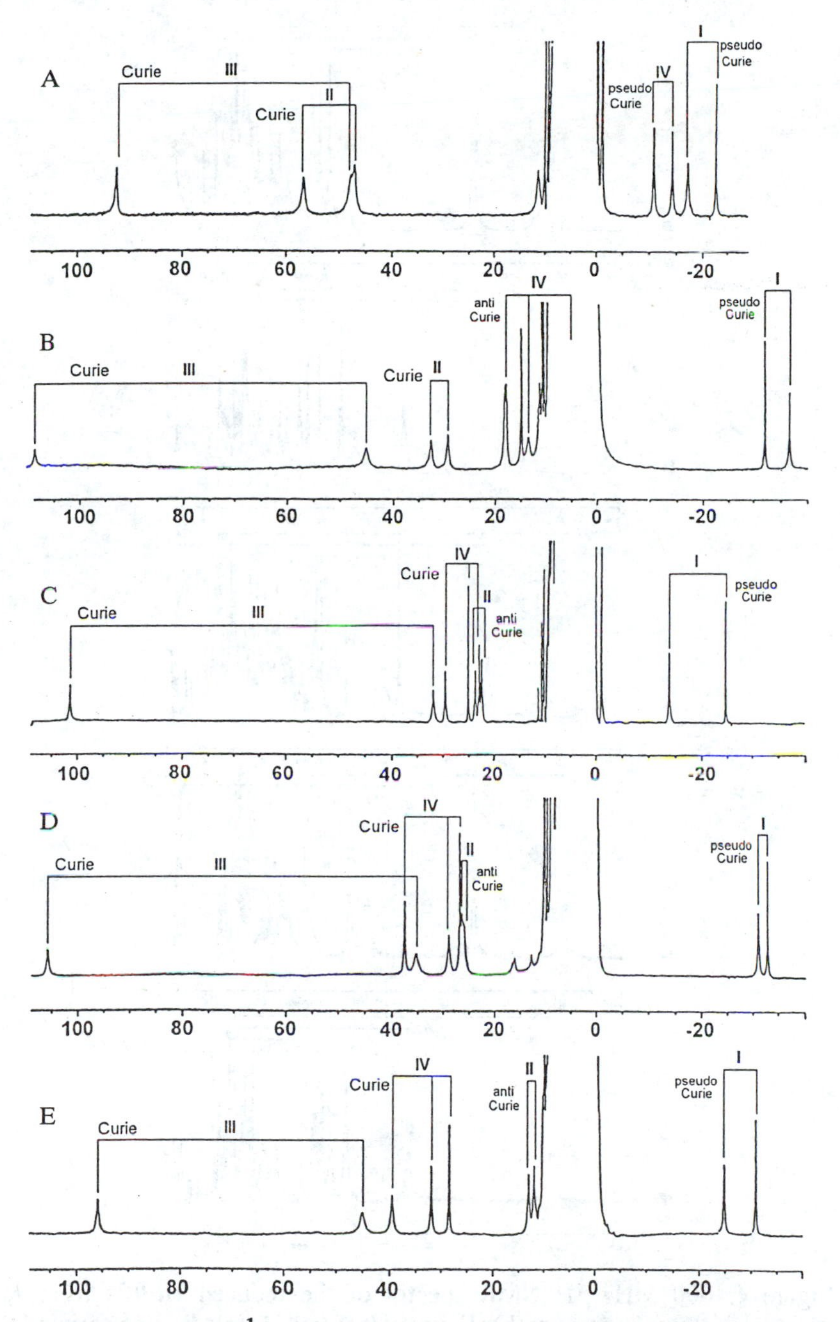

Figure 3. 600 MHz ^1H NMR spectra of the oxidized HiPIPs from *E. halophila* (iso II) (A) (*5*), *R. globiformis* (B) (*22*), *E. vacuolata* (iso II) (C) (*23*), *C. vinosum* (D) (*11*) and *R. gelatinosus* (E) (*17*). Roman numerals indicate the sequence-specific assignment of cysteine β-CH_2 protons.

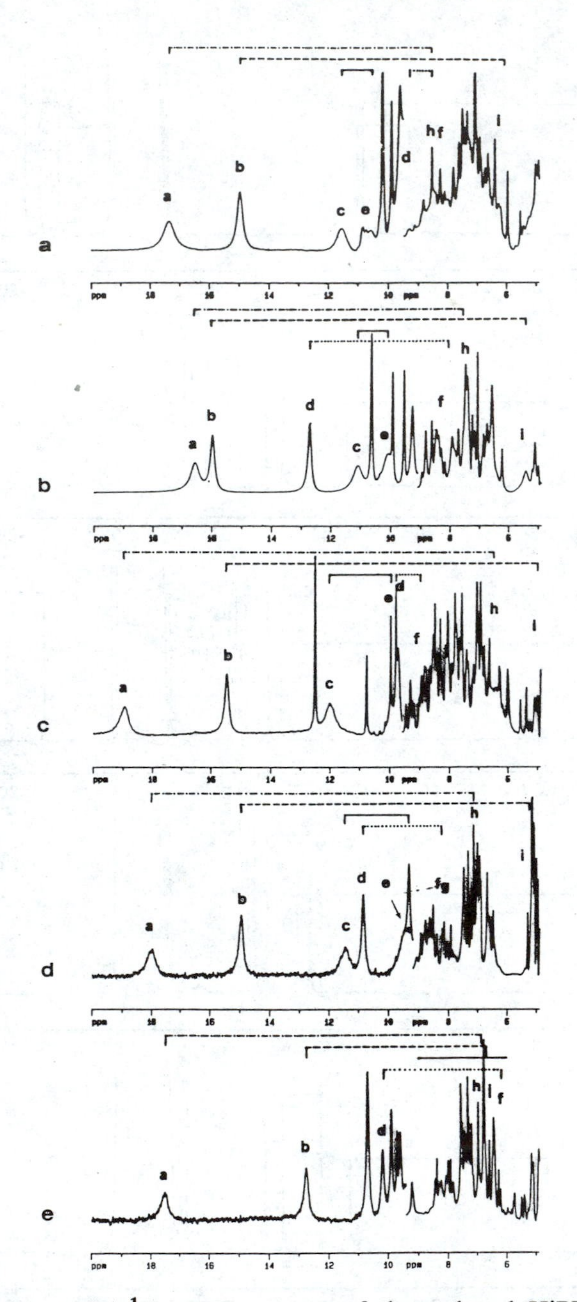

Figure 4. 600 MHz ^1H NMR spectra of the reduced HiPIPs from *R. gelatinosus* (a), *C. vinosum* (b), *E. vacuolata* (iso II) (c), *R. globiformis* (d) and *E. halophila* (iso II) (e). Lines refer to sequence-specific assignment of cysteine β-CH_2 protons to Cys I (-.-.-), Cys II (_____), Cys III (- - -) and Cys IV (.....). (Reproduced with permission from ref. *22*. Copyright 1993 FEBS).

suggested that the mixed valence character involves three iron ions (Figure 5) (*17*). Then, an equilibrium between two isomers as depicted in the upper part of Figure 6 was proposed (*23*). If the equilibrium is fast on the NMR time scale, then the protons of two cysteines will experience intermediate shifts between the upfield and the downfield limits. If the shifts of the fully Fe^{3+} and $Fe^{2.5+}$ are taken as the limiting shifts, the intermediate shifts provide an estimate of the equilibrium constant (*24*).

Is There a Theoretical Requirement for the Electronic Structure?

Let us consider an ideal $(RS)_4[Fe_4^{III}S_4]$ cluster of T_d symmetry containing four iron(III) ions. The question can be asked as to whether a theoretical prediction is possible for the electronic structure of this system after addition of one electron. Four different scenario's are possible:
A) the extra electron is completely localized on one of the four iron ions, which then displays an oxidation number 2+. Its bond lengths and angles, as well as its J values with the neighboring irons, will change accordingly. The resulting symmetry is C_{3v}. The cluster has still T_d symmetry in a time-averaged sense, and the system will display four equivalent energy minima (one for each iron becoming iron(II)), separated by equal energy barriers (Figure 7A).
B) The extra electron is completely delocalized on the four iron ions. Only one energy minimum exists, and all iron ions have oxidation number 2.75+. T_d symmetry is retained (Figure 7B).
C) The electron is delocalized over two iron ions, which then display an oxidation number of 2.5+ while the other two iron ions retain oxidation number 3+. The resulting symmetry is C_{2v}. Six of such equivalent situations exist, again separated by equal energy barriers (Figure 7C) and averaging T_d symmetry.
D) The electron is delocalized over three iron ions, which then display an oxidation number of 2.67+. As in case A, the symmetry is C_{3v} and the time-averaged T_d symmetry is restored by equilibrium over four equivalent states (Figure 7D).

The protein data discussed above clearly point to case C as the only experimentally observed. A reason for this could be that proteins are intrinsically dissymmetric, and may thus favor case C which is the one of lowest symmetry. However, symmetric model compounds also show spectroscopic evidence of C_{2v} symmetry, provided that the time scale of the spectroscopy is short enough, and the temperature is low enough to prevent overcoming the energy barriers (*25*). Therefore, it appears that case C would also be favored on theoretical grounds.

The electronic structure of the symmetric system has been theoretically analyzed in terms of the following driving forces (*26*): a) the Heisenberg antiferromagnetic exchange coupling between each pair of iron ions; b) the exchange integrals that tend to delocalize the extra electron over more than one iron center, favoring a ferromagnetic ordering of the spins; c) the vibronic coupling that tends to stabilize and "trap" each possible electronic configuration by instantaneously lowering the symmetry of the cluster. The result of this

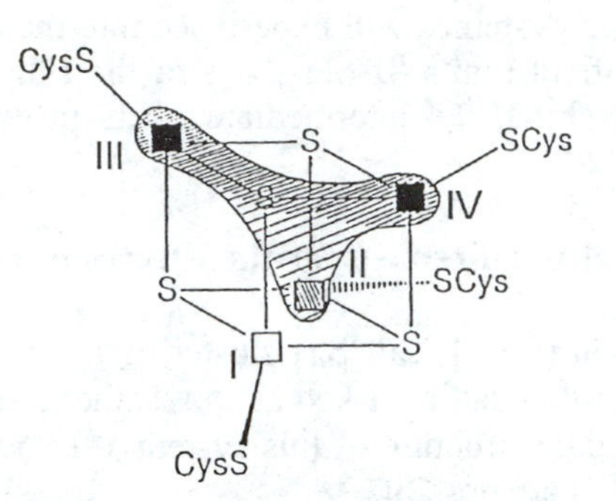

Figure 5. Pictorial representation of the partial involvement of the iron bound to cysteine II in the electron delocalization over the mixed valence pair of iron ions III and IV (*17,23*) (Reproduced from ref. *23*. Copyright 1993 ACS).

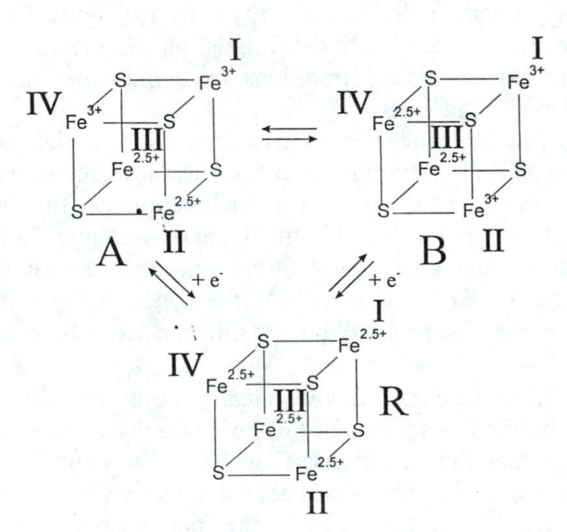

Figure 6. Equilibrium between the two electronic isomers having the mixed valence pair on irons III and II (A) or on irons III and IV (B), and their redox equilibria with the reduced species (R) (Reproduced from ref. *34*. Copyright 1996 ACS).

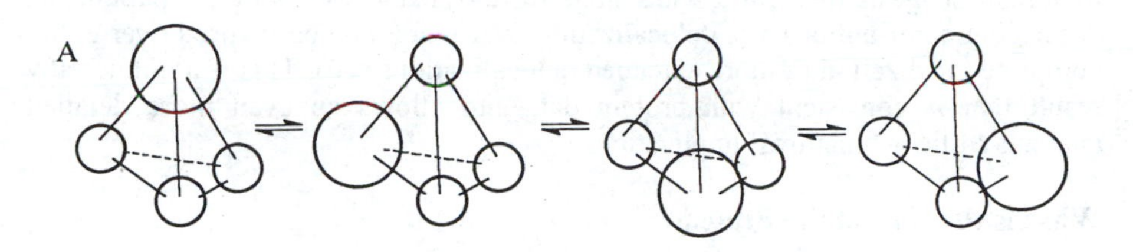

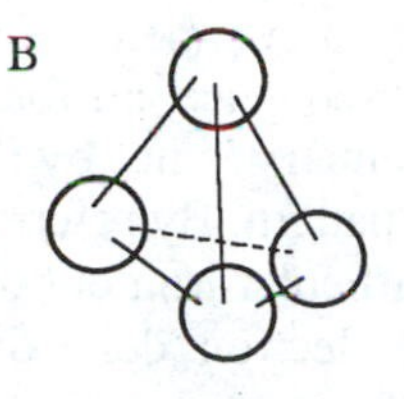

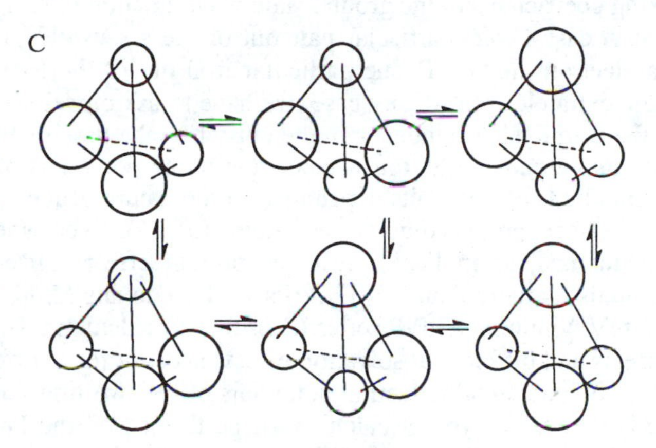

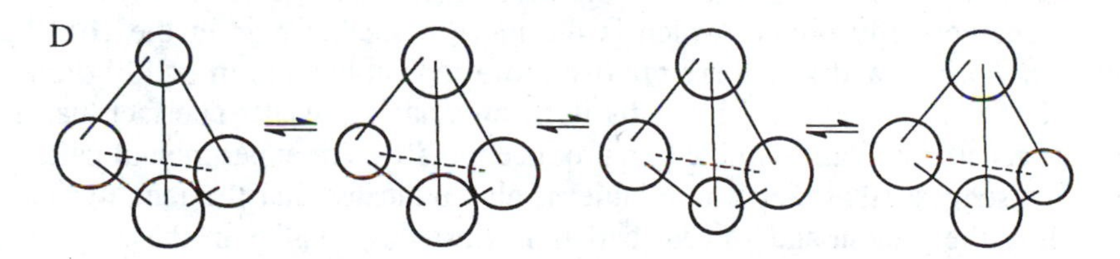

Figure 7. Possible electronic isomers in $[Fe_4S_4]^{3+}$ clusters. Full electron localization on one iron (A), full electron delocalization (B), electron delocalization over pairs (C), and electron delocalization over three of the four irons (D). The diameters of the spheres qualitatively reflect the percent of the extra electron.

treatment suggests that, for a wide range of reasonable values of the parameters describing contributions a-c, delocalization over pairs may be favored over either complete localization or more extended delocalizations (*26*). This is a noteworthy result that is consistent with protein data and allows an even more detailed analysis of their functional implications.

What is the Role of the Protein?

By definition, a cluster embedded in a protein, except for the special case where it bridges two subunits related by a symmetry element, has only C_1 symmetry (i.e. no symmetry at all). As a consequence, all four irons are inequivalent, as are all six iron pairs. The lack of symmetry is not by itself an argument in favor of a particular electron delocalization pattern. By referring to the preceding discussion and to Figure 7, in case A, one particular iron out of the four would preferentially become ferrous; in case B the electron delocalization would be unbalanced (different mixing coefficients in the ground state wavefunction (*27*)) towards that particular iron; in case C one particular pair out of the six would preferentially bear the extra electron; in case D one particular iron out of the four would not share any of the extra electron. In any case, we have to ask ourselves what is the magnitude of the inequivalence induced in the cluster by the protein environment. Protein effects can be quite large: for instance, the redox potentials of heme iron ions differ by hundreds of millivolts depending on the nature of the axial ligands and electrostatic effects produced by protein atoms (*28,29*). Even when all metal ligands are maintained, as in Fe_4S_4 systems, protein effects differentiate the reduction potentials by more than 1 V (HiPIPs *vs.* ferredoxins (*1,30,31*)) and by more than 300 mV within the HiPIP series (*32*). Recent calculations have shown that protein effects and differential solvation effects account for a difference of at least 100 mV in the individual reduction potentials of the two iron ions in Fe_2S_2 ferredoxins (*33*). The same type of calculations performed on the Fe_4S_4 cluster of HiPIPs yield very small energy differences when the extra electron is delocalized over each of the six possible iron pairs (*33*). However, the calculations correctly predict which is the more reducible pair in the HiPIP II from *E. halophila*, and which are the two more reducible pairs in equilibrium in the HiPIP from *C. vinosum*. This is by itself an amazing result. The fact that the calculations are qualitatively able to reproduce the different experimental patterns in HiPIPs suggests that they are reliable; it also indicates that the inequivalence induced by the protein and solvent part is intrinsically smaller in these systems than it is, for instance, in Fe_2S_2 systems. What is then the reason for the experimentally observed marked preference for the extra electron to delocalize over one (or two) of the six possible pairs? In the light of the theoretical analysis described in the preceding section we can suggest that the cluster environment in Fe_4S_4 systems induces just a small energy preference for one (or two) pairs; this then "triggers" a further distortion in bond lengths and angles that deepens the energy well corresponding to that particular distortion. The preference for pairwise delocalization is thus intrinsic, but the choice of the actual pair(s) is dictated by the environment.

Can We Learn about the Microscopic Reduction Potentials?

With the above discussion in mind, we now turn back to the experimental protein data, summarized in Figure 3, and interpret them on the basis of the equilibrium depicted in the upper part of Figure 6. In the HiPIP II from *E. halophila*, the equilibrium can be assumed to be virtually 100% to the left. Then we learn that, by starting from the hypothetical $[Fe_4S_4]^{4+}$ cluster, the iron pair bound to cysteines sequentially numbered II and III (Cys 42 and Cys 55 in HiPIP II from *E. halophila*)) has the highest reduction potential, whereas the iron pair bound to Cys I and IV (Cys 39 and Cys 71 in HiPIP II from *E. halophila*) has the lowest reduction potential. We have no means of discriminating between the individual reduction potentials of Fe_{II} and Fe_{III} on one side and of Fe_{I} and Fe_{IV} on the other. As we move to the other proteins of the series, we notice that the equilibrium of Figure 6 is progressively shifted (up to about 70%) to the right. This is illustrated in Figure 8. We then learn that the individual reduction potential of Fe_{III} (the one involved in both pairwise electron delocalizations A and B of Figure 6) remains the highest, while that of Fe_{II} decreases and that of Fe_{IV} increases along the series. In all situations the reduction potential of Fe_{I} remains the smallest. An order of the individual reduction potentials starting from four Fe^{3+} ions is thus obtained (*34*). The order is:

$$Fe_{III} > Fe_{II} \cong Fe_{IV} > Fe_{I}$$

where $Fe_{II} > Fe_{IV}$ for the HiPIPs I (*24*) and II (*5*) from *E. halophila* and the HiPIP from *R. globiformis* (*22*), and $Fe_{II} < Fe_{IV}$ for the HiPIPs I (35) and II (*23*) from *E. vacuolata* and for the HiPIPs from *C. vinosum* (*11*), *R. gelatinosus* (*36*) and *R. fermentans* (*37*).

We have recently corroborated the whole picture by mutating one of the four cysteine ligands of the Fe_4S_4 cluster to a serine (*34*). The idea was to selectively perturbing the individual reduction potential of one iron ion. We have chosen a HiPIP with intermediate equilibrium position (*C. vinosum* HiPIP) and mutated one of the two cysteines bearing an iron that shows a variable individual reduction potential along the series (Fe_{IV}). NMR studies have shown that the overall structure remains the same (*38*). The overall reduction potential decreases by 25 mV, the reduction potential for the A→R and B→R processes (Figure 6) decrease by 40 and 10 mV, respectively, and the position of the equilibrium between species A and B changes from 45%-55% to 60%-40%, as determined by NMR (*34*). This change is largely outside the experimental uncertainty, and is particularly meaningful as it crosses the 50%-50% situation, where the individual reduction potentials of Fe_{IV} and Fe_{II} cross. In other words, by a single mutation we have reverted the order of the individual reduction potentials of Fe_{IV} and Fe_{II} and caused this *C. vinosum* mutant to become a HiPIP of the same type as the

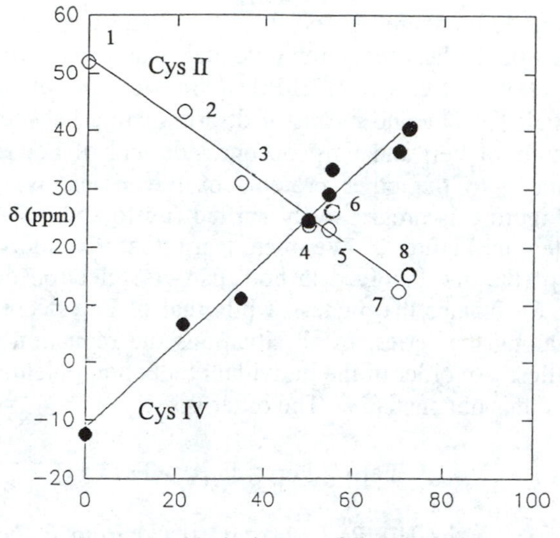

Figure 8. Correlation between average hyperfine shifts of the β-CH_2 cysteine protons (O = Cys II, ● = Cys IV) and % ferric character of the iron bound to Cys II in the HiPIPs from *E. halophila*, iso II (1), *E. halophila*, iso I (2), *R. globiformis* (3), *E. vacuolata*, iso I (4), *E. vacuolata*, iso II (5), *C. vinosum* (6), *R. gelatinosus* (7) and *R. fermentans* (8) (Reproduced from ref. *24*. Copyright 1995 ACS).

HiPIPs I and II from *E. halophila* and as the HiPIP from *R. globiformis* (Figure 8).

The overall picture that clearly emerges from these experiments, again by referring to the addition of one electron to the hypothetical $[Fe_4S_4]^{4+}$ cluster, is that in all HiPIPs Fe_{III} is the most reducible and Fe_I the least reducible. As these properties of Fe_I and Fe_{III} are common to all proteins of the series, they presumably depend on global rather than local environment features. Fe_I is by far the innermost iron of the cluster, and all four of its sulfur donors are buried inside the protein. Its environment is rather hydrophobic, and should thus stabilize the zero local charge achieved by a ferric iron and its negative sulfur ligands (-1 + -3*2/3) rather than a mononegative charge of a ferrous iron. The other three iron ions are also buried, but they share a bridging sulfide ion that is partially exposed to the solvent. Furthermore, of their three cysteine ligands, Cys III is partly exposed to the solvent: this could make Fe_{III} the most reducible. Discrimination between Fe_{II} and Fe_{IV} is less obvious, and in fact is more likely to depend on local features that vary from one protein to another.

Concluding Remarks

The combined application of Mössbauer, NMR and other physico-chemical techniques (*25,39-44*) has provided an example of deep understanding of the electronic isomerism in oxidized HiPIPs. There is now general agreement that each geometrical isomer contains two Fe^{3+} and two $Fe^{2.5+}$ ions. The geometrical isomers are generally two, and the particular isomerism seems dictated by the electrostatic properties of the protein and solvent. The electronic ground state seems to result from antiferromagnetic coupling between a largely $S = 9/2$ subspin of the mixed valence pair and a largely $S = 4$ subspin of the ferric pair. It is possible that the asymmetry of the polymetallic center requires the mixing of the above subspins with other subspin levels. Covariance between the decrease in the antiferromagnetic coupling constant within the mixed valence pair and the double exchange parameter within the same pair prevents an independent estimate of the two parameters. The estimate through NMR of the equilibrium constant between the two electronic isomers, together with the overall reduction potential, permits the attainment of a deeper insight into the redox properties of the individual iron ions.

Literature Cited

1. Carter, C. W. J.; Kraut, J.; Freer, S. T.; Alden, R. A.; Sieker, L. C.; Adman, E. T.; Jensen, L. H. *Proc. Natl. Acad. Sci. USA* **1972**, *69*, 3526-3529.
2. Middleton, P.; Dickson, D. P. E.; Johnson, C. E.; Rush, J. D. *Eur. J. Biochem.* **1980**, *104*, 289-296.
3. Noodleman, L. *Inorg. Chem.* **1988**, *27*, 3677-3679.
4. Papaefthymiou, V.; Girerd, J. -J.; Moura, I.; Moura, J. J. G.; Munck, E. *J. Am. Chem. Soc.* **1987**, *109*, 4703-4710.

5. Banci, L.; Bertini, I.; Capozzi, F.; Carloni, P.; Ciurli, S.; Luchinat, C.; Piccioli, M. *J. Am. Chem. Soc.* **1993**, *115*, 3431-3440.
6. Dunham, W. R.; Palmer, G.; Sands, R. H.; Bearden, A. J. *Biochim. Biophys. Acta* **1971**, *253*, 373-384.
7. Banci, L.; Bertini, I.; Luchinat, C. *Struct. Bonding* **1990**, *72*, 113-135.
8. Bertini, I.; Capozzi, F.; Luchinat, C.; Piccioli, M.; Vila, A. J. *J. Am. Chem. Soc.* **1994**, *116*, 651-660.
9. Karplus, M.; Fraenkel, G. K. *J. Chem. Phys.* **1961**, *35*, 1312-1323.
10. Bertini, I.; Campos, A. P.; Luchinat, C.; Teixeira, M. *J. Inorg. Biochem.* **1993**, *52*, 227-234.
11. Bertini, I.; Capozzi, F.; Ciurli, S.; Luchinat, C.; Messori, L.; Piccioli, M. *J. Am. Chem. Soc.* **1992**, *114*, 3332-3340.
12. Schulz, C.; Debrunner, P. G. *J. Phys. (Paris)* **1976**, *37*, C6-153.
13. Mouesca, J. -M.; Noodleman, L.; Case, D. A.; Lamotte, B. *Inorg. Chem.* **1995**, *34*, 4347-4359.
14. Belinskii, M. I.; Bertini, I.; Galas, O.; Luchinat, C. *Z. Naturforsch.* **1995**, *50a*, 75-80.
15. Bertini, I.; Luchinat, C. *JBIC* **1996**, *1*, 183-185.
16. Noodleman, L. *Inorg. Chem.* **1991**, *30*, 246-256.
17. Banci, L.; Bertini, I.; Briganti, F.; Luchinat, C.; Scozzafava, A.; Vicens Oliver, M. *Inorg. Chem.* **1991**, *30*, 4517-4524.
18. Belinskii, M. I. *Chem. Phys.* **1993**, *176*, 37-45.
19. Belinskii, M. I.; Bertini, I.; Galas, O.; Luchinat, C. *Inorg. Chim. Acta* **1996**, *243*, 91-99.
20. Mouesca, J. M.; Chen, J. L.; Noodleman, L.; Bashford, D.; Case, D. A. *J. Am. Chem. Soc.* **1994**, *116*, 11898-11914.
21. Bertini, I.; Ciurli, S.; Luchinat, C. In *Structure and Bonding*; Springer-Verlag, Vol. 83: Berlin Heidelberg, 1995; pp. 1-54.
22. Bertini, I.; Capozzi, F.; Luchinat, C.; Piccioli, M. *Eur. J. Biochem.* **1993**, *212*, 69-78.
23. Banci, L.; Bertini, I.; Ciurli, S.; Ferretti, S.; Luchinat, C.; Piccioli, M. *Biochemistry* **1993**, *32*, 9387-9397.
24. Bertini, I.; Capozzi, F.; Eltis, L. D.; Felli, I. C.; Luchinat, C.; Piccioli, M. *Inorg. Chem.* **1995**, *34*, 2516-2523.
25. Gloux, J.; Gloux, P.; Lamotte, B.; Mouesca, J. M.; Rius, G. J. *J. Am. Chem. Soc.* **1994**, *116*, 1953-1961.
26. Bominaar, E. L.; Borshch, S. A.; Girerd, J. -J. *J. Am. Chem. Soc.* **1994**, *116*, 5362-5372.
27. Ding, X. -Q.; Bill, E.; Trautwein, A. X.; Winkler, H.; Kostikas, A.; Papaefthymiou, V.; Simopoulos, A.; Beardwood, P.; Gibson, J. F. *J. Chem. Phys.* **1993**, *99*, 6421-6428.
28. Moore, G. R.; Pettigrew, G. W.; Rogers, N. K. *Proc. Natl. Acad. Sci. USA* **1986**, *83*, 4998-4999.

29. Moore, G.R.; Pettigrew, G.W. *Cytochromes c; Evolutionary, Structural and Physicochemical Aspects*; Springer-Verlag: Berlin, 1990;
30. Cammack, R.; Dickson, D.; Johnson, C. In *Iron Sulfur Proteins*; Lovenberg, W., Ed.; Academic Press: New York, 1977; pp. 283-330.
31. Jensen, G. M.; Warshel, A.; Stephen, P. J. *Biochemistry* **1994**, *33*, 10911-10924.
32. Luchinat, C.; Capozzi, F.; Borsari, M.; Battistuzzi, G.; Sola, M. *Biochem. Biophys. Res. Commun.* **1994**, *203*, 436-442.
33. Banci, L.; Bertini, I.; Gori Savellini, G.; Luchinat, C. *Inorg. Chem.* **1996**, in press.
34. Babini, E.; Bertini, I.; Borsari, M.; Capozzi, F.; Dikiy, A.; Eltis, L. D.; Luchinat, C. *J. Am. Chem. Soc.* **1996**, *118*, 75-80.
35. Bertini, I.; Gaudemer, A.; Luchinat, C.; Piccioli, M. *Biochemistry* **1993**, *32*, 12887-12893.
36. Bertini, I.; Capozzi, F.; Luchinat, C.; Piccioli, M.; Vicens Oliver, M. *Inorg. Chim. Acta* **1992**, *198-200*, 483-491.
37. Ciurli, S.; Cremonini, M. A.; Kofod, P.; Luchinat, C. *Eur. J. Biochem.* **1996**, *236*, 405-411.
38. Bentrop, D.; Bertini, I.; Capozzi, F.; Dikiy, A.; Eltis, L. D.; Luchinat, C. *Biochemistry* **1996**, in press.
39. Thompson, A.J. In *Metalloproteins*; Harrison, P., Ed.; Verlag Chemie: Weinheim, FRG, 1985; pp. 79.
40. Backes, G.; Mino, Y.; Loehr, T. M.; Meyer, T. E.; Cusanovich, M. A.; Sweeney, W. V.; Adman, E. T.; Sanders-Loehr, J. *J. Am. Chem. Soc.* **1991**, *113*, 2055-2064.
41. Dunham, W. R.; Hagen, W. R.; Fee, J. A.; Sands, R. H.; Dunbar, J. B.; Humblet, C. *Biochim. Biophys. Acta* **1991**, *1079*, 253-262.
42. Czernuszawicz, R. S.; Macor, K. A.; Johnson, M. K.; Gevirth, A.; Spiro, T. S. *J. Am. Chem. Soc.* **1987**, *109*, 7178-7187.
43. Przysiecki, C. T.; Meyer, T. E.; Cusanovich, M. A. *Biochemistry* **1985**, *24*, 2542-2549.
44. Moss, T. H.; Petering, D.; Palmer, G. *J. Biol. Chem.* **1969**, *244*, 2275-2277.

Chapter 4

Redox Metalloenzymes Featuring S-Donor Ligands Hydrogenase: A Case Study

Michael J. Maroney, Christian B. Allan, Balwant S. Chohan, Suranjan B. Choudhury, and Zhijie Gu

Department of Chemistry, University of Massachusetts, Amherst, MA 01003–4510

Many metalloenzymes have active sites composed of metals coordinated to S-donor ligands. Among the examples are Ni, Fe hydrogenases. These enzymes catalyze the two-electron redox chemistry of H_2 and are believed to contain a heterodinuclear active site composed of a Ni center bridged to an Fe center by cysteinate ligands. The possible roles of the thiolate ligands in constructing the active site, in the redox chemistry of the active site, and in the binding of H^+ are discussed in the context of an overview of the results of physical studies of the enzyme and dinuclear model compounds.

A large number of redox metalloproteins that employ thiolate (cysteinate) and/or sulfide as metal ligands is known (see Chapter 1). These proteins play key roles in biological redox processes including respiration and photosynthesis (*e.g.*, ferredoxins (*1,2*), blue Cu proteins (*3*), Cu_A in cytochrome oxidase(*4-6*)) and in the oxidation or reduction of substrates (*e.g.*, nitrogenase (*7*), hydrogenase (*8-11*), sulfite reductase (*12*)). Among these metalloproteins are a group of enzymes that contain Ni (*13*). Nickel containing redox metalloenzymes include methylcoenzyme M reductase (*14*), carbon monoxide dehydrogenase (COdH) (*15*) and most hydrogenases (H_2ases) (*8,16*).

Methylcoenzyme M reductase is found in methanogenic bacteria where it catalyzes the last step in methanogenesis:

$$\underset{\text{Me-coenzyme M}}{CH_3SCH_2CH_2SO_3^-} + \text{HS-HTP} \longrightarrow CH_4 + {}^-O_3SCH_2CH_2S\text{-S-HTP}$$

It contains a unique Ni-containing tetrahydrocorphin cofactor, F_{430}, which is involved in the reduction of the methyl group from methylcoenzyme M (a methylthioether) coupled with the formation of a disulfide involving coenzyme M and *N*-7-mercaptoheptonyl-*O*-phospho-L-threonine (thiols), a two-electron redox process. The exact role of the Ni in this enzyme is not well-known (*17-19*), although it is clear that the net redox chemistry involves thiolate oxidation.

Carbon monoxide dehydrogenase catalyzes nature's version of water-gas shift chemistry--the two-electron redox chemistry of CO:

0097–6156/96/0653–0074$16.75/0

$$CO + H_2O \rightleftharpoons CO_2 + 2H^+ + 2e-$$

The enzyme is found in bacteria in both relatively simple enzymes (*e.g.*, *Rhodospirillum rubrum*) and more complex multicomponent enzymes (*e.g.*, *Clostridium thermoaceticum*) that also catalyze acetate biosynthesis at a separate site containing Ni and Fe. Vibrational spectroscopy has been used to address the role of the Ni center in the *C. thermoaceticum* enzyme. These studies reveal that CO binds to an Fe site (*20*), but that the Ni site does serve to bind the methyl group in acetate biosynthesis (*21*). Both types of enzymes contain Ni ligated to S-donor ligands as well as Fe-S clusters. Analysis of EXAFS data obtained on the Ni site in *R. rubrum* COdH indicate that the Ni is coordinated to ~2 S-donor ligands at a distance of 2.23 Å, and to ~3 O,N-donor atoms at 2.00 Å (*22*). Analysis of EXAFS data obtained from the α-subunit of *C. thermoaceticum* COdH (acetate biosynthesis) also reveals that the Ni features S-donor ligation (*23*).

Hydrogenases are redox enzymes that catalyze the two-electron redox chemistry of dihydrogen:

$$H_2 \rightleftharpoons 2H^+ + 2e^-$$

These enzymes play a central role in anaerobic metabolism and may be grouped by the metal content of the enzyme (*16*). The first group are quasi-H_2ases (the pure protein cannot activate H_2) represented by enzymes from *Methanobacterium* and contain no known metals (*24*). In place of a metal, an organic redox cofactor, methylenetetrahydromethanopterin, serves as a source of hydride ion (*25,26*). With this exception, all other hydrogenases contain Fe-S clusters. A second class of H_2ases contains only Fe and features an Fe-S cluster that has unique spectral properties (the H-cluster). This cluster is believed to constitute the active site in the Fe-only enzymes (*11*). By far the largest group of hydrogenases contain a Ni atom in addition to various Fe-S clusters (*8*). Although it is clear that Ni is not a requirement for an active catalyst, much attention has been focused on the function of the Ni in this enzyme. Like the other Ni-containing enzymes, EXAFS has consistently shown that the Ni in H_2ases is bound by S-donor ligands (*27-29*). In the case of the enzyme from *Thiocapsa roseopersicina*, the ligand environment is remarkably similar to that found for the Ni site in *R. rubrum* COdH (~2 S @ 2.23 Å + ~3 O,N @ 2.00 Å) (*28*).

The use of Ni in redox proteins appears to be a curious choice given the fact that there is only one common oxidation state (II) of Ni due to the normally extreme potentials of the Ni(II/III) and Ni(II/I) couples (*30-32*). This is particularly striking in the known Ni redox enzymes, all of which catalyze two-electron processes and must therefore involve two relatively uncommon oxidation states, if all of the redox activity is attributed to Ni. Two alternatives are possible: First, the redox chemistry might be due in part to S redox chemistry. All of the Ni sites are associated with S-donor ligands or with the catalysis of S-redox chemistry. Model studies have contributed to the understanding of what effect thiolate ligation has on Ni redox chemistry. In general, thiolate ligation stabilizes the higher oxidation states of transition metals and thus makes the Ni(III/II) couple more accessible to biological systems (*10,33-38*). At the same time, thiolate ligation generally makes the potentials associated with lower oxidation states (*i.e.*, the Ni(II/I) couple) even more inaccessible. However, at least one system involving thiolate ligation is able to produce both stable Ni(III) and Ni(I) species (*39*).

The composition of COdH and H_2ase suggest a second possibility: That some or all of the chemistry catalyzed by these enzymes may involve Fe, a metal noted for facile redox chemistry that is widely distributed in redox metalloproteins.

Structural Aspects

Hydrogenases containing Ni are typically isolated as heterodimers composed of subunits with molecular weights of ca. 30 and 60 kDa. The structural genes that code for these two subunits have been sequenced in a number of cases and reveal that two cys-X-X-cys sequences are conserved in the larger subunit, one near the N-terminus and one near the C-terminus (*40,41*). In some enzymes (*e.g.*, those from *Desulfomicrobium baculatum* and *Methanococcus voltae*) the first cysteine residue in the C-terminal Cys-X-X-Cys sequence is substituted by selenocysteine, which has been shown to be a Ni ligand by analysis of EXAFS data (*42*) and by the observation of ^{77}Se hyperfine coupling in the epr spectra of these enzymes (*43,44*).

The understanding of the relationship of the cofactors to each other and the structure of the Ni site has been greatly enhanced by the recent publication of a crystal structure at 2.8 Å resolution of the H_2ase from *Desulfovibrio gigas* (*45*). The crystal structure shows that the two subunits of the heterodimer are intimately associated with each other, and that the various redox cofactors are widely separated. The Fe-S clusters are located entirely within the small subunit, while the Ni site lies entirely within the large subunit. The Fe-S clusters found in the *D. gigas* enzyme are comprised of 2 Fe_4S_4 clusters and a Fe_3S_4 cluster. These clusters are arranged in a linear fashion spaced ca. 10 Å apart with the Fe_3S_4 cluster in the middle. The Ni center lies about 10 Å from the proximal Fe_4S_4 cluster. The Fe_4S_4 cluster furthest from the Ni site (the distal cluster) is unique in that one of the Fe ligands is a histidine imidazole. The surface-exposed histidine plus the linear arrangement of the Fe-S clusters suggest an electron transfer pathway leading to or from the Ni center.

One of the most interesting aspects of the crystal structure is the revelation that the Ni center is actually a dimetallic cluster composed of Ni and Fe (Figure 1). The assignment of the second metal center as an Fe center was based on the metal analysis (only Ni and Fe were found), the strong anomalous scattering of Cu-Kα radiation (distinct from Ni) and the electron density associated with the metal center (appropriate for a first row transition metal) (*45*). This assignment is supported by recent data collected at other x-ray wavelengths (*46*). The dimetallic cluster is ligated by the four conserved cysteines found in the large subunit. Two of these cysteines are bound as terminal ligands to the Ni center. The remaining two cysteines bridge between the Ni and Fe atoms. No other protein ligands were identified. The structure of the Ni site can be described as a highly distorted trigonal-pyramidal arrangement of the four cysteine residues. Three of the cysteine ligands have average Ni-S bond lengths of 2.25 Å with the remaining S at ~2.6 Å. The Fe center lies ~2.7 Å from the Ni atom and appears to be five-coordinate with three exogenous ligands. These exogenous ligands were modeled as H_2O molecules, but recent IR data identified bands in the 2000 cm^{-1} region of the spectrum that belong to triply bonded species such as CN^- or CO (*47*). These features are linked to the Ni site by the fact that their frequencies shift in response to the redox state of the enzyme. Further examination of the crystallographic data are consistent with diatomic ligands (*46*), but cannot differentiate between the various possibilities.

The information obtained from the crystal structure suffers from the fact that the crystals employed were a mixture of at least three different redox states of the enzyme (50% SI, 36% A and 14% B, as defined below). This fact may have contributed to a large amount of disorder in the Ni-Fe site. The structure of the Ni site that is emerging from the crystallographic studies can be compared with information obtained from XAS on redox poised enzymes from a number of different species(*27,28,42*). A recent comparison involving enzymes from photosynthetic bacteria (*T. roseopersicina.* and *Chromatium vinosum*), sulfate reducing bacteria (*D. gigas* and *D. desulfuricans*), and *E. coli* has been performed (*29*). The Ni K-edge XANES data provides a useful predictor of coordination number/geometry (*48*). The data obtained from all of the enzymes lack the 1s -> $4p_z$ transition that is typical of

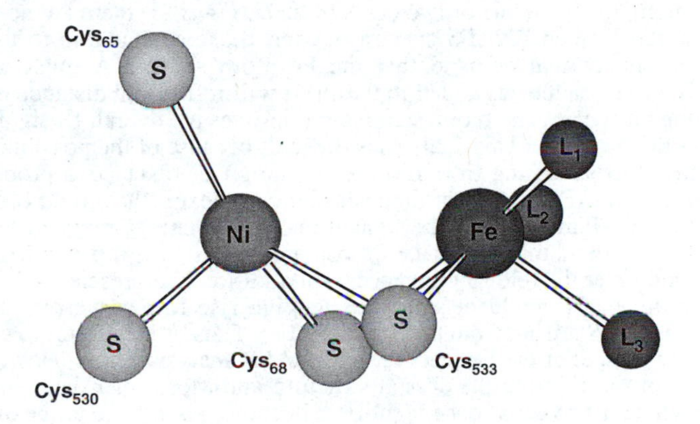

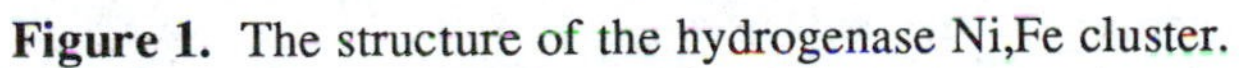
Figure 1. The structure of the hydrogenase Ni,Fe cluster.

planar Ni complexes, and exhibits peak areas for the 1s -> 3d transitions that are consistent with either a five-coordinate (particularly for oxidized forms) or a six-coordinate geometry (particularly for *T. roseopersicina* H_2ase and reduced forms). The arrangement of the ligands in the crystal structure is not inconsistent with these predictions, since placing an exogenous fifth ligand (*e.g.*, H_2O, O_2, CO, H^-) into the vacant axial position of the Ni center in the crystal structure would likely produce a distorted trigonal bipyramidal site, and place the additional ligand into a bridging position between the Ni and Fe centers.

The EXAFS data from all of the H_2ases is dominated by the presence of S-scattering atoms at 2.22(1) Å. In most cases, additional long Ni-S (2.4 - 2.9 Å) and/or Ni-Fe (2.4 - 2.9Å) interactions can also be accommodated by the data (*29,49,50*). The number of S-scatterers, as well as the presence of additional O,N donors appears to be variable among the enzymes from various species and is not well-determined by EXAFS analysis. The trend is that the oxidized enzymes frequently exhibit evidence for O,N ligation, but the reduced enzymes are most often fit best by a set of four S-donor ligands (*29*). One exception is the enzyme from *T. roseopersicina.*, which exhibits evidence of O,N ligation at all redox levels (*28*).

Although it has long been suspected that the Ni center might be associated with an Fe atom, in all but one or two cases (*e.g., D. gigas* SI) there are no features in the Fourier-transformed EXAFS spectra beyond those attributable to the primary coordination sphere that demand that the Fe atom at ~2.7 Å must be present. However, if an Fe scatterer is added to the fit, it will refine to a distance of 2.4 - 2.9 Å (depending on species and redox state) and improves the overall fit significantly in at about half of the cases. This analysis is difficult because of the possibility of large Debye-Waller factors arising from disorder in the Ni-Fe distance, a problem that is observed in the EXAFS spectra of dinuclear Ni complexes, that could obliterate the expected features. Further, given the signal to noise and energy range of the data sets on the enzymes (often limited by trace Cu contamination), it is difficult to distinguish between S and Fe at the relevant distances. In addition, the presence of S and Fe at comparable distances would be expected to give rise to destructive interference between the EXAFS arising from the S and the Fe. Thus, it is difficult to confirm the presence (or absence) of the Fe scatterer by EXAFS analysis alone. However, given the presence of the Fe from the crystal structure and Mössbauer data, the EXAFS arising from the Ni-Fe vector, once identified, becomes a sensitive probe of structural changes in the dimetallic site (*vide infra*).

The structure of the Ni-Fe cluster may also be compared to other M_2S_2 units found in redox metalloproteins and is unique in two ways: First, it is a heterodimetallic cluster, and second it does not feature a flat M_2S_2 core. The Fe_2S_2 centers found in ferredoxins feature planar Fe_2S_2 units (*51*), though the S-donors involved are sulfide ions and not cysteine thiolate ligands. A planar Cu_2S_2 unit involving cysteinate ligation is found in the recently characterized Cu_A site in cytochrome oxidase (*4-6*). The tendency of Ni(II) alkylthiolate complexes to polymerize via the formation of μ-dithiolato groups is well-established.(*35,52-54*) In most of these complexes the Ni_2S_2 cores are not flat, but bent along the vector containing the bridging S donor atoms. One example where polymerization was stopped at the dimer stage by incorporation of an N-donor ligand is shown in Figure 2. This dimer (and others like it) have several gross structural features in common with the H_2ase Ni-Fe cluster (*53*). The dihedral angle formed between the planes defined by the MS_2 fragments (the fold angle) is typically ~90-120° and compares with a fold angle of ~96° calculated for the H_2ase active site. The folding places the Ni centers closer together at a distance of 2.68 Å, a distance that is indistinguishable from the Ni-Fe distance in the H_2ase structure (~2.7 Å). The folding is in large part a consequence of the use of p-orbitals on the S-donor atoms, which dictate ~90° bond angles around the S atom. Thus, the S center can choose to either position the alkyl

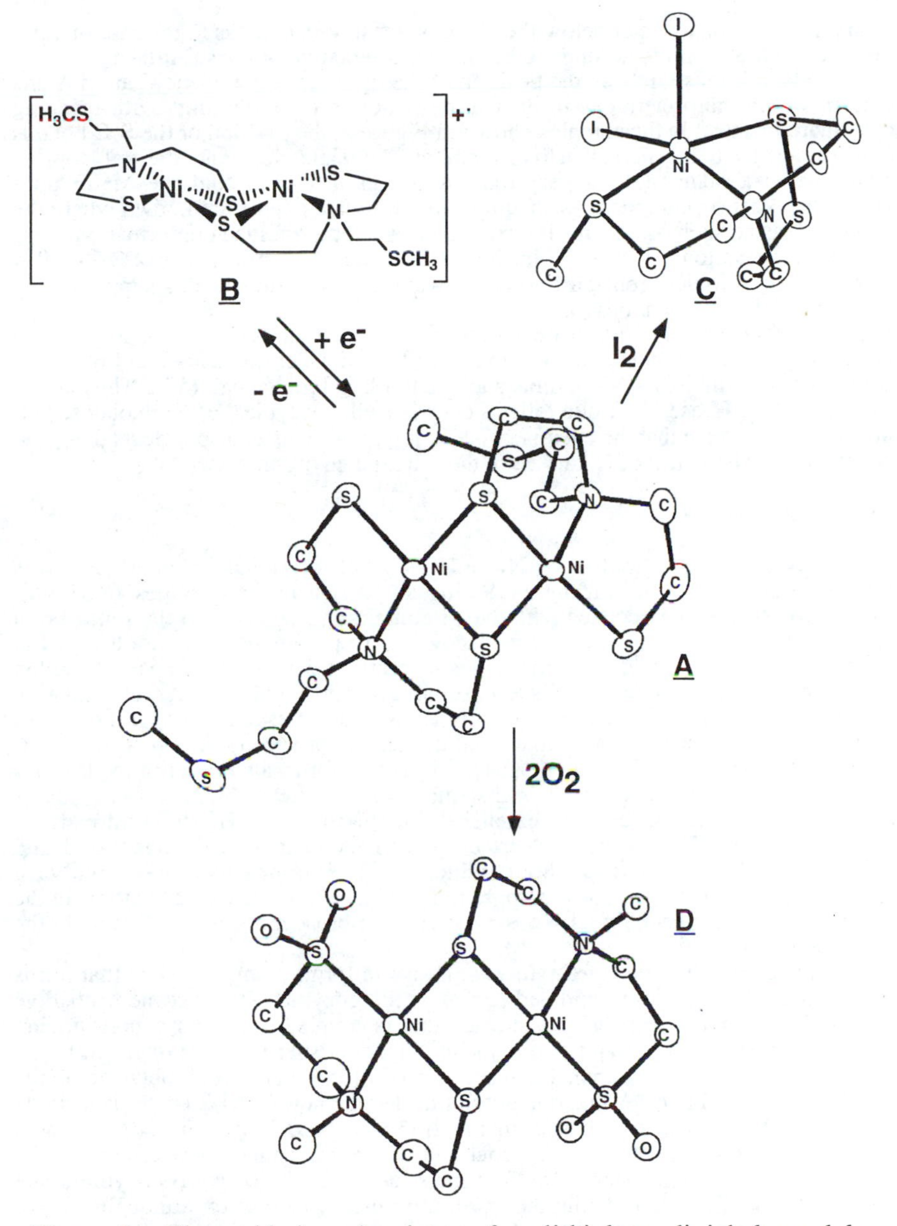

Figure 2. The oxidative chemistry of μ-dithiolato dinickel models, $\{Ni[RN(C_2H_4S)_2]\}_2$. A. The structure of a Ni(II,II) dimer, $\{Ni[MeSC_2H_4N(C_2H_4S)_2]\}_2$. B. The one-electron oxidation complex characterized by spectroscopic techniques. C. The structure of a disulfide from the two-electron oxidation of each Ni center. D. The O_2 oxidation product of the Ni(II,II) dimers. (The structure shown for the oxidation of $\{Ni[MeN(C_2H_4S)_2]\}_2$).

or aryl substituent above or below the plane of a flat M_2S_2 unit as in the case of Cu_A, or bend the M_2S_2 unit to accommodate another orientation of the substituent.

Placing constraints on the position of the cysteine β-methylene C atom is one way in which the protein can influence the structure of M_2S_2 units with bridging cysteinate ligands. In the complex shown in Figure 2, the position of the S-C bond is constrained by being part of a five-membered chelate ring, and the only option available to a planar (albeit very distorted) complex is to fold the M_2S_2 unit. However, other Ni complexes with simple alkylthiolate donors are known where the complex has no such constraints but nevertheless adopts a folded conformation. One possible explanation is that Ni-Ni interactions occurring over distances of 2.6 - 2.9 Å might favor a folded configuration. However, this notion is not supported by qualitative M.O. calculations (*55*).

The tendency of Ni thiolates to bridge to another metal extends to heteropolynuclear complexes as well (*56,57*). The Ni_3Fe cluster shown in Figure 3 contains two μ–dithiolato Ni-Fe units and one thiolato-bridged pair (*57*). Thus, the structure of the H_2ase Ni-Fe site reflects chemistry that is typical of Ni thiolates, and one might anticipate that the chemistry occurring at this cluster also reflects the chemistry that is typical of Ni and Fe in these ligand environments.

Redox Chemistry

The redox chemistry exhibited by Ni-Fe H_2ases is exceptionally rich and involves the Ni-Fe cluster and the various Fe-S clusters present in the enzymes. (*8,58-66*). The redox chemistry associated with the Ni-containing site has been the subject of a great deal of investigation since the discovery of epr signals that are observed at temperatures > 77 K and display hyperfine interactions with ^{61}Ni (*67,68*). The redox chemistry associated with the Ni site is summarized in Figure 4. Aerobic isolation and purification of Ni-Fe H_2ases generally leads to samples containing a mixture of two catalytically inactive forms that exhibit distinct epr spectra (Form A: g = 2.31, 2.23, 2.02; Form B: g = 2.33, 2.16, 2.01). These two forms also differ in the kinetics of reductive activation. Form B is instantaneously reduced by H_2 (it is ready or active) while form A requires extensive incubation with H_2 (it is unready or inactive). Form B may be converted to form A in a poorly understood and irreversible slow process (without reducing the enzyme) that may involve a conformational change and/or incorporation of O_2. The presence of O atoms in the oxidized forms has been directly observed via line broadening of epr signals in the presence of $^{17}O_2$ (*69*).

Oxidative titrations of *C. vinosum* H_2ase in forms A and B reveal that forms A and B are not the fully oxidized forms of the enzyme (*70*). At more positive potentials, a new S = 1/2 center is created that is coupled to the epr signals arising from Ni in forms A and B and also to the epr signal arising from the oxidized Fe_3S_4 cluster that exists in this enzyme. An analysis of Mössbauer spectra obtained on the oxidized samples indicates that one possibility for the new S = 1/2 center is a single Fe atom with an unusually small isomer shift (δ = 0.05 - 0.15 mm/s). The weakness of the coupling (6 mT) to the Ni signal and the large distance between the Ni-Fe cluster and the Fe_3S_4 cluster (~18 Å) in the structure of the *D. gigas* enzyme argue against this S = 1/2 center being the Fe atom in the Ni-Fe cluster, although it is the only possibility suggested by the 2.85 Å crystal structure.

Both forms A and B are reduced by one-electron to produce an epr-silent intermediate (SI) (*8,58*). The existence of active and inactive conformations of the enzyme is supported by a study of the reductive interconversion of forms A and B (*8*). When form A is reduced at 4° to the SI level and reoxidized, only form A is observed as a product. However, when the reduction is carried out and the resulting SI state is allowed to equilibrate at 30°, form B is also produced upon oxidation. A similar temperature dependence is observed for SI samples produced from form B.

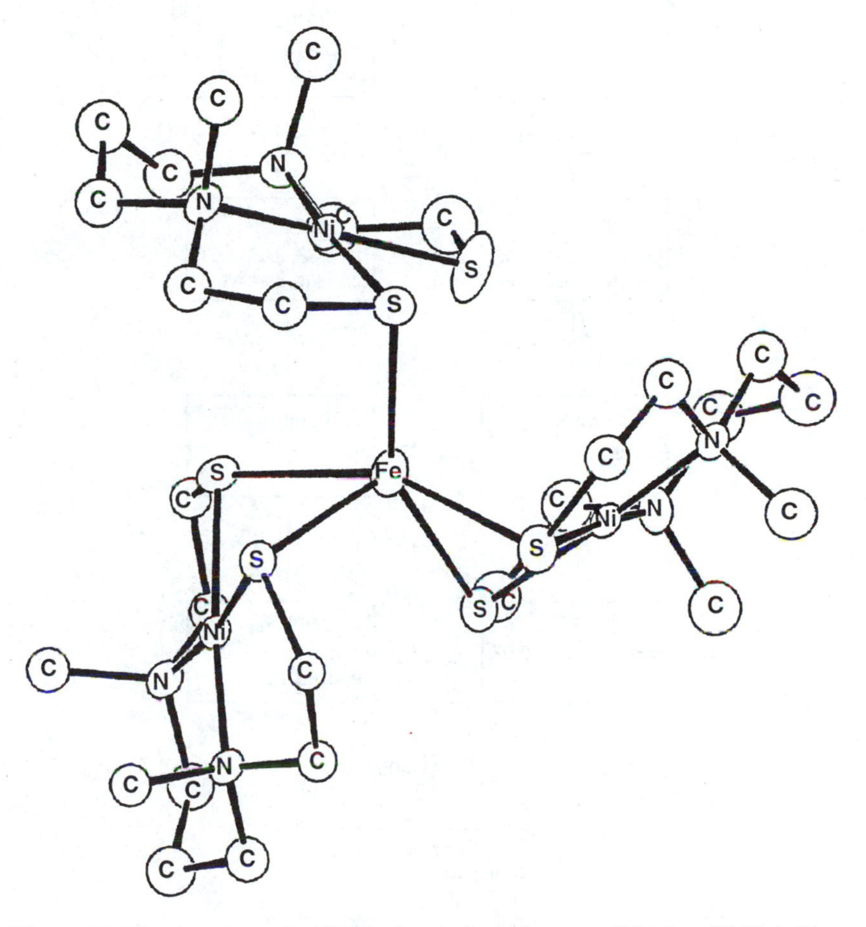

Figure 3. The structure of a Ni_3Fe cluster featuring two dithiolato Ni,Fe bridges and one monothiolato bridge. (Adapted from ref. 57).

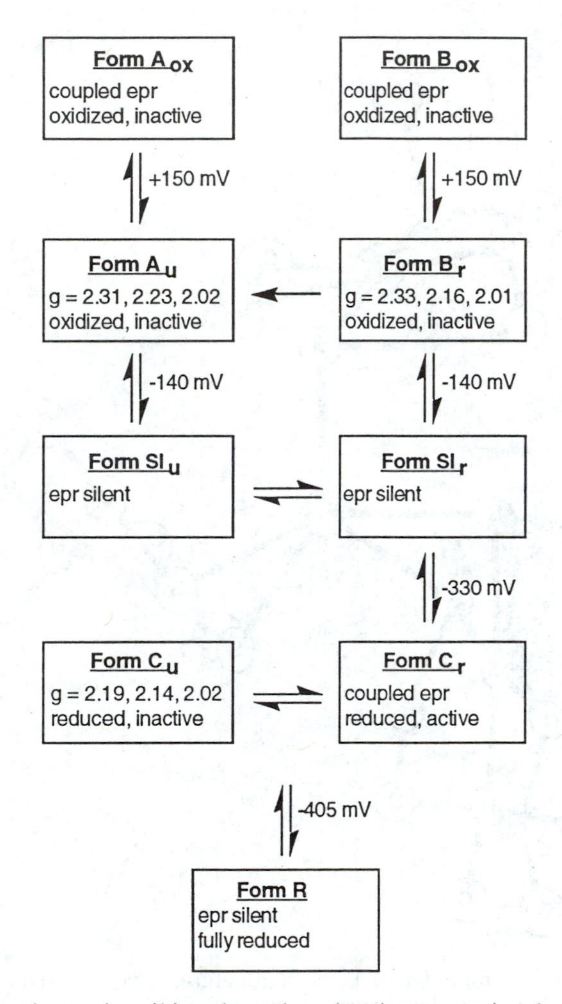

Figure 4. A scheme describing the redox chemistry associated with the Ni,Fe cluster in hydrogenases.

The SI redox states may also be reduced by one-electron to yield another epr-active redox state, form C (g = 2.19, 2.14, 2.02) (*8,58*). This state has long been thought to constitute a catalytic intermediate such as an enzyme/substrate complex (*e.g.*, $Ni-H^-$ or $Ni-H_2$). However, recent results have indicated that this state is stable indefinitely in the absence of H_2, and thus cannot be a catalytic intermediate (*60*). Form C may be further reduced to another epr inactive redox level where all of the metal cofactors are in their fully reduced state (R). This reduction, and the stability of the Form C in *C. vinosum* has been studied in detail under a variety of conditions by Coremans *et al.* (*61*).

All of the redox chemistry associated with Ni occurs at potentials between -100 and -400 mV. This is a range that is not typical of one-electron Ni redox chemistry, and is unprecedented for two-electron Ni redox chemistry. Nonetheless, several schemes based on Ni redox chemistry have been proposed, examples of which are shown in Table I. The chief advantage of these schemes is that they

Table I. Redox Schemes

Sample	*77 K epr Signal*	*Scheme A*	*Scheme B*	*Scheme C*	*Scheme D*
Form A	g = 2.31, 2.23, 2.02	Ni(III)	Ni(III)	Ni(I), Fe(II)	Ni(III), Fe(IV)
Form B	g = 2.33, 2.16, 2.01	Ni(III)	Ni(III)	Ni(I), Fe(II)	Ni(III), Fe(IV)
SI	none	Ni(II)	Ni(II)	Ni(I), Fe(I)	Ni(III), Fe(III)
Form C	g = 2.19, 2.14, 2.02	Ni(III)	Ni(I)	Ni(I), Fe(II)	Ni(III), Fe(II)
R	none	Ni(II)	Ni(0)	Ni(I), Fe(I)	Ni(II), Fe(II)

provide simple explanations for the sequential appearance and disappearance of S=1/2 epr signals that are associated with the Ni site via the observation of ^{61}Ni hyperfine couplings in labeled samples. In both schemes, the epr-active oxidized forms of the protein are assigned to Ni(III) centers. Scheme A utilizes one-electron redox chemistry at the Ni center by reducing Ni(III) to Ni(II) in SI and then reoxidizing it to Ni(III) in form C and reducing it again to Ni(II) in R. Scheme B reduces Ni through four redox states (III->II->I->0). Scheme A has the advantage that it utilizes only two oxidation states for Ni, and that the unusual oxidation state involved (III) is the one favored by thiolate ligation. However, scheme A needs a structural change in the Ni site in order to account for the reoxidation of the Ni(II) SI form at a lower potential. Scheme B implies unprecedented redox chemistry for a single Ni center that must involve considerable structural change in order to compress the Ni(III/II) and Ni(II/I) couples that are normally separated by about 2 V into an ~ 250 mV range. A baseline for the kinds structural change that might be expected is provided by a model system where Ni-S bonds lengthen by 0.14 Å upon reduction of Ni(III) to Ni(II) (*37*).

The redox activity of the Ni center has been examined by using XAS. Measurement of the differences in the Ni K-edge energy between the various redox levels provides a sensitive method for detecting changes in the charge density on the Ni center (*48*). The technique has been applied to many redox enzymes and has proven useful in detecting redox reactions that are ligand-centered rather than metal-centered, provided that the ligand environment of the metal is relatively constant (*10*). The edge energies determined from the Ni K-edge spectra in Figure 1 are shown in Table II. For a one-electron metal-centered redox process, a shift of ~2 eV in the edge energy is expected. In none of the cases is the edge energy shift between fully oxidized and fully reduced samples greater than 1.5 eV, ruling out scheme B as a realistic description of the redox chemistry. This view is also supported by a Ni L-edge study of *P. furiosus* H_2ase (*71*). Scheme A cannot be ruled out entirely, but is overly simplistic in that the analysis of the EXAFS data reveals no significant changes in the Ni-S distances as a function of redox poise. Thus, the redox chemistry associated with at least three electrons that are required to go from oxidized to reduced enzyme must either be localized elsewhere, or delocalized over several atomic centers in the Ni-Fe cluster.

Table II. Ni K-edge and XANES Data for Ni,Fe Hydrogenases

Sample	*Source*	*Edge Energy (eV)*	*1s->3d Area (10^2 eV)*
Form A	*C. vinosum*	8340.0(2)	7.6(5)
	D. gigas	8340.5(2)	5.7(5)
	E. coli	8340.3(2)	4.4(5)
	T. roseopersicina	8340.4(2)	1.5(5)
Form B	*C. vinosum*	8340.4(2)	6.3(5)
	D. desulfuricans	8340.3(2)	4.3(5)
	T. roseopersicina	8339.8(2)	1.6(5)
SI	*D. gigas*	8339.2(2)	2.2(5)
	T. roseopersicina	8339.8(2)	1.4(5)
Form C	*C. vinosum*	8339.4(2)	2.4(5)
	D. gigas	8339.6(2)	2.5(5)
	T. roseopersicina	8339.6(2)	<0.1(5)
R	*C. vinosum*	8339.0(2)	1.1(5)
	D. desulfuricans	8339.6(2)	1.4(5)
	D. gigas	8339.0(2)	2.9(5)
	T. roseopersicina	8339.5(2)	0.7(5)

Another possibility is that the Fe atom in the Ni-Fe center is primarily responsible for the redox chemistry. Iron-centered redox schemes are also summarized in Table I (*46*). These schemes feature Ni(I) (Scheme C) or Ni(III) (Scheme D) to account for the epr signal that is associated with the Ni site, but do not vary the Ni oxidation state. Instead, the S=1/2 Ni centers are antiferromagnetically coupled to S =1/2 Fe centers in order to produce the epr silent states. The choice of the favored Fe and Ni oxidation states is determined by the hypothetical choice of CO (lower oxidation states) or CN^- (higher oxidation states) as the terminal Fe ligands in the cluster. These mechanisms have several advantages over the Ni redox schemes. First, Fe has facile redox chemistry and is widely utilized as a redox center in biology. Second, the schemes are consistent with the XAS data in that the structure and the charge density of the Ni center need not change much. The schemes are also consistent with a role for Fe as a CO binding site (*vide infra*). However, both schemes still utilize unusual oxidation states for Ni and have many of the problems associated with the Ni-centered schemes. In addition, a stable Ni(I) species is inconsistent with an oxidized protein and a thiolate-rich Ni ligand environment. The Ni(III) center favored by the thiolate environment in the oxidized protein is unlikely to be stable in a reducing environment. The Fe-centered schemes also provide no compelling explanation for the absence of ^{57}Fe hyperfine splittings or line broadening in the epr spectra associated with the Ni-Fe cluster in labeled H_2ase samples (*72*). In essence, these schemes simply transfer the redox processes from Ni to Fe.

A third possibility is that the thiolate ligands are intimately involved in the redox process and that the redox chemistry is a property of the cluster and not a particular metal center. This possibility is supported by studies of μ-dithiolato dinickel(II) model complexes that also feature terminal thiolate ligation. The products of various oxidations of $\{Ni[RN(CH_2CH_2S)_2\}_2$ dimeric complexes reveal evidence of S-oxidation in every case (figure 4) (*10,33,53,73,74*).

One-electron electrochemical or chemical oxidations of the dimeric systems lead to species that exhibit rhombic S = 1/2 epr spectra that are similar to those characteristic of H_2ase. The potentials measured for the one-electron oxidation of the dimers using electrochemical methods can be low (*e.g.*, +30 mV vs. NHE), indicating that the thiolate ligation has stabilized the *formal* Ni(II/III) couple, as expected. The Ni(II,II) dimers have no accessible reductive chemistry in CH_2Cl_2/0.1mM $Bu_4N(ClO_4$ or $PF_6)$ solution. Although the oxidized species have proven to be too reactive to crystallize and structurally characterize by x-ray diffraction, the assignment of the oxidation product as a *formal* Ni(II/III) dimer is consistent with the coulometry (one-electron/dimer), with structural precedent in a similar system, with the quasi-reversible electrochemistry exhibited by the dimers, and with the analysis of epr spectra in dimers labeled with ^{61}Ni where the hyperfine is best fit by including two Ni centers. The electronic structure of the one-electron oxidation product has been examined by using epr, ^{1}H-ENDOR, electronic absorption and theoretical techniques (*10,55*). The epr spectra obtained from the ^{61}Ni-labeled dimers with N-substitutents that do not contain donor atoms exhibit very small hyperfine interactions that average for the two Ni centers to 10.5, 0.0, and 4.1 G for g = 2.20, 2.14 and 2.02, respectively (for N-substituent = $-CH_2Ph$). The magnitude of the hyperfine couplings observed in the models is about half that found in H_2ases (*75*), which is about half of that is observed in Ni(III) complexes with O- or N-donor ligands (*30,31*). The ^{1}H-ENDOR spectrum of the S=1/2 species reveals hyperfine couplings to $S-CH_2$- protons with a coupling constant of 12-14 MHz, demonstrating that substantial spin density resides on some of the S-donor atoms (*10,55*). Similar ^{1}H-ENDOR resonances that arise from solvent nonexchangeable protons are observed in the ^{1}H-ENDOR spectra of *D.gigas* H_2ase in forms A, B and C and in form C of *T. roseopersicina* H_2ase (*10,50,76*). These nonexchangeable protons were assigned to cysteine β-CH_2 protons, and provide evidence of substantial spin density on one or more S donor atoms in the enzyme. Further, the magnitude of the coupling to the cysteine β-CH_2 protons is essentially identical in all of the H_2ase samples. This is strong evidence that forms A, B, and C are electronically similar species and are derived from *the same oxidation state* of the Ni-Fe cluster. This notion is also consistent with XAS data that show the charge density on the Ni atom in form A, B and C to be essentially the same (*28,29,71*), and with redox state dependent IR results that show similar shifts in the vibrations tentatively assigned to the ligands of the Fe center in forms A, B, and C (*47*).

It is easy to envision how spin density might find its way to the $S-CH_2$ protons in these oxidized complexes, either by viewing the product as containing an oxidized S center or as S covalently bonded to a Ni(III) center. However, it is not so easy to understand how spin density of this magnitude could be transmitted to an S-donor ligand with a full valence shell from a reduced (*e.g.*, Ni(I)) center. This difficulty, the stabilization of the oxidized dimers by the thiolates, plus the similarity of the H_2ase ENDOR spectra to those of an oxidized model compound are strong evidence that forms A, B, and C contain an oxidized Ni-Fe center.

Qualitative *ab initio* theoretical calculations predict that the HOMO of the Ni(II,II) dimers is largely a combination of p_z orbitals on the terminal thiolate ligands (Figure 5) (*55*), a situation that is predicted to be maintained in the oxidation product if no major structural changes accompany the oxidation. The assumption that large structural changes do not occur upon oxidation is supported by the structure of the Ni(II/III) dimeric complex $\{Ni[P(o-C_6H_4S)_3]\}_2$, which exhibits a very similar Ni_2S_2 core to that found in the Ni(II,II) dimers of the $RN(CH_2CH_2S)_2^{2-}$ ligands, and by the results of XAS structural comparisons between the oxidized and reduced dimers.

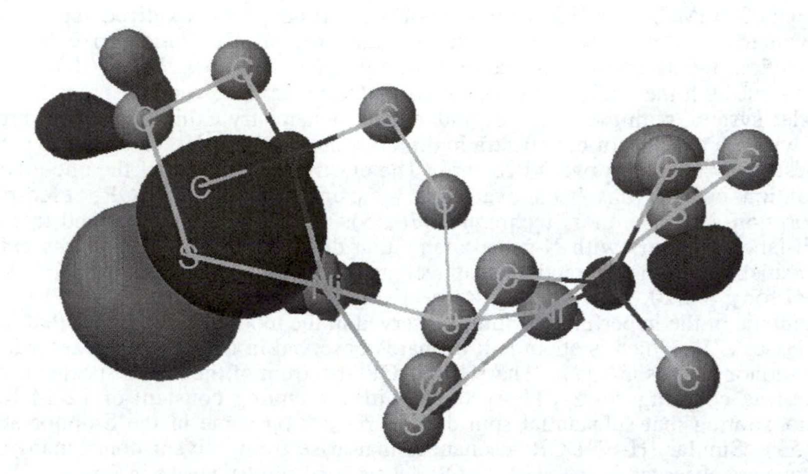

Figure 5. The HOMO calculated for $\{Ni[MeN(C_2H_4S)_2]\}_2$ from an HF-level *ab initio* calculation using the LANL2-DZ basis set.

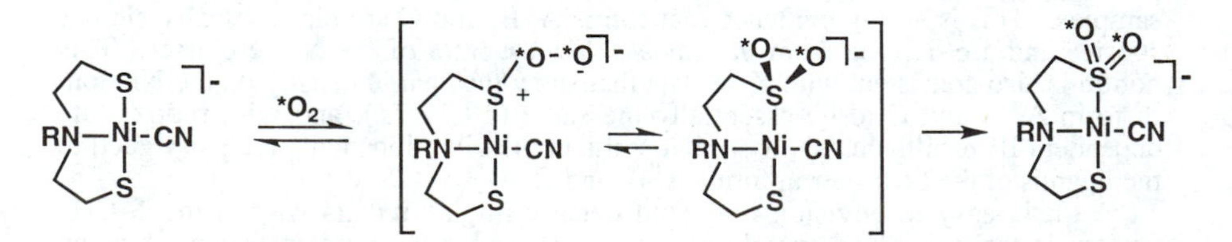

Figure 6. A proposed mechanism for the oxidation of Ni(II) thiolato ligands by O_2 (Reproduced with permission from ref. 74. Copyright 1993 ACS.).

Thus, on the basis of this model, the oxidized Ni-Fe center might be more appropriately viewed as involving oxidation of one or both of the terminal Ni cysteinate ligands. The localization of the spin density on these terminal ligands and the Ni center provides one possible explanation for why no hyperfine coupling is observed in the epr spectrum of H_2ase samples labeled with ^{57}Fe.

The fact that the g-values in the epr spectra are all greater than 2.0 might be interpreted as evidence that a Ni(III) center has been produced. However, electrons localized on S also can experience spin-orbit coupling of sufficient magnitude to lead to the observed g-values. Epr spectra of thiyl radicals (*e.g.*, cysteinyl radical) taken in frozen MeOH solutions reveal axial spectra with $g_{\|} = 2.3$ and $g_{\perp} = 2.0$ ($g_{ave} = 2.1$) (*77*). The deviation of the g-values from g = 2.0 in H_2ase and in the model system are within this range and have average values of about 2.1 (*10,55*). The epr spectra of the thiyl radicals are typically axial due to the degeneracy of the p-orbitals perpendicular to the S-C bond. This degeneracy could be lifted in a metal complex because of the additional S-M interaction, giving rise to a rhombic spectrum. Thus, the observed g-values and small hyperfine interactions are also consistent with largely S-centered oxidation.

The involvement of the S-donor atoms in the oxidative chemistry of the Ni alkylthiolate dimeric complexes is also apparent in the products from two- and four-electron oxidations. When I_2 is used as a two-electron chemical oxidant, the thiolate ligands are oxidized to disulfides (*33*). The stoichiometry of the reaction indicates that each Ni center is oxidized by two-electrons and quantitaively converts all of the thiolates to disulfides. In the case where R = $-CH_2CH_2SMe$, the product has been isolated and crystallographically characterized (Figure 2) (*33*). It is found to be a five-coordinate, mononuclear high-spin Ni(II) complex with a monodentate disulfide. The remaining coordination positions on Ni are occupied by the tertiary amine N-donor, the thioether, and two I^- ligands. If substoichiometric I_2 is used, the product of the reaction is a mixture of the disulfide complex and unreacted dimer.

Nickel complexes of alkyl thiolates have also been shown to be susceptible to oxidation by O_2, leading to the formation of S-bound sulfinate ligands, a four-electron S-centered oxidation. (*73,74,78-80*). Reaction of the dimer leads to the oxidation of only the terminal thiolates, consistent with their greater nucleophilicity and the mechanistic proposal that the reaction involves nucleophilic attack by thiolate S on O_2 (*74*). Studies of various planar dithiolates reveal that only one of the thiolates is oxidized by O_2. The second thiolate is not converted to a sulfinate, an observation that has been traced to changes in the electronic structure of the complexes that occurs upon oxidation. Qualitative *ab initio* calculations of planar cis- and trans-dithiolate complexes reveal that the filled frontier M.O.s of the dithiolates are dominated by S p-orbitals, as was the case in the dimeric system (*81*). Upon oxidation, the filled frontier M.O.s of the monosulfinato complexes are dominated by contributions from the O atoms of the sulfinate group. (*81*). Calculations of the electrostatic potential surfaces of the molecules show that the S centers in the dithiolates are nucleophilic, but that the remaining thiolate S-donor in the monosulfinate complexes is not (*81*).

Kinetic investigations of the O_2 oxidations of planar Ni thiolates indicates that the reaction is quite slow ($t_{1/2}$ = hours - days) and is first order in [Ni] and in [O_2] (*74,78*). These studies also provide no evidence for stable intermediates, radicals, or free $^1O_2$. Cross-over experiments performed with mixtures of $^{18}O_2$ and $^{16}O_2$ reveal that the reaction follows a dioxygenase-like mechanism, proceeding via the incorporation of both atoms of O from a single O_2 molecule (*74,80*). A mechanism analogous to that proposed for the oxidation of thioethers by $^1O_2$ has been proposed (Figure 6) (*74*). This mechanism, which involves a thiadioxirane/persulfoxide intermediate, is consistent with the known features of the reaction mechanism.

All of the known examples of the O_2 oxidation of Ni thiolate complexes involve planar (S = 0) complexes. Efforts to test the reaction on high-spin systems were stymied by the paucity of five- and six-coordinate thiolate complexes featuring chelating alkyl thiolate ligands. This situation was overcome when a five-coordinate, high-spin, Ni(II) dithiolate was synthesized by Kovacs and coworkers (*82*). This complex has been examined for reactivity with O_2, and preliminary work indicates that a reaction similar to the low-spin systems occurs. As in the planar systems, the reaction can be followed spectrophotometrically and proceeds through isosbestic points, an observation that is consistent with the formation of a single product (Figure 7). This product has not been crystallographically characterized, but its IR spectrum features bands in the 1000 - 1200 cm^{-1} region that are characteristic of sulfinate groups (*74*). The greatest difference lies in the rate of reaction ($t_{1/2}$ ~14 min) and indicates that either the thiolates in the high-spin complex are much more nucleophilic, or the reaction of the planar system is inhibited by the spin-forbidden nature of the process (the complexes are singlets and O_2 is a triplet species).

The ubiquitous nature of the oxidation of alkyl thiolate ligands in Ni complexes suggests that the same chemistry should occur in the H_2ase Ni-Fe site unless the thiolates are specifically protected from oxidation. One possibility is that the loss of reactivity that is associated with the conversion of form B to form A may be due to the oxidation of a thiolate ligand. The 2.85 Å crystal structure provides no support for this hypothesis, but it would be difficult to rule it out entirely, since only 1/3 of the sample is in form A. Another aspect of the oxidation of thiolates to sulfoxy species is that they are not known to be easily reversed (*83*), which is a requirment for H_2ase. One exception to the oxidative deactivation of H_2ases by O_2 involves enzymes that incorporate selenocysteine as a Ni ligand. The H_2ase from *D. baculatum* may be isolated aerobically in an epr silent state that requires no reductive activation. (*16*). Amino acid sequence homology with the structurally characterized *D. gigas* enzyme reveals that it is cys-530, a terminal Ni thiolate ligand in the Ni-Fe site, that is substituted by selenocysteine (*84*). Studies of the oxidative chemistry of the Se containing analog, $\{Ni[MeN(CH_2CH_2Se)]\}_2$, show that it is readily oxidized by one-electron but is very unreactive toward O_2 (*85*).

Several H_2ases that do not contain Ni have been characterized (*11*). Compared to the Fe-only enzymes, the Ni-Fe enzymes are typically only 1-10% as active (*10*). The Ni-Fe enzymes usually function as H_2 uptake enzymes (H_2 oxidation), suggesting that one possible role for Ni is to produce an active site that favors H_2 oxidation under biological conditions. However, this is not the case for *Pyrococcus furiosus* H_2ase, which contains Ni and preferentially catalyzes H_2 production (*86*). The most outstanding feature associated with the Ni-containing enzymes is their O_2 tolerance and the reversibility of the oxidatively inactivated enzymes. Thus, one view of the redox role of Ni that emerges from the model chemistry presented here is that it is involved in modifying the reactivity of thiolates or Fe-thiolate complexes to make them more stable to oxidation. This property would be even more developed by replacing key cysteinate groups with selenocysteinate.

The Role of Ni as a Binding Site

Another aspect of many hypothetical reaction mechanisms for H_2ase is the use of Ni as a binding site for substrate (H_2 or H^-) and inhibitors (*e.g.*, CO). Such a role is suggested by two properties of the enzyme: the photochemical reactivity of form C (*87*) and the observation of ^{13}C hyperfine coupling in the 77 K epr signal of form C in the presence of ^{13}CO (*69*). However, recent investigations of these properties suggest that Ni may not serve as a binding site for H_2, or at the very least is not the only binding site.

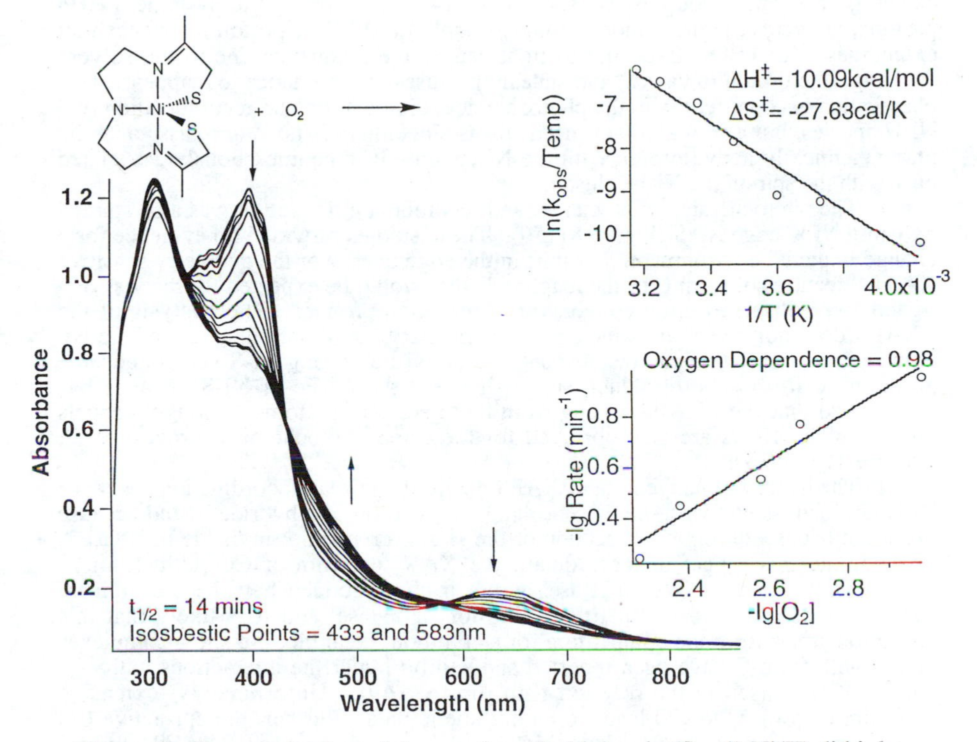

Figure 7. Characterization of the reaction of a high-spin (S = 1) Ni(II) dithiolate with O_2: spectral changes, activation parameters, and O_2 dependence.

When form C is exposed to light, it is converted to another epr-active form, form C* (g = 2.29, 2.13, 2.05) (*50,87*). At temperatures below ~200 K the process is irreversible but it can be reversed by annealing the sample at temperatures above ~200 K. Thus, the photochemistry has the characteristics of a ligand photodissociation and recombination reaction. One possible mechanism that would explain the photochemistry involves the photodissociation and recombination of an S = 1/2 $Ni-H^-$ complex. This proposal is consistent with the observation of a large ^2H-isotope effect on the photochemistry (*88*). The photodissociation of a Ni-H species has been examined by a combination of ENDOR and XAS studies of the reaction using H_2ase from *T. roseopersicina* (*50*) The ^1H-ENDOR spectrum reveals two sets of resonances in the enzyme corresponding to protons that were not solvent exchangeable with a coupling constant of 12-14 MHz (assigned to cysteine β-CH_2 protons, *vide supra*), and a more strongly coupled (20 MHz) proton resonance that exchanges with D_2O. Exposure to light causes the resonance due to the solvent exchangeable proton to vanish, and annealing causes the resonance to reappear. This observation is consistent with the photochemical dissociation and recombination of a Ni-H species, but also with other mechanisms since there is no way to associate the photochemically active proton with the Ni center. It is unambiguously associated only with the spin in the Ni-Fe cluster.

The photochemical dissociation and recombination reaction was also studied by using Ni K-edge XAS (Figure 8) (*50*). These studies provided no evidence for a change in the Ni environment. No shift in the edge energy or the geometry sensitive XANES features of form C of the magnitude that would be expected for the loss of a ligand were observed upon conversion to the photoproduct. The analysis of the EXAFS does not reveal any changes in the primary coordination sphere of the Ni. However, a feature that can be attributed to the Ni-Fe or long Ni-S vectors become pronounced in the photoproduct, suggesting that the Ni-Fe or Ni-S distance has changed so that the EXAFS arising from these scattering atoms no longer cancels out. Similar effects are also observed for the same photochemical process in *C. vinosum* H_2ase (*89*).

The lack of a change in the EXAFS due to the primary coordination sphere of Ni is not inconsistent with the presence and dissociation of a hydride ligand because the small backscattering cross-section of the H atom renders it invisible in EXAFS. To examine this aspect in more detail, the XAS spectrum of CO adducts of *T. roseopersicina* and *C. vinosum* H_2ases and their photoproducts have been examined. Carbon monoxide is a competitive inhibitor of H_2ase, and it is likely that this inhibition arises from the competition for a common metal binding site. Complexes of CO with form C have been reported and exhibit hyperfine interactions with ^{13}C when ^{13}CO is used in the reaction with form C (*69*). Unfortunately, extensive exposure of form C to CO leads to an epr silent state. Further, the epr-active CO adduct is formed in low yield (~25%) and is unsuitable for EXAFS studies. However, the epr silent species that is formed corresponds to the CO complex of the SI redox level, as shown by IR spectroscopy (*90*). The IR spectrum of this CO adduct displays a single vibration at 2060 cm^{-1} that was assigned to a terminal Ni-CO complex based on the stretching frequency. This complex is still light-sensitive and was shown to reversibly photodissociate CO.

In contrast to H^-, CO is a ligand that gives rise to a rich EXAFS spectrum that arises from a short M-C distance and a second coordination sphere O atom. Examination of the EXAFS spectrum of a crystallographically characterized Ni-CO complex with a ligand environment similar (with respect to EXAFS analysis) to that found for Ni in H_2ase, $Ni(NP_3)CO$ (ave. crystallographic distances for Ni-P = 2.215(3) Å; Ni-C = 1.74(2) Å) (*91*), clearly shows the short Ni-C interaction and second coordination sphere O atom (Figure 9). No such features are observed in the Ni K-edge EXAFS spectrum of H_2ase (Figure 10) (*89*). Further, photodissociation of the CO from the enzyme does not lead to significant changes in either the Ni K-

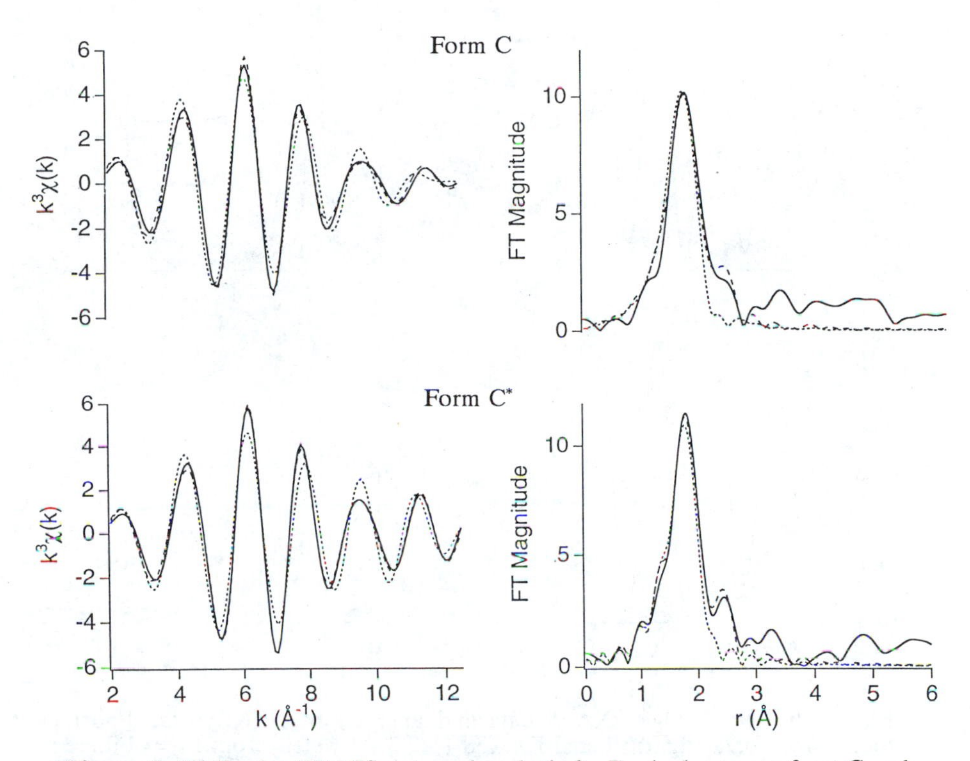

Figure 8. Ni K-edge EXAFS data and analysis for *T.r.* hydrogenase form C and its photoproduct. Data is shown as a solid line as Fourier transforms (right) and filtered data (1 - 3 Å) (left). Fits shown for Form C: 1N @ 1.92 Å + 3 S @ 2.22 Å (--------); 1 N @1.92 Å + 3S @ 2.22 Å + 1 S @ 2.75 Å (-----). Fits shown for Form C*: 1 N @ 1.92 Å + 4 S @ 2.22 Å (--------); 1 N @1.94 Å + 3S @ 2.23 Å + 1 S @ 2.70 Å (-----). (Adapted from ref. 57).

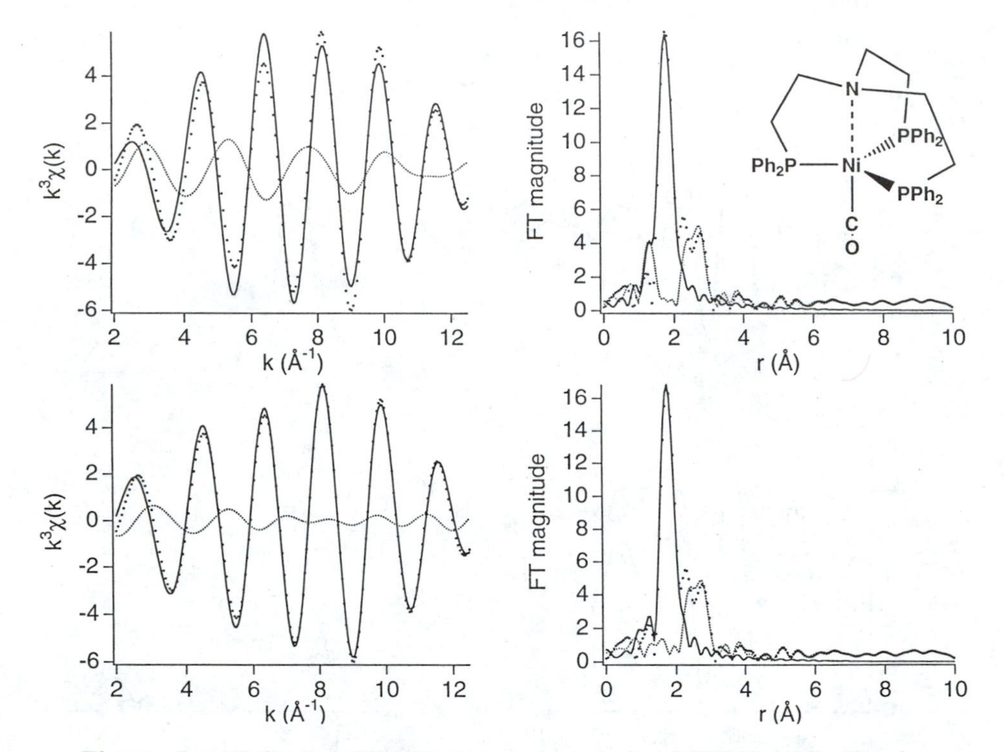

Figure 9. Ni K-edge EXAFS data and analysis of $Ni(NP_3)CO$. Fourier transformed data (right) filtered EXAFS (1 - 3 Å) (left). Top: fit = Filtered EXAFS (1 - 2.5 Å) (dots) and a fit (solid line) = 3 P @ 2.19 Å. The difference (data - fit) is shown as a dotted line. Bottom: The EXAFS data (dots) and a fit (solid line) with 3 P @ 2.20 Å + 1 C @ 1.72 Å. The difference (data - fit) is shown as a dotted line. The addition of the C scatterer improves the goodness of fit by a factor of 2. The short C scattering atom is clearly seen as giving rise the residual EXAFS wave in the top figure. Features at distances > 2Å in the FT spectrum are due to scattering from the semi-coordinated N-donor and the second coordination sphere CO O-atom.

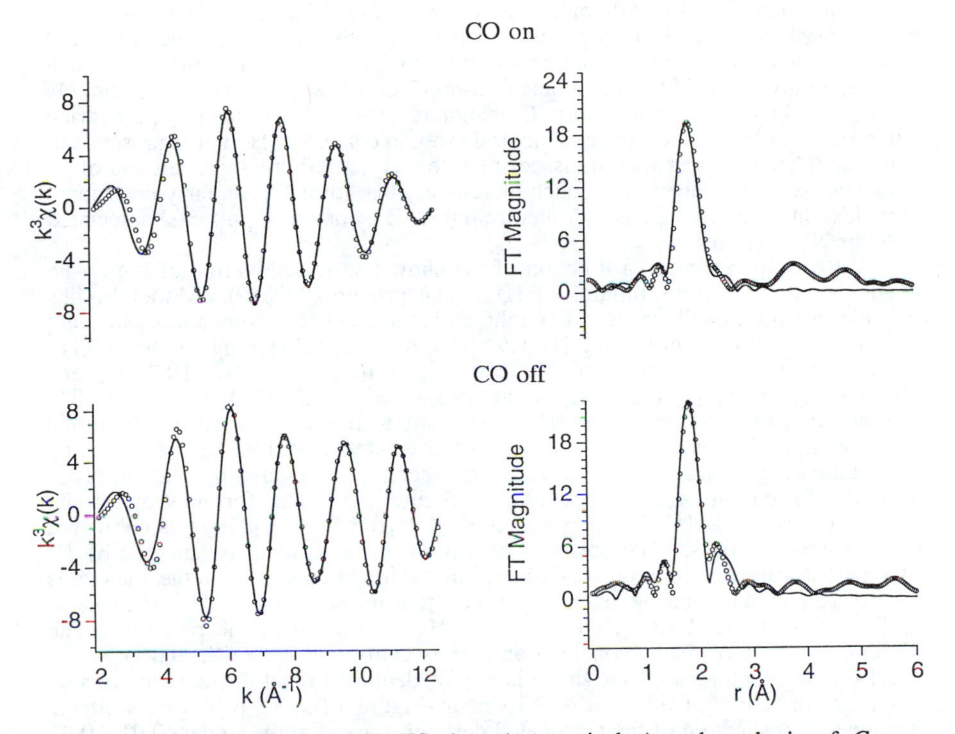

Figure 10. Ni K-edge EXAFS data (open circles) and analysis of *C.v.* hydrogenase + CO (top) and its photoproduct where CO is dissociated (bottom). Note the absence of features attributable to a short Ni-C bond or to a second coordination sphere O atom. Fits shown (solid lines) for the CO complex and its photoproduct are: 4 S @ 2.24 Å + 1 Fe @ 2.57 Å and 4 S @ 2.24 Å + 1 Fe @ 2.63 Å, respectively.

edge energy, the geometry-sensitive XANES features, or the EXAFS spectrum. Thus, the Fe center in the Ni-Fe site is implicated as the binding site for the terminal CO ligand. This conclusion is consistent with recent Mössbauer studies of reduced *C. vinosum* H_2ase (form C and R), which reveal the spectrum of a diamagnetic Fe center (*e.g.*, low-spin Fe(II)) that is sensitive to the presence of CO (*92*).

Because CO is a competitive inhibitor, the Fe atom is also a likely binding site for H_2 or H^-. This suggestion is supported by the discovery that the unique IR features associated with the Fe site are also found in Fe-only H_2ases (*93*), indicating that the Fe and the ligands giving rise to these unusual spectral features are part of an active site found in all metal-containing hydrogenases. An Fe-H_2(H^-) complex is also consistent with ^2H isotope effect on the photochemical dissociation and recombination. However, it is unlikely that Form C is an H_2 or H^- complex. When H_2 is carefully removed from a sample of form C, the epr signal is stable indefinitely (*60*). This is inconsistent with form C being a reactive intermediate (*e.g.*, a hydride complex) or an H_2 adduct, since the hydride should convert to H_2 in the presence of water and the H_2 adduct should dissociate at low H_2 partial pressure. Instead of an enzyme/substrate complex, it is much more likely that form C is an enzyme/product complex. In other words, that it differs from form B by having a proton(s) associated with the Ni-Fe cluster.

Redox titrations as a function of pH show that the reduction of the Ni-Fe cluster is coupled to the binding of H^+ to the protein (*8,58,59*). Model studies suggest that the cysteinate ligands might serve as a site for protonation and yield compounds capable of producing H_2 (*9,94-96*). An example employing the dinickel model complexes discussed here is provided by the titration of $\{Ni[RN(CH_2CH_2S)_2]\}_2$ with 1 equivalent of $[H(OEt_2)_2]^+[B(3,5\text{-}Me_2C_6H_3)_4]^-$ (*97*). The addition of 1 equivalent of H^+/dimer leads to the protonation of a terminal thiolate ligand, which may be deprotonated by addition of base. The protonation/deprotonation may be followed spectrophotometrically and proceeds with the formation of isosbestic points (Figure 11). The formulation of the protonated species as $[\{Ni[RN(CH_2CH_2S)_2]\}_2H]^+[B(3,5\text{-}Me_2C_6H_3)_4]^-$ is confirmed by elemental analysis. The presence of a thiol in the product is detected by IR spectra that feature a vibration at 2690 cm^{-1} that shifts to 2630 cm^{-1} in the analogous selenolate complex, and by ^2H-NMR spectra that reveal a single ^2H resonance at 1.95 ppm when $[^2H(OEt_2)_2]^+([B(3,5\text{-}Me_2C_6H_3)_4]^-$ is used in the reaction. The dimeric structure of the protonated complex is confirmed by XAS. The EXAFS spectrum of the protonated derivative is nearly identical to that of the unprotonated complex, and clearly exhibits a Ni-Ni vector (Figure 12). All of the Ni-scatterer distances characteristic of the protonated derivative are slightly longer (0.01 - 0.02 Å) than in the unprotonated dimer. The energy of the Ni K-edge shifts by 0.8(2) eV to higher energy, reflecting the changes in the charge density on Ni when a thiolate ligand is protonated and demonstrates the sensitivity of these measurements to small changes in charge density.

A similar protonated complex can be envisioned for form C of H_2ase. The thiol proton would pick up spin density by being associated with the Ni-Fe cluster, leading to the observation of a relatively weakly coupled proton (20 MHz) in the ENDOR spectrum. This proton could be lost from the ENDOR spectrum by photolytic cleavage of either the Ni-S bond or the S-H bond, processes that are likely given the dominance of the electronic absorption spectrum by S -> M LMCT transitions.

Summary

The remarkable crystal structure obtained from *D.gigas* H_2ase reveals a heterodimetallic Ni-Fe cluster that serves as the active site. The structure of this cluster appears to be consistent with common Ni thiolate chemistry, which is

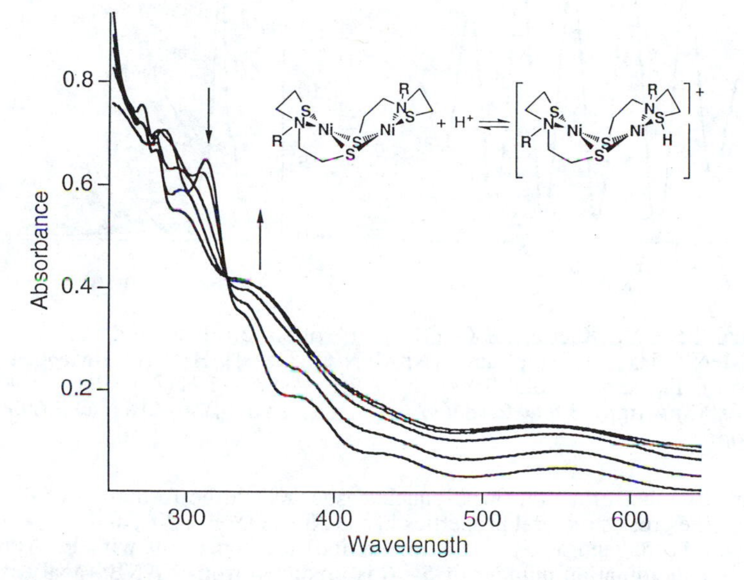

Figure 11. Electronic spectral changes accompanying the reversible protonation of $\{Ni[MeN(C_2H_4S)_2]\}_2$.

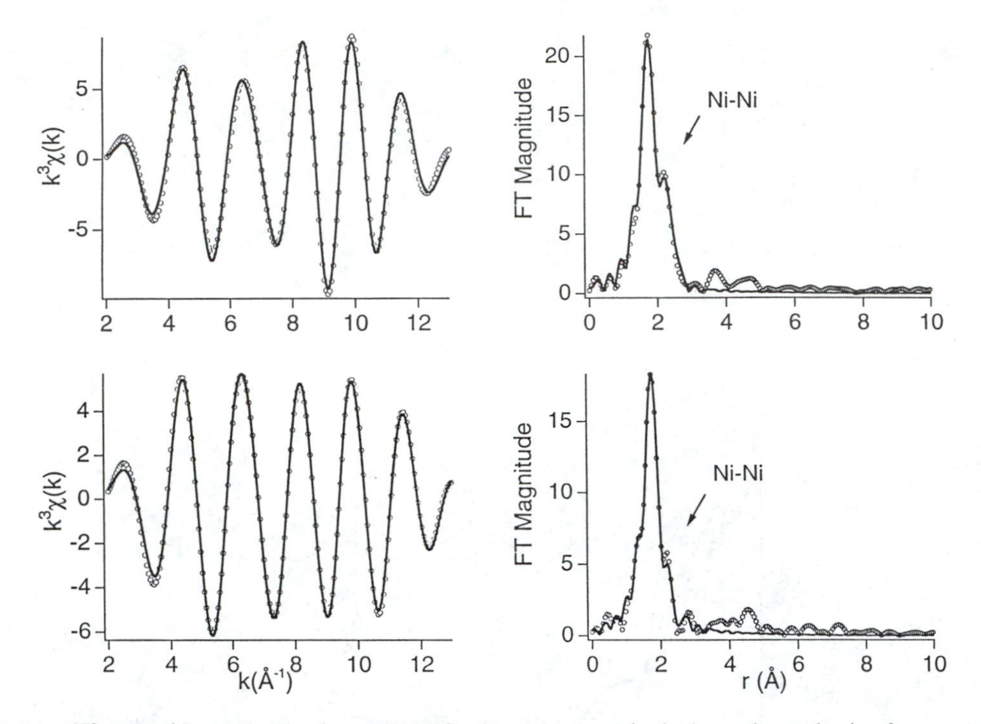

Figure 12. Ni K-edge EXAFS data (open circles) and analysis for $\{Ni[MeN(C_2H_4S)_2]\}_2$ (top) and $[\{Ni[MeN(C_2H_4S)_2]\}_2H]^+[B(3,5\text{-}Me_2C_6H_3)_4]^-$ (bottom). Fits shown (solid lines): 3S @ 2.164(1) Å + 1 N @ 1.896(5) Å + 1 Ni @ 2.680(3) Å (top), 3 S @ 2.184(1) Å + 1 N @ 1.901(4) Å + 1 Ni @ 2.690(3) Å (bottom).

dominated by the formation of polynuclear species via the formation of dithiolato bridges. The gross structural properties of the Ni site (S at 2.2 Å, and the possibility of a Ni-Fe vector and a long Ni-S interaction) are consistent with EXAFS data, although a coordination number of 5 - 6 is predicted from XANES analysis. The basic structural features are similar in several H_2ases obtained from a variety of bacteria. The largest differences involve the presence of O,N ligation in many enzymes that is not obvious in the published crystal structure.

The redox chemistry of the Ni-Fe cluster has been probed by several techniques, none of which provide unambiguous evidence for Ni-centered redox chemistry. XAS studies reveal edge energy shifts that are too small to accommodate more than two oxidation states for Ni in the enzyme. ENDOR spectroscopy reveals that the resonances associated with cysteinate methylene protons exhibit the same coupling constant in forms A, B, and C, thus demonstrating that all three of these forms feature the same oxidation state of the Ni-Fe cluster, in agreement with the XAS experiments.

Several explanations for the redox chemistry of the cluster are possible and include redox activity at the Fe center and redox activity involving the thiolate ligands. A role for Fe in the redox chemistry is suggested by IR studies of the redox-dependent frequencies of bands that have been associated with the Fe ligands in the cluster. Mechanisms involving thiolate redox chemistry are supported by the redox chemistry exhibited by a series of dinickel model complexes that feature μ–dithiolato and terminal thiolate ligation. The redox chemistry associated with these models is dominated by products that reflect S-centered oxidation. No reductive chemistry is observed at modest potentials. The electronic structure of the thiolate complexes, as examined by qualitative *ab initio* theory, shows that S p-orbitals dominate the occupied frontier orbitals of the Ni alkylthiolate complexes. These results support the notion that the Ni center is not the primary redox center, suggest that the thiolate ligands are intimately involved in the redox chemistry, and confirm that higher oxidation states should be supported by thiolate ligation.

Studies aimed at providing evidence for the role of Ni as a substrate and/or inhibitor binding site fail to provide unequivocal evidence of Ni-H or Ni-CO complexes, and recent studies of the CO complexes formed with H_2ase clearly implicate Fe as a potential binding site for this competitive inhibitor. Thus, there is little evidence to indicate that Ni serves as a binding site.

Given that one of the outstanding properties of the Ni-containing H_2ases is their stability to irreversible oxidation by O_2, it seems possible that one of the functions of Ni is to modify the Fe and/or S catalytic sites so that the site is not irreversibly inactivated by exposure to O_2. This notion is supported by the fact that H_2ases containing only Fe are much better catalysts of H_2 redox chemistry, but are irreversibly deactivated by O_2. The trend in O_2 stability seems to be furthered by the substitution of a terminal Ni-cysteinate ligand by selenocysteinate.

Acknowledgments

Support for this work by the NIH (MJM, GM38829) is gratefully acknowledged. We thank Dr. Simon S. P. J. Albracht of the E. C. Slater Institute, University of Amsterdam, for samples of H_2ase from *Chromatium vinosum* and a critical review of this manuscript.

Literature Cited

(1) Thomson, A. J. In *Metalloproteins*; P. Harrison, Ed.; Verlag Chemie: Deerfield Beach, FL, 1985; pp 79-120.
(2) Beinert, H. *FASEB J.* **1990**, *4*, 2483-2491.
(3) Adman, E. T. In *Advances in Protein Chemistry*; C. B. Anfinsen, J. T. Edsall, D. S. Eisenberg, F. M. Richards, Eds.; Academic Press: San Diego, 1991; Vol. 42; pp. 145 - 198.
(4) Iwata, S.; Ostermeier, C.; Ludwig, B.; Michel, H. *Nature* **1995**, *376*, 660-669.
(5) Tsukihara, T.; Aoyama, H.; Yamachita, E.; Tomizaki, T.; Yamaguchi, H.; Shinzawa-Itoh, K.; Nakashima, R.; Yaono, R.; Yoshikawa, S. *Science* **1995**, *269*, 1069-1074.
(6) Wilmanns, M.; Lappalainen, P.; Kelly, M.; Sauer-Eriksson, E.; Saraste, M. *Proc. Natl. Acad. Sci. USA* **1995**, *92*, 11955-11959.
(7) Kim, J.; Rees, D. C. *Nature* **1992**, *360*, 553-560.
(8) Albracht, S. P. J. *Biochim. Biophys. Acta* **1994**, *1188*, 167-204.
(9) Maroney, M. J. *Comments on Inorganic Chemistry* **1995**, *17*, 347-375.
(10) Maroney, M. J.; Pressler, M. A.; Mirza, S. A.; Whitehead, J. P.; Gurbiel, R. J.; Hoffman, B. M. *Adv. Chem. Ser.* **1995**, *246*, 21-60.
(11) Adams, M. W. W. *Biochim. Biophys. Acta* **1990**, *1020*, 115-145.
(12) Crane, B. R.; Siegel, L. M.; Getzoff, E. D. *Science* **1995**, *270*, 59-67.
(13) Hausinger, R. P. *Sci. Total Environ.* **1994**, *148*, 157-166.
(14) Won, H.; Olson, K. D.; Summers, M. F.; Wolfe, R. S. *Comments Inorg. Chem.* **1993**, *15*, 1-26.
(15) Ferry, J. G. *Annu. Rev. Microbiol.* **1995**, *49*, 305-333.
(16) Fauque, G.; Peck, H. D., Jr.; Moura, J. J. G.; Huynh, B. H.; Berlier, Y.; DerVartanian, D. V.; Teixeira, M.; Przybyla, A. E.; Lespinat, P. A.; Moura, I.; LeGall, J. *FEMS Microbiol. Rev.* **1988**, *54*, 299-344.

(17) Berkessel, A. *Bioorg. Chem.* **1991**, *19*, 101-115.
(18) Maroney, M. J. In *Encyclopedia of Inorganic Chemistry*; R. B. King, Ed.; Wiley: Sussex, 1994; pp 2412-2427.
(19) Kovacs, J. A. *Adv. Inorg. Biochem.* **1994**, *9*, 173-218.
(20) Qiu, D.; Kumar, M.; Ragsdale, S. W.; Spiro, T. G. *J. Am. Chem. Soc.* **1995**, *117*, 2653-2654.
(21) Kumar, M.; Qiu, D.; Spiro, T. G.; Ragsdale, S. W. *Science* **1995**, *270,* 628-630.
(22) Tan, G. O.; Ensign, S. A.; Ciurli, S.; Scott, M. J.; Hedman, B.; Holm, R. H.; Ludden, P. W.; Korszun, Z. R.; Stephens, P. J.; Hodgson, K. O. *Proc. Natl. Acad. Sci. USA* **1992**, *89*, 4427-4431.
(23) Xia, J.; Dong, J.; Wang, S.; Scott, R. A.; Lindahl, P. A. *J. Am. Chem. Soc.* **1995**, *117*, 7065-7070.
(24) Zirngibl, C.; Van Dongen, W.; Schworer, B.; Von Bunau, R.; Richter, M.; Klein, A.; Thauer, R. K. *Eur. J. Biochem.* **1992**, *208*, 511-520.
(25) Berkessel, A.; Thauer, R. K. *Angew. Chem., Int. Ed. Engl.* **1995**, *34*, 2247-2250.
(26) Klein, A. R.; Fernandez, V. M.; Thauer, R. K. *FEBS Lett.* **1995**, *368*, 203-206.
(27) Eidsness, M. K.; Sullivan, R. J.; Scott, R. A. In *The Bioinorganic Chemistry of Nickel*; J. R. Lancaster, Jr., Ed.; VCH: New York, 1988; pp. 73-91.
(28) Bagyinka, C.; Whitehead, J. P.; Maroney, M. J. *J. Am. Chem. Soc.* **1993**, *115*, 3576-3585.
(29) Gu, Z.; Dong, J.; Allan, C. B.; Choudhury, S. B.; Franco, R.; Moura, J. J. G.; Moura, I.; LeGall, J.; Przybyla, A. E.; Roseboom, W. R.; Albracht, S. P. J.; Axley, M. J.; Scott, R. A.; Maroney, M. J. **1996**, manuscript in preparation.
(30) Nag, K.; Chakravorty, A. *Coord. Chem. Rev.* **1980**, *33*, 87-147.
(31) Haines, R. I.; McAuley, A. *Coord. Chem. Rev.* **1981**, *39*, 77-119.
(32) Lappin, A. G.; McAuley, A. *Adv. Inorg. Chem.* **1988**, *32*, 241-294.
(33) Kumar, M.; Day, R. O.; Colpas, G. J.; Maroney, M. J. *J. Am. Chem. Soc.* **1989**, *111*, 5974-5976.
(34) Fox, S.; Wang, Y.; Silver, A.; Millar, M. *J. Am. Chem. Soc.* **1990**, *112*, 3218-3220.
(35) Krüger, H. J.; Holm, R. H. *Inorg. Chem.* **1989**, *28*, 1148-1155.
(36) Krüger, H.-J.; Peng, G.; Holm, R. H. *Inorg. Chem.* **1991**, *30*, 734-742.
(37) Krüger, H.-J.; Holm, R. H. *J. Am. Chem. Soc.* **1990**, *112*, 2955-2963.
(38) Franolic, J. D.; Wang, W. Y.; Millar, M. *J. Am. Chem. Soc.* **1992**, *114*, 6587-6588.
(39) Marganian, C. A.; Vazir, H.; Baidya, N.; Olmstead, M. M.; Mascharak, P. K. *J. Am. Chem. Soc.* **1995**, *117*, 1584-1594.
(40) Przybyla, A. E.; Robbins, J.; Menon, N.; Peck, H. D. J. *FEMS Microbiol. Rev.* **1992**, *88*, 109-135.
(41) Wu, L. F.; Mandrand, M. A. *FEMS Microbiol. Rev.* **1993**, *104*, 243-269.
(42) Eidsness, M. K.; Scott, R. A.; Prickril, B. C.; DerVartanian, D. V.; Legall, J.; Moura, I.; Moura, J. J. G.; Peck, H. J. *Proc. Natl. Acad. Sci. USA* **1989**, *86*, 147-151.
(43) He, S. H.; Teixeira, M.; LeGall, J.; Patil, D. S.; Moura, I.; Moura, J. J. G.; DerVartanian, D. V.; Huynh, B. H.; Peck, H. D., Jr. *J. Biol. Chem.* **1989**, *264*, 2678-2682.
(44) Sorgenfrei, O.; Klein, A.; Albracht, S. P. J. *FEBS Lett.* **1993**, *332*, 291-297.
(45) Volbeda, A.; Charon, M. H.; Piras, C.; Hatchikian, E. C.; Frey, M.; Fontecilla-Camps, J. C. *Nature* **1995**, *373*, 580-587.
(46) Fontecilla-Camps, J. C. private communication.
(47) Bagley, K. A.; Duin, E. C.; Roseboom, W.; Albracht, S. P. J.; Woodruff, W. H. *Biochemistry* **1995**, *34*, 5527-5535.

(48) Colpas, G. J.; Maroney, M. J.; Bagyinka, C.; Kumar, M.; Willis, W. S.; Suib, S. L.; Mascharak, P. K.; Baidya, N. *Inorg. Chem.* **1991**, *30*, 920-928.
(49) Whitehead, J. P.; Colpas, G. J.; Bagyinka, C.; Maroney, M. J. *J. Am. Chem. Soc.* **1991**, *113*, 6288-6289.
(50) Whitehead, J. P.; Gurbiel, R. J.; Bagyinka, C.; Hoffman, B. M.; Maroney, M. J. *J. Am. Chem. Soc.* **1993**, *115*, 5629-5635.
(51) Fukuyama, K.; Hase, T.; Matsumoto, S.; Tsukihara, T.; Katsube, Y.; Tanaka, N.; Kakudo, M.; Wada, K.; Marsubara, H. *Nature* **1980**, *286*, 522.
(52) Tremel, W.; Kriege, M.; Krebs, B.; Henkel, G. *Inorg. Chem.* **1988**, *27*, 3886-3895.
(53) Colpas, G. J.; Kumar, M.; Day, R. O.; Maroney, M. J. *Inorg. Chem.* **1990**, *29*, 4779-4788.
(54) Melnik, M.; Sramko, T.; Dunaj-Jurco, D.; Sirota, A.; Holloway, C. E. *Rev. Inorg. Chem.* **1994**, *14*, 2 - 346.
(55) Maroney, M. J.; Pressler, M. A.; Gurbiel, R.; Hoffman, B. M. unpublished results.
(56) Mills, D. K.; Hsiao, Y. M.; Farmer, P. J.; Atnip, E. V.; Reibenspies, J. H.; Darensbourg, M. Y. *J. Am. Chem. Soc.* **1991**, *113*, 1421-1423.
(57) Colpas, G. J.; Day, R. O.; Maroney, M. J. *Inorg. Chem.* **1992**, *31*, 5053-5055.
(58) Roberts, L. M.; Lindahl, P. A. *J. Am. Chem. Soc.* **1995**, *117*, 2565-25672.
(59) Roberts, L. M.; Lindahl, P. A. *Biochemistry* **1994**, *33*, 14339-14350.
(60) Barondeau, D. P.; Roberts, L. M.; Lindahl, P. A. *J. Am. Chem. Soc.* **1994**, *116*, 3442-3448.
(61) Coremans, J. M. C. C.; Van, G. C. J.; Albracht, S. P. J. *Biochim. Biophys. Acta* **1992**, *1119*, 148-156.
(62) Huynh, B. H.; Patil, D. S.; Moura, I.; Teixeira, M.; Moura, J. J. G.; DerVartanian, D. V.; Czechowski, M. H.; Prickril, B. C.; Peck, H. D. J.; LeGall, J. *J. Biol. Chem.* **1987**, *262*, 795-800.
(63) Teixeira, M.; Moura, I.; Xavier, A. V.; Moura, J. J. G.; LeGall, J.; DerVartanian, D. V.; Peck, H. D., Jr.; Huynh, B.-H. *J. Biol. Chem.* **1989**, *264*, 16435-16450.
(64) Teixeira, M.; Moura, I.; Fauque, G.; Dervartanian, D. V.; Legall, J.; Peck, H. D., Jr.; Moura, J. J. G.; Huynh, B. H. *Eur. J. Biochem.* **1990**, *189*, 381-386.
(65) Moura, J. J. G.; Teixeira, M.; Moura, I.; LeGall, J. In *The Bioinorganic Chemistry of Nickel*; J. R. Lancaster, Jr., Ed.; VCH: New York, 1988; pp. 191-226.
(66) Cammack, R.; Fernandez, V. M.; Schneider, K. In *The Bioinorganic Chemistry of Nickel*; J. R. Lancaster, Jr., Ed.; VCH: New York, 1988; pp. 167-190.
(67) Lancaster, J. R., Jr. *FEBS Lett* **1980**, *115*, 285-288.
(68) Graf, E. G.; Thauer, R. K. *FEBS Lett.* **1981**, *136*, 165-169.
(69) van der Zwaan, J. W.; Coremans, J. M. C. C.; Bouwens, E. C. M.; Albracht, S. P. J. *Biochim. Biophys. Acta* **1990**, *1041*, 101-110.
(70) Surerus, K. K.; Chen, M.; van der Zwaan, J. W.; Rusnak, F. M.; Kolk, M.; Duin, E. C.; Albracht, S. P. J.; Muenck, E. *Biochemistry* **1994**, *33*, 4980-4993.
(71) van Elp, J.; Peng, G.; Zhou, Z. H.; Adams, M. W. W.; Baidya, N.; Mascharak, P. K.; Cramer, S. P. *Inorg. Chem.* **1995**, *34*, 2501-2504.
(72) Maroney, M. J., unpublished results; Huynh, B. H. private communication; Albracht, S. P. J. , private communication.
(73) Kumar, M.; Colpas, G. J.; Day, R. O.; Maroney, M. J. *J. Am. Chem. Soc.* **1989**, *111*, 8323-8325.
(74) Mirza, S. A.; Pressler, M. A.; Kumar, M.; Day, R. O.; Maroney, M. J. *Inorg. Chem.* **1993**, *32*, 977-987.
(75) Moura, J. J. G.; Teixeira, M.; Xavier, A. V.; Moura, I.; LeGall, J. *J. Mol. Cat.* **1984**, *23*, 303.
(76) Fan, C.; Teixeira, M.; Moura, J.; Moura, I.; Huynh, B. H.; Le Gall, J.; Peck, H. D., Jr.; Hoffman, B. M. *J. Am. Chem. Soc.* **1991**, *113*, 20-24.

(77) Gilbert, B. C. In *Sulfur-Centered Reactive Intermediates in Chemistry and Biology*; C. a. A. Chatgilialoglu, K.-D., Ed.; Plenum Press: New York, 1990; Vol. A197; pp. 135-154.
(78) Mirza, S. A.; Day, R. O.; Maroney, M. J. *Inorg. Chem.* **1996**, *34*, in press.
(79) Darensbourg, M. Y.; Farmer, P. J.; Soma, T.; Russell, D. H.; Solouki, T.; Reibenspies, J. H. In *Act. Dioxygen Homogeneous Catal. Oxid., [Proc. Int. Symp.], 5th*; D. H. R. M. Barton, Arthur Eael; Sawyer, Donald T, Ed.; Plenum: New York, 1993; pp. 209-223.
(80) Farmer, P. J.; Solouki, T.; Mills, D. K.; Soma, T.; Russell, D. H.; Reibenspies, J. H.; Darensbourg, M. Y. *J. Am. Chem. Soc.* **1992**, *114*, 4601-4605.
(81) Maroney, M. J.; Choudhury, S. B.; Bryngelson, P. A.; Mirza, M. A., Sherrod, M.J. *Inorg. Chem.* **1996**, *35*, 1073-1076.
(82) Shoner, S. C.; Olmstead, M. M.; Kovacs, J. A. *Inorg. Chem.* **1994**, *33*, 7-8.
(83) Farmer, P. J.; Verpeaux, J. N.; Amatore, C.; Darensbourg, M. Y.; Musie, G. *J. Am. Chem. Soc.* **1994**, *116*, 9355-9356.
(84) Voordouw, G.; Menon, N. K.; LeGall, J.; Choi, E. S.; Peck, H. J.; Przybyla, A. E. *J. Bacteriol.* **1989**, *171*, 2894-2899.
(85) Choudhury, S. B.; Pressler, M. A.; Mirza, S. A.; Day, R. O.; Maroney, M. J. *Inorg. Chem.* **1994**, *33*, 4831-4839.
(86) Bryant, F. O.; Adams, M. W. W. *J. Biol. Chem.* **1989**, *264*, 5070-5079.
(87) van der Zwaan, J. W.; Albracht, S. P. J.; Fontijn, R. D.; Slater, E. C. *FEBS Lett.* **1985**, *179*, 271-277.
(88) Chapman, A.; Cammack, R.; Hatchikian, C. E.; McCracken, J.; Peisach, J. *FEBS Lett.* **1988**, *242*, 134-138.
(89) Gu, Z.; Maroney, M. J.; Roseboom, W.; Albracht, S. P. J. unpublished results.
(90) Bagley, K. A.; Van Garderen, C. J.; Chen, M.; Woodruff, W. H.; Duin, E. C.; Albracht, S. P. J. *Biochemistry* **1994**, *33*, 9229-9236.
(91) Ghilardi, C. A.; Sabatini, A.; Sacconi, L. *Inorg. Chem.* **1976**, *15*, 2763-2767.
(92) Münck, E.; Albracht, S. P. J. unpublished results.
(93) van der Spek, T. M.; Arendsen, A. F.; Happe, R. P.; Yun, S.; Bagley, K. A.; Stufkens, D. J.; Hagen, W. R.; Albracht, S. P. J. *Eur. J. Biochem.* **1996**, *237*, 629-634.
(94) Sellmann, D.; Kaeppler, J.; Moll, M. *J. Am. Chem. Soc.* **1993**, *115*, 1830-5.
(95) Sellmann, D.; Geck, M.; Moll, M. *J. Am. Chem. Soc.* **1991**, *113*, 5259-64.
(96) Waden, H.; Seak, W.; Haase, D.; Pohl, S. *Angew. Chem.* **1996**, in press.
(97) Allan, C. B.; Maroney, M. J. unpublished results.

Chapter 5

Challenges of Biocatalysts: Modeling the Reactivity of Nitrogenases and Other Metal–Sulfur Oxidoreductases

Dieter Sellmann and Jörg Sutter

Institut für Anorganische Chemie der Universität Erlangen-Nürnberg, Egerlandstrasse 1, D–91058 Erlangen, Germany

In the quest for complexes that model the reactivity of nitrogenases, hydrogenases, and other oxidoreductases containing metal sulfur units as active centers, the coordination of transition metals by sulfur donor ligands is regarded as a basic and essential structural feature that governs the catalytic functions. Evidence for structure-function relationships is found a) in electronic, structural, and redox changes caused by protonation and alkylation of the thiolate donors in $[Fe(CO)_2('S_4')]$ and $[Fe(CO)('N_HS_4')]$, b) in the reactions of $[Rh(H)(CO)('S_4')]$, which catalyzes the H_2/D^+ exchange reaction typical of hydrogenases, and c) in the stabilization and reactivity of diazene, HN=NH, bound to [FeS] complex fragments.

For many reasons, catalysis is a major topic of chemical research (*1*). One reason is that nature has created catalysts, the enzymes, of whose activity and efficiency chemists can only dream. A few of these biocatalysts catalyze reactions of molecules such as dinitrogen, dihydrogen, dioxygen, or carbon dioxide; molecules that represent the building blocks of all living matter. Some of these molecules, e.g., dinitrogen are very inert under standard conditions; others, e.g., dihydrogen may play a role as a future energy source (and carrier). The challenges for chemists are evident: It is the synthesis of compounds that are able to model enzymatic reactions under similarly mild conditions in order to yield perspectives for the synthesis of compounds which can ultimately serve as competitive catalysts.

Numerous efforts have been made to model the active centers of metal enzymes using low molecular weight complexes. Many of the resultant 'models' have been criticized for being too unlike the active centers. Frequently, this criticism results from the misconception that a model has to reduplicate every property of the active center, e.g., structure, spectroscopy, magnetism, and reactivity. A (chemical) model, however, is *per definitionem* an abstraction that, even if it models only a very few features of the real thing, can nevertheless serve to elicit a better understanding of the enzyme (*2*).

In search of such model compounds, another problem is encountered. Often it is known or it can be predicted that the active centers will be either inactive

0097–6156/96/0653–0101$15.00/0

or even unstable in the absence of proteins reflecting the fact that active centers of enzymes need their proteins and the fulfillment of many other conditions in order to work properly. For instance, the FeMo cofactor isolated from FeMo nitrogenases does not catalyze N_2 reduction, and, in addition, it has only a relatively short lifetime in aqueous media ($\tau_{1/2}$ ~ 2h) [The properties of the native FeMoco have been reviewed comprehensively (*3*)]. Thus, even if it were possible to synthesize in vitro such a [FeMoS] cluster as the FeMoco, this cluster can not be expected to be the catalyst that is sought. Consequently, in the quest for competitive catalysts a way has to be chosen that is different from the attempts to rebuild in vitro the active centers of enzymes as precisely as possible. Nevertheless, the enzymes may show in which direction to proceed.

Our approach is based on the generally accepted presumption that structure-function relationships exist not only for the complete enzymes but also for their active centers. The most basic structural feature of a metal complex, however, is the type of donor atoms surrounding the metal center. In this context it is noted that typical oxidases such as methane monoxygenase or the water oxidizing complex (WOC) of photosystem II employ hard oxygen or nitrogen donors for coordinating the transition metals of their active centers. In contrast, typical reductases such as nitrogenases, hydrogenases or CO dehydrogenase appear to prefer soft sulfur donors for their metals iron, molybdenum, vanadium, or nickel (*4*).

This could be one of the above mentioned structure-function relationships. Our general efforts aim at elucidating the characteristic features connected with the combination of transition metals and sulfur donors. Our particular interest focuses on compounds that are able to model reactions of nitrogenases, hydrogenases, and CO dehydrogenase. Such compounds ought to be stable in the absence of proteins, and as will be shown below, they must also be robust enough to withstand the attack of species such as protons. These requirements demand special ligands. A few of these ligands and their complexes are surveyed in Scheme I.

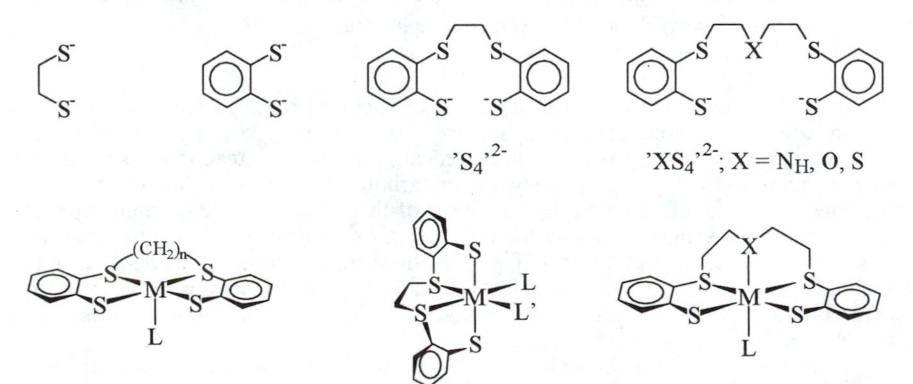

M = Fe, Ru, Os; Cr, Mo, W; Ni, Pd, Pt; Co, Rh; L or L' = N_2H_2, N_2H_3, N_2H_4, NH, NH_2, NH_3; NO^+, $NO^\cdot$, NH_2O, NH_2OH; (H^+), {H_2}, H^-;(CH_3^+), CH_3^-, CO, {CH_3CO-SR}; Cl^-, N_3^-, O^{2-}; H_2S, S^{2-}, S_2^{2-}, {S_2^-}, S_2^0

Scheme I. Sulfur Ligands, Complexes and Coligands

Starting from bidentate dithiolates such as ethane- or benzene-dithiolate, tetra- and pentadentate ligands of the 'S_4'$^{2-}$ and 'XS_4'$^{2-}$ type have been synthesized that contain thioether and thiolate donors and in the case of 'XS_4'$^{2-}$ additionally amine N or ether O donors (*5*). These ligands coordinate to biorelevant metals such as Fe, Mo, W, Ni, Co or their group homologues, and the resultant complex fragments are able to bind a large variety of coligands L or L'. It may be noted that L and L' comprise species being relevant to N_2 fixation, the nitrate-ammonia interconversion, hydrogenases, and CO dehydrogenases and the biological sulfur cycle. The coligands, which range from soft σ-π ligands such as CO to hard ligands such as chloride or oxide further indicate that these metal sulfur fragments can exhibit a remarkable electronic flexibility. Scheme II suggests a few reasons for this flexibility and characteristics of [MS] fragments that potentially result from this flexibility.

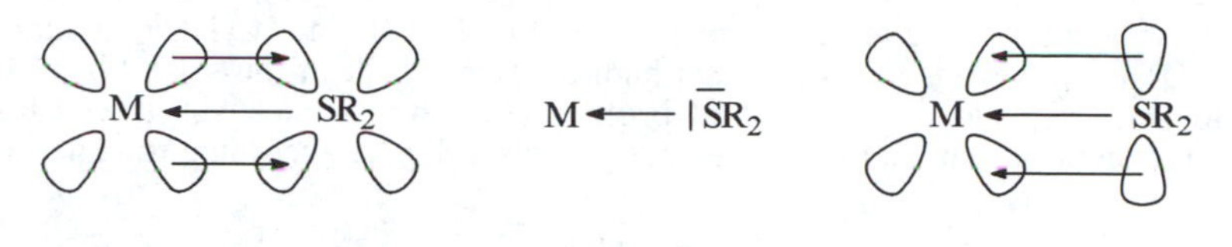

σ-donor-π-acceptor σ-donor σ-donor-π-donor

redox activity — Brönsted acid-base — vacant sites

Scheme II. Bonding Modes of Thioether and Thiolate Ligands and Potentially Resultant Characteristics of [MS] Fragments

From a biological, but also from a theoretical point of view, it can be anticipated that ligands with biorelevant sulfur donors, i.e., sulfide, thiolate and thioether donors, differ from ligands, e.g., such as phosphine or carbon ligands. For instance, thioether, thiolate or sulfide S atoms can be expected to act as σ-donor, as σ-donor-π-acceptor, or due to the remaining lone pairs, as σ-donor-π-donor ligands. The actual bonding mode will depend on the occupation of the metal d orbitals. The variability of the bonding modes also plausibly explains why metal sulfur fragments so frequently exhibit redox activity, Brönsted acid-base behaviour, and vacant sites of coordination. These three properties are essentials if a molecule A is to be activated or stabilized by coordination to a metal sulfur site in order to be transformed subsequently into B by addition of protons and electrons according to equation 1.

$$A + x\,H^+ + y\,e^- \longrightarrow B \qquad (1)$$

Equation 1 is the most general expression of the typical reductase reaction stressing the coupling of proton and electron transfer.

In the following sections, general and, with regard to hydrogenase and nitrogenase reactions, specific aspects of this equation will be discussed.

Protonation and Alkylation of Metal Sulfur Complexes: Structural, Electronic and Redox Effects

Little is known of the coupling of the transfer of protons and electrons, for instance, the order of the transfer steps and the way in which they influence each other (*6*). In order to shed light upon the question of how primary protonation of metal sulfur sites may influence a) the metal sulfur cores, b) small molecules such as CO, N_2 (or hydride) bound to them, and c) the subsequent transfer of electrons, we have investigated protonation, alkylation and redox reactions of complexes such as $[Fe(CO)_2('S_4')]$ and $[Fe(CO)('N_HS_4')]$.

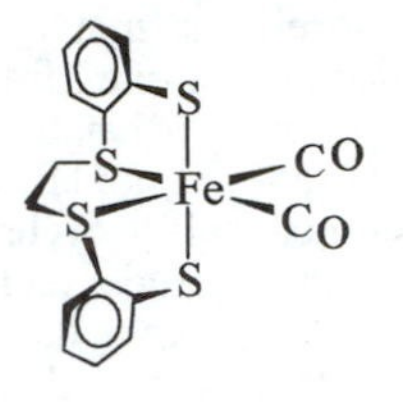

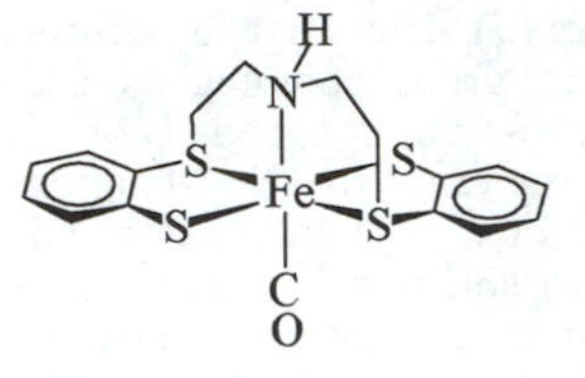

[Fe(CO)$_2$('S$_4$')] [Fe(CO)('N$_H$S$_4$')]

Stepwise addition of one and two equivalents of HBF_4 to CH_2Cl_2 solutions of [Fe(CO)$_2$('S$_4$')] leads to a significant high-frequency shift of the ν(CO) bands by 35 to 40 cm^{-1} per added equivalent of protons (*7*). This shift can be traced back to protonation of one and both Brönsted basic thiolate donors according to equation 2.

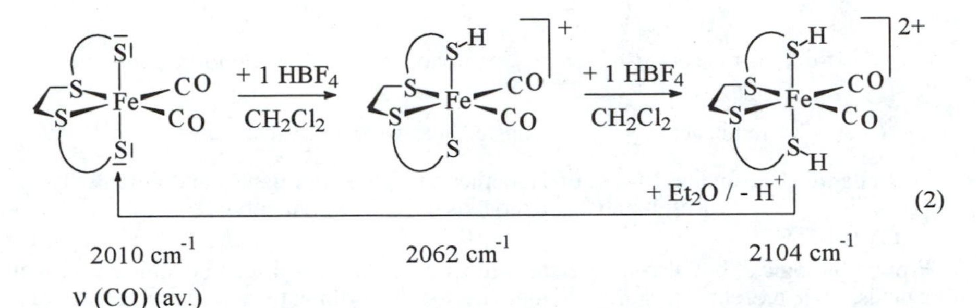

Protonation of the Brönsted basic thiolate donors causes a decrease of electron density at the iron centers and consequently a weakening of the Fe–CO π back-bonding. Upon addition of ether, the starting complexes are regenerated showing that the protonations are reversible. Also for this reason, the thiol complexes could not be isolated. In order to more strongly implicate the thiolate as the site of attack, [Fe(CO)$_2$('S$_4$')] has been also alkylated by stepwise treatment with Et_3OBF_4 yielding the isoelectronic salts [Fe(CO)$_2$('S$_4$-Et')]BF$_4$ and [Fe(CO)$_2$('S$_4$-Et$_2$')]-(BF$_4$)$_2$, which could be characterized by X-ray structure analyses (*7*).

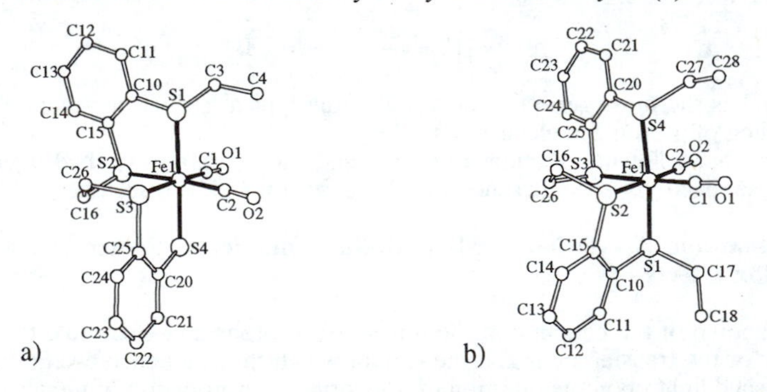

Figure 1. Molecular structures of the cations of a) [Fe(CO)$_2$('S$_4$-Et')]BF$_4$ and b) [Fe(CO)$_2$('S$_4$-Et$_2$')](BF$_4$)$_2$.

The same increase of ν(CO) frequencies by some 40 cm^{-1} per alkylation took place, and the molecular structures clearly confirm the thiolate donors as sites of electrophilic attack.

The increase of ν(CO) frequencies is remarkably large. It compares to the difference observed for couples of homoleptic carbonyl metal complexes such as $[V(CO)_6]^-$ (1860 cm^{-1}) and $[V(CO)_6]$ (1976 cm^{-1}) (*8*) whose oxidation states differ by a whole unit. We therefore expected a correlation between the ν(CO) frequencies and the Fe–S distances. However, the data of Table I show that such a correlation does not exist.

Table I. ν(CO) frequencies versus invariance of FeS distances of $[Fe(CO)_2('S_4')]$ and its alkyl derivatives $[Fe(CO)_2('S_4\text{-}Et')]^+$ and $[Fe(CO)_2('S_4\text{-}Et_2')]^{2+}$

	$\nu(CO)_{av.}$ cm^{-1}	$\bar{d}$ (Fe–S) pm
$[Fe(CO)_2('S_4')]$	2010	228.4(2)
$[Fe(CO)_2(Et\text{-}'S_4')]^+$	2049	228.9(2)
$[Fe(CO)_2(Et_2\text{-}'S_4')]^{2+}$	2091	228.1(2)

The (averaged) Fe–S distances in the neutral starting complex and the mono- and dialkylated derivatives remain invariant. (In this context it is to be noted that in the starting complex even the Fe–S(thiolate) and Fe–S(thioether) distances are identical within standard deviations).

In order to explain the invariance of Fe–S distances which strongly contrasts with the difference in electron density at the Fe centers as indicated by the large ν(CO) shifts, we suggest the following bonding scheme.

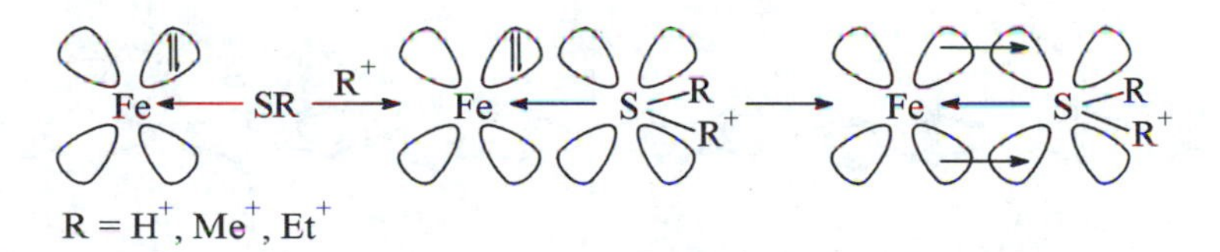

The Fe–S(thiolate) bonds are assumed to have predominantly σ-donor bond character. Protonation, and likewise alkylation, of the thiolate donors leads to a weakening of the respective S→Fe σ-bonds and an inductive withdrawal of electron density from the Fe centers. The thiolate donors that turned into thiol or thioether donors, however, gain π-acceptor properties such that partial Fe→S π-back bonds form, which leads to a further decrease of electron density at the Fe centers. The Fe–S distances, on the other hand, remain invariant because weakening of the S→Fe σ-donor bond and formation of the Fe→S π-back bond compensate each other.

The $[Fe(CO)_2('S_4')]$ complex and its derivatives proved electrochemically inactive such that conclusions on the redox consequences could not be drawn. This was possible for $[Fe(CO)('N_HS_4')]$ and its derivatives. In this case, cyclic voltammograms of even the protonated derivatives could be obtained.

Like $[Fe(CO)_2('S_4')]$, $[Fe(CO)('N_HS_4')]$ can be protonated reversibly and stepwise, it can also be alkylated, and the ν(CO) frequencies exhibit a similar shift of 35–40 cm^{-1} per step of protonation or alkylation. Comparison of the corresponding distances in $[Fe(CO)('N_HS_4']$ and its dialkyl derivative $[Fe(CO)('N_HS_4\text{-}Me,Et')]$ $(BF_4)_2$ shows that again the Fe–S (and Fe–N) core distances remain invariant within standard deviations (Table II) (Sellmann, D.; Becker, T.; Knoch, F. *Chemistry*, submitted.).

Table II. Comparison of corresponding distances [pm] in the neutral parent complex [Fe(CO)('N_HS_4')] and its dialkylated derivative [Fe(CO)('N_HS_4-Me,Et')]$(BF_4)_2$

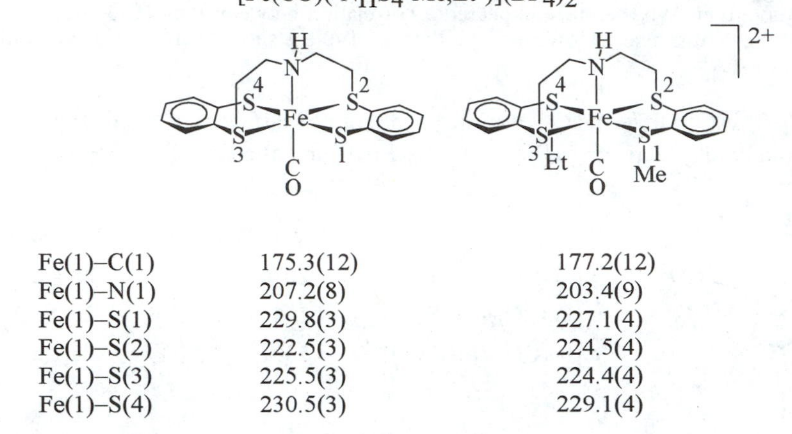

Fe(1)–C(1)	175.3(12)	177.2(12)
Fe(1)–N(1)	207.2(8)	203.4(9)
Fe(1)–S(1)	229.8(3)	227.1(4)
Fe(1)–S(2)	222.5(3)	224.5(4)
Fe(1)–S(3)	225.5(3)	224.4(4)
Fe(1)–S(4)	230.5(3)	229.1(4)

The electrochemical data of Table III demonstrate a dramatic effect upon the redox potentials of corresponding redox couples.

Table III. Redox potentials of [Fe(CO)('N_HS_4')] and its protonated or alkylated derivatives (in [V] vs. NHE, in CH_2Cl_2, v = 100mV/s, q = quasi reversible, i = irreversible, n.o. = not observed)

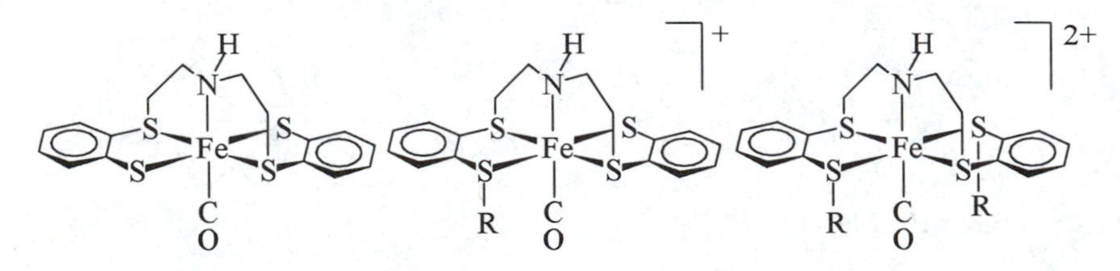

redox couple			
Fe(II) / Fe(III)	+ 0.35 q	R = H : 0.98 q	R = H : 1.46 i
		R = Et : 1.02 q	R = Et : n.o.
Fe(II) / Fe(I)	n.o.	R = Et : -1.34 i	R = Et : -0.54 q
Fe(I) / Fe(0)	n.o.	n.o.	-1.44 i

The results can be summarized as follows.

1. Protonation or alkylation have identical consequences. They shift the redox potential of corresponding redox pairs ca. 500–700 mV per step of protonation or alkylation. Nearly identical shifts of protonated and alkylated derivatives further justify consideration of the alkylated derivatives as a kind of 'frozen-in' protonated species.
2. A logical consequence of the redox potential shifts is that protonated or alkylated species are more difficult to oxidize and easier to reduce. This in turn allows species with 19 or 20 valence electrons (VE), which cannot be observed electrochemically in the case of the neutral parent complex, to become accessible when one or both thiolate donors are alkylated.

The shift of redox potentials and simultaneous invariance of Fe–S distances may also reveal a fundamental structure-function relationship for [MS] oxidoreductases. The [MS] cores can be anticipated to facilitate the uptake and the release of electrons because no activation barriers due to the rearrangement of atoms, distances, or angles impede the electron transfer steps.

Hydrogenase Reactions

Hydrogenases contain [NiFeS] or only [FeS] centers (*9*). The X-ray structure analysis of the hydrogenase isolated from *Desulfovibrio gigas* (*10*) shows that the active center contains a nickel atom in a coordination sphere of soft cysteine thiolate donors (Figure 2). Two of them and an as yet undefined X bridge link the nickel atom to a second metal atom M that probably is iron which additionally coordinates three diatomic YZ molecules. These YZ molecules, too, could not yet be identified and possibly are CO ligands

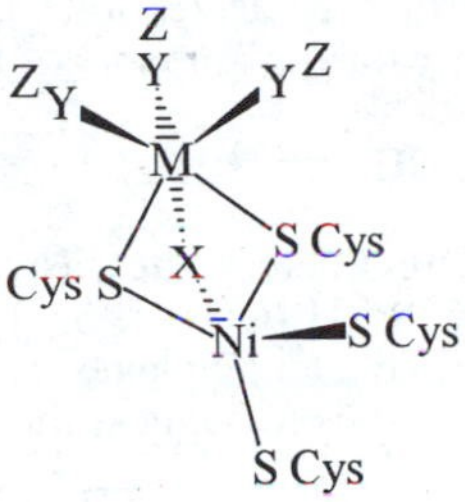

Z Y Z Y Y Z M X Cys S S Cys Ni S Cys S Cys

Figure 2. Molecular structure of the active center of hydrogenase isolated from *Desulfovibrio gigas* (*10*).

Hydrogenases catalyze the H^+/H_2 redox equilibrium and the exchange between molecular hydrogen and the protons of water (equations 3a and 3b).

$$2\,H^+ + 2\,e^- \rightleftharpoons H_2 \quad (3a)$$

$$H_2 + D_2O \rightleftharpoons HD + HDO \quad (3b)$$

From a technological point of view reaction 3a is of interest with regard to H_2 production from renewable sources. From the chemical point of view, however, reaction 3b is more interesting. It requires the heterolytic cleavage of the strong H–H bond (ΔH = -436 kJ/mole = -104 kcal/mole)) under mild conditions without producing any strongly exergonic products. If such a heterolytic cleavage of H_2 is catalyzed by [MS] centers, it can be expected that the resultant protons bind to sulfur donors while the hydride ion gives a metal hydride (*11*) (equation 4).

$$\begin{matrix} M \\ | \\ S \end{matrix} + H_2 \longrightarrow \begin{matrix} M\text{---}H \\ | \quad\; | \\ S\text{---}H \end{matrix} \longrightarrow \begin{matrix} M\text{—}H^- \\ | \\ S\text{—}H^+ \end{matrix} \quad (4)$$

For that reason, metal sulfur hydride complexes gain considerable importance. However, the number of such complexes that have been fully characterized and catalyze the reaction of equation 3b is very small. Nickel sulfur hydride complexes such as $[Ni(H)\{SRC_2H_4)_3N\}]^+$ (*11*) and $[Ni(H)(S_2C_6H_4)(PMe_3)]^-$ (Sellmann, D.; Häußinger, D., unpublished results.), which are of primary interest, do not catalyze the H_2/D^+ exchange. Therefore, we have extended our research to metals other than nickel (or iron) and found the rhodium complex [Rh(H)(CO)('$^{bu}S_4$')]. It forms when $[RhCl(CO)_2]_2$ is treated with the neutral thioether thiol '$^{bu}S_4$'-H_2. CO and hydrogen chloride are removed, and one SH bond oxidatively adds to the rhodium center, which becomes formally a Rh(III) (equation 5) (*12*).

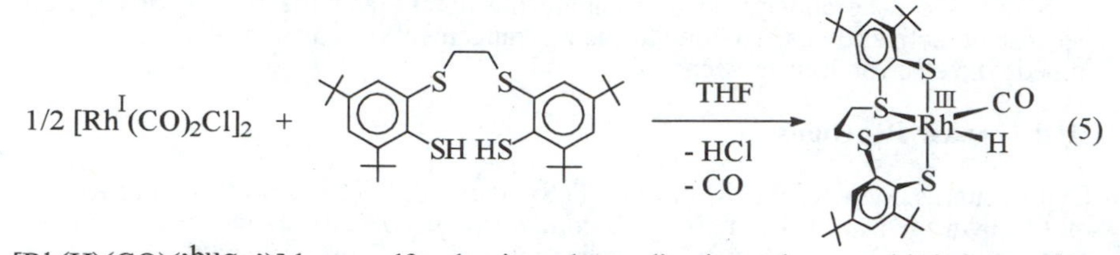

[Rh(H)(CO)('^{bu}S$_4$')] has a sulfur dominated coordination sphere, and it is interesting to investigate how the reactivity of this complex differs from that of common hydride complexes. The most significant result has been that [Rh(H)(CO)('^{bu}S$_4$')] exchanges its hydride ligand with deuterium upon reaction with molecular D_2 only in the presence of catalytic amounts of protons from Brönsted acids such as hydrochloric acid. [Rh(H)(CO)('^{bu}S$_4$')] further catalyzes the D_2/H^+ exchange according to equation 6, which again requires catalytic amounts of HCl/H_2O.

$$D_2 + EtOH \longrightarrow HD + EtOD \qquad (6)$$

These reactions strongly implicate a heterolytic cleavage of H_2 (or D_2) by a species that results from the reaction of [Rh(H)(CO)('^{bu}S$_4$')] and protons. The isolation and full characterization of the H and D complexes and the essential requirement of catalytic amounts of protic acids led us to suggest the mechanism of Scheme III.

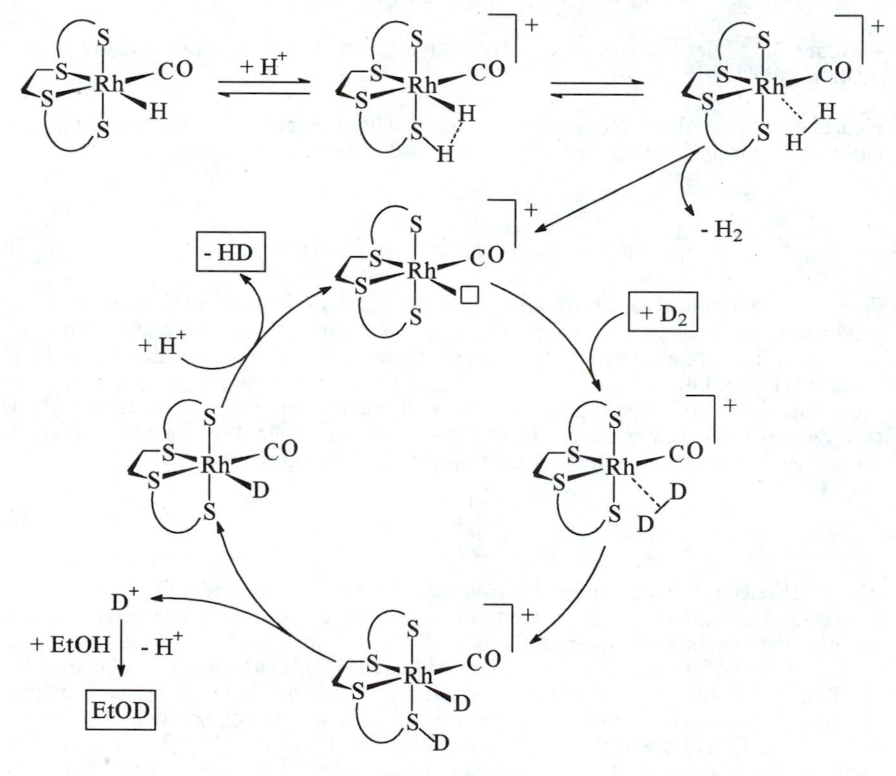

Scheme III.

The Brönsted basic thiolate donors are protonated. Reaction between H^+ (from the resulting thiol) and H^- (from the hydride ligand) yields a nonclassical dihydrogen complex (*13*), which releases H_2 to give the coordinatively unsaturated species $[Rh(CO)('^{bu}S_4')]^+$. Its vacant site is occupied by D_2 which gets cleaved into D^+ and D^-. D^+ is released and yields EtOD plus H^+, which subsequently reacts with the D complex to regenerate the vacant site at the rhodium center.

Thus, the essential key step, the heterolytic cleavage of D_2 is achieved by the concerted action of the Brönsted basic thiolate donor and the Lewis acidic Rh center upon the D_2 molecule. The vacant site of the coordinatively unsaturated species, which is a five coordinate 16 valence electron Rh(III) complex, is evidently long-lived enough in order to allow for the release of H_2 and the addition of D_2. The stabilization of this vacant site can be traced back to the π-donor capacity of the two thiolate donors (and it does not need the protective wrapping by a protein!).

In order to further substantiate the suggested mechanism of Scheme III, protonation reactions of $[Rh(H)(CO)('^{bu}S_4')]$ and its phosphine derivative $[Rh(H)(PCy_3)('^{bu}S_4')]$ have been investigated at temperatures between +20 °C and -50 °C. Monitoring these reactions by IR and ^1H NMR spectroscopy yielded conclusive evidence (ν(SH) bands and δ(SH) ^1H NMR signals) for the formation of the hydride thiol complexes $[Rh(H)(L)('^{bu}S_4'\text{-}H)]\,BF_4$ (L = CO, PCy_3) according to equation 7 (Sellmann, D.; Rackelmann, G., unpublished results.).

+ HBF_4 / CH_2Cl_2 ; - H^+ (L = PCy_3) ; BF_4 ; - H_2 ; + H_2 ; BF_4

L = CO, PCy_3 L = PCy_3 (7)

When CH_2Cl_2 solutions of the PCy_3 derivative $[Rh(H)(PCy_3)('^{bu}S_4'\text{-}H)]BF_4$ are evaporated at ambient temperature, H_2 is released and the coordinatively unsaturated $[Rh(PCy_3)('^{bu}S_4')]BF_4$ forms. $[Rh(PCy_3)('^{bu}S_4')]BF_4$ could be isolated in the solid state, and proved highly reactive toward H_2O, which added reversibly, and toward CO, which yields the stable CO complex $[Rh(CO)(PCy_3)('^{bu}S_4')]BF_4$. Most importantly, $[Rh(PCy_3)('^{bu}S_4')]BF_4$ also adds H_2. The H_2 molecule is heterolytically cleaved and yields the thiol hydride complex $[Rh(H)(PCy_3)('^{bu}S_4'\text{-}H)]BF_4$. Upon treatment with bases such as solid Na_2CO_3 or water, $[Rh(H)(PCy_3)('^{bu}S_4'\text{-}H)]BF_4$ regenerates the hydride $[Rh(H)(PCy_3)('^{bu}S_4')]$. These results demonstrate the heterolytic cleavage of H_2 at a metal-sulfur site according to equation 8, firmly establish the reversibility of the reactions according to equation 7, and lend further support to the mechanism suggested in Scheme III.

$$\text{—Rh—H (S)} + H^+ \rightleftharpoons [\text{—Rh—H (S—H)}]^+ \rightleftharpoons [\text{—Rh—}\square\text{ (S)}]^+ + H_2 \quad (8)$$

It is to be mentioned that $[Rh(PCy_3)('^{bu}S_4')]BF_4$ also catalyzes the exchange reaction between EtOH protons and D_2 (equation 6). In this case, however, no catalytic amounts of protic acids are required, which in the case of $[Rh(H)(CO)('^{bu}S_4')]$ are necessary in order to generate a vacant coordination site at the rhodium center.

N_2 Fixation

Biological N_2 fixation is catalyzed by Mo/Fe, V/Fe or 'Fe-only' nitrogenases and represents another one of the biological challenges that have not yet been met by chemists. The extremely different reaction conditions of the biological and the Haber-Bosch processes of N_2 reduction, i.e., of standard temperature and pressure and biological redox potentials on the one hand, red-hot temperatures and high-pressures on the other hand, make this challenge particularly fascinating (*14*).

Over the last 30 years, chemists have discovered a number of reactions of N_2 that occur at ambient temperature and pressure. Breakthroughs comprise the discovery of metal N_2 complexes (*15*). A few reactions have been found converting the N_2 ligands, for example, in $[Mo(N_2)_2(dppe)_2]$ (*16*), $[Fe(N_2)(dmpe)_2]$ (*17*) and $[CpMn(CO)_2N_2]$ (*18*), into lower valent nitrogen species by electrophilic (H^+ or CH_3^+), radical ($CH_3\cdot$) or nucleophilic ($C_6H_5^-$, CH_3^-) attack. Most recently, it was shown that 'simple' compounds such as the tris(amido) Mo(III) complex $[Mo(NRR')_3]$ ($R = C(CD_3)_2CH_3$, $R' = 3,5\text{-}C_6H_3(CH_3)_2$)) are able to react with N_2 and completely break the N_2 triple bond by forming of the molybdenum (VI) nitrido complex $[Mo(VI){\equiv}N(NRR')_3]$ (*19*). Aqueous systems, too, such as Schrauzer's (*20*) and Shilov's (*21*) molybdenum or vanadium based systems, have been found to reduce N_2 to N_2H_4 or NH_3 (*22*).

But even the most efficient of these systems has not yet matched the nitrogenase challenge. Otherwise we might have cheap fertilizers nowadays. One major problem is that usually strong abiological reductants such as alkaline metals are needed, if not for the actual reduction of N_2 then in one of the preceeding steps, when the N_2 complexes are being synthesized. The other problem is that none of these systems is truly catalytic. Thus, numerous hopes focussed on the X-ray structure analysis of nitrogenase expecting that all these problems would become resolvable by elucidating the molecular structure of the FeMo cofactor (Figure 3) (*23*).

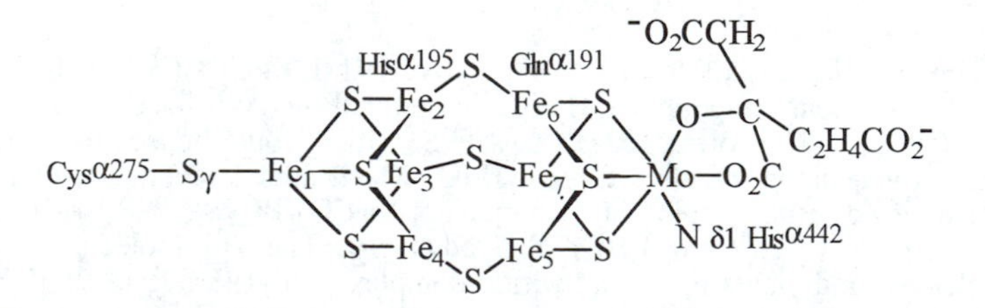

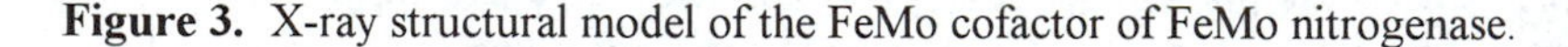

Figure 3. X-ray structural model of the FeMo cofactor of FeMo nitrogenase.

Such expectations have not yet come true. Additionally, as outlined above, even if the challenge can be met by synthesizing such a cluster in vitro, the cluster may very well prove not to be a catalyst for N_2 reduction under mild conditions. On the other hand, thermodynamics clearly shows that the reaction according to equation 9 needs a

$$N_2 + 6\,H^+ + 6\,e^- \xrightarrow{H_2O} 2\,NH_3 \qquad E_0 = -280\text{ mV (pH 7)} \tag{9}$$

reduction potential of only -280 mV at pH 7. This is a lower potential than that required for the reduction of protons to dihydrogen (-410 mV at pH 7)! Hence, there is in principle no reason that rules out the possibility of reducing N_2 under mild conditions.

Iron in the coordination sphere of sulfur donors plays a major role in the FeMoco and certainly does so in the active centers of 'Fe-only' nitrogenases. It has been shown above that protonation dramatically influences the redox potentials of complexes containing [MS] cores. These two facts constitute the basis of a working hypothesis for N_2 reduction which we want to explain by discussing the chemistry of diazene complexes containing [Fe–S] fragments.

Diazene, HN=NH, can be assumed to be the first key intermediate of N_2 fixation which presumably proceeds via $2e^-/2H^+$ steps (*22b,24*). In the free state, however, diazene has such a high positive enthalpy of formation (ΔH_f = +212 kJ/mole = +51 kcal/mole) that it rapidly decomposes above temperatures of ~ -180 °C (*25*). The question arises as to how this unstable molecule can be stabilized, especially in the coordination sphere of iron sulfur complexes, in order to avoid unsurmountably high energy barriers on the reaction coordinate from N_2 to NH_3.

Scheme IV shows a model cycle for the reduction of N_2 in the coordination sphere of [Fe–S] complexes that contain the [Fe('N_HS_4')] fragment (*26*).

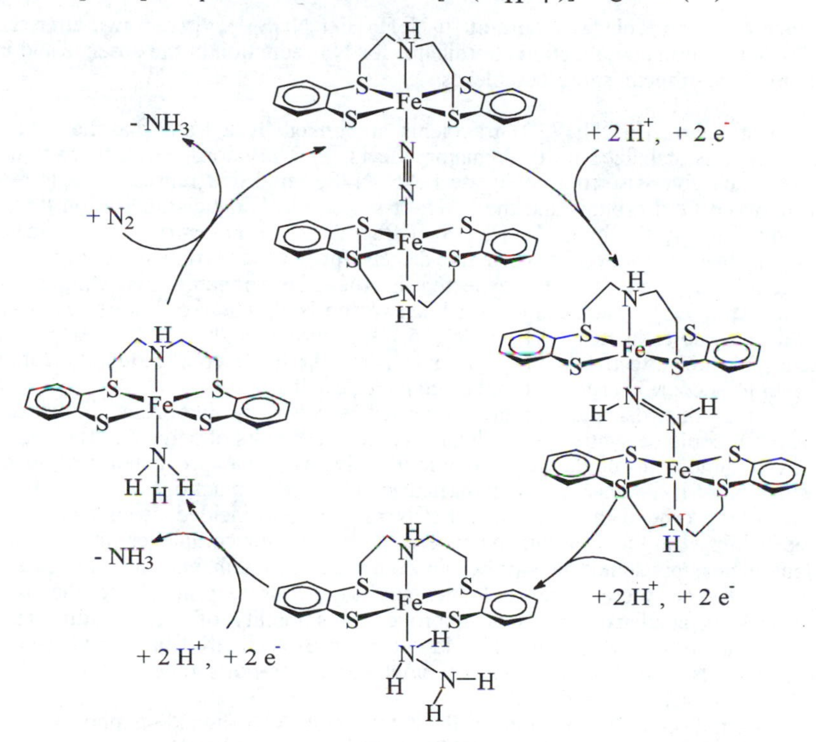

Scheme IV. Hypothetical Cycle for N_2 Fixation Catalyzed by [Fe('N_HS_4')] Fragments

Three of the four key intermediates, i.e., the N_2H_2, N_2H_4 and NH_3 complexes, have been isolated and characterized. Significantly the cycle contains no biologically unusual Fe oxidation states such as Fe(0) or Fe(I), but only Fe(II). These Fe(II) centers, however, are high-spin in the N_2H_4 and NH_3 species and low-spin in the N_2H_2 species. Thus, when going through the cycle, changes of spin-states have to be envisaged that influence the individual redox steps.

The molecular structure of the diazene complex is shown in Figure 4 (*27*).

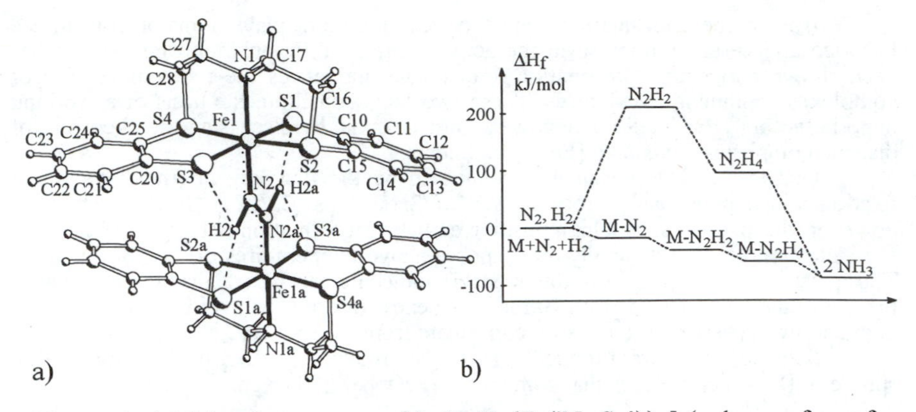

Figure 4. a) Molecular structure of $[\mu\text{-}N_2H_2\{Fe('N_HS_4')\}_2]$ (redrawn after ref. (*27*)) and b) potential reaction coordinates for N_2 reduction in the absence and in the presence of metal sulfur complexes.

Two enantiomeric $[Fe('N_HS_4')]$ fragments are linked by a trans diazene, and the N_2H_2 ligand is stabilized by three major effects: a) The coordination to two metal centers which gives rise to a conjugate Fe∸N∸N∸Fe bond that represents a 4c-6e π-bond between Fe d-orbitals and the N_2H_2 π-system. b) Kinetic stabilization through steric shielding by the bulky $[Fe('N_HS_4']$ fragments. c) Tricentric (bifurcated) N–H··$(S)_2$ hydrogen bridges between the diazene protons and thiolate donors. Such hydrogen bridges can not be expected in the corresponding N_2 complex $[\mu\text{-}N_2\{Fe('N_HS_4')\}_2]$, which has not yet been synthesized. These H-bridges alone may 'neutralize' some estimated 70 kJ/mole (16.1 kcal/mole) of the positive ΔH_f of free diazene. All three effects taken together may allow the first step of reduction from N_2 to N_2H_2 to become exergonic as indicated in Figure 4b.

Thus, this diazene complex revealed some very important information. However, it could be synthesized only in very small amounts of some 50–100 mg, and it proved almost insoluble in all common solvents. This prevented systematical investigation of its chemistry and, in particular, of its redox reactions.

On the other hand, $[\mu\text{-}N_2H_2\{Fe('N_HS_4'\}_2]$ had yielded proof that diazene can be stabilized in the coordination sphere of metal sulfur complexes not containing any abiological phosphine ligands which could act as stabilizing and thus essential coligands. Hence, there were no longer any objections in principle to the use of phosphines as auxiliary ligands to improve the solubility of metal-sulfur diazene complexes in (organic) solvents. This idea proved exceedingly fruitful and gave not only high yields of well soluble diazene complexes in one-pot syntheses, but also what we call our 'miracle of substituents'.

When $FeCl_2$, the dianion of the tetradentate $'S_4'$-ligand, a phosphine and excessive hydrazine are combined in THF according to equation 10,

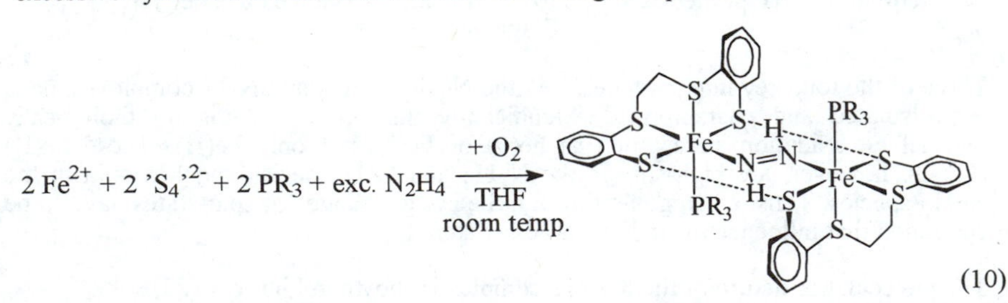

olive-green to brown solutions form which presumably contain the very labile $[Fe(N_2H_4)(PR_3)('S_4')]$ complexes. When air is bubbled through these solutions, they instantly turn dark blue-green, and the diazene complexes $[\mu\text{-}N_2H_2\{Fe(PR_3)\text{-}('S_4')\}_2]$ form in high yields. The 'miracle of substituents' refers to the fact that this procedure does not work with PMe_3, PEt_3 or PPh_3, however, with $P(nPr)_3$ and $P(nBu)_3$, it yields up to 10 g of diazene complex in one run. The resultant diazene complexes feature the same characteristics as $[\mu\text{-}N_2H_2\{Fe('N_HS_4')\}_2]$. Figure 5 shows the molecular structure of the $P(Pr)_3$ derivative, which is schematically drawn in equation 10 for R = Pr (*28*).

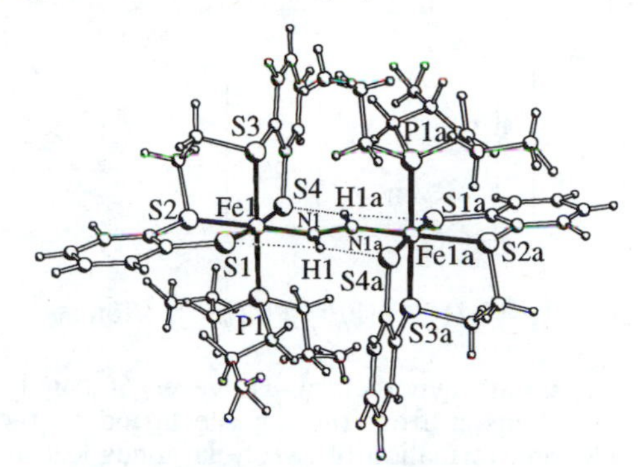

Fe(1)–N(1)	190.0(9) pm
N(1)–N(1a)	128.8(15) pm
S(1)–H(1)	281.4 pm
S(4a)–H(1)	235.8 pm

Figure 5. Molecular structure and selected structural parameters of $[\mu\text{-}N_2H_2\{Fe(PPr_3)('S_4')\}_2]$.

The molecular parameters and spectroscopic properties of $[\mu\text{-}N_2H_2\{Fe(PPr_3)\text{-}('S_4')\}_2]$ clearly show that the same effects stabilize the diazene ligand in $[\mu\text{-}N_2H_2\{Fe('N_HS_4')\}_2]$ and $[\mu\text{-}N_2H_2\{Fe(PPr_3)('S_4')\}_2]$: steric shielding, a Fe∴N∴N∴Fe 4c-6e π-system, and N–H···$(S)_2$ bridges. In contrast to $[\mu\text{-}N_2H_2\{Fe('N_HS_4')\}_2]$, $[\mu\text{-}N_2H_2\{Fe(PPr_3)('S_4')\}_2]$ allowed cyclic voltammograms (CV) to be recorded (Figure 6).

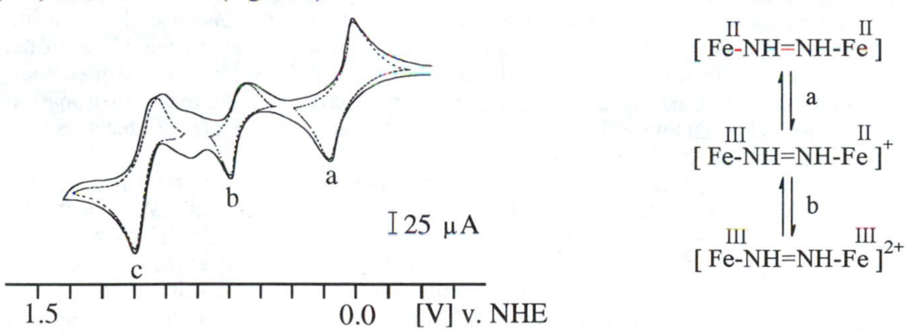

Figure 6. Cyclic voltammogram of $[\mu\text{-}N_2H_2\{Fe(PPr_3)('S_4')\}_2]$ (in CH_2Cl_2, 20 °C, v = 100 mV/s) and assignment of redox waves and Fe oxidation states.

The CV of $[\mu\text{-}N_2H_2\{Fe(PPr_3)('S_4')\}_2]$ exhibits three well resolved and nearly reversible redox waves a, b and c in the anodic region. At -78 °C, even a fourth (irreversible) wave at 1.32V can be observed. The redox waves can be assigned to the formation of the corresponding mono-cation (a), di-cation (b), and tri-cation (c). As shown in Figure 6, assignment of formal oxidation states indicates that the neutral diazene parent complex, the monocation, and the dication contain Fe(II)/Fe(II), Fe(II)/Fe(III), and Fe(III)/Fe(III) pairs, respectively.

The formation of the dication and the reversibility of the corresponding redox process may have particular significance for the requirements that have to be fulfilled in order to reduce N_2 in the coordination sphere of metal sulfur complexes under likewise mild conditions as in nitrogenases. Redrawing of the relevant core atoms of the dication (Fe, N_2H_2, S(thiolate) donors) shows that the dicationic diazene complex B and the diprotonated N_2 complex C are redox isomers (or valence tautomers) (Scheme V).

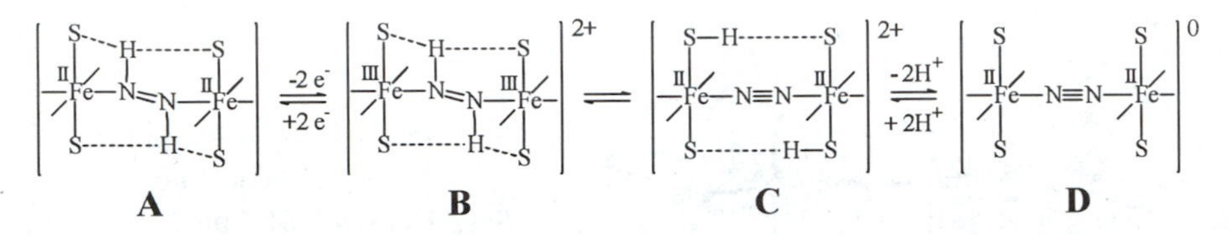

Scheme V. Redox Isomerism of the $[\mu\text{-}N_2H_2\{Fe(PPr_3)('S_4')\}_2]^{2+}$ Ion

It is stressed that species C and D are as yet hypothetical. However, it can be envisaged that an intramolecular electron transfer from the diazene ligand to the Fe(III) centers of B, cleavage of both NH and formation of two S–H bonds lead to the diprotonated N_2 complex C. It yields the neutral N_2 complex D upon loss of the two protons. The last step can be expected to take place reversibly and in a stereocontrolled way such that formation of S–H···S hydrogen bridges may even stabilize the N_2 complex.

Two alternatives can be discussed for the key reaction that transforms the species C into the neutral diazene complex A:

An equilibrium exists between B and C such that protonation of D is sufficient in order to reach the stage of B which can be reduced to give A as shown by the reversible redox waves b and a in Figure 6. This is considered the less likely alternative because the diprotonated N_2 complex C is an 18 valence electron (VE) complex.

Complexes with 18 VE are usually difficult to reduce because the entering electrons are required to be in antibonding orbitals. However, in species C all atoms that are necessary to form the neutral diazene complex A have already assumed their correct position, and, as has been shown above for carbonyl complexes, protonation can be expected to facilitate the reduction of the N_2 complex. Thus the decisive step might be the reduction of C to give (directly) A.

The distinction between these two alternatives requires unequivocal proof for the neutral N_2 complex D or $[\mu\text{-}N_2\{Fe(PPr_3)('S_4')\}_2]$, respectively. We did obtain evidence for such a species when $[\mu\text{-}N_2H_2\{Fe(PR_3)('S_4')\}_2]$ is oxidized by $[Cp_2Fe]PF_6$ at -78 °C. The resultant purple species, however, is extremely labile. It slowly releases N_2 at -78 °C, and very rapidly releases N_2 upon warming to yield a mixture of $[Fe(PPr_3)_2('S_4')]$ and other species that no longer contain nitrogen and that have not been characterized in detail (Sellmann, D.; Hennige, A., unpublished results.).

The importance of Brönsted basic thiolate donors and their protonation for a reduction of dinitrogen complexes such as D is underlined by the results of *Collman et al.* (*29*). The dinuclear $[Ru_2(DPB)(Im)_2]$ complex (DBP = diporphyrinato-biphenylentetraanion, Im = 1-tert-butyl-5-phenylimidazol) contains two cofacial [Ru(II)(porphyrinato]-fragments and can bind $\mu\text{-}N_2$, $\mu\text{-}N_2H_2$, $\mu\text{-}N_2H_4$ and (terminal) NH_3 ligands. Reversible $2e^-/2H^+$ redox reactions transform the N_2H_2, N_2H_4 and NH_3 complexes into each other. The N_2H_2 can be oxidized to give the N_2 complex plus 2 H^+, but this process is irreversible. There may be several reasons why just this step is irreversible (see above), but one reason could be that the [Ru(porphyrinato)] fragments have no Brönsted basic centers whose protonation could facilitate the uptake of electrons.

Concluding Discussion

Realization of structural principles but not the precise reduplication of structural details of the active centers is the basis for the synthesis of complexes intended to model the reactivity of enzymes and to yield perspectives for future competitive catalysts. With regard to the active centers of oxidoreductases such as nitrogenases and hydrogenases, surrounding transition metals by sulfur donors is a key structural principle. Structure-function relationships have been elucidated for such metal sulfur centers by investigation of $[Fe(CO)_2('S_4')]$, $[Fe(CO)('N_HS_4')]$ and $[Rh(H)(CO)('S_4')]$ in which preferably thiolate and thioether donors surround the metal centers.

Protonation or alkylation of the thiolate donors in the Fe complexes has been shown to cause a strong decrease in electron density at the Fe centers, and a drastic increase of redox potentials by some 500–700 mV per step of protonation (or alkylation) that makes the protonated (or alkylated) derivatives easier to reduce. In contrast to these remarkable electronic changes, the Fe–S distances remain invariant in the neutral parent complexes and their derivatives. This is traced back to the electronic flexibility of metal thiolate bonds, which can assume σ-donor, σ-donor-π-acceptor, or σ-donor-π-donor features.

Protonation of $[Rh(H)(CO)('buS_4')]$ leads to similar effects, and beyond those to release of H_2. The resultant coordinatively unsaturated species $[Rh(CO)('^{bu}S_4')]^+$ catalyzes the heterolytic cleavage of molecular dihydrogen, which is one of the two key hydrogenase reactions. Heterolysis of H_2 is achieved by the concerted attack of the Lewis acidic Rh center and the Brönsted basic thiolate donors upon the H_2 molecule. A mechanism for the catalysis is proposed and corroborated by isolation of the phosphine derivative $[Rh(PCy_3)('^{bu}S_4')]BF_4$. This species is also catalytically active and reversibly gives, upon reaction with H_2, $[Rh(H)(PCy_3)('^{bu}S_4'\text{-}H)]BF_4$, which is deprotonated by Na_2CO_3 to yield $[Rh(H)(PCy_3)('^{bu}S_4')]$.

The relationship between primary protonation and subsequent electron transfer is potentially essential for the reduction of N_2 in the coordination sphere of metal sulfur complexes. Cyclic voltammetry shows that the diazene complex $[\mu\text{-}N_2H_2\{Fe(PPr_3)('S_4')\}_2]$ can be reversibly oxidized to give the dication $[\mu\text{-}N_2H_2\{Fe(PPr_3)('S_4')\}_2]^{2+}$, which is a redox isomer of the diprotonated dinitrogen complex $[\mu\text{-}N_2\{Fe(PPr_3)('S_4'\text{-}H)_2]^{2+}$. This dication and the influence of protonation upon reduction potentials point out a way in which reduction of N_2 may be achieved under conditions as mild as by nitrogenases and close to the theoretical thermodynamic limit of E_o = -280 mV.

Acknowledgment

D.S. thanks his coworkers who are cited in the references for their dedication to obtain the results described here. Support of our research by the Deutsche Forschungsgemeinschaft, Fonds der Chemischen Industrie, Bundesministerium für Forschung und Technologie, and Fa. Degussa is gratefully acknowledged.

Literature Cited

(1) Pimentel Report, *Chem. Eng. News.* **1993**, *71* (May 31), 45.
(2) a) Ibers, J. A.; Holm, R. A. *Science* **1980**, *209,* 223. b) Wieghardt, K. *Nachr. Chem. Tech. Lab.* **1985,** *33,* 961.
(3) Burgess, B.K. *Chem. Rev.* **1990**, *90*, 1377.

(4) *Bioinorganic Catalysis*, Reedijk, J. Ed., Marcel Dekker, Inc., New York, **1993**.
(5) cf. Sellmann, D. In *Molybdenum Enzymes, Cofactors and Model Systems*; Stiefel, E. I.; Coucouvanis, D.; Newton, W. E., Eds.; *ACS Symposium Series*, Washington, DC, **1993,** *535*, 332.
(6) Kramarz, K. K.; Norton, J. R. *Progr. Inorg. Chem.* **1994**, *42*, 1.
(7) Sellmann, D.; Mahr, G.; Knoch, F.; Moll, M. *Inorg. Chim. Acta* **1994,** *224*, 59.
(8) Elschenbroich, C.; Salzer, A. *Organometallchemie*, Teubner, Stuttgart, **1988**, p. 280.
(9) a) Kolodziej, A. F. *Progr. Inorg. Chem.* **1994**, *41*, 493. b) Adams, M. W. W. Biochim. Biophys. Acta **1990,** *1020*, 115.
(10) Volbeda, A.; Charon, M. H.; Piras, C.; Hatchikian, E. C.; Frey, M.; Fontecilla-Camps, J. C. *J. Inorg. Biochem*. **1995**, *59,* 637.
(11) Stavropoulos, P.; Carrié, M.; Muetterties, M. C.; Holm, R. H. *J. Am. Chem. Soc.* **1990**, *112,* 5385; ibid. **1991**, *113*, 8485.
(12) Sellmann, D.; Käppler, J.; Moll, M. *J. Am. Chem. Soc.* **1993**, *115*, 1830.
(13) Kubas, G. J. *Acc. Chem. Res.* **1988**, *21*, 120.
(14) Eady, R. R.; Leigh, G. J. *JCS Dalton Trans.* **1994**, 2739.
(15) Allen, A. D.; Senoff, C.V. *JCS Chem. Comm.* **1965**, 621.
(16) Chatt, J.; Heath, G. A.; Richards, R. L. *JCS Chem. Comm.* **1972**, 1010.
(17) Hills, A.; Hughes, D. L.; Jimenez-Tenorio, M.; Leigh, G. J.; Rowleigh, A.T. *JCS Dalton Trans.* **1993**, 3041.
(18) Sellmann, D.; Weiss, W. *Angew. Chem. Int. Ed.* Engl. **1977**, *16*, 880.
(19) Laplaza, C. E.; Cummins, C. C. *Science* **1995**, *268*, 861.
(20) Weathers, B. J.; Schrauzer, G. N. *J. Am. Chem. Soc.* **1979**, *101*, 917.
(21) Denisov, N. T.; Efimov, O. N.; Shuvalova, N. I.; Shilova, A. K.; Shilov, A. E. *Zhur. Fiz. Khim.* **1970**, *44*, 2694.
(22) a) Henderson, R. A.; Leigh, G. J.; Pickett, C. J. *Adv. Inorg. Chem. Radiochem.* **1983**, *27*, 197. b) Leigh, G. J. *Eur. J. Biochem.* **1995**, *229*, 14.
(23) a) Kim J.; Rees, D. C. *Science* **1992**, *257*, 1653. b) Kim, J.; Rees, D. C. *Nature* **1992**, *360*, 553.
(24) Hardy, R. W. F. In *A Treatise on Dinitrogen Fixation*, Hardy, R. W. F.; Bottomley, F.; Burns, R. C. Eds.; Wiley, New York, **1979**, p. 515.
(25) a) Foner, S. N.; Hudson, R. L. *J. Chem. Phys.* **1978**, *68*, 3162. b) Pople, J. A. *J. Chem. Phys.* **1991**, *95*, 4385. c) Back, R. A. *Rev. Chem. Intermed.* **1984,** *5*, 293.
(26) Sellmann, D.; Soglowek, W.; Knoch, F.; Ritter, G.; Dengler, J. *Inorg. Chem.* **1992**, *31*, 3711.
(27) Sellmann, D.; Soglowek, W.; Knoch, F.; Moll, M. *Angew. Chem. Int. Ed.* Engl. **1989**, *28*, 1271.
(28) Sellmann, D.; Friedrich, H.; Knoch, F.; Moll, M. *Z. Naturforsch.* **1994**, *49b*, 660.
(29) Collman, J. P.; Hutchison, J. E.; Ennis, M. S.; Lopez, M. A.; Guilard, R. *J. Am.Chem. Soc.* **1992**, *114*, 8074.

Chapter 6

Catalytic Multielectron Reduction of Hydrazine to Ammonia and Acetylene to Ethylene with Clusters That Contain the MFe_3S_4 Cores (M = Mo, V)

Relevance to the Function of Nitrogenase

D. Coucouvanis, K. D. Demadis, S. M. Malinak, P. E. Mosier, M. A. Tyson, and L. J. Laughlin

Department of Chemistry, University of Michigan, Ann Arbor, MI 48109–1055

Clusters with the $[MFe_3S_4]^{n+}$ core, (M = Mo, n=3; M = V, n=2). are used as catalysts for the reduction of substrates relevant to nitrogenase function. Substrates such as hydrazine and acetylene, are catalytically reduced by $(NEt_4)_2[(Cl_4\text{-cat})(CH_3CN)MoFe_3S_4Cl_3]$, **I**, to ammonia and ethylene respectively, in the presence of added protons and reducing equivalents. Hydrazine also is catalytically reduced by the $(NEt_4)[(DMF)_3VFe_3S_4Cl_3]$ cubane under similar conditions. Catalysis in excess of 100 turnovers (for hydrazine reduction) and in excess of 15 turnovers (for acetylene reduction) has been observed over a period of 24 hours. Kinetic studies of the acetylene reduction reaction have been carried out. Considerable evidence has been amassed which directly implicates the Mo and V atoms as the primary catalytic sites. The reduction of hydrazine is accelerated in the presence of carboxylate ligands bound to the Mo atom in **I** and this effect is interpreted in terms of a proton delivery shuttle involving the carboxylate group and the substrate during reduction. The possible role of the homocitrate ligand in the nitrogen cofactor is analyzed in terms of these findings.

In recent single crystal X-ray structure determinations of the Fe-Mo protein of nitrogenase from *Clostridium pasteurianum* (*1,2*) and *Azotobacter vinelandii* (*2*) the structure of the Fe/Mo/S site has been revealed to near atomic resolution. This site catalyzes the biological reduction of dinitrogen (*3*) to ammonia under ambient temperature and pressure and its exact role and mechanism of action have been subject to intense interest. The Fe/Mo/S cluster contains two cuboidal subunits, Fe_4S_3 and $MoFe_3S_3$, bridged by three S^{2-} ions. The cluster is anchored to the protein matrix by a cysteinyl residue coordinated to an Fe atom at one end of the cluster and by an imidazole group from a histidine residue that is bound to the Mo atom at the other end of the cluster, Fig. 1. The Mo atom also is coordinated by a homocitrate molecule that serves as a bidentate chelate. Unusual structural features include the unprecedented trigonal planar coordination geometry for the six μ-S-bridged iron atoms, and the unusually short Fe-Fe distances across the two subclusters (2.5-2.6Å).

0097–6156/96/0653–0117$15.00/0

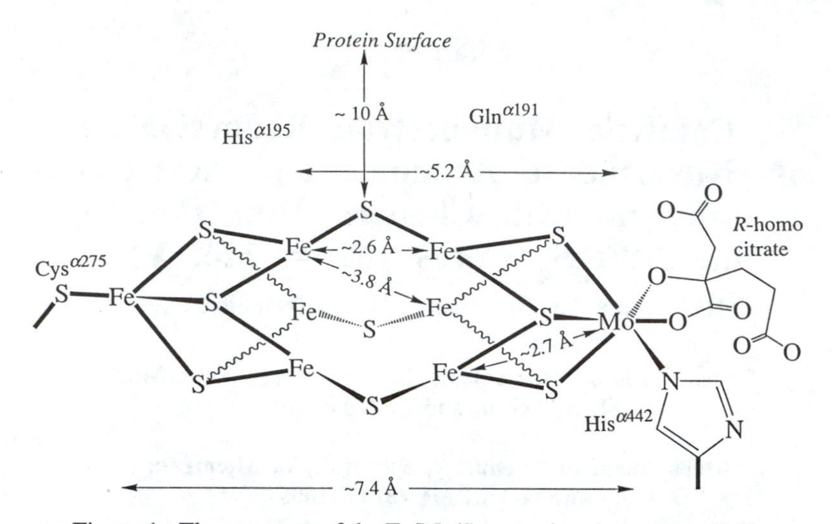

Figure 1. The structure of the Fe/Mo/S center in nitrogenase (*1,2*).

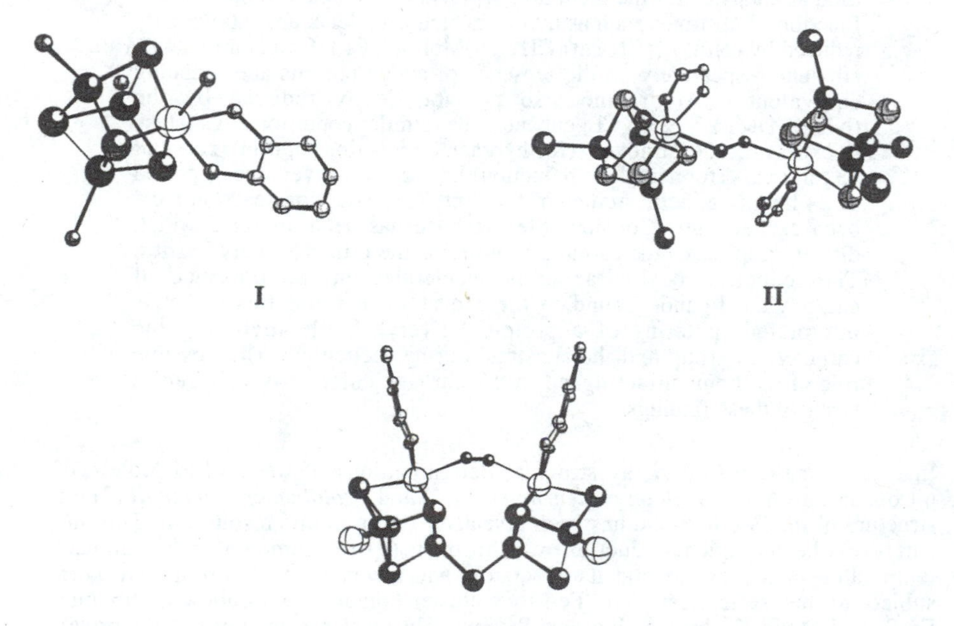

Figure 2. Molecular structures of : **I**, the $[(cat)MoFe_3S_4(Cl)_3L]^{n-}$ cubane clusters (L= a neutral, n=2 or anionic n=3 ligand, cat = a substituted catecholate dianion; (*15, 16*). **II**, the structure of the $\{[(Cl_4\text{-cat})MoFe_3S_4(Cl)_3]_2(\mu\text{-}N_2H_4)\}^{4-}$, singly -bridged double cubane (*20*). **III**, the structure of the $\{[(Cl_4\text{-cat})MoFe_3S_4(Cl)_2]_2(\mu\text{-}N_2H_4)(\mu\text{-}S)\}^{4-}$ (*21*).

The location of the Mo atom at the periphery of the $MoFe_7S_{8-9}$ cluster and the apparent coordinative saturation of the Mo atom have been interpreted as indications that the Mo atom may not be directly involved in N_2 fixation.

The revealed structural features of the nitrogenase Fe/Mo/S center have not led to a clear understanding of the mode of substrate activation and reduction and many questions regarding the catalytic function of nitrogenase remain unanswered. One of these questions concerns the role of the Mo atom (and of the V atom in alternate nitrogenase) in the function of the "M" centers in nitrogenase and may be explored by reactivity studies on appropriate Fe/M/S model complexes. Numerous models for the Fe/Mo/S center in nitrogenase have been proposed over the years (*4,5*) and in each case their constitution and structure was based on the spectroscopic and analytical information available at the time of their intellectual conception.

Until recently the Fe/M/S synthetic analog clusters were not known to mediate chemistry related to dinitrogen reduction (*6*). Nevertheless fundamentally important chemistry has been developed in this area mainly with molybdenum and tungsten complexes of limited *direct* biological significance. Features of this chemistry include: a) the protonation-assisted reduction of dinitrogen complexes of low-valent molybdenum and tungsten to the hydrazido(-2) state (*7*); b) the catalytic reduction of N_2 to NH_3 with molybdocyanide catalysts (*8*); c) the proposed bimetallic activation of N-N bonds in protic environments and the reduction of N_2 to the hydrazine level (*9*); d) the catalytic reduction of hydrazine to ammonia on pentamethylcyclopentadienyl-trimethyl tungsten hydrazine complexes (*10*); and e) the catalytic reduction and disproportionation of hydrazine on certain Mo(IV) sulfur-ligated complexes(*11*).

Single metal-atom site mechanisms have been proposed for these reactions with pathways involving: a) the stoichiometric reduction of N_2 to ammonia with $M{=}NNH_2$ as intermediates (*7*); b) the side-on (η^2) metal binding of N_2 followed by protonation and disproportionation; (*8*) and c) reduction of N_2 by one-electron one-proton steps where an intermediate with coordinated hydrazine is proposed to lie on the pathway to ammonia. (*10*) For some of the model reactions, (*5,9,11,12*) binuclear activation mechanisms also have been proposed. The latter are not as firmly supported as the single metal-atom mechanisms.

The reductions of nitrogenase substrates by electrochemically reduced, synthetic Fe/Mo/S (*4,5*) and Fe/S clusters (*13*) {such as $[Mo_2Fe_6S_8(SPh)_9]^{3-}$ and $[Fe_4S_4(SPh)_4]^{2-}$} have been reported. These reactions (*14*) are mostly non-catalytic, heterogeneous, and not well defined in terms of the actual nature of the reactant clusters.

The molecules that we have chosen to examine in substrate reduction studies contain as a commom structural feature the MFe_3S_4 "cubane" unit (M = Mo, V) and their chemistry has been reviewed. (*4,5*) They are: a) the $([(L)MoFe_3S_4Cl_3(CH_3CN)]^{n-}$ single cubane clusters, **SC**, (L= Cl_4-catecholate (*15*), n = 2, **Ia** ; L = citrate (*16,17*) , citramalate (*17*), methyl-iminodiacetate (*17*), nitrilo-triacetate (*18*), thio-diglycolate (*19*), n=2,3); $[MoFe_3S_4Cl_3(\text{thiolactate})]_2^{4-}$, (*19*) $(MoFe_3S_4Cl_4)_2\mu\text{-oxalate})]^{4-}$, (*17*), $\mathbf{Ib_n}$); b) the singly-bridged double-cubane, (*20*) $[(Cl_4\text{-cat})MoFe_3S_4Cl_3]_2(\mu\text{-}NH_2NH_2)]^{4-}$ clusters, **SBDC**, **II** ; and c) the doubly-bridged double-cubanes, (*21*) , $\{[(Cl_4\text{-cat})MoFe_3S_4Cl]_2(\mu\text{-L})(\mu\text{-S})]\}^{4-}$, **DBDC**, **III** (μ-L = NH_2NH_2 and CN^-). The MFe_3S_4 cores in these model complexes, Fig. 2,3, structurally are quite similar to the MFe_3S_3 cuboidal subunits in the nitrogenase cofactor and show nearly identical first and second coordination spheres around the Mo atom.

Catalytic Reduction of Hydrazine

Three different processes were investigated (*22*). a) stoichiometric reactions with cobaltocene as a reducing agent and 2,6-lutidine hydrochloride, Lut·HCl, as a source of protons; b) *catalytic reductions* with various $N_2H_4/MoFe_3S_4$ cubane ratios in the presence of cobaltocene and Lut·HCl, (eq.1);

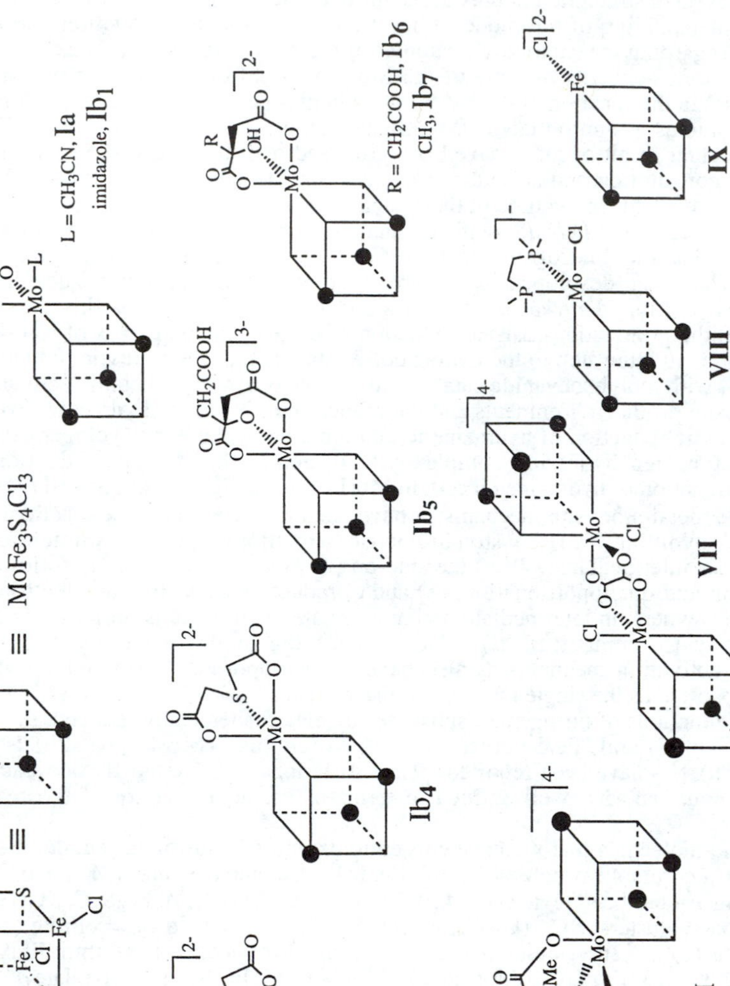

Figure 3. Schematic structures of the [(L)$MoFe_3S_4$ $(Cl)_3$] cubanes used in this study with L = carboxylate group containing ligands.

$$N_2H_4 + 2e^- + 2H^+ \rightarrow 2NH_3 \quad \text{(eq.1)}$$

and c) *catalytic disproportionation* reactions where N_2H_4 serves as both a reducing agent and a proton source (eq.2)

$$3N_2H_4 \rightarrow 4NH_3 + N_2 \quad \text{(eq.2)}$$

It was established that ammonia did not form in appreciable amounts from N_2H_4 in the presence of $Co(Cp)_2$ and Lut·HCl alone. Also, it was demonstrated that CH_3CN (a poor substrate of nitrogenase (*23*)) used as a solvent did not undergo reduction under the reaction conditions. Of the Fe/Mo/S clusters studied, the doubly-bridged double-cubane, **III**, (Fig.2) and the $[(Cl_4\text{-cat})Mo(O)(\mu\text{-}S)_2FeCl_2]^{2-}$ dimer (*24*) (one possible oxidative degradation product of the $[(Cl_4\text{-cat})MoFe_3S_4(Cl)_3{\cdot}CH_3CN]^{2-}$ cubane) did not promote the reduction or disproportionation of hydrazine to ammonia. Very slow activity was detected when **II** (*20*) was used as a catalyst. By far the most active catalysts in these reactions were the $[(L)MoFe_3S_4(Cl)_3]^{2-}$ single cubanes (*15*), **Ia** (*16*), and $\mathbf{Ib_n}$ (*17-19*).

The use of **Ia** in all of the reactions outlined above at ambient temperature resulted in the catalytic disproportionation and the stoichiometric or catalytic reduction of N_2H_4 to NH_3 (Tables I,II). In all *catalytic disproportionation* reactions (in the absence of $Co(Cp)_2$ or Lut·HCl) the concentration of **Ia** after 12h of reaction time was found to be nearly the same as it was in the begining of the reaction (*25*), the production of ammonia was reproducible, and the yields for repetitive experiments fell within a relatively narrow range.

The *catalytic reduction* of N_2H_4 with $Co(Cp)_2$ as a source of electrons and Lut·HCl as a source of protons stops before all of the hydrazine is reduced to ammonia. This is attributed to the precipitation of the **Ia** anion by counterions that are generated as the reaction proceeds (*26*) including: $LutH^+$, $[Co(Cp)_2]^+$, NH_4^+ and $N_2H_5^+$. The onset of precipitation within ~0.5h is not exactly predictable and, as a result, the yields of ammonia between successive runs show some variation. In DMF solution, the onset of precipitation is longer (~12h). However, the substitution of the DMF ligand by hydrazine in the coordination sphere of the molybdenum atom is not as facile as the substitution of CH_3CN (*27*) and consequently the catalytic reduction is slower.

Qualitative comparative studies show that the rates of ammonia formation with **II** as a catalyst were much slower than those observed with **I** while **III** was totally ineffective as a catalyst. These results indicate that *the hydrazine molecule is activated by coordination to only one $MoFe_3S_4$ cubane and the addition of an additional equivalent of cubane (which is known (20) to give **II**) inhibits the reduction.* Moreover, the results suggest that the availability of an uncoordinated NH_2 group (and the lone pair of electrons needed for protonation) is essential for the reduction of N_2H_4 to ammonia. The lack of reactivity of **III** almost certainly derives from the robust nature of the double cubane (*21*) which precludes rupture of the N_2H_4 bridge and consequently prevents the generation of an available lone pair. The reactivity of **II** (Fig. 2) may well be attributed to **I** that very likely exists in small amounts in equilibrium with **II**. Indeed, addition of N_2H_4 to a solution of **II** leads to the formation of the $[(Cl_4\text{-cat})MoFe_3S_4(Cl)_3(N_2H_4)]^{2-}$ cubane. Additional evidence that the interaction of hydrazine with a single cubane is necessary *and sufficient* for catalytic reduction is available in studies with phenyl hydrazine, $PhHNNH_2$. The replacement of the CH_3CN molecule in **I** by $PhHNNH_2$, a known substitution reaction (*15*), occurs readily. The product, **IV**, which for steric reasons does not interact further with another cubane molecule to form a bridged double cubane similar to **II**, has been structurally characterized(*28*). The stoichiometric or catalytic reductions of the terminally coordinated $PhHNNH_2$ in **IV** proceed with the formation of NH_3 and aniline(*22*).

Table I. Production (%) of NH_3[a] from the Catalytic Reduction of NH_2NH_2[b] by Various $MoFe_3$-Cubanes Using $Co(Cp)_2$ as the Reducing Agent and Lut·HCl as the Proton Source. The $[NH_2NH_2]$:[Catalyst] Ratio was 100:1.[c]

Cluster Catalyst[d]	No. of Trials	5 min	30 min	1 h	12h
$[MoFe_3S_4Cl_3(Cl_4\text{-cat})(CH_3CN)]^{2-}$ (**Ia**)	3	30	34	38	61
$[MoFe_3S_4Cl_3(Cl_4\text{-cat})(im)]^{2-}$ (**Ib$_1$**)[e]	1	31	34	35	48
$[MoFe_3S_4Cl_3(mida)]^{2-}$ (**Ib$_2$**)	2	55	64	70	79
$[MoFe_3S_4Cl_3(Hnta)]^{2-}$ (**Ib$_3$**)	3	48	53	59	65
$[MoFe_3S_4Cl_3(tdga)]^{2-}$ (**Ib$_4$**)	3	38	45	47	65
$[MoFe_3S_4Cl_3(Hcit)]^{3-}$ (**Ib$_5$**)	4	83	92	96	98
$[MoFe_3S_4Cl_3(H_2cit]^{2-}$ (**Ib$_6$**)	3	80	86	94	95
$[MoFe_3S_4Cl_3(Hcmal)]^{2-}$ (**Ib$_7$**)	3	62	66	68	78
$[MoFe_3S_4Cl_3(tla)]_2^{4-}$ (**VI**)[f]	2	58	58	66	71
$[(MoFe_3S_4Cl_4)_2(\mu\text{-ox})]^{4-}$ (**VII**)[f]	2	63	71	77	85
$[MoFe_3S_4Cl_4(dmpe)]^{-}$ (**VIII**)	1	12	71	77	85
$(Fe_4S_4Cl_4)^{2-}$ (**IX**)	2	0	0	0	7

[a]The experiments were performed at ambient temperature in CH_3CN solution. [b]Periodically, N_2H_4 quantification with the *p*-(dimethylamino)-benzaldehyde method (*30*) confirmed complete N atom balance. In all cases, there was a balance of 100 ± 5%. [c]Sampling of the reaction mixture and treatment of the samples were performed as described before(22). [d]Reactions of catalyst with $Co(Cp)_2$ and Lut·HCl (absence of N_2H_4), or of N_2H_4 with $Co(Cp)_2$ and Lut·HCl (absence of catalyst) gave NH_3 within background limits (< 3 %). [e]Abbreviations used: im = imidazolate, H_2cit = citrate dianion, Hcmal = citramalate, mida = methyl-iminodiacetate, Hnta = nitrilo-triacetate dianion, tdga = thio-diglycolate, tla = thio-lactate, ox = oxalate, dmpe = dimethylphosphinoethane), [f]For clusters **VI** and **VII** two molecules of N_2H_4 per double cubane (or one N_2H_4 / Mo) were used in the experiments.

Table II. Catalytic disproportionation of N_2H_4, ($3N_2H_4 \rightarrow 4NH_3 + N_2$) in CH_3CN by$[(Cl_4\text{-cat})MoFe_3S_4Cl_3(CH_3CN)]^{2-}$ at ambient temperature

N_2H_4/catalyst[a]	Time (min)	NH_3[b]	% reduction[c]
10	30	6.2	46
20	30	7.1	27
40	30	11.0	21

a) Cubane concentration 2 x 10^{-4} M. b) Ammonia yields reported as equivalents of catalyst were quantified by the indophenol method. c) Maximum yield is 4/3 NH_3 per N_2H_4

Importance of the Mo Atom

With the effectiveness of **I** in the catalytic reduction of hydrazine to NH_3 established, the question regarding the identity of the metal site involved in catalysis remained to be answered. Toward this goal the reduction of N_2H_4 was attempted with the $[(Cl_4\text{-cat})MoFe_3S_4(Cl)_3(PEt_3)]^{2-}$, **V** (*29*), and $[(L)MoFe_3S_4(Cl)_3]^{3-}$, L = citrate (*16*), citramalate (*17*), methyl-iminodiacetate (*17*), nitrilo-triacetate (*18*), thio-diglycolate (*19*), (n =2,3); $[MoFe_3S_4Cl_3(\text{thiolactate})]_2^{4-}$ (*19*), $[(MoFe_3S_4Cl_4)_2(\mu\text{-oxalate})]^{4-}$ (*17*), **Ib$_n$**, clusters as potential catalysts. The PEt_3 ligand in **V** is known (*29*) to be substitutionally inert as are the carboxylate ligands in **Ib$_n$**. The results show that under identical conditions the hydrazine *disproportionation* reactions (reactions that do not need the addition of external H^+ and are not complicated by catalyst precipitation) catalyzed with either **V** or **Ib$_n$** are very much slower than the same reactions catalyzed with **Ia**. In the *catalytic* reactions, however, Table I, (with Lut·HCl added as a source of H^+) the **Ib$_n$** are better catalysts than **Ia.** The Mo-bound tridentate citrate ligand, in the citrate-cubane catalyst **Ib$_5$**, (the best of the **Ib$_n$** catalysts) rather than blocking access of hydrazine (and inhibiting subsequent reduction) promotes the catalytic process. One can speculate that protonation of the citrate ligand causes dissociation of one of the oxygen donor atoms and allows the N_2H_4 molecule access to the Mo coordination sphere. Further the protonated carboxylate group can deliver the proton following substrate reduction. *In general it appears that the availability of carboxylate ligands, which can undergo protonation, may facilitate the transfer of protons to the hydrazine molecule as the reduction takes place. The results suggest that the homocitrate ligand bound to the Mo atom in the nitrogenase cofactor may play a similar, proton transfer role.*

The catalytic *disproportionation* of N_2H_4, Table II, is not observed when substitution-inert ligands are coordinated to the Mo atom in the $[(Cl_4\text{-cat})MoFe_3S_4Cl_3(L)]^{n-}$ cubane catalyst (L = CN^-, Cl^-, CH_3NH_2). In contrast the *catalytic reduction* by the same cubanes is significantly affected but not completely eliminated by the substitutionally inert ligands.

The results show, Table III, that in the *disproportionation* reaction, where no external source of protons is added, the Mo atom is effectively blocked and the coordination/reduction of N_2H_4 cannot take place to a significant extent. The *catalytic reduction* of N_2H_4, however, is possible with the same clusters because now the externally supplied acidic protons may be added to the O atoms of the Mo-bound Cl_4-catecholate ligand and in the process generate a coordination site on the Mo atom, thus allowing the N_2H_4 molecule to coordinate and undergo reduction. The site of N_2H_4 reduction also is the site of H^+ reduction. Indeed N_2H_4 is an inhibitor of H^+ reduction, Table IV. This inhibition is less pronounced with substituted hydrazines that are bound to the Mo atom with the R substituted N atom. Steric interactions of the R group with the catecholate ligand promote dissociation of the $RNHNH_2$ ligand and allow for the reduction of H^+, which undoubtedly hydrogen-bonds to the Mo-bound O atoms of the catecholate ligand. The importance of the Mo atom, rather than the Fe atoms, in catalysis is supported further by the observation that the $[Fe_4S_4Cl_4]^{2-}$ cluster (*13*) is not active as a N_2H_4 reduction or disproportionation catalyst over a period of 12h. After extended periods of time (~ 36h) some hydrazine is converted to NH_3. However, the electronic spectrum of the solution at this stage differs from the expected spectrum of the $[Fe_4S_4Cl_4]^{2-}$ cluster and very likely the reaction is catalyzed by a different, as yet unidentified, species. The structural similarity of the $MoFe_3S_4$ clusters (Figs. 2 and 3) to the Fe/Mo/S center in nitrogenase, Fig.1, and the competence of the former in the catalytic reduction of N_2H_4, a nitrogenase substrate, (*31*) raise the possibility that the activation and reduction of dinitrogen by the Fe/Mo/S center in nitrogenase takes place on the Mo atom by a single metal-site mechanism.

Table III. Catalysis of hydrazine disproportionation and reduction reactions with the $[(Cl_4\text{-cat})MoFe_3S_4\ Cl_3(L)]^{n-}$ cubanes

Catalyst (L)[a]	N_2H_4 /catalyst	Reductions[b,c]	Disproportion[b,c]
CH_3CN	10	14.6(73%)	8.6(64%)
CN^-	10	8.1(40%)	1.1(8%)
CH_3NH_2	10	8.6(43%)	1.2(9%)
Cl^-	10	7.2(36%)	2.4(18%)

a) Concentration of catalyst was 2 x 10^{-4} M. b) All reactions were analyzed for NH_3 after 60 min. c) Ammonia quantified by the indophenol method.

Table IV. Proton reduction/hydrogen evolution catalyzed by $[(Cl_4\text{-cat})MoFe_3S_4Cl_3(CH_3CN)]^{2-}$ (I) in CH_3CN in the presence of $Co(Cp)_2$ and Lut· HCl as sources of e^- and H^+, respectively

Catalyst	NH_2NHR	$RNHNH_2$/ catalyst	NH_3[a]	H_2[b]
no	no	0	0	1.0 (5%)
(Ia)	no	0	0	~20 (100%)
(Ia)	H	10	19 (95%)	1.0(5%)
(Ia)	CH_3[c]	10	10 (50%)	10.0(50%)

a) Yields reported as equivalents of ammonia per equivalent of **I**. b) Hydrogen yields are reported as equivalents per equivalent of I and were quantified by gas chromatography. c) The crystal structure of the $(CH_3)HNNH_2$-**SC** cluster (*34*) has been determined and shows the methyl hydrazine ligand coordinated to the Mo atom by the CH_3-substituted (more basic) N atom.

Inhibition Reactions with PEt_3 and CO.

To further verify that the Mo site is involved in hydrazine binding and reduction we performed two sets of inhibition experiments, which are summarized in Table V.
(a) Inhibition by PEt_3. Triethylphosphine is known to bind exclusively and irreversibly to the Mo site of the $[MoFe_3S_4]^{3+}$ cubanes.(*32*) Blocking the sixth coordination site on the Mo with PEt_3 greatly reduces the NH_3 yields. The low (above background) levels of NH_3 production can be explained by the possible loss of some of the bound PEt_3 and also by the protonation and dissociation of one of the Cl_4-cat ligand oxygen donors. Either of these events will generate a vacant site and allow coordination of hydrazine. Large excess of PEt_3 suppresses hydrazine reduction (see experiment with compound $\mathbf{Ib_7}$).
(b) Inhibition by CO. Carbon monoxide is known to inhibit nitrogenase reactivity(*33*). In addition, CO can bind to the $MoFe_3S_4$ cubanes *only* at their reduced (2+) oxidation level(*32b*). Nevertheless, the precise metal site of coordination (Fe, Mo, or both) is not clear. A hydrazine reduction experiment using cluster **1a** as a catalyst was performed under a CO atmosphere. The ammonia yields, although slightly above background, were much lower than the corresponding yields under N_2 atmosphere. The above result suggests that the $[MoFe_3S_4]^{3+}$ core goes through a reduced (2+) oxidation level during the catalytic proccess, at which it is attacked and irreversibly inactivated by CO (very likely at the Mo site). The above results are graphically depicted in Fig 4.

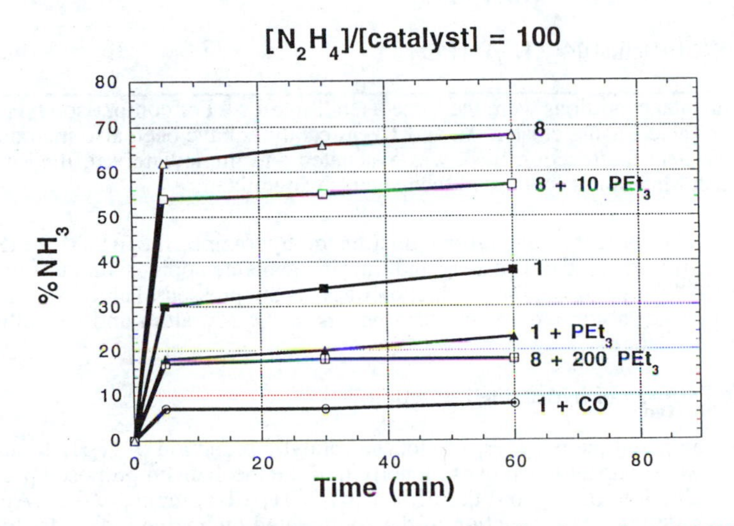

Figure 4. Catalysis and inhibition of N_2H_4 reduction by selected $MoFe_3S_4$ cubanes.

Table V. Production (%) of NH_3 by the Catalytic Reduction of NH_2NH_2 by Various $MoFe_3$-carboxylate Cubanes[a] in the Presence of Inhibitors PEt_3 or CO. Experimental Conditions Were the Same as in Table I. The $[NH_2NH_2]$:[Catalyst] Ratio was 100:1.

Cluster Catalyst	No. of Trials	5 min	30 min	1 h	12h
$[MoFe_3S_4Cl_3(Cl_4\text{-cat})(CH_3CN)]^{2-}$ (**Ia**)[b]	3	30	34	38	61
$[MoFe_3S_4Cl_3(Cl_4\text{-cat})(CH_3CN)]^{2-}$ (**Ia**)+ PEt_3	2	18	20	23	37
$[MoFe_3S_4Cl_3(Cl_4\text{-cat})(CH_3CN)]^{2-}$ (**Ia**) + CO[c]	2	7	7	8	9
$[MoFe_3S_4Cl_3(Hcmal)]^{2-}$ (**Ib$_7$**)[b]	3	62	66	68	78
$[MoFe_3S_4Cl_3(Hcmal)]^{2-}$ (**Ib$_7$**) +PEt_3 (1:10)	1	54	55	57	66
$[MoFe_3S_4Cl_3(Hcmal)]^{2-}$ (**Ib$_7$**) +PEt_3 (1:200)	1	17	18	18	19

[a] Experimental conditions were the same as in Table I. [b] For comparison NH_3 yields (%) from Table I using clusters **1a** and **1b$_7$** as catalysts have been also included. [c] In this experiment the reaction flask was evacuated and immediately refilled with CO (Johnson & Mathey) as soon as all the reagents were mixed.

In parallel with the catalytic reduction of hydrazine, the $[(L)MoFe_3S_4Cl_3]^{n-}$ cubanes in the presence of $Co(Cp)_2$ and Lut·HCl generate copious amounts of H_2. In the presence of hydrazine, the production of H_2 is suppressed (Table IV). The data show that the catalytic reduction of H^+ occurs at the Mo atom and is inhibited by hydrazine reduction.

Possible Pathway of hydrazine reduction

A simple proposed pathway, Fig. 5, for the catalytic reduction of N_2H_4 to ammonia by the Fe_3MoS_4 cluster is similar to a portion of the mechanism proposed previously for the reduction of N_2 on the $(Me_5Cp)W(CH_3)_3$ fragment (*10*). An initial protonation step prior to reduction of the coordinated hydrazine molecule, Fig. 5, is supported by the synthesis and reactivity of the $N_2H_5^+$ **SC** cluster. The latter has been isolated and characterized and upon addition of $Co(Cp)_2$ (in the absence of Lut·HCl) affords ammonia.

Catalysis of Hydrazine Reduction with the V/Fe/S clusters.

The catalytic behavior of synthetic Fe/V/S clusters that structurally resemble the Fe/V/S site of the alternate V-nitrogenase (*35,36*) was investigated (*37*) with the $[(L)(L')(L'')VFe_3S_4Cl_3]^{n-}$ clusters (*37,38*) (L, L', L" = DMF, **X**; Fig. 6; L = PEt_3, L', L" = DMF, **XI**; L, L' = 2,2'-bipyridyl, L" = DMF, **XII**; n = 1). These clusters which contain the $[VFe_3S_4]^{2+}$ cuboidal core, also are effective catalysts in the reduction of hydrazine (a nitrogenase substrate) to ammonia in the presence of cobaltocene and 2,6-lutidine hydrochloride as sources of electrons and protons, respectively, Table VI. The catalytic reduction of phenylhydrazine to ammonia and aniline also is effected by **X, XI,** and **XII**.

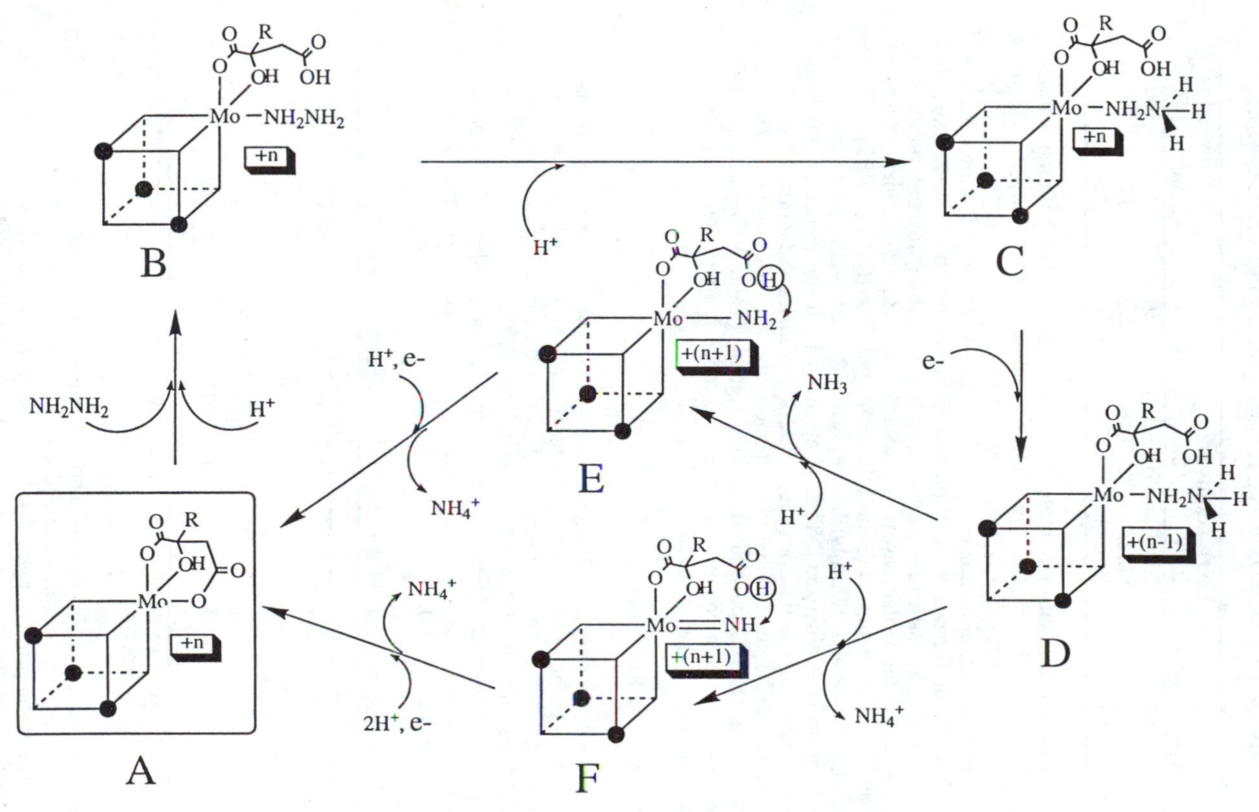

Figure 5. A proposed mechanism for the catalytic reduction of N_2H_4 to NH_3 by the $[(Citr)MoFe_3S_4(Cl)_3]^{3-}$ (**Ib$_5$**, **Ib$_6$**) cubanes with emphasis on a possible H^+ delivery role of the carboxylate function. The intra- and inter-molecular electron transfer steps arbitrarily are depicted by changes in the formal oxidation level of the Mo atom.

Table VI. Production (%) of NH_3 from the Catalytic Reduction of N_2H_4 by the (L)(L')(L")$VFe_3S_4(Cl)_3$-Cubanes Using $Co(Cp)_2$ as a Reducing Agent and Lut·HCl as a Source of Protons. **X**, L=L'=L"=DMF; **XI**, L=L'=DMF, L"=PEt_3; **XII**, L=DMF, L', L" = bipy).

Catalyst	N_2H_4/ Catalyst	NH_3 (Equiv) 0.5h	2.0h	$NH_{3\,(max)}$	% Conversion 0.5h	2.0h
X	10	15.2(3)	20.0(3)	20	76	100
XI	10	5.6(2)	9.4(2)	20	28	47
XII	10	---	3.4(2)	20	trace	17

The ability of hydrazine-like substrate molecules to interact directly with the V atom, has been demonstrated by the synthesis and characterization of the $(Me_4N)[(PhHNNH_2)(bpy)VFe_3S_4Cl_3]$ single cubane (*39*).

The catalytic reduction of hydrazine by **X**, **XI**, and **XII** (Table VI) shows that the V-coordinated terminal ligands have a profound effect on the relative rates of hydrazine reduction. Specifically, as the number of labile solvent molecules coordinated to the V atom decreases, the relative rate of hydrazine reduction decreases (Table VI). This behavior reaches the limit with the $[(HBpz_3)VFe_3S_4Cl_3]^{2-}$ cubane (L, L', L" = hydrotris(pyrazolyl)borate, **XIII**; n = 2), where all coordination sites on the V atom are "blocked." The latter, for which the structure has been determined (*37*), shows no catalytic or stoichiometric hydrazine reduction.

Unlike clusters **X**, **XI** and **XII** that do not show significant changes in the 1-/2- reduction potential as DMF is substituted by other ligands (Table VII), **XIII** shows a less negative 2-/3- reduction potential and this difference in reduction potential may indeed be the reason for its lack of catalytic activity.

Catalysis of Acetylene Reduction

Previous studies on the abiological reduction of C_2H_2 (*40-49*) suffer from major drawbacks including incomplete catalyst identification, (*38,39,41*) the use of mononuclear or binuclear Mo complexes of little relevance to the nitrogenase problem, (*42-48*) and sub-stoichiometric, non-catalytic, substrate reduction (*49*). Our studies have provided substantial evidence that identifies the $[MoFe_3S_4]^{3+}$ cubanes as catalysts in the reduction of acetylene to ethylene and implicates both the Mo and Fe sites in acetylene reduction (*50*).

Table VII. Electrochemical data of the substituted $[VFe_3S_4]^{2+}$ cores

Cluster	Oxidation[a]	Reduction	Ref.
[**X**]	+0.17(qr)[b,c]	-1.20(irr)	38,37
[**XI**]	+0.13(qr)	-1.20(irr)	37
[**XII**]	+0.30(qr)	-1.22(qr)	37
[**XIII**]	0.00(qr)	-0.68(irr)	37
[**X**]-Br	+0.12(qr)	-1.18(irr)	37
[**X**]-I	+0.14(qr)	-1.12(irr)	37

a) vs Ag/AgCl in CH_3CN solution b) qr = quasireversible process, irr = irreversible process. c) E_{pc} or E_{pa} is reported.

Catalytic reductions of C_2H_2 to C_2H_4 and traces of C_2H_6 were carried out at 20°C using **Ia** as the catalyst and cobaltocene and 2,6-lutidine hydrochloride, Lut·HCl, as sources of electrons and protons, respectively. The initial-rate method,(*51*) whereby [catalyst] << [substrate] was used to obtain reaction velocities, v_o(M/min), at less than 5% substrate consumption,Table VIII. The data obey saturation kinetics, as found for enzyme catalysis, and reaction rates remain constant for C_2H_2:cubane ratios < 30:1, indicating zero-order substrate dependence. A double reciprocal plot, v_o^{-1} vs $[C_2H_2]^{-1}$, is linear at optimum substrate concentration from which K_m = 17.9 mM and V_{max} =

1.1 x 10^{-4} M/min are calculated. Catalyst **Ia** reduces acetylene with a turnover number of 0.11 mol C_2H_2/mol catalyst/min, which is approximately 0.08% the enzymatic rate of acetylene reduction by nitrogenase.

The reaction also shows first-order dependence on proton concentration and zero-order dependence on reductant concentration A study of the reaction at five temperatures indicates a moderate activation energy [E_{act} = 9(1) kcalmol^{-1}] but a large entropy of activation [$\Delta S^{\ddagger}$ = -32(2) cal K^{-1} mol^{-1}] which extrapolates to a significant Gibbs free energy of activation [$\Delta G^{\ddagger}$ = 19(1)].

The characteristic EPR spectrum of **Ia** (S = 3/2) remains quantitatively unchanged as the reaction progresses and indicates that the $[MoFe_3S_4]^{3+}$ cubane core remains structurally intact during the reaction.

The stereochemistry of addition across the substrate triple bond was investigated using gaseous FT-IR spectroscopy using deuterated acetylene (C_2D_2) as a substrate. The formation of solely *cis*-1,2-$C_2D_2H_2$ was revealed by the characteristic IR absorption, n_7 = 842 cm^{-1}.(*52*) This result is consistent with a transition state intermediate in which substrate acetylene is side-on bonded to the Mo atom of **Ia**. Nitrogenase catalyzed reduction of acetylene also proceeds with a high degree of stereoselectivity resulting in *cis* addition across the triple bond(*53*). In attempts to block the Mo site from catalysis, it was discovered that the Fe atoms on cubane **Ia** also are competent in substrate reduction albeit at a slower rate (Table VIII). Typically, all the Fe sites combined completed 2-3 turnovers in a 24 hr period, while the entire cluster, with the Mo site free for substrate binding, was capable of at least 16 turnovers in the same time. The use of $[NBu_4]_2[Fe_4S_4Cl_4]$ as the catalyst effected no appreciable increase in reaction rate or overall yield compared to cubane **Ia** with the Mo atom "blocked" from substrate binding/reduction. The rate of acetylene reduction varies considerably depending on the reaction site, Mo *vs* Fe, the former being decidedly more competent in catalyzing the reaction.

Table VIII. Initial Velocities (v_0) for the Reduction of Acetylene to Ethylene Catalysed by Various Cubanes [a]

Cubane Catalyst	v_o (M/min) [b] (x 10^5)	v_o normalized to Cubane **I**.
Ia	6.6(3)	1.0
$[(\eta^3\text{-citrate})MoFe_3S_4Cl_3]^{3-}$	6.4(2)	0.97
$[Fe_4S_4Cl_4]^{2-}$	1.3(3)	0.20
$[(Cl_4\text{-cat})MoFe_3S_4Cl_3(CN)]^{3-}$	1.6(1)	0.24
$[(CO)_3MoFe_3S_4Cl_3]^{3-}$	1.5(2)	0.23
$[(CO)_3MoFe_3S_4R_3]^{3-}$ R = *p*-Cl-SPh	1.1(3)	0.17
Ia + PEt_3 (1:2) [c]	1.3(2)	0.20
Blank (no catalyst) [d]	0.011	0.002

[a] Experiments are performed in DMF at 20°C using $CoCp_2$ and Lut.HCl as sources of electrons and protons, respectively. The initial substrate:catalyst ratio used is 20:1.

[b] Initial velocity, v_o, is obtained as the slope from $[C_2H_4]$ (M) *vs* t(min), which obeys a straight-line relationship during the initial stages of the reaction.

[c] Acetylene reduction performed using cubane **I** as the catalyst in the presence of PEt_3 such that Mo:PEt_3 = 1:2.

[d] Typical acetylene reduction conditions except for the absence of catalyst.

The involvement of the Mo atom in acetylene reduction has been confirmed by modification of the coordination sphere around the heterometal. The rate of ethylene production varies considerably depending on the availability of Mo coordination sites. Replacing the bidentate Cl_4-cat ligand in **Ia** with η^3-citrate(*16*) effects a slight reduction of reaction rate. This is in contrast with the reduction of hydrazine with the same catalyst (vide supra). A significant decrease in the reaction rate is observed when CN^- is used as the terminal ligand in place of CH_3CN or DMF (Figure 6). The non-labile cyanide ligand precludes binding of substrate acetylene to the Mo atom thereby restricting substrate reduction to the Fe sites and slowing ethylene production to a rate comparable with $[Fe_4S_4Cl_4]^{2-}$ cubane. A similar result is obtained when the $[(CO)_3MoFe_3S_4R_3]^{3-}$ (R = Cl, *p*-Cl-SPh) cubanes (*54*) are used as catalysts.

Catalytic C_2H_2 reduction by cubane **Ia** is effectively prevented if the reaction is carried out under a saturating CO atmosphere and the overall yield of C_2H_4 and C_2H_6 is considerably diminished to approximately that of the background level (no catalyst present). This suggests that CO precludes C_2H_2 binding to the Mo atom of **Ia**, probably by virtue of its own binding to the heterometal. Interestingly, a saturated CO atmosphere also appears to block substrate reduction at the secondary (Fe) catalytic sites on **Ia**. An inhibition study of the Mo and Fe substrate binding sites using PEt_3 revealed that both centers are adversely affected by the phosphine. The affinity of PEt_3 for the heterometal in cubane **I** is well documented(*29*) and our work now shows that at sufficiently high concentrations PEt_3 also serves to block the Fe sites on **I** from reducing acetylene. Recently, we have reported the synthesis and structural characterization of the new $Mo_2Fe_6S8(PR3)_6(Cl_4\text{-cat})_2$ (*55*) and $Fe_4S_4Cl(^tBu_3P)_3$ (*56*) clusters that contain PR_3 ligands terminally bound to the Fe atoms.

An important contrast can be made between the catalytic reduction of acetylene and the catalytic reduction of N_2H_4 by **Ia**. In the latter the $[Fe_4S_4Cl_4]^{2-}$ cubanes were shown to be totally ineffective and no ammonia was found. The results imply that the Mo and Fe sites in the Fe/Mo/S cubanes have varying affinities for different nitrogenase substrates and also that both metal sites may be important at specific times during the step-wise reduction of a given substrate.

Summary and Conclusions

The results of our studies demonstrate that the molybdenum atom in a multitude of $MoFe_3$ cubanes **Ia** , and the V atom in **X, XI** and **XII**, in environments very similar to those in the Fe/M/S centers of the nitrogenases, are catalytically active in the reduction of N_2H_4 to ammonia. At present we have been unable to show any reactivity of **Ia** with N_2 and, as reported earlier, the CH_3CN ligand in **Ia** is not replaced by N_2 although the possibility still exists that **I** may bind to N_2 through the Mo atom at a different (lower) oxidation level.

As suggested recently,(*2, 57,58*) for the nitrogenase cofactor, the direct involvement of the unique coordinatively unsaturated Fe atoms cannot be ruled out. Indeed, the early stages of N_2 reduction on the Fe/Mo/S center of nitrogenase may involve binuclear (Fe-Fe or Fe-Mo) activation prior to reduction to the hydrazine level (Fig. 7). In support of the possible direct involvement of the Mo atom, at least in the reduction of hydrazine, is the observation that the Fe atoms in the $[Fe_4S_4Cl_4]^{2-}$ or $[(L)MoFe_3S_4Cl_3]^{2-}$ cubanes do not show any affinity towards hydrazine.

Our studies thus far have shown that the synthetic Fe/Mo/S clusters are versatile in the activation and reduction of nitrogenase substrates. Within these clusters distinct, substrate specific, reactivity sites have been identified in catalytic reductions of N_2H_4 and C_2H_2 .

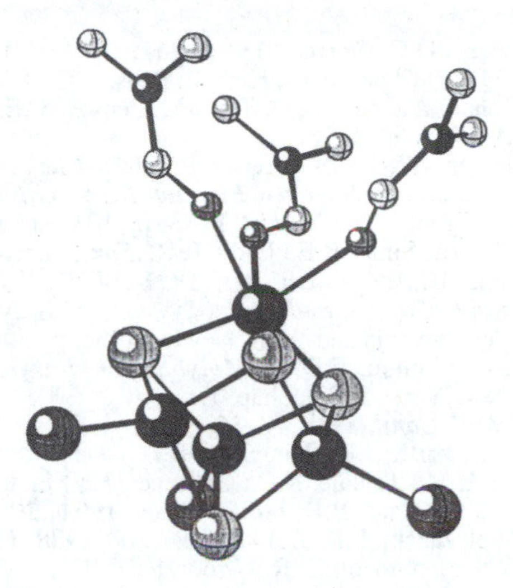

Figure 6. The structure of the $[(DMF)_3VFe_3S_4Cl_3]^-$ anion (*38*).

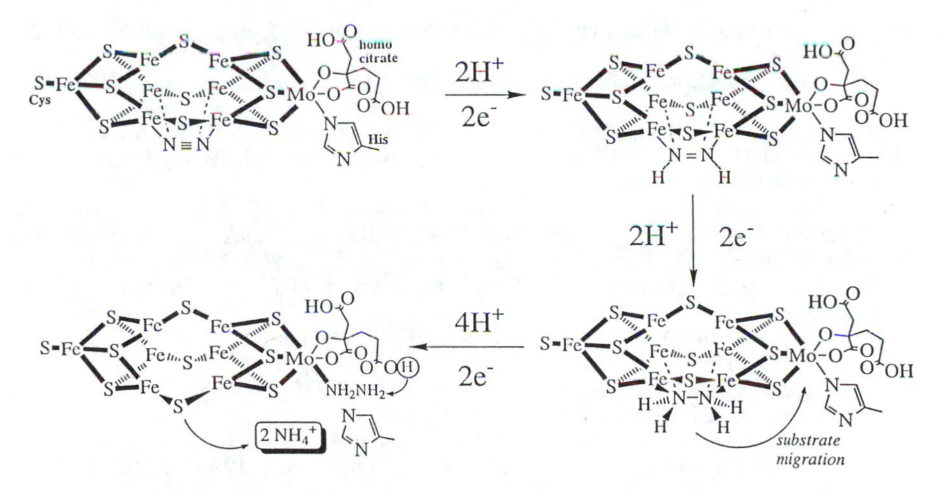

Figure 7. A possible multisite scheme for the activation and reduction of N_2 by the nitrogenase cofactor. The activation of N_2 on the Fe_6 core of the cofactor has been suggested previously (*57, 58*).

Acknowledgment. This work was made possible by funding from the National Institutes of Health (GM-33080).

References

1) (a) Kim, J.; Rees, D.C. *Science* **1992**, *257*, 1677. (b) Kim, J.; Rees, D.C. *Nature* **1992**, *360*, 553. (c) Chan, M.K.; Kim, J.; Rees, D.C. *Science* **1993**, *260*, 792.
2) Bolin, J.T.; Ronco, A.E.; Morgan, T.V.; Mortenson, L.E.; Xuong, N.H. *Proc. Natl. Acad. Sci. U.S.A.* **1993**, *90*, 1078.
3) (a) Orme-Johnson, W.H. *Ann. Rev. Biophys. Chem.* **1985**, *14*, 419. (b) Burgess, B.K. In *Advances in Nitrogen Fixation Research;* Veeger, C.; Newton, W.E.; Eds.; Martinus-Nijhoff: The Hague, **1983**; pp 103. (c) Burris, R.H. *J. Biol. Chem.* **1991**, *266*, 9339. (d) Smith B.E.; Eady, R.R. *Eur. J. Biochem.* **1992**, *205*, 1.
4) (a) Coucouvanis, D. *Acc. Chem. Res.* **1991**, *24*, 1. (b) Coucouvanis, D. In *Molybdenum Enzymes, Cofactors, and Model Systems,* ACS Symposium Series 535 Eds. E.I.Stiefel, D. Coucouvanis and W. E. Newton **1993**. pp 304-331.
5) (a) Holm, R.H.; Simhon, E.D. In *Molybdenum Enzymes*, T.G. Spiro, Ed. (Wiley Interscience, New York, **1985**), chap. 1.
6) Leigh, G.J. *J. Mol. Catal.* **1988**, *47*, 363.
7) (a) Chatt, J.; Dilworth, J.R.; Richards, R.L. *Chem. Rev.* **1978**,*78*, 589. (b) George, T.A.; Koczon, L.M.; Tisdale, R.C.; Gebreyes, K.; Ma, L. *Inorg. Chem.* **1991**, *30*, 883. (c) George, T.A; Kaul, B.B. *Inorg. Chem.* **1991**, *30*, 883. (d) Cummins, C.C.; Baxter, S.M.; Wolczanski, P.K. *J. Am. Chem. Soc.* **1988**, *110*, 8731.
8) Schrauzer, G.N.; Robinson, P.R.; Moorehead, E.L.; Vickrey, T.M *J. Am. Chem. Soc.* **1976**, *98*, 2815 (and references therein).
9) (a) Shilov, A.E. *J. Mol. Catal.* **1987**, *41*, 221. (b) Bazhenova, T.A.; Kachapina, L.M.; Shilov, A.E.; Antipin, Yu. M.; Struchkov, T. *J. Organomet. Chem*, **1992**, *428*, 107.
10) Schrock, R.R.; Glassman, T.E.; Vale, M.G.; Kol, M. *J. Am. Chem. Soc.* **1993**, *115*, 1760.
11) Block, E.; Ofori-Okai, G.; Kang, H.; Zubieta, J. *J. Am. Chem. Soc.* **1992**, *114*, 758.
12) Hardy, R.W.F.; Burns, R.C.; Parshall, G.W *Adv. Chem. Ser.* **1971**, *100*, 219 .
13) Berg J.M.; Holm, R.H. *Iron Sulfur Proteins*, T. Spiro, Ed. (Wiley Interscience, New York, 1982), p. 1.
14) (a) Tanaka, M.; Tanaka, K.; Tanaka, T. *Chem. Lett.* **1982,** 767. (b) Tanaka, K.; Hozumi, Y.; Tanaka, T. *Chem. Lett.* **1982**, 1203. (c) Tanaka, K.; Imasaka, Y; Tanaka, M.; Honjo, M.; Tanaka, T. *J. Am. Chem. Soc.* **1982**, *104*, 4258.
15) Palermo, R.E; Singh, R.; Bashkin, J.K.; Holm, R.H *J. Am. Chem. Soc.* **1984**, *106*, 2600.
16) Coucouvanis, D.; Demadis, K.D.; Kim, C.G.; Dunham, R.W.; Kampf, J.W. *J. Am. Chem. Soc.* **1993**, *115*, 3344.
17) Demadis, K.D.; Coucouvanis, D. *Inorg. Chem.* **1995**, *34*, 436.
18) Demadis, K.D.; Coucouvanis, D. *Inorg. Chem.* Submitted.
19) Demadis, K.D.; Coucouvanis, D. *Inorg. Chem.* **1995**, *34*, 3658
20) Mosier, P.E.; Kim, C.G.; Coucouvanis, D. *Inorg. Chem.* **1993**, *32,* 3620.
21) (a) Challen, P.R.; Koo, S.-M.; Kim, C.G; Dunham, W.R.; Coucouvanis, D. *J. Am. Chem. Soc.* **1990**, *112,* 8606. (b) Challen, P.R. Thesis, University of Michigan **1990.** (c) Coucouvanis, D. et al. *Inorg. Chem.* **1989**, *28*, 4181.
22) Coucouvanis, D.; Mosier, P.E.; Demadis, K.D.; Patton, S.; Malinak, S.M.; Kim, C.G.; Tyson, M.A. *J. Am. Chem. Soc.* **1993**, *115*, 12193.
23) Hardy, R.W.F; Jackson, E.K. *Federation, Proc.* **1967,** *26*, 725.
24) Coucouvanis, D.; Al-Ahmad, S.; Kim, C.G; Mosier, P.E; Kampf, J.W. *Inorg. Chem.* **1993**, *32*, 1533. Cl_4-cat = the tetrochlorocatecholate dianion.

25) In a catalytic disproportionation reaction with a N_2H_4 to **Ib** ratio of 20, the conversion of N_2H_4 to NH_3 after 72 hours was found to be 32%. At this time a spectroscopic determination of the amount of **Ib** in solution showed that the concentration of the cluster was lower by 40%. The results indicate that, over extended periods of time, **Ib** decomposes.
26) In the reactions with **Ia** the suspension of the insoluble precipitate in CH_3CN solution dissolves upon the addition of Bu_4NI and shows an electronic spectrum nearly identical to that of **Ia**. This solution was found active in the catalytic generation of NH_3 upon the addition of N_2H_4, LutHCl and $Co(Cp)_2$.
27) Mascharak, P.K; Armstrong, W.H.; Mizobe, Y.; Holm, R.H. *J. Am. Chem. Soc.* **1983**, *105*, 475
28) Coucouvanis, D.; Patton, S.; Kim, C.G. to be published.
29) Palermo, R.E.; Holm, R.H. *J. Am. Chem. Soc.* **1983**, *105*, 4310.
30) Watt, G. W.; Chrisp, J. D. *Anal. Chem.* **1952**, *24*, 2006.
31) (a) Thorneley, R.N.F.; Eady, R.R.; Lowe, D.J. *Nature* **1978**, *272*, 557. (b) Thorneley, R.N.F.; Lowe, D.J. *Biochem. J.* **1984**, *224*, 887. (c) Davis, L.C. *Arch. Biochem. Biophys.* **1980**, *204*, 270.
32) (a) Holm, R.H. in *Biomimetic Chemistry;* Yoshida, Z.; Ise, N. Eds;Elsevier; N.Y. **1983**, pp79-99.
(b) Mascharak, P.K; Armstrong, W.H.; Mizobe, Y.; Holm, R.H. *J. Am. Chem. Soc.* **1983**, *105*, 475.
33) (a) Smith, B.E.; Lowe, D.J.; Bray, R.C. *Biochem. J.* **1973**, *135*, 331. (b) Hawkes, T.R.; Lowe, D.J.; Smith, B.E. *Biochem. J.* **1983**, *211*, 495. (c) Hwang, J.C.; Chen, C.H.; Burris, R.H. *Biochim. Biophys. Acta* **1973**, *292*, 256. (d) Davis, L.C.; Shah, V.K.; Brill, W.J. *Biochim. Biophys. Acta* **1975**, *403*, 67. (e) Orme-Johnson, W.H.; Davis, L.C.; Henzl, M.T.; Averill, B.A.; Orme-Johnson, N.R.; Münck, E.; Zimmerman, R. in *Recent Developments in Nitrogen Fixation*, Newton, W.E.; Postgate, J.R.; Rodriguez-Barrueco, C. Eds.; Academic Press, London, **1977**, pp 131.
34) Mosier P., Ph.D Thesis University of Michigan **1995**.
35) (a) Hales, B.J.; Langosch, D.J; Case, E.E. *J. Biol. Chem.* **1987**, *261*, 15301. (b) Eady, R.R.; Robson, R.L; Richardson, T.H.; Miller, R.W.; Hawkins, M. *J. Biochem.* **1987**, *244*, 197.
36) (a) George, G.N.; Coyle, C.L.; Hales, B.J.; Cramer, S.P. *J. Am. Chem. Soc.* **1988**, *110*, 4057. (b) Arber, J.M. et al. *Nature* **1987**, *325*, 372
37) Malinak, S.M.; Demadis, K.D.; Coucouvanis, D., *J. Am. Chem. Soc.* **1995**, *117*, 3126-3133.
38) (a) Kovacs, J.A; Holm, R.H. *J. Am. Chem. Soc.* **1986**, *108*, 340. (b) Carney, M.J. et al. *Inorg. Chem.* **1987**, *26*, 719.
39) Coucouvanis, D; Malinak, S.M. work in progress.
40) Tanaka, K.; Honjo, M.; Tanaka, T. *J. Inorg. Biochem.* **1984**, *22,* 187.
41) Tanaka, K.; Nakamoto, M.; Tashiro, Y.; Tanaka, T. *Bull. Chem. Soc. Jpn.* **1985**, *58,* 316.
42) Schrauzer, G.N.; Schlesinger, G. *J. Am. Chem. Soc.* **1970**, *92,* 1808.
43) Schrauzer, G.N.; Doemeny, P.A. *J. Am. Chem. Soc.* **1971**, *93,* 1608.
44) Ledwith, D.A.; Schultz, F.A. *J. Am. Chem. Soc.* **1975**, *97,* 659.
45) Corbin, J.L; Pariyadath, N.; Stiefel, E.I. *J. Am. Chem. Soc.* **1976**, *98,* 7862.
46) Moorehead, E.L.; Robinson, P.R.; Vickery, T.M.; Schrauzer, G.N. *J. Am. Chem. Soc.* **1976**, *98,* 6555.
47) Moorehead, E.L.; Weathers, V.J.; Ufkes, E.A.; Robinson, P.R.; Schrauzer, G.N. *J. Am. Chem. Soc.* **1977**, *99,* 6089.
48) Rubinson, J.F.; Behymer, T.D.; Mark Jr., H.B. *J. Am. Chem. Soc.* **1982**, *104,* 1224.
49) McMillan, R.S.; Renaud, J.; Reynolds, J.G.; Holm, R.H. *J. Inorg. Biochem.* **1979**, *11*, 213.

50) Laughlin, L.J.; Coucouvanis, D. *J. Am. Chem. Soc.* **1995**, *117*, 3118-3125.
51) Segel, I.H. *Enzyme Kinetics*; Wiley-Interscience: New York, **1975**.
52) Crawford Jr., B.L.; Lancaster, L.E.; Inskeep, R.G. *J. Chem. Phys.* **1953**, *21*, 678.
53) Hardy, R.W.F.; Holsten, R.D.; Jackson, E.K.; Burns, R.C. *Plant Physiol.* **1968**, *43*, 1185.
54) (a) Coucouvanis, D.; Al-Ahmad, S.; Salifoglou, A.; Dunham, W.R; Sands, R.H. *Angew. Chem. Int. Ed. Engl.* **1988**, *27,* 1353. (b) Coucouvanis, D. et al. *J. Am. Chem. Soc.* **1992**, *114*, 2472.
55) Demadis, K.D.; Campana, C.F.; Coucouvanis, D. *J. Am. Chem. Soc.* **1995**, *117*, 7832.
56) Tyson, M.A.; Demadis, K.D.; Coucouv anis, D. *Inorg. Chem .* **1995,** 34, 4519.
57) Dance, I.G. *Aust. J. Chem.* **1994**, *47*, 979.
58) Deng, H.; Hoffman, R. *Angew. Chem. Int. Ed. Engl.* **1993**, *32*, 1062.

Chapter 7

Computational Methods for Metal Sulfide Clusters

Ian Dance

School of Chemistry, University of New South Wales, Sydney 2052, Australia

The applicability of contemporary techniques in computational chemistry for large molecules containing metal and sulfur atoms is considered, with emphasis on density functional (DF) and force-field methods. Non-local density functional methodology has been evaluated for three different metal sulfide clusters and shown to yield accurate geometries by optimization of the total electronic energy. This method is then applied to calculation of the structures (and related reactivities) of new copper sulfide clusters $[Cu_xS_y]^-$ for which the available experimental data comes from the gas phase. DF calculations applied to metallocarbohedrenes inspired postulation of a mechanism for the binding and reduction of N_2 at the Fe_7MoS_9 cluster of the nitrogenase active site. The N_2 is bound to an Fe_4 face, and the N–N bond weakened by torsion of the cluster. The pathway for protonation of the reducing N_2, and the egress of the reduction products, are explained in this concerted mechanism, which involves the surrounding protein and is supported by DF calculations. Force-field calculations are used to model the molecular biomineral with composition $Cd_{80}S_{62}\{(\gamma\text{-glu–cys})_3\text{gly}\}_{22}$ which is formed when yeasts are grown with a burden of Cd^{2+}.

The progress we make in understanding the chemistry of metal-sulfur compounds depends on the perspectives we take and the questions we ask. In this account, in which computations are the investigative tool, I adopt two perspectives on metal sulfide clusters. Most molecular metal sulfide clusters contain the metal and sulfur atoms in a core structure, surrounded by terminating ligands, which may be elaborated sulfur (i.e., thiolate) or heteroligands. One perspective strips away the terminating ligands, and asks fundamental questions about the metal-sulfur core of the cluster, and its existence, stability, and structure in the absence of the coating of ligands. The other perspective focuses on the coating and the environment, and enquires about the influences of the environment on the properties of the molecular cluster. This is the domain of supramolecular inorganic chemistry (1,2)

The significance of both of these approaches is illustrated by the Fe-Mo-S cluster at the active site of nitrogenase (*3-11*). This Fe_7MoS_9 cluster (described further below) has a remarkable structure, unprecedented amongst the plethora of metal sulfide molecules synthesised in mimicry, and in comparison with the standard clusters is remarkably underligated with connections to the protein by only two amino

0097–6156/96/0653–0135$15.00/0

acids at two of the eight metal atoms. Further, for this Fe_7MoS_9 cluster to perform its remarkable function of catalyzing the reduction of triply-bonded N_2 to NH_3 under ambient conditions there is strong dependence on the protein environment. Thus the core of the cluster is distinctive and the environment essential, and the two research perspectives are demonstrated. The mechanism of action at the active site of nitrogenase has been partly revealed by computational methods, as explained below.

Questions about unligated cluster cores, and about environmental influences, are not so readily approached experimentally (except where the environment is a crystal lattice). One technique allowing experimental access to metal sulfide cores is gas phase synthesis in the cooling energized plume formed by laser ablation of solids. By coupling this synthesis with Fourier transform ion-cyclotron resonance mass spectrometry we have revealed the existence of many hundreds of metal sulfide clusters, and been able to describe the reactivities of some of them (*12-18*). More often than not the compositions of these binary metal sulfide molecules are unexpected and unprecedented. However, these experiments are performed under high vacuum conditions, and direct measurement of structure is not yet possible.

Computational methods (i.e., theory in practice) in inorganic chemistry are now powerful and reliable, and can be used profitably in investigations of the questions just raised. Figure 1 shows the principal methods and their features and relationships. In compounds where the bonds and stereochemistry are well defined, mechanical models and force-field methods are most applicable (*19-22*), particularly where the atoms are numerous. Where the bonding is unconventional, uncertain, or variable, calculations of electronic structure through the molecular orbitals are most suitable. In compounds where the bonding has a high degree of ionic character, electrostatic models are useful (*23*).

Calculations of electronic structure are parametrized (semi-empirical) or *ab initio*. Conventional *ab initio* methods are the widely used Hartree-Fock (HF) calculation, preferably supplemented with a procedure such as configurational interaction (CI) to account for electron correlation, which is significant in inorganic molecules. The alternative to HF-CI is the density functional (DF) method in which the electron exchange and correlation are built in through the use of "functionals", which describe exchange and correlation as functions of electron density (*24-36*). There are various functionals, derived from the properties of an electron gas, and corrections (known as gradient corrections, or non-local density functionals) that take account not only of the electron density but also its curvature.

Finally, a promising development in computational inorganic chemistry involves the embedded cluster methodology (see Fig. 1), in which the core structures with unknown bonding are treated by electronic structure calculations while the surroundings (such as solvent, or protein) are treated at the same time by force-field methods.

While unbiased calculations of electronic structure (and geometry optimization by minimization of electronic energy) are clearly desirable for inorganic molecules of all types, there is the practical issue of computational effort and demand for resources. This effort is roughly proportional to the number of orbitals (basis functions) raised to an exponent n:

$$\text{computational effort} \approx (\text{number of orbitals})^n$$

When DF methods are compared with HF-CI methods of comparable accuracy, it is generally found that the exponent n for DF is at least one integer less than n for HF-CI. Therefore with density functional methods there is a saving of computational time/resources by a factor that is of the order of the number of orbitals, which for typical inorganic clusters could be of order 1000. Thus, with density functional methods it is possible to tackle big molecules with big (i.e., inorganic) atoms, containing significantly correlated electrons.

The accuracy of density functional methods for metal sulfide clusters

Expediency must be accompanied by accuracy. What is the accuracy of DF methods for metal sulfide clusters? Accuracy can be assessed against various observables, each of which has an energy scale to be considered in the evaluation of the agreement between calculated and observed properties. By far the most widely available observable is molecular geometry, from diffraction analyses of crystals. Here the flatness of the geometry–energy hypersurface is a factor, which can be assessed experimentally by examining the variation of "equivalent" dimensions in the molecule, and if available, the variation of dimensions over symmetry-inequivalent molecules in the crystal, or molecules in different crystals. (The accuracy of molecular geometries from crystal structures is usually less than the quoted precision.) Another common observable for these clusters is spin state, which is often dependent on relatively small differences in the energies of orbitals that are neither strongly bonding or antibonding, and on energies as low as kT. In the following, the DF methodology is evaluated by its reproduction of observed geometry in selected compounds (37). Accurate reproduction of geometry is a prerequisite for accurate calculation of other properties.

The implementation of DF methods in the program DMol was used (*30, 38-40*). Various local density functionals and non-local density functionals were tested, and the best results were obtained with Becke's 1988 version (*41*) of the gradient corrected exchange functional combined with the Lee-Yang-Parr (*42*) correlation functional which includes both local and non-local terms. This combination is labeled the "blyp" functional. The non-local corrections were applied after the self consistent field convergence.

The basis sets are expressed numerically rather than analytically, and are generated by solution of the DF equations (for each element) with the same functionals as used for the complete cluster. The basis sets employed were double numerical, augmented with polarization functions (basis type DND in DMol). Core orbitals can be frozen without significant loss of accuracy.

Clusters were evaluated in their even-electron charge state, with spin restricted calculations. The electronic state was the aufbau ground state. In molecules where there are closely spaced orbitals at the Fermi level or degenerate partially occupied orbitals, and thus the possibility of non-singlet spin states and low-lying excited states, the calculation of electronic structure was facilitated by a smearing of the occupancy of orbitals within about 0.1 eV at the Fermi level. This is equivalent to calculation of the average of the accessible states. The geometry optimization, by minimization of the total energy, was commenced with a geometry close to that observed. Tertiary phosphine and organo-thiolate ligands were simplified to PH_3 and SH, respectively.

$[Fe_6S_6(PEt_3)_6]^{2+}$. A significant structure type in FeS cluster chemistry is the "basket" isomer of $[Fe_6S_6(PEt_3)_6]^{2+}$ (*43*). Figure 2 shows the structure of $[Fe_6S_6(PH_3)_6]^{2+}$ as optimised in symmetry C_{2v}, and defines the atom types and notation. The best results were obtained with the blyp functional, and are are compared in Table I with the observed distances in $[Fe_6S_6(PEt_3)_6]^{2+}$. The calculated and observed structures are virtually superimposable, with agreement between the

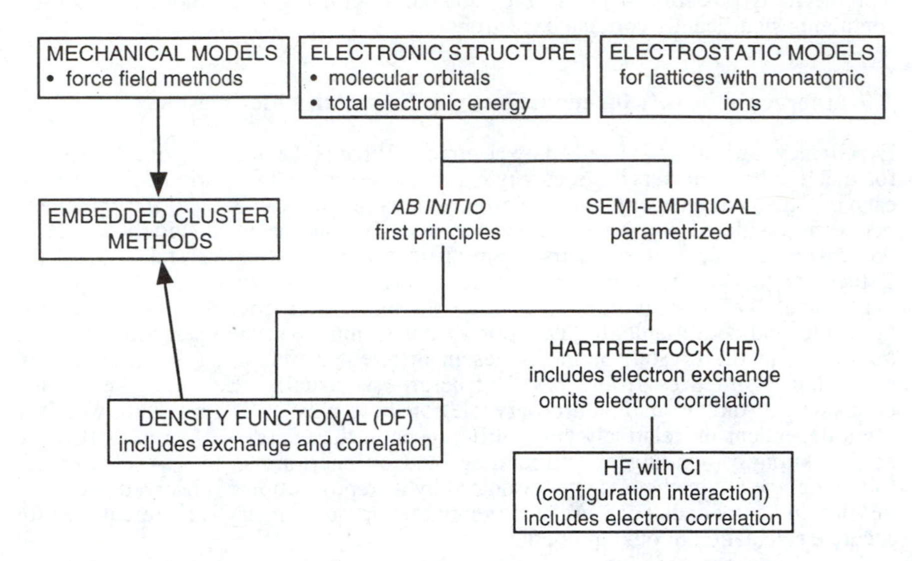

Figure 1. The methodologies of computational inorganic chemistry.

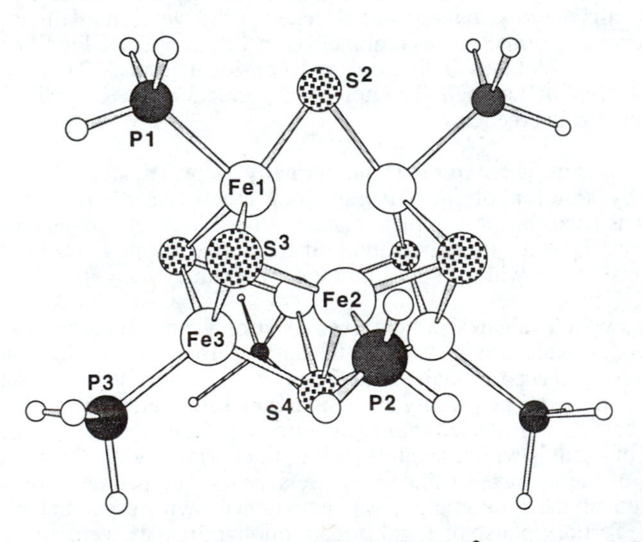

Figure 2. The optimized structure of $[Fe_6S_6(PH_3)_6]^{2+}$, with the definitions of atom types: symmetry C_{2v} applies. In the notation S^n, n also represents the coordination number of the sulfur atom. Fe–Fe bonds are not drawn, for clarity.

calculated and observed Fe–S bond distances and Fe–Fe distances better than 0.04Å in all cases. The Fe1–S^2 distance involving the unusual doubly bridging sulfide is calculated exactly. The two different types of Fe–Fe bond at 2.63Å are reproduced very well, and the longer non-bonded Fe2--Fe2 distance of 2.97Å is calculated within 0.03Å. The terminal Fe–P bond distances are observed both slightly shorter and slightly longer than those calculated.

$[VFe_4S_6(PEt_3)_4(SPh)]^-$. The second test cluster is $[VFe_4S_6(PEt_3)_4(SPh)]^-$ (*44*). The cluster core approximates threefold symmetry, although this is quite strongly disrupted by the thiolate substituent. The core can be regarded as a trigonal bipyramid of metal atoms with V axial, bridged by μ_2–S between $Fe^{equatorial}$ and V and by μ_3–S between $Fe^{equatorial}$ and Fe^{axial}. Again the best results come with use of the blyp functional (see Table II). Note that the bonds involving bridging S atoms are all calculated to better than 0.04Å, as are the V–Fe^e and Fe^a–Fe^e bonds. The Fe^e–Fe^e distance is calculated to be ca 0.10Å longer than the observed range. The occurrence of this range of 0.07Å in the crystal is indicative of the relatively weak, perturbable nature of this bond. The calculated bond lengths to the terminal phosphine and thiolate ligands are in very good agreement with those observed.

$[Nb_6S_{17}]^{4-}$. The third test involves an unusual metal sulfide cluster involving a second transition series metal, $[Nb_6S_{17}]^{4-}$, which approaches C_{5v} symmetry and contains five different types of S atom (*45*). The structure is shown in Figure 4. A pentagonal pyramid of Nb atoms (Nb^a axial, Nb^e equatorial) is triply-bridged on its triangular faces by S^3, doubly-bridged on its equatorial edges by S^2, and sextuply-bridged at its center by S^6: Nb^a and Nb^e carry terminal sulfur atoms S^a and S^e respectively. The geometry calculated using the blyp functional (see Table III) calculates Nb–S^3 distances up to 0.1Å longer than observed, while the Nb^e–S^2 distances are calculated longer by 0.07Å. The unique Nb^a–S^6 bond is reproduced exactly, although the longer non-bonding Nb--Nb distances are calculated up to 0.2Å longer than observed.

Other non-local density functionals can be used, with slightly different results, and for some metal sulfide clusters not reported here the other functionals perform better than blyp. Local density functionals, without the gradient corrections, consistently calculate tighter bonding with shorter bonds and excessive binding energies. While improvements in DF methods are expected and further evaluations are required and are in progress (*37*), it can be concluded that the current DF procedures provide reliable information about metal sulfide clusters.

As a first application of DF methods to metal sulfide systems I describe copper sulfide clusters generated and investigated in the gas phase.

Pristine copper sulfide clusters

Laser ablation of solid copper-sulfide compounds yields a collection of clusters $[Cu_xS_y]^-$, with the compositions plotted in Figure 5 (*14,18*). Using a Fourier transform ion cyclotron resonance (FTICR) mass spectrometer these ions can be collected in an ion trap. Then a particular cluster ion can be isolated in the ion trap by removal of all others, and the selected cluster allowed to react with gaseous reagents, or dissociated (*17,18,46*). Figure 6 summarizes some of these results, showing the most abundant clusters, the least reactive clusters, and the dissociation pathways. $[Cu_6S_4]^-$ is abundant, unreactive, and the common product of dissociation of larger ions, and clearly is very stable. $[Cu_3S_3]^-$ and $[Cu_{10}S_6]^-$ are similarly stable

Table I. Bond distances observed in $[Fe_6S_6(PEt_3)_6]^{2+}$ in comparison with those in $[Fe_6S_6(PH_3)_6]^{2+}$ optimized (C_{2v}) with the functional blyp. See Fig. 2.

Bond	Obs (Å) (ref 43)	Calc (Å)
Fe1 – S^2	2.15	2.15
Fe1 – S^3	2.22, 2.23	2.20
Fe2 – S^3	2.19, 2.20	2.20
Fe3 – S^3	2.22	2.18
Fe3 – S^4	2.28, 2.30	2.25
Fe2 – S^4	2.23	2.19
Fe1 – Fe1	2.63	2.62
Fe2 – Fe1	2.72	2.73
Fe2 – Fe2	2.97	3.00
Fe2 – Fe3	2.63, 2.65	2.61
Fe1 – P	2.27, 2.29	2.32
Fe2 – P	2.28	2.31
Fe3 – P	2.36, 2.38	2.31

Table II. Bond distances in $[VFe_4S_6(PEt_3)_4(SPh)]^-$ in comparison with those optimized for structure $[VFe_4S_6(PH_3)_4(SH)]^-$ (symmetry C_s, functional blyp): see Fig. 3 for definitions of atom types.

Bond	Obs (Å) (ref 44)	Calc (Å)
Fe^e – S^3	2.15–2.19	2.17
Fe^a – S^3	2.19–2.22	2.17
Fe^e – S^2	2.22–2.24	2.25, 2.26
V – S^2	2.16–2.20	2.22
Fe^a – S^a	2.25	2.23
Fe^e – Fe^e	2.61–2.68	2.72, 2.77
V – Fe^e	2.58–2.61	2.57, 2.61
Fe^a – Fe^e	2.62–2.67	2.59, 2.63
V – P^a	2.48	2.53
Fe^e – P^e	2.26–2.29	2.29, 2.32

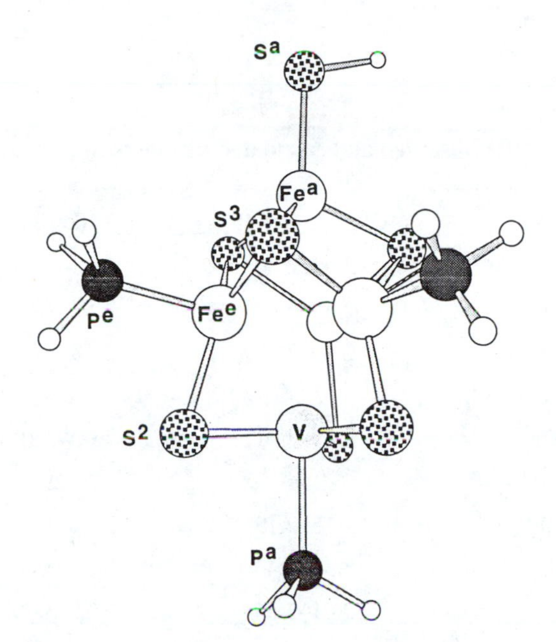

Figure 3. The structure of $[VFe_4S_6(PR_3)_4(SR')]^-$, as observed (R=Et, R'=Ph) and calculated (R=R'=H). Fe^a, P^a and S^a lie on the molecular axis, Fe^e and P^e are equatorial, and S^3, S^2 refer to the triply- and doubly-bridging sulfur atoms respectively (threefold symmetry assumed in atom labelling). Bonds between metal atoms are not drawn, for clarity.

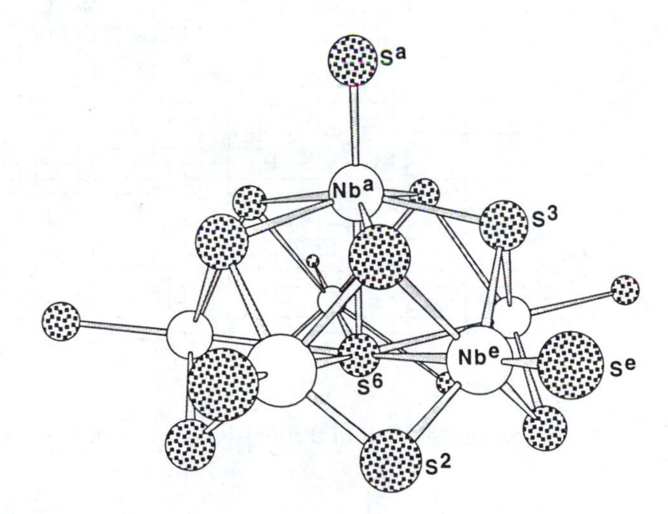

Figure 4. The structure and atoms types of $[Nb_6S_{17}]^{4-}$: five-fold symmetry is assumed in the atom labelling.

Table III. Observed and calculated distances in $[Nb_6S_{17}]^{4-}$

Bond	Obs (Å) (ref 45)	Calc (Å)
$Nb^a - S^3$	2.58–2.60	2.68–2.70
$Nb^e - S^3$	2.49–2.53	2.57–2.63
$Nb^e - S^2$	2.40–2.42	2.48–2.49
$Nb^a - S^6$	2.64	2.64
$Nb^e - S^6$	2.89–3.01	3.00–3.18
$Nb^a - S^a$	2.20	2.29
$Nb^e - S^e$	2.15–2.19	2.27–2.28
$Nb^e - Nb^e$	3.38–3.44	3.52–3.66
$Nb^a - Nb^e$	3.61–3.65	3.75–3.80

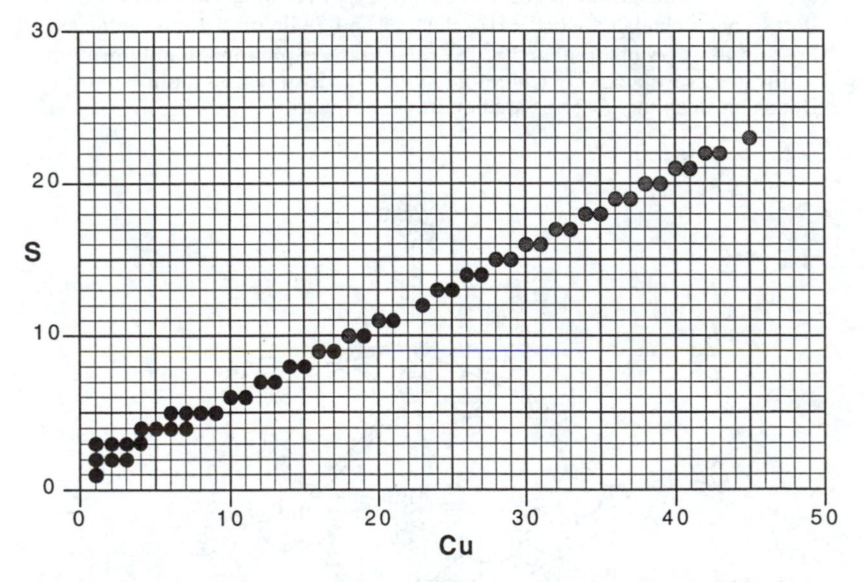

Figure 5. Map of the compositions of the ions $[Cu_xS_y]^-$ generated by laser ablation of KCu_4S_3.

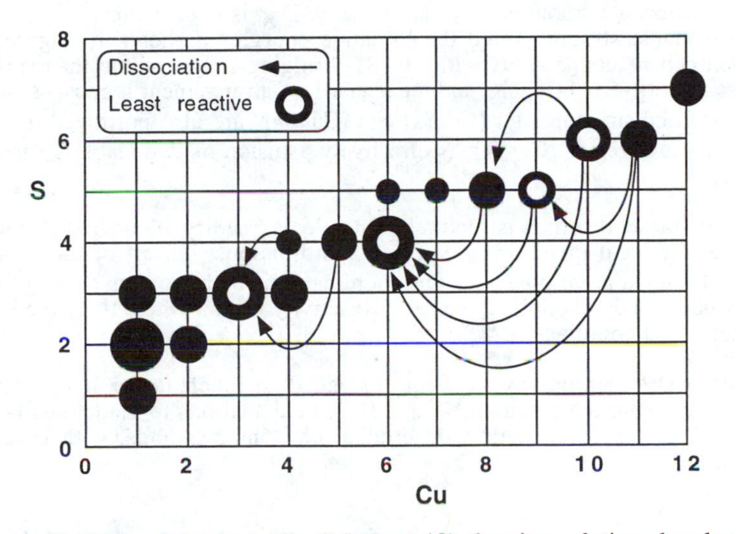

Figure 6. Map of the ions $[Cu_xS_y]^-$ ($x \leq 12$) showing relative abundance (circle size), those least reactive with thiols and H_2S (open circle), and the products of collisionally induced dissociation (arrows).

compositions, as is $[Cu_9S_5]^-$ even though its abundance is lower. In contrast, the other ions are generally less abundant and more reactive.

These findings provoke many questions. For example, what are the geometrical and electronic structures of these clusters? What factors stabilize the four special compositions? Since these experiments are performed at 10^{-8} mbar there are no spectroscopic data (and it is not likely that such data could be interpreted unambiguously to yield structure), and so DF methods can be deployed. The following results were obtained with the progam DMol, as above, using the blyp functional, DND basis, spin restricted calculations of even–electron species, with geometry optimizations usually in symmetry lower than ideal.

Figure 7 shows the most stable isomer for $[Cu_3S_3]^-$, the D_{3h} structure, and the next most stable geometry, $[(\mu_3\text{–}S)(\mu_2\text{–}S)_2Cu_3]$, which is 47 kcal mol^{-1} higher, and which transforms without energy barrier to the D_{3h} isomer. For $[Cu_4S_4]$ the most and next stable structures and their relative energies a shown in Figure 7: the common cubane connectivity with $(\mu_3\text{–}S)_4$ bridging is less stable than square Cu_4 isomers with $(\mu\text{–}S)_4$ bridging, and the planar D_{4h} arrangement is most stable. The best calculated structures for $[Cu_6S_4]$ and $[Cu_9S_5]^-$ are also portrayed in Figure 7. The best structure for $[Cu_9S_5]^-$ is effectively a fusion of two stable T_d isomers of $[Cu_6S_4]$.

From these results a structural principle is readily identified: stability is conferred by local quasi-linear S–Cu–S coordination. In all of the most stable isomers each Cu atom possesses this coordination: both isomers of $[Cu_9S_5]^-$ have this property, and are similarly stable. Reactivity occurs where there are Cu atoms with non-linear coordination (*18*).

These DF calculations are being extended to much larger $[Cu_xS_y]$ clusters. Ahlrichs and Fenske have used HF and HF-CI calculations to model similar copper sulfide and selenide clusters with terminal phosphine ligands, with good results (*47,48*).

Nitrogenase

Elucidation of the structure and mechanism of the enzyme nitrogenase has been a longstanding goal of research. The crystal structure of the protein, first reported in 1992 (*4-11*), uncovered the special structure of the iron-molybdenum-sulfide cluster at the active site, but did *not reveal* (*49-53*) the location or mode of binding of N_2 at this site. Computational methods have now provided considerable insight into the probable location and stereochemistry of the binding of N_2, and elucidated the mechanism for the weakening of the N–N bond and the pathway for proton transfer to the reducing N_2 (*54*).

Figure 8 shows the (cysteine)Fe_7S_9Mo(histidine)(homocitrate) cluster at the active site, together with some significant surrounding protein. A central trigonal prism of Fe atoms is capped on the upper triangular face by $(\mu_3\text{–}S)_3$Fe(S-cysteine) and on the lower triangular face by $(\mu_3\text{–}S)_3$Mo(N-histidine)(O,O-homocitrate). The three vertical edges of the Fe_6 trigonal prism each carry doubly-bridging sulfide. A significant feature is that the cluster is bound to the protein only at the Mo and the upper Fe atom: the six-coordinate Mo provides an anchor point at the base of the cluster, while the upper section should be free to rotate about the Fe–S-cysteine bond, allowing torsion in the central region of the cluster. In the center, there is trigonal prism of six under-coordinated Fe atoms. This trigonal prism presents three Fe_4 faces around the equator, and one of these is sufficiently unobstructed by protein as to

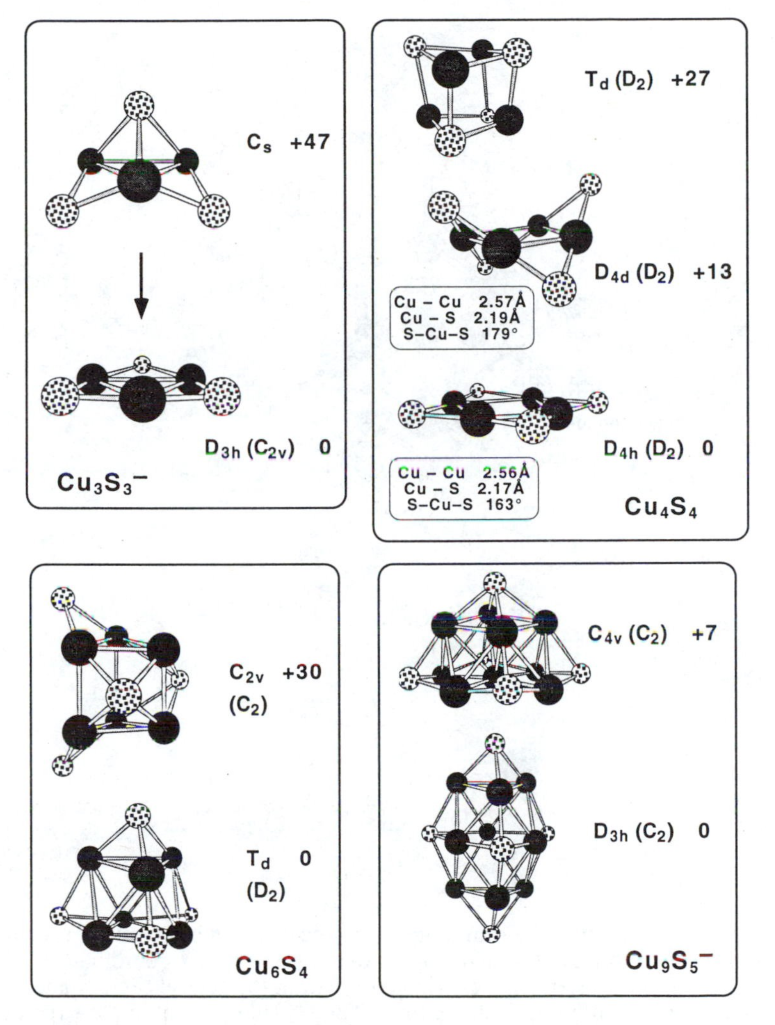

Figure 7. Isomers of $[Cu_xS_y]^{0,-}$, with relative energies in kcal mol^{-1} (symmetries in parentheses are those imposed during geometry optimization).

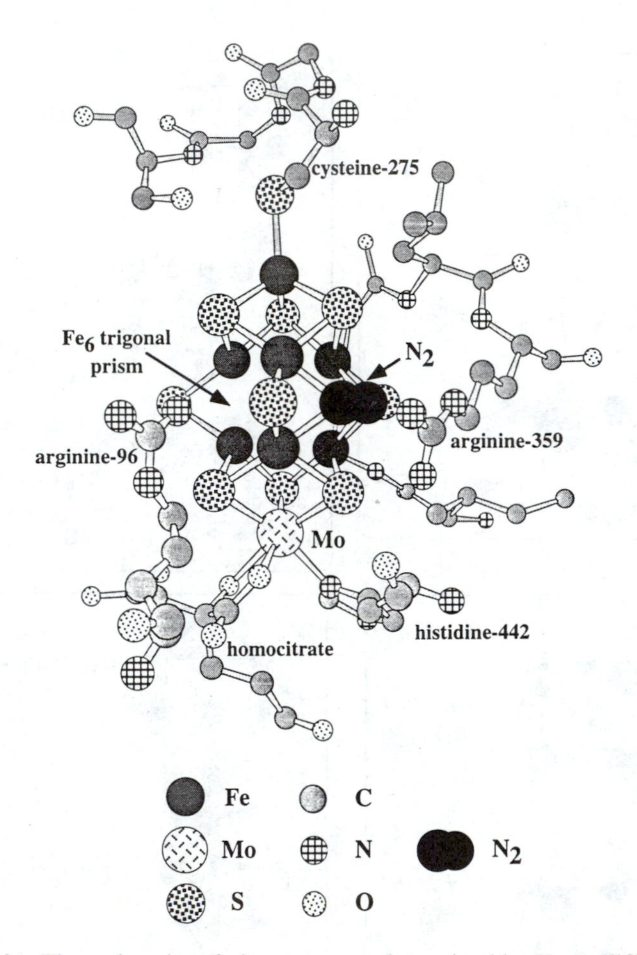

Figure 8. The active site of nitrogenase as determined by X-ray diffraction (*55*), showing the histidine and homocitrate ligands for six-coordinate Mo at the base of the cluster, the Fe_6 trigonal prism in the central region, and the cysteine coordination of Fe at the top. The N_2 (*not* present in the structure determination) is shown at the proposed binding site, enclosed by arginine-359, in the parallel binding conformation (see text). A hydrogen bond exists between arginine-96 and the μ_2–S ligand in the center-front of the picture.

offer a site for binding of N_2 (*54*). Arginine-359 provides a cover for this binding site.

Clues about the geometry of binding and weakening of N_2 came from computations on the metallocarbohedrenes (binary metal-carbon clusters, M_xC_y) (*56-59*), in which C_2^{2-} (iso-electronic with N_2) binds to a quadrilateral of metal atoms (*60-66*). The two relevant conformations (labelled *diagonal* and *parallel*) are shown in Figure 9, for N_2 over Fe_4, analogous to C_2 over M_4. More stable binding occurs for N_2 diagonal to an Fe_4 rhombus, with favourable overlap of the N_2 π-bonding orbitals with Fe 3d. The alternative binding of N_2 parallel to the edges of an Fe_4 rectangle is less stable overall, and involves overlap of Fe 3d orbitals with N_2 π-antibonding orbitals, thus weakening the N–N bond. This parallel conformation can alternatively be regarded as four Fe–N σ-bonds to N_2^{2-} in the doubly reduced state. The key concept is that the N_2 can be initially trapped in the diagonal conformation, and then its N–N bond weakened by conversion to the parallel conformation. The interconversion of these two Fe_4 shapes could be achieved by rotation of the upper section of the cluster, causing torsion of the Fe_6 trigonal prism. In Fig. 8 the N_2 is shown bound in the parallel conformation.

These hypotheses have been substantiated and quantitated by DF calculations (*54*), which also reveal another significant characteristic of the active site. Increased electron population during the reduction is calculated to have largest effect on atomic charges at the doubly-bridging sulfide ligands which flank the binding site. These μ_2-S ligands, unusual in polymetallic sulfide clusters (*16*), are thus postulated to play a significant role in the proton transfer components of the total reduction of N_2. This proposal is supported by the fact that both of the flanking μ_2–S ligands accept hydrogen bonds from behind, as marked in Figure 10 (and portrayed in more detail in ref 54). Thus, during reduction of bound N_2 there is also an increase in basicity of the flanking μ_2–S ligands, facilitating transfer of H^+ to them via protein. Inversion of the hydrogen on the resulting Fe_2SH (a facile process) places it immediately adjacent to the reducing N_2, and subsequent transfer of the proton to the increasingly basic nitrogen atom is stereochemically facile (see Figure 10). This proton transfer stage has been evaluated with DF calculations (*54*). While the μ_2–S ligands were unexpected in the Fe-Mo-S cluster, they are seen to have a mechanistic function that could not be achieved with μ_3-S.

Finally, another role for the obligatory homocitrate ligand is recognised. After reduction, the amine product must leave the active site via the opening near the guanidine terminus of arginine-359. The homocitrate ligand at the base of the egress route provides the required hydrophilic pathway. The other role of homocitrate is to anchor the coordinatively saturated Mo atom, as required for the torsion of the upper part of the cluster. It is proposed that this torsion is induced by hydrogen bonds with the surrounding protein.

Thus a concerted mechanism is proposed for the function of the active site of nitrogenase, and supported computationally (*54*). This proposal emphasises and illustrates nicely the significance of the two questions raised in the introduction, concerning details of fundamental cluster core structure, and the role of the cluster environment.

Biomineralization

Plants and fungi detoxify heavy metals such as Cd^{2+}, Hg^{2+}, Pb^{2+} by inducing the synthesis of cysteine-rich oligopeptides known as phytochelatins (PC), which then

Diagonal binding of N_2 over Fe_4 rhombus

Fe N N Fe Fe Fe

Fe overlap with N_2 π-bonding orbitals

Parallel binding of N_2 over Fe_4 rectangle

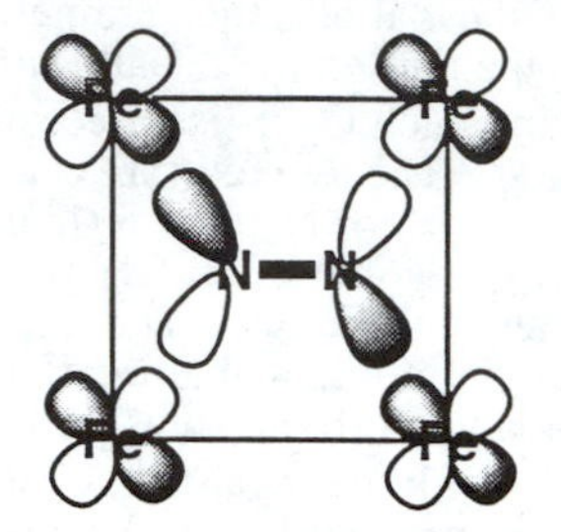

Fe overlap with N_2 π-antibonding orbitals

Figure 9. The diagonal and parallel conformations of N_2 bonded over an Fe_4 quadrilateral. Adapted from reference 54.

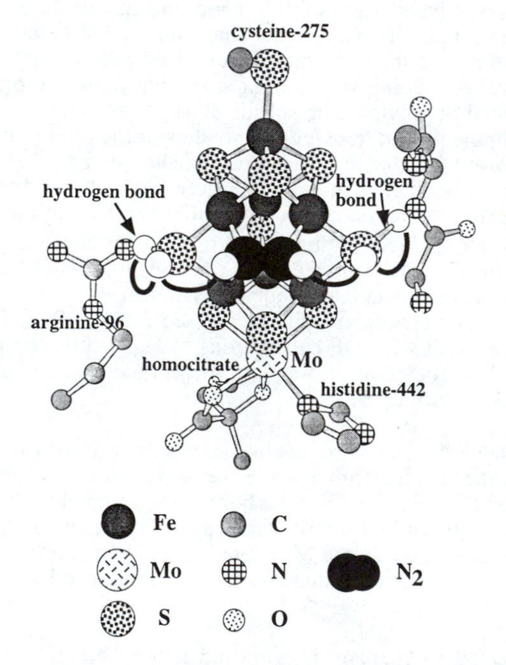

Figure 10. Representation of the hypothesis for proton transfer from the hydrogen bonds behind the two μ_2–S ligands which flank the binding site, around μ_2–S by inversion, and to reducing N_2: hydrogen atoms (open circles) are drawn in each of the locations along these two flanking pathways. Arginine-359 which covers this site has been removed for clarity.

sequester the metal ions in innocuous complexes. The phytochelatins are usually (γ-glutamic acid–cysteine)$_n$glycine with n varying between 2 and 8, symbolised PC_n.

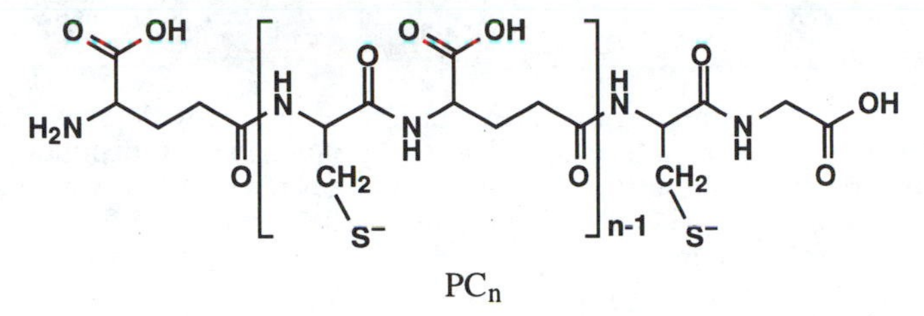

PC_n

Organisms such as the yeasts *Candida glabrata* and *Schizosaccharomyces pombe* have increased the efficacy of this protection by further generating sulfide, S^{2-} (presumably from glutathione), which initiates formation of a metal sulfide biomineral together with the metal–phytochelatin complexes. Thus, when these yeasts are burdened with Cd^{2+} they grow a crystallite of CdS, coated with phytochelatins. This is an unusual example of biomineralization, or biochemically-controlled mineral formation. The dimensions of the CdS nanocrystallite are restricted, to *ca* 20 Å diameter: the thermodynamically favoured growth (Ostwald ripening) of nanocrystalline CdS to bulk CdS which would otherwise occur is controlled and restricted by a coating of phytochelatin peptides. The formation (*in vivo* and *in vitro*) and properties of these have been described by Dameron, Winge et al. (*67-69*).

These biomineral materials are significant in fields other than metal detoxification, since CdS is a size-quantized photoresponsive semiconductor. That is, the band gap for absorption of radiation, and the energy separation of the reducing electrons in the valence band and oxidizing holes in the conduction band, are tunable by variation of the size of the crystallite in the nm size domain (*70*). Thus, control of nanocrystallite size and avoidance of Ostwald ripening are important objectives, and the yeasts provide a natural solution to a technological problem in materials science.

Using knowledge of many smaller cadmium-sulfide-thiolate clusters with known structures in crystals and in solution, I have developed a computer model of a representative CdS bio-nanocrystallite, with composition $Cd_{80}S_{62}\{(\gamma\text{-glu–cys})_3\text{gly}\}_{22}$, chosen to fit the compositional, diffraction, and size data (*71*). This model and the modelling process were very informative about structural possibilities, which are in fact limited as a consequence of the unusual coordination properties of the phytochelatins. A chelate ring formed by the cys–γ-glu–cys sequence contains 14 atoms, which is anomalously large and conformationally demanding.

In the absence of data interpretable at the atomic level, computational methods are valuable. The refinement of this model (which has a total of 1990 atoms) employed force-field methods, using parameters determined for Cd-S-SR clusters (*72*). The significant characteristic of this structure is the close packing of the 22 peptides over the surface; surface packing is tight, and the core is completely covered by peptide. There is a preponderance of carboxylate oxygen atoms on the surface, with a total charge of –96 if all are deprotonated. The close packing of peptides over the surface provides the clue to the mechanism for size control. The diagrammatic representation (Fig 11) of the separation of peptides anchored with *fixed spacing* over the surface of small and large mineral cores indicates how the peptides come together as the core grows. At the limiting size, any addition to the core would decrease the general curvature of the surface, force the peptides together, prevent the ingress of Cd^{2+} and S^{2-}, and prevent further growth.

Figure 11. Diagrammatic representation of the growth of the CdS biomineral core, with surface coordination by phytochelatin peptides. As the core grows and the surface curvature decreases the peptides (which have fixed separations at the connection points on the core) are forced together and prevent further growth.

Conclusions

I conclude that contemporary computational techniques, especially the density functional calculations, enable expedient insight into key questions at the frontiers of metal sulfide cluster chemistry, providing information not yet accessible experimentally, and providing direction for experimental programs.

Acknowledgments. This research is supported by the Australian Research Council. I thank Professor Doug Rees for provision of information on the structure of nitrogenase, and Professors Ahlrichs and Fenske for information in advance of publication. Computational resources provided by the Australian Nuclear Science and Technology Organisation are gratefully acknowledged.

References

1. Dance, I.G. in *Perspectives in Supramolecular Chemistry*, G. Desiraju, ed. John Wiley and Sons, **1995**, *2*, 137-233.
2. Müller, A; Reuter, H; Dillinger, S. *Angew. Chem. Int. Ed. Engl.*, **1995**, *34*, 2328-2361.
3. Burgess, B.K. *Chem. Rev.*, **1990**, *90*,1377-1406.
4. Georgiadis, M.M.; Komiya, H.; Chakrabarti, P.; Woo, D.; Kornuc, J.J.; Rees, D.C *Science*, **1992**, *257*, 1653-1659.
5. Kim, J.; Rees, D.C. *Science*, **1992**, *257*, 1677-1682.
6. Kim, J.; Rees, D.C. *Nature*, **1992**, *360*, 553-560.
7. Rees, D.C.; Chan, M.K.; Kim, J. *Adv. Inorg. Chem.*, **1993**, 40, 89-118.
8. Chan, M.K.; Kim, J.; Rees, D.C. *Science*, **1993**, *260*, 792-794.
9. Bolin, J.T.; Ronco, A.E.; Morgan, T.V.; Mortenson, L.E.; Xuong, N.H. *Proc. Nat. Acad. Sci. USA*, **1993**, *90*, 1078-82.
10. Rees, D.C; Kim, J.; Georgiadis, M.M.; Komiya, H.; Chirino, A.J.; Woo, D.; Schlessman, J.; Chan, M.K.; Joshua-Tor, L.; Santillan, G.; Chakrabarti, P.; Hsu, B.T. *ACS Symposium Series*, **1993**, *535*, 171 - 185.
11. Bolin, J.T.; Campobasso, N.; Morgan, T.V.; Muchmore, S.W.; Mortenson, L.E. in *Molybdenum Enzymes, Cofactors and Models*, Stiefel, E.I.; Coucouvanis, D.; Newton, W.E. Eds, *ACS Symposium Series*, **1993**, *535*, 186-215.
12. El-Nakat, H.J.; Dance, I.G.; Fisher, K.J.; Willett, G.D. *J. Chem. Soc., Chem. Comm.*, **1991**, 746-748.
13. El Nakat, J.H.; Dance, I.G.; Fisher, K.J.; Rice, D.; Willett, G.D. *J. Am. Chem. Soc.*, **1991**, *113*, 5141-5148
14. El Nakat, J.H.; Dance, I.G.; Fisher, K.J.; Willett, G.D. *Inorg. Chem.*, **1991**, *30*, 2957-2958
15. El Nakat, J.H.; Fisher, K.J.; Dance, I.G.; Willett, G.D. *Inorg. Chem.*, **1993**, *32*, 1931-1940.

16. Dance, I.G.; Fisher, K.J. *Prog. Inorg. Chem.*, **1994**, *41*, 637-803.
17. Dance, I.G.; Fisher, K.J. *Materials Science Forum*, **1994**, *152-153*, 137-142.
18. Fisher, K.J.; Dance, I.G.; Willett, G.D.; Yi, MaNu *J. Chem. Soc., Dalton Trans.*, **1996**, 709-718.
19. Zimmer, M. *Chem. Rev.*, **1995**, *95*, 2629-2649.
20. Comba, P.; Hambley, T.W. *Molecular Modeling of Inorganic Compounds*, VCH, **1995**, 300p.
21. Hay, B.P. *Coord. Chem. Rev.*, **1993**, *126*, 177-236.
22. Rappé, A.K.; Casewit, C.J.; Colwell, K.S.; Goddard, W.A.; Skiff, W.M. *J. Am. Chem. Soc.*, **1992**, *114*, 10024-10035.
23. Catlow, C.R.A.; Price, G.D. *Nature*, **1990**, *347*, 243-248.
24. Jones, R.O.; Gunnarsson, O. *Review. Modern Phys.*, **1989**, *61*, 689-746.
25. Parr, R.G.; Yang, W. *Density-Functional Theory of Atoms and Molecules*, **1989**.
26. Labanowski, J.K.; Andzelm, J.W. *Density Functional Methods in Chemistry*, **1991**, 443pp.
27. Ziegler, T. *Chem. Rev.*, **1991**, *91*, 651-667
28. Fitzgerald, G.; Andzelm, J. *J. Phys. Chem.*, **1991**, *95*, 10531-10534.
29. Ziegler, T. *Pure and Appl. Chem.*, **1991**, *63*, 873-878
30. Delley, B.; Wrinn, M.; Lüthi, H.P. *J. Chem. Phys.*, **1994**, *100*, 5785-5791.
31. Schwerdtfeger, P.; Boyd, P.D.W.; Fischer, T.; Hunt, P.; Liddell, M. *J. Am. Chem. Soc.*, **1994**, *116*, 9620-9633.
32. Special issue on the applications of density functional theory: *Int. J. Quantum Chem.*, **1994**, *52*, 693-1121.
33. Van Leeuwen, R.; Baerends, E.J. *Int. J. Quantum Chem.*, **1994**, *52*, 711.
34. Andzelm, J.; Sosa, C.; Eades, R.A. *J. Phys. Chem.*, **1993**, *97*, 4664-4669.
35. Ziegler, T. *Energetics of Organometallic Species,* J.A. Martinho Simoes, Ed., Kluwer Academic Publishers, **1992**, 357-385
36. Jones, R.O. *Angew. Chem. Int. Ed. Engl.*, **1991**, *30*, 630-640.
37. Dance, I.G. Results to be published.
38. DMol v 2.36, Biosym-MSI, San Diego, USA.
39. Delley, B. *J. Chem. Phys.*, **1990**, *92*, 508-517.
40. Delley, B. *J. Chem. Phys.*, **1991**, *94*, 7245.
40. Delley, B. *New J. Chem.*, **1992**, *16*, 1103-1107.
41. Becke, A.D. *Phys. Rev. A*, **1988**, *38*, 3098.
42. Lee, C.; Yang, W.; Parr, R.G. *Phys. Rev. B*, **1988**, *37*, 785-789.
43. Snyder, B.S.; Holm, R.H. *Inorg. Chem.*, **1990**, *29*, 274-279.
44. Cen, W.; MacDonnell, F.M.; Scott, M.J.; Holm, R.H. *Inorg. Chem.*, **1994**, *33*, 5809-5818.
45. Sola, J.; Do, Y.; Berg, J.M.; Holm, R.H. *Inorg. Chem.*, **1985**, *24*, 1706-1713.
46. Fisher, K.J.; Dance, I.G.; Willett, G.D. *Rapid Comm. Mass Spect.*, **1996**, *10*, 106-109.
47. Dehnen, S.; Schafer, A.; Fenske, D.; Ahlrichs, R. *Angew. Chem. Int. Ed. Engl.*, **1994**, *33*, 746-749.
48. Dehnen, S.; Schafer, A.; Ahlrichs, R.; Fenske, D. *Chem. Eur. J.*, **1996**, *2*, 429-435.
49. Orme-Johnson, W.H. *Science*, **1992**, *257*, 1639-1640.
50. Deng, H.; Hoffmann, R. *Angew. Chem. Int. Ed. Engl.*, **1993**, *32*, 1062-1065
51. Sellman, D. *Angew. Chem., Int. Ed. Engl.*, **1993**, *32*, 64.
52. Kim, J.; Rees, D.C *Biochemistry*, **1994**, *33*, 389-397.
53. Leigh, G.J. *Science*, **1995**, *268*, 827-828.
54. Dance, I.G. *Aust. J. Chem.*, **1994**, *47*, 979-990.
55. From atomic coordinates provided by Professor D. Rees.
56. Guo, B.C.; Kerns, K.P.; Castleman, A.W. *Science*, **1992**, *255*, 1411-1413.
57. Bowers, M.T. *Acc. Chem. Res.*, **1994**, *27*, 324-332.

58. Pilgrim, J.S.; Duncan, M.A. *Adv. Metal and Semiconductor Clusters*, **1995**, *3*, 181-221.
59. Pilgrim, J.S.; Duncan, M.A. *J. Am. Chem. Soc.*, **1993**, *115*, 6958-6961.
60. Dance, I.G. *J. Chem. Soc., Chem. Comm.*, **1992**, 1779-1780.
61. Grimes, R.W.; Gale, J.D. *J. Phys. Chem.*, **1993**, *97*, 4616-4620.
62. Rohmer, M-M; Benard, M.; Henriet, C.; Bo, C.; Poblet, J-M *J. Chem. Soc., Chem. Comm.*, **1993**, 1182-1185.
63. Lin, Z.; Hall, M.B. *J. Am. Chem. Soc.*, **1993**, *115*, 11165-11168.
64. Dance, I.G. *J. Chem. Soc., Chem. Comm.*, **1993**, 1306-1308.
65. Dance, I.G. *J. Am. Chem. Soc.*, **1993**, *115*, 11052-11053.
66. Rohmer, M.-M.; Benard, M.; Bo, C.; Poblet, J-M *J. Am. Chem. Soc.*, **1995**, *117*, 508-517.
67. Dameron, C.T.; Reese, R.N.; Mehra, R.K.; Kortan, A.R.; Carrol, P.J.; Steigerwald, M.L.,;Brus, L.E.; Winge, D.R. *Nature*, **1989**, *338*, 596-597.
68. Dameron, C.T., Smith, B.R.; Winge, D.R. *J. Biol. Chem.*, **1989**, *264*, 17355-17360.
69. Dameron, C.T.; Winge, D.R. *Inorg. Chem.*, **1990**, *29*, 1343-1348.
70. Wang, Y.; Herron, N. *J. Phys. Chem.*, **1991**, *95*, 525-532.
71. Dameron, C.T.; Dance, I.G. Biogenic CdS Semiconductors, in *Biomimetic Approaches in Materials Sciences*, ed S. Mann, **1995**, VCH.
72. Gizachew, D.; Dance, I.G. unpublished data.

Hydrodesulfurization and Related Systems

Chapter 8

Hydrodesulfurization Catalysts and Catalyst Models Based on Mo–Co–S Clusters and Exfoliated MoS_2

M. David Curtis

Willard H. Dow Laboratory, Department of Chemistry, University of Michigan, Ann Arbor, MI 48109–1055

Results from the author's laboratory on the desulfurization activity of $Cp_2Mo_2Co_2S_3(CO)_4$ (**1**) and $Cp_2Mo_2Co_2S_4(CO)_2$ (**2**) clusters, as well as the HDS activity of catalysts derived from exfoliated MoS_2, are reviewed. Cluster **1** abstracts the sulfur atom from a variety of organic sulfides to give cluster **2** and a desulfurized organic product. With thiols, the mechanism involves C-S bond homolysis. Supporting these clusters on Al_2O_3 gives HDS catalysts with activity nearly identical to that of conventional Mo/Co/S catalysts. New HDS catalysts derived from exfoliated MoS_2 are described.

Hydrodesulfurization (HDS) of fossil fuels is a process important to the preservation of environmental quality. When combusted, sulfur in fossil fuels forms sulfur oxides that combine with atmospheric water and contribute to acid rain and other undesirable environmental contamination. Sulfur-containing organic molecules are also potent poisons for noble metal-based catalysts, e.g.,, those used for reforming or hydrocracking in fossil fuel refining. As increasingly sulfur-rich, residual crude oil fractions are converted to lighter MW products by hydrocracking, it becomes increasingly important to remove sulfur from crude oil fractions.

The commercial catalyst most commonly used for this process is based on the sulfides of molybdenum and cobalt supported on a high surface area oxide, e.g., γ-Al_2O_3 (*1-3*). Other metal combinations, e.g., Mo/Ni, W/Co, and W/Ni are also effective. The late transition metal, Co or Ni, is usually referred to as the "promoter" element although Mo can also "promote" Co-based catalysts (*1*). The nature of these complex, supported sulfided catalysts has been the subject of investigation and controversy for nearly 20 years.

These supported metal sulfide catalysts are prepared by impregnating the refractory oxide support 127. "date and the promoter ion (the impregnation steps may be simultaneous or sequential). The impregnated support is then dried and calcined at 400-500 °C. This step converts the supported metal species to various disordered oxide phases (*4-6*). These oxides are then reduced with a mixture of H_2 and H_2S at 350-500 °C. This sulfiding treatment converts the oxides to sulfides. It is the structure of these mixed metal sulfides (and/or oxysulfides) that has engendered debate and controversy since the mid-1980's. Some authors claim that sulfidation is never complete and that up to 30% of the molybdenum remains bonded to oxygen in the initially prepared (most active) catalysts (*7*). Other workers see a continuous increase

0097–6156/96/0653–0154$15.50/0

in sulfur coordination to Mo, with these molybdenum sulfides becoming more ordered into MoS_2-like structures as the sulfiding temperature is increased (*8,9*).

At this time, the most widely accepted model of the HDS catalyst structure is the "CoMoS phase" model proposed by Topsøe and coworkers (*5,6*) (Figure 1.). In this model, the catalyst consists of small crystallites of MoS_2 supported on the Al_2O_3 surface. The Co promoter occupies sites on the edges of the basal planes of the MoS_2 crystallites; this is the "CoMoS" phase. The Co is also known to be present in other phases, e.g., Co in tetrahedral holes in the Al_2O_3, and as the cobalt sulfide, Co_9S_8. These latter phases are believed to have very low activity for HDS catalysis.

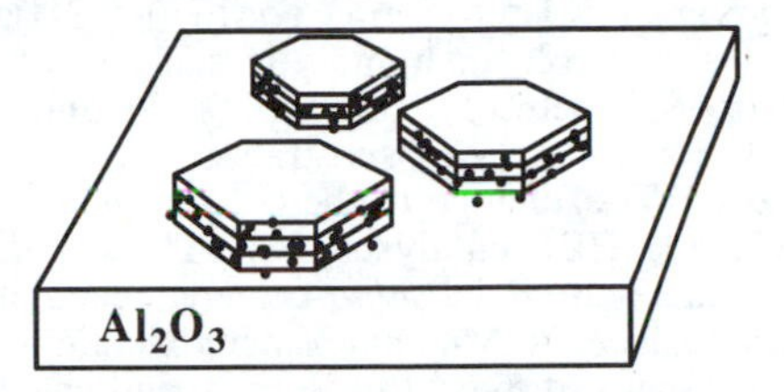

Figure 1. Active "CoMoS" phase for HDS catalysis. Dots represent Co atoms decorating the edges of the MoS_2 crystallites.

Although this physical model is widely accepted, it does not specify a role for the Co as a promoter element. How does the presence of the Co increase the activity of the MoS_2, a catalyst in its own right? In fact, the argument has been put forward that it is actually the Co that is the most competent catalyst and that the MoS_2 is merely acting as a support to effect the optimum dispersion of the Co active sites (*10*). In order to answer these questions, one needs to answer even more fundamental questions. For example, what is (are) the mode(s) of coordination of the organosulfur substrates to the catalyst surface or active sites? What are the mechanistic details of the C-S bond breaking? How is H_2 activated on metal sulfide surfaces,? Does C-S activation occur at the same site(s) as H_2 activation? What are the mobilities of the various species on the catalyst surface?

Attempts to answer these fundamental questions pose some exceedingly thorny experimental problems, even over "simple" surfaces, e.g., clean metals in UHV. On the complex, sulfided catalyst surface, these questions may be next to impossible to answer. In such cases, modeling studies can often come to the rescue. The more faithfully the model resembles the object being modeled, the more confidence one has in transferring conclusions reached on the basis of model behavior to the real system of interest.

Organometallic Models

Organometallic chemistry has long been associated with catalysis. This association arises from the fact that organometallic compounds are often catalysts in their own right. Examples abound; hydrogenation, hydroformylation, olefin oligomerization and polymerization, olefin metathesis, hydrosilation, and cyclopropanation, are reactions with well-developed homogeneous catalytic processes (*11,12*). Furthermore, the elementary steps of many heterogeneously catalyzed reactions may be modeled or imitated by well characterized organometallic complexes. At the most fundamental level, organometallic structures provide models for possible surface species (*13*). Secondly, the spectroscopic properties of organometallic compounds often resemble those of corresponding surface species, e.g., ν_{CH} of μ_3 - ethylidyne, M-NO or M-CO vibrational frequencies, etc. Thirdly, the reactivity of ligands in organometallic complexes can, and often does, mimic that of adsorbate molecules on catalyst surfaces. Finally, organometallics may serve as precursors to supported catalysts that show different activities or selectivities from those prepared by the conventional impregnation procedures. Even if a catalyst prepared from organometallic precursors exhibits the same reactivity as one prepared by the conventional impregnation route, the surface of the former is often "cleaner", i.e., free of extraneous phases, so that interpretation of spectroscopic data is often facilitated. This particular advantage of organometallic precursors seems to be little appreciated.

Until fairly recently there were no good organometallic models for reactivity patterns on sulfided surfaces, or for binding modes of thiophene compounds, for example. The latter situation has changed dramatically with the detailed and thorough work of Rakowski DuBois, Angelici, Rauchfuss, Jones, Adams, Bianchini, and others (*14,15*). Rakowski DuBois has made an extensive study of the reactivity of dimeric molybdenum sulfides, e.g., $Cp_2Mo_2S_4$ and $Cp_2Mo_2(SR)_2(S)_2$ (R=H, alkyl, etc.) (*16*). Of particular relevance here are her observations that the complex $Cp_2Mo_2(\mu_2\text{-}S_2)(\mu_2\text{-}S)_2$ splits H_2 to form $Cp_2Mo_2(\mu_2\text{-}SH)_2(\mu_2\text{-}S)_2$, i.e., the sulfide ligands may play an active role in homolytic scission of H_2, that $Cp_2Mo_2(\mu_2\text{-}S_2CH_2)(\mu_2\text{-}SMe)(\mu_2\text{-}S)^+$ heterolytically splits H_2 and that these species are hydrogenation catalysts for a variety of substrates. The complex, $Cp_2Mo(S_2CH_2)(SH)(S)^+$ also catalyzes H/D exchange of the α-hydrogens on thiophene, a reaction also observed over CoMoS HDS catalysts (*17*). The significance of this body of work is that it demonstrates substantial *sulfur*-centered reactivity. The role of the metal is nonetheless important, since the Mo can change oxidation states to support changes in the formal oxidation states of the sulfur ligands, and vacancies in the metal coordination sphere facilitate desulfurization reactions (*18*).

Rauchfuss has made important contributions to understanding the reactivity of polysulfide ligands, e.g., extrusion of S_x (*19*), interconversion of binding modes of disulfide ligands ($\mu\text{-}\eta^1,\eta^1$, $\mu\text{-}\eta^2,\eta^2$) (*20*), their reactivity with alkynes (*21*), and the coordination chemistry of thiophene (*22*). In connection with the latter, these workers have postulated the importance of the bimetallic, multihapto coordination of thiophenes to the desulfurization process and have also demonstrated a remarkable reactivity of the coordinated thiophenes with the hydroxide ion or the hydrogen ion (*23*). These latter demonstrations are especially important since both base (MS^-, MO^-, OH^- etc.) or Bronsted acids (SH) are present on the surface of MoS_2 catalysts and may serve as agents for thiophene ring opening.

Angelici and co-workers have conducted thorough investigations of the coordination chemistry of thiophenes and have discovered several transformations relevant to desulfurization reactions over heterogeneous catalysts (*14*). In particular, their demonstration that the deuterium exchange pattern on π-bound thiophene is similar to that observed on heterogeneous catalysts (*24*), of H^-/H^+ addition to C=C bonds (*25*) and C-S bond cleavage in π–bound thiophene (*26*), the greatly increased basicity of the sulfur atom in η^4-bound thiophene, and interconversions between η^4(C) and η^2(C,S) bonding modes (*27*) are especially noteworthy. The measurements of relative binding strengths of methyl substituted thiophenes in π- and S-bound coordination modes has also been extremely useful in the interpretation of adsorption and relative HDS rates of substituted thiophenes over supported HDS catalysts (*28,29*). Benzothiophene coordination chemistry and reactivity have also received attention (*21,30-32*).

Most of the recent investigations of reactions of organic sulfur compounds with transition metal complexes have involved mononuclear derivatives, but several reports of C-S bond activation at dinuclear and polynuclear centers have appeared. Rauchfuss et al. showed that a thiaferracyclohexadiene is an intermediate in the desulfurization of thiophene with $Fe_3(CO)_{12}$ (*21*). Boorman et al. studied the cleavage of a C-S bond of sulfides coordinated to two metal centers (*33*), and Rakowski DuBois recently reported that a dinuclear molybdenum complex can desulfurize thiirane (*18*). Jones et al. have observed cleavage of the C-S bond of thiophene in dinuclear Co and Rh complexes (*34-37*). An especially interesting result was obtained with the reaction of $Cp^*{}_2Ir_2H_6$ with thiophene. The sulfur atom was abstracted and formed a μ–S bridge while the C_4H_4 fragment was reduced to butadiene and was also incorporated as a bridging ligand (*38*). The sequence of reactions that were proposed to account for the formation of the final product are very similar to those we have proposed for the heterogeneously catalyzed HDS of thiophene (*39*). Bianchini et al. have shown that an Ir complex inserts into the C-S bond of benzothiophene or thiophene, and the ring-opened complexes react with acids, e.g., PhSH, to give H_2S, ethylphenyl thiol, or butenyl thiols (*40-43*). In fact, Bianchini and coworkers have demonstrated for the first time a homogeneous *catalysis* of HDS with (triphos)IrH_2Et complex (*44*).

Adams et al. have studied the reactions of organic sulfides with osmium clusters (*45-47*). In an extensive series of papers, they have isolated new clusters resulting from S-H, C-S, and C-H bond insertions, as well as reductive elimination of alkanes or arenes, dehydrogenation of the organic moiety and oligomerization of the organic sulfide. All these reaction types have been observed over heterogeneous HDS catalysts or clean metal surfaces.

At this point a fair question is: Have these organometallic studies helped the interpretation of data obtained on the real catalyst systems? I believe the answer to this question is most certainly in the affirmative. The review articles in references (*1*) and (*6*) both refer extensively to organometallic model systems. Recent studies on clean and sulfided metal surfaces under ultrahigh vacuum conditions are beginning to uncover aspects of the reactivity of surface bound thiolates that have direct parallels with reaction pathways seen on organometallic centers. Thus, elimination of H-atoms in the β-position with respect to the thiol sulfur atom has been observed on Au (*48*) as has the extrusion of butadiene from 2,5-dihydrothiophene over Mo(110) (*49*) (similar reactions have been seen in organometallic reactions). Stiefel has also drawn attention to the correspondence of the structural features found in metal sulfido clusters and the MoS_x species believed to exist at the edges of the basal planes of MoS_2 (*50*). Hence, there exist organometallic models for the structural features as well as the kinetic or reaction pathways postulated to be important in the catalysis of the HDS reaction.

Model studies on well ordered surfaces (*6*), the EXAFS, MES, and other spectroscopic studies by Topsøe et al. on sulfided Co-Mo-S catalysts, and the organometallic studies described above have led to a greatly increased understanding of the nature of the catalyst and possible coordination modes of organic sulfides with species likely to be present on metal sulfide catalysts. However, there are many questions yet unanswered. The role of the promoter metals, e.g., Co or Ni, in the catalytic cycle is still not established. Is it electronic (modifying the Mo-S bond strength) (*48,49*), physical (promoting *more* active sites), or kinetic (enhancing the *quality* of the active sites or acting as a center for dissociative adsorption of H_2) in nature? Is the cobalt, as some have suggested, the real active site with the MoS_2 acting merely as a support or modifier to the Co-active site(s)? How are the C-S bonds activated on the catalyst surface? What are the kinetically important steps?

It is universally accepted that the oxidation state of Mo in the active HDS catalysts is +4 in the MoS_2 phases, although very low concentrations of +3 and/or +5 sites are detected by ESR spectroscopy (*51*). Harris has concluded that only very electron rich complexes are expected to insert into the C-S bond of thiophene (*52*). Thus, model studies with low-valent organometallics may show reaction paths that are not relevant on the actual HDS catalysts.

In the remainder of this chapter, we present a summary of our work with the Mo/Co/S clusters **1** and **2** as models for the putative "CoMoS" phase on heterogeneous HDS catalysts. These clusters have the metal atoms in an average oxidation states of +2 and +2.5 for **1** and **2**, respectively; but oxidation states of +3 to +4 for Mo are reasonable in these clusters. Hence, clusters **1** and **2** are in many ways, ideal models for the "CoMoS" phase in HDS catalysts.

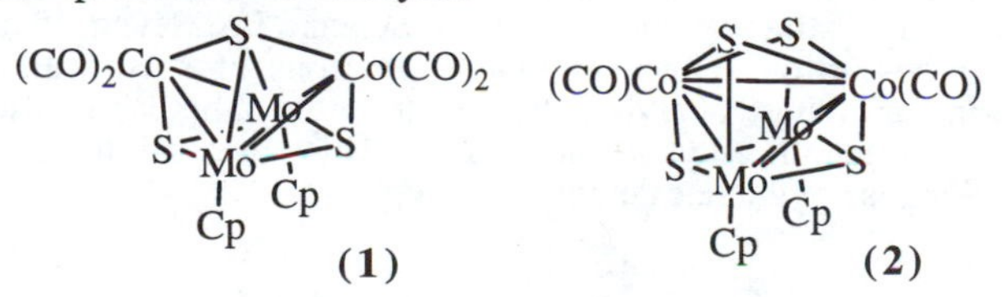

Electronic Structure of the Clusters

A series of bimetallic sulfido clusters of the type, $Cp_2Mo_2M'_2S_nL_m$ (M' = Fe, Co, Ni; n = 2-4, L_m = ligands, e.g., CO, Cp) have been synthesized and structurally characterized in our laboratories (*53-57*). For the most part, these clusters obey the "18 electron rule". When a cluster obeys the 18 electron rule, the number of valence shell electrons (VSE) in a four-metal cluster would be 4x18=72 VSE in the absence of metal-metal bonding. However, each pair of electrons engaged in metal-metal bonding gets "counted" twice, once for each metal. Hence, the number of electrons required to reach the 18-electron count is 18M-2n, where M = number of metal atoms and n = number of metal-metal bonds. Thus, an M_4 cluster with six M-M bonds requires 72-12 = 60 VSE.

To arrive at the number of VSE, one adds the numbers of electrons contributed by each metal atom plus those contributed by the ligands. For clusters of the type considered here, counting is easier if all atoms are considered to be neutral. Thus, Cp is a five-electron donor; and CO, R_3P, RNC, etc. are two electron donors. Bridging groups, especially, sulfur, can present some problems in electron counting. A μ_2-sulfide, counted as a neutral, will donate two VSE; a μ_3-S will donate four VSE. A terminal RS or halogen is a one-electron donor as the neutral atom. As a μ_2-ligand, these groups donate 3 VSE.

With the above counting scheme, cluster **2** is a 60 VSE, electron precise cluster. But what about cluster **1**? How many electrons does the μ_4-sulfur contribute? If the μ_4-S also contributes 4 electrons, then cluster **1** is also a 60 VSE cluster; but **1** has only five M-M bonds and thus requires 62 VSE to be electron precise. Cluster **1** is therefore electron deficient in the same sense as are boron hydrides.

EHMO calculations support the idea that the μ_4-S ligand in **1** contributes the same number of electrons to the overall bonding as does the μ_3-S ligand (*58*). The calculated Mulliken overlap populations are: Mo-S(μ_3) = 0.56, Mo-S(μ_4) = 0.46, Co-S(μ_3) = 0.35, and Co-S(μ_4) = 0.31 e^-. The total overlap population for a μ_3-S atom is (2Mo-S + Co-S) = 1.47 while for the μ_4-S it is (2Mo-S + 2Co-S) = 1.54. Thus, the μ_4-S has about the same number of bonding electrons spread over four bonds as the μ_3-S has spread over 3 bonds. The Mo-S overlap populations are larger than the corresponding Co-S populations by about 50%, and the M-S(μ_4) values are less than those of the M-S(μ_3) bonds.

Thus, the EHMO calculations are consistent with counting the bridging sulfur atoms as four-electron donors, and thus ascribing a 60 VSE electron count to cluster **1**. The electronic structure of **1** may be important in relation to its desulfurization activity (see below). Since **1** is electron deficient, it can be viewed as possessing a *latent* vacancy in the metal coordination sphere, i.e., the cluster can accept an additional ligand with its two electrons. This concept helps to relate the desulfurization activity of cluster **1** to the requirement of vacancies (or latent vacancies!) on the surfaces of heterogeneous catalysts.

Homogeneous S-Abstraction Reactions of Mo/Co/S Clusters

The Mo/Co/S cluster, **1**, reacts with a variety of S-containing organic molecules to give quantitative yields of cluster **2** and the corresponding desulfurized organic molecule (*59*). A summary of the reactions is given in Scheme I. Several features of these reactions are noteworthy. The rates of the reactions of **1** with alkane thiols are qualitatively similar to those of aryl thiols, in spite of the rather large differences in C-S bond energies between the alkyl and aryl series. This observation would suggest that C-S bond breaking is not the rate determining step.

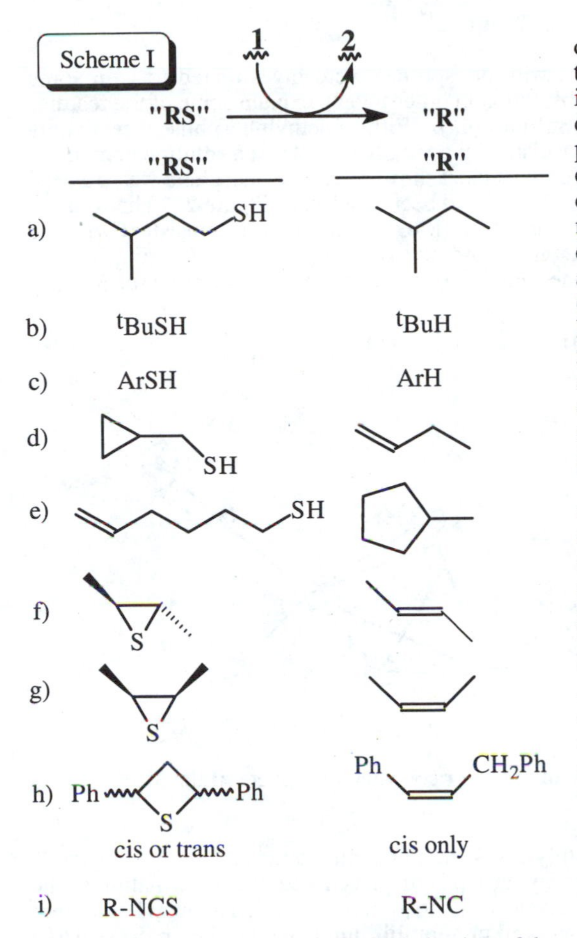

Another striking observation is the simplicity of the products. When conducted in a sealed NMR tube, the desulfurization of alkyl thiols produces only the corresponding alkane and cluster **2**. No skeletal rearrangements or H^+ elimination to give isomeric alkanes or alkenes are observed (entries "a" and "b" in Scheme I). This suggests that the sulfur is not removed as the SH^- anion, which would produce carbocations. The latter are expected to lead to alkenes and/or rearranged alkanes.

The desulfurizations of *cis*- or *trans*-2,3-dimethylthiirane were stereospecific: only *cis*- or *trans*-2-butene were seen as the products (entries "f" and "g", Scheme I) (*59,60*). The sulfur abstraction from thietanes was *stereoselective*, however. Both *cis*- and *trans*-2,4-diphenylthietane gave *cis*-1,3-diphenylpropene as the only observed product (*60*) (entry "h" in Scheme I). The C-S bond strengths are decreased in the strained, cyclic sulfides, and thiiranes are known to donate sulfur to a wide variety of low valent metal complexes, so the desulfurization of these cyclic sulfides is not particularly novel.

However, even the very strong C=S double bonds in organic isothiocyanates and in COS are rapidly desulfurized by **1** (*59*). In the case of RNCS (entry "i", Scheme I), the organic product is the isocyanide, a very good ligand. The released isocyanide displaces CO from **1**, as well as from product **2**, to give isonitrile-substituted clusters, $Cp_2Mo_2Co_2S_3(CO)_{4-x}(RNC)_x$ (x = 1-3) and $Cp_2Mo_2Co_2S_4(CO)_{2-x}(RNC)_x$ (x = 1, 2) (*61*).

In contrast to the desulfurization reactions of simple alkyl thiols, the reactions of cyclopropylmethyl thiol, and 6-mercaptohexene with cluster **1** lead to rearranged products: 1-butene and methylcyclopentane, respectively (entries "d" and "e", Scheme I) (*62-63*). These products are typical of rearrangements of the cyclopropylmethyl and 1-hexene-6-yl radicals, respectively (*64*).

Reactions of Nucleophiles with 1 and 2

The reactions of clusters **1** and **2** with phosphines were investigated to gain some insight into the mechanisms of substitution in order to help explain some of the features of the reactions of **1** with organic sulfides (*65*). With trimethylphosphine, **1** reversibly forms a deep red adduct that was isolated by crystallization from a solution containing excess Me_3P. A crystal structure of the adduct showed that the phosphine was attached to a Co-vertex and that the Co-S(μ_4) bond had been displaced (Figure 2). The solution NMR of the adduct **3** (L = Me_3P, Cp' = C_5H_4Me, Scheme II) was consistent with the solid state structure: one Cp-Me signal and one ABCD pattern for the ring protons show the presence of a mirror plane containing the Co-atoms and bisecting the Mo-Mo bond.

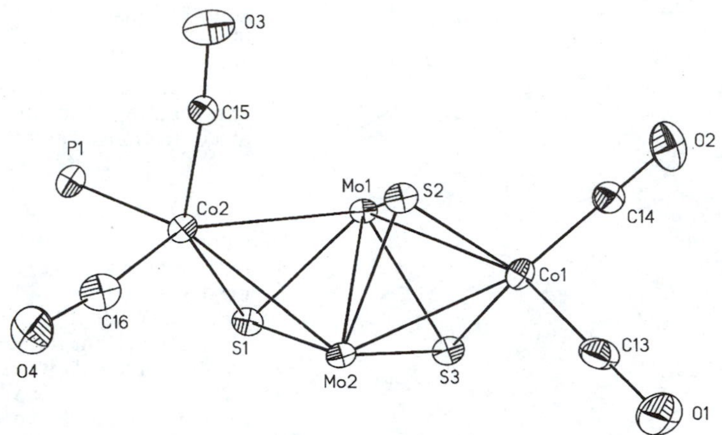

Figure 2. ORTEP drawing of the inner coordination sphere of the Me_3P adduct of cluster **1**.

The Me_3P adduct, **3**, readily loses the phosphine and regenerates cluster **1**. With less basic phosphines, a carbonyl group is displaced with the re-formation of the Co-S(μ_4) bond (Scheme II) (*65*). Even with phosphines for which adduct **3** could not be directly detected, the kinetics showed pre-equilibrium behavior, i.e., plots of 1/[L] vs. $1/k_{obs}$ were linear as required by the kinetic pathway shown in Scheme II. Measurement of the rates as a function of temperature allowed us to determine the values of the pre-equilibrium constant, K_{eq}, and CO displacement rate, k, indicated in Scheme II. For (n-Bu_3)P, $\Delta H = -12.2 \pm 0.2$ kcal/mol, and $K_{eq}^{298} = 91$, and the activation parameters for k, were found to be $\Delta H^{\ddagger} = 24 \pm 3$ kcal/mol and $\Delta S^{\ddagger} = -0.5 \pm 8$ eu. A typical graph of energy vs. reaction coordinate is shown in the insert to Scheme II.

These studies showed that cluster **1** does behave as an electron-deficient cluster in that **1** can accept two more electrons to form the 62 VSE, saturated cluster, **3**. This adduct is sufficiently stable to be isolated at low temperatures. Cluster **2** also reacts with nucleophiles in an associative mechanism, but in no case have we been able to detect measurable concentrations of the 62 VSE intermediate, only the final, CO-substitution product may be isolated.

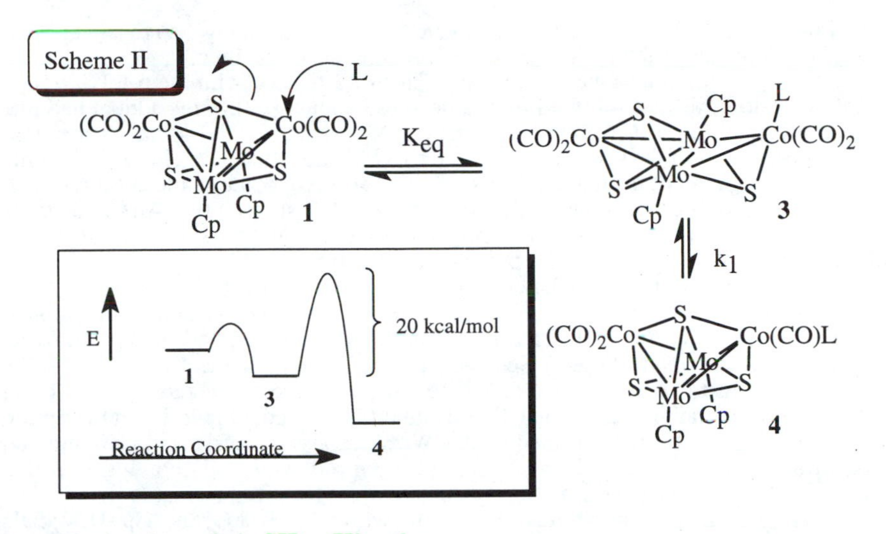

Desulfurization of ArSH: Kinetics

The kinetics of the desulfurization reactions shown in equation 1 were measured for a variety of substituents, X. The reaction followed the 2nd-order rate law, rate = k[ArSH][**1**], and the presence of 1 atm of CO had no effect on the rate of the desulfurization reaction. The entropy of activation, $\Delta S^{\ddagger}$ ranged from -5 eu (X = H) to -22 eu (X = MeO), and $\Delta H^{\ddagger}$ ranged from +27 kcal/mol to +19 kcal/mol, respectively. Electron releasing substituents led to faster S-atom abstraction.

These results were interpreted to mean that the rate-determining step (rds) of the desulfurization reaction was the attachment of the thiol to the cluster (adduct formation, **3**, L = ArSH). The subsequent desulfurization was rapid and no detectable concentrations of intermediate products were observed (*62*).

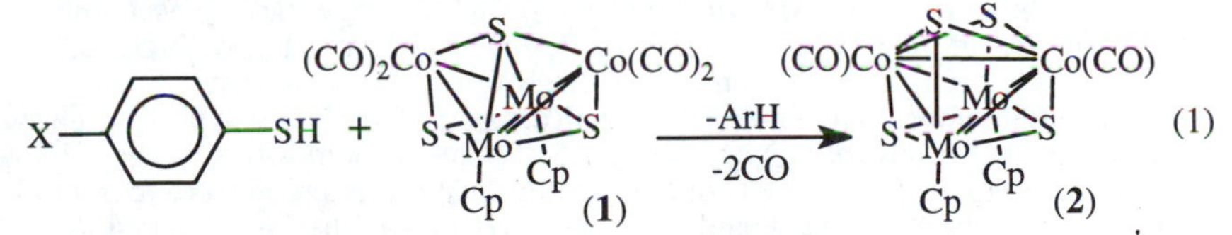

Regardless of the exact mechanism of C-S bond cleavage, the $\Delta H^{\ddagger}$ of the desulfurization reaction puts an upper bound on the C-S bond dissociation energy (BDE) of the aryl thiol. In aromatic thiols, the C-S BDE is near 80 kcal/mol (*66*). Hence, the C-S BDE is reduced by coordination to cluster **1** by a factor of 3-4! One need search no further to find the source of C-S bond activation on Mo/Co/S catalysts. However, we would still like to know the intimate mechanism whereby the C-S BDE is drastically reduced.

Desulfurization of Thiolate Anions, ArS^-

To find the intimate details of the C-S bond activation, we investigated the reactions of thiolate anions. Our reasoning was that, if the rate determining step for the desulfurization of thiols was the initial adduct formation, then the thiolate anion would, by virtue of its increased nucleophilicity, move the rate determining step to a later stage in the reaction, thus enabling us to view the C-S bond cleavage directly.

These expectations were borne out. Cluster **1** reacted with the p-tolylthiolate anion in acetonitrile at ≤ -40°C to form a deep red adduct (*62*). This adduct had one ABCD multiplet for the Cp'-protons and one Cp'-Me signal, and the spectrum appeared very similar to that of the structurally characterized Me_3P-adduct (see cluster **5**, Scheme III). As the solution of the thiolate adduct, **5**, was allowed to warm, the color changed from red to olive green and the NMR spectrum at -24°C showed two ABCD patterns and 2Cp-Me resonances. This green compound thus has only C_1 symmetry. Cooling the solution to -40°C did not restore the original spectrum.

There are two likely candidates for the green material with C_1 symmetry. One corresponds to a cluster with one of the four carbonyl groups displaced by the thiolate (cf: compound **7**, Scheme III). The other likely product corresponds to a μ_2-thiolate, μ_2-sulfido tetracarbonyl, cluster **6**, Scheme III. The latter assignment was shown to be correct by the fact that the addition of acid to a solution of **6** regenerated cluster **1** and the thiol in quantitative yield. The CO-substituted cluster could give **1** in a maximum yield of only 75% if **1** were formed at all. We were also successful in isolating and structurally characterizing the CO-substitution product **7** and showing that this compound does not have the spectroscopic signature of the green adduct **6**.

As the green solution was allowed to warm to -10 °C, the Cp'-H signals broadened and the Cp'-Me signals began to coalesce. The NMR spectra were simulated with the program DNMR5, and it was apparent that the fluxional process responsible for the observed spectral changes could be accounted for by a concerted movement of the μ_2-sulfido and μ_2-thiolato through the plane containing the Co vertices. This fluxional, wind-shield wiper motion establishes an effective mirror plane that bisects the Mo-Mo bond, makes the two Cp'-rings equivalent, and results in one ABCD pattern for the Cp'-H resonances (step "a", Scheme III).

Increasing the temperature further to +22 °C results in continued evolution of the Cp'-H signals and the complete coalescence of the Cp'-Me signals. Detailed simulation was consistent with the ABCD pattern evolving to a single A_2B_2 pattern, i.e., toward the spectrum that would be associated with a species with C_{2v} symmetry. We therefore suggest that in the higher energy fluxional process the μ_2-thiolato and μ_2-sulfido ligands traverse the Mo-Mo bond (step "b", Scheme III). The combined fluxional motions of steps "a" and "b" impart effective C_{2v} symmetry to the adduct, and the μ_2-S and μ_2-SAr ligands are mobile over the "surface" of the cluster.

The high-temperature, limiting spectrum expected for the C_{2v}-symmetry cluster was not observed because at temperatures ≥ 35 °C, the decomposition of the thiolate adduct became rapid. The products of the "decomposition" reaction were found to be the anion of the cubane cluster, **2**, and the arene, ArH. When the "decomposition" was conducted in CD_3CN solvent, the organic product was the monodeuterio arene, ArD. Similarly, desulfurization of the cyclopropylmethylthiolate anion in CD_3CN gave 1-butene-d_1 as the sole organic product. Thus, the thiolate-cluster adducts react by C-S bond homolysis (eq. 2). The identity of the anionic, paramagnetic cluster **8** was established unambiguously by preparing it from the one-electron reduction of cubane **2** and structurally characterizing the anion (*62*). Adams et al. have previously observed homolytic C-S bond scission in the reactions of thiols and thietanes with Os_3 clusters (*45,47,67*). In these reactions, multi-point bonding of the thiolate to the cluster is also observed, so that the mechanism of C-S bond activation proposed here applies to the Os_3 cluster-mediated reactions as well.

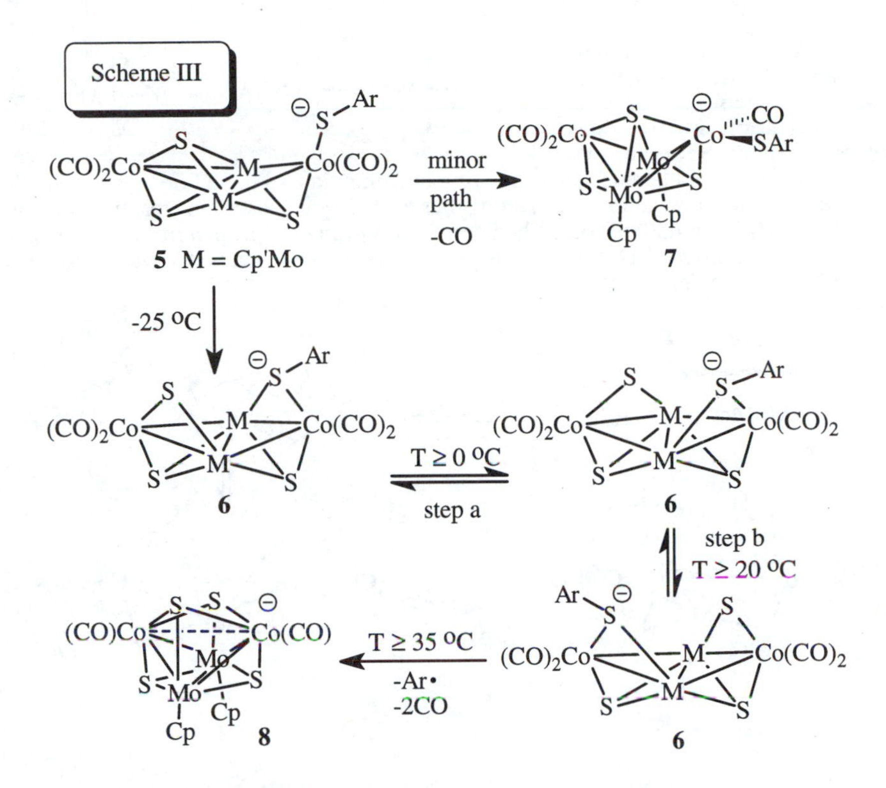

Proposed Desulfurization Mechanism

These results allow us to piece together a self-consistent mechanism for the desulfurization of thiols by cluster **1**. The rate-determining step is the nucleophilic attack of the thiol to give an intermediate, **9** (Scheme IV). The terminal thiol can then bend over and bind to the Mo-atom in an μ_2-mode (intermediate **10**). Intermediate **10** has a quaternary sulfur that should render its attached H-atom acidic. Transfer of this proton to the μ_2-sulfido ligand gives intermediate **11**. Now the thiolate sulfur can bind in a μ_3-fashion. We believe it is this μ_3-thiolate binding mode that activates the C-S bond to homolytic cleavage. C-S bond homolysis, followed by H-atom abstraction leads to intermediate **14**. Loss of CO and "folding up" of the cluster gives the ultimate cubane cluster, **2**.

A comment about the C-S bond homolysis is perhaps useful. A bond energy is the difference between two thermodynamic states. Thus, if we compare the C-S bond energy in the two reactions depicted in eqs. 3 and 4, the respective C-S bond energies will depend on the relative energies of HS• radical and the M_3S fragment (formally oxidized with respect to the $ArSM_3$ cluster). Since the S-atom in M_3S is in a stable bonding mode as compared to the unstable SH radical, the apparent Ar-S bond energy is greatly decreased in cluster-bound thiolate and C-S homolysis is thermally accessible.

$$\text{Ar-SH} \longrightarrow \text{Ar}\bullet + \bullet\text{SH} \quad (3)$$

A slight variation on the mechanism shown in Scheme IV can account for the desulfurization of thietanes (entry "h", Scheme I).

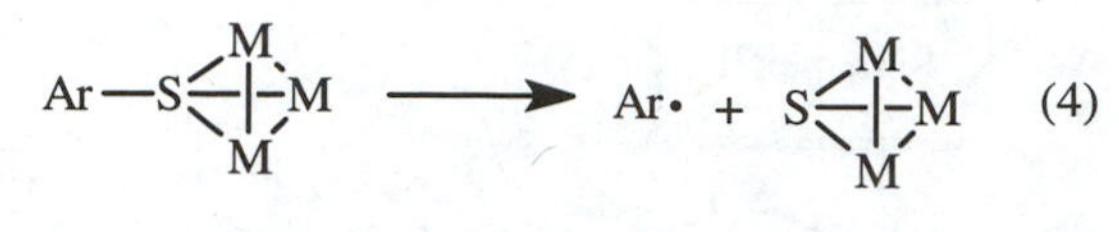

Coordination of the thietane can lead to ring opening via C-S bond homolysis. The S-tethered radical can bond to a metal atom, thus effecting an "insertion" reaction into the C-S bond. β-Elimination could then form a coordinated propenylthiolato cluster (Scheme IV, compound **11** with R = propenyl), after which the desulfurization follows the path shown in Scheme IV.

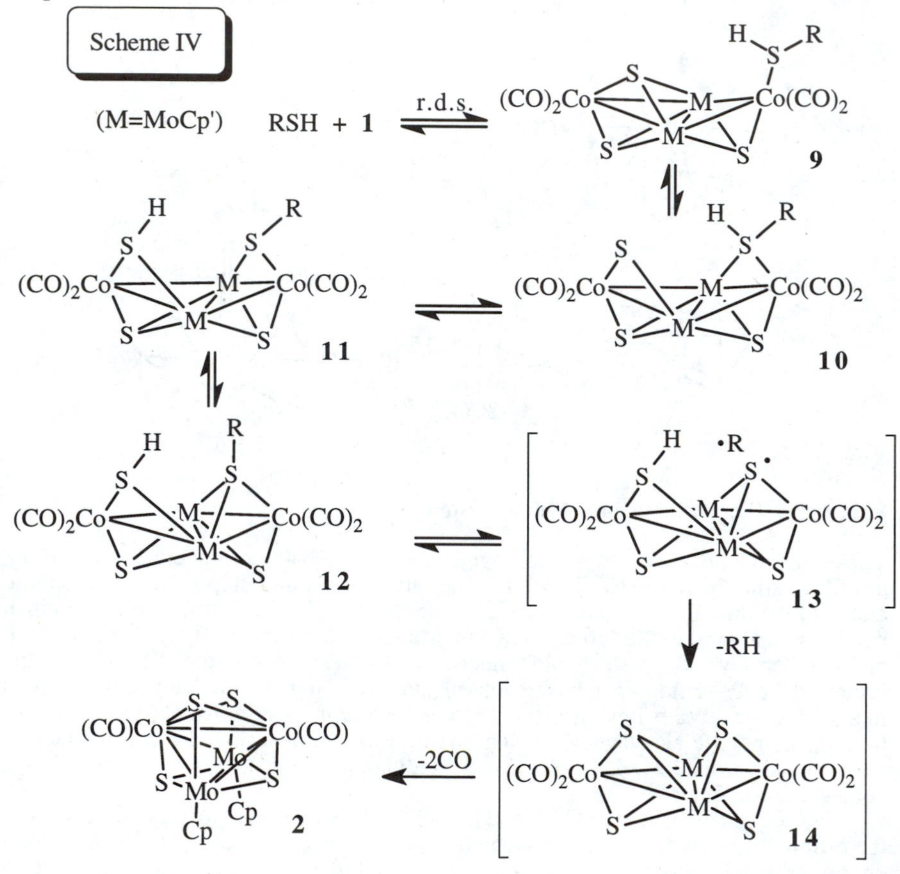

Desulfurization Reactions with Electron Deficient Clusters

In the course of the study of the desulfurization reactions with cluster **1**, we observed that cluster **2** reacted with good sulfur atom donors, e.g., thiiranes or thietanes, to form a black, insoluble powder that contained no carbonyl groups. This solid analyzed well for $Cp'_2Mo_2Co_2S_5$ and the far-IR spectrum of the black solid showed a prominent absorption at 388 cm^{-1}, a peak characteristic of all the $Mo_2M_2'S_4$ cubane clusters that we have examined. We therefore proposed that the structure of the black solid is a chain of cubane clusters linked by sulfide bridges as shown in equation 5 (*68*).

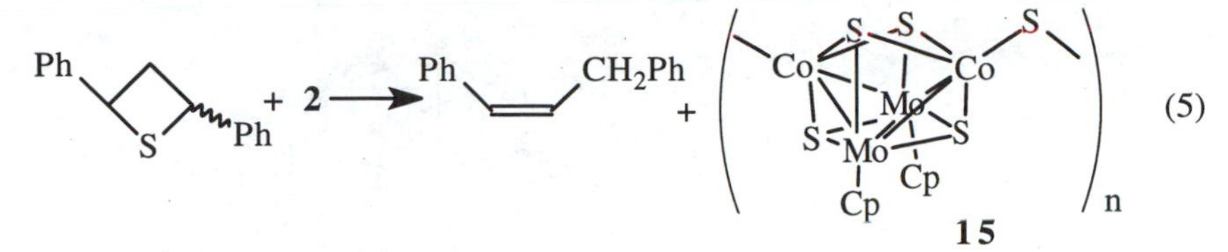

The cluster cores in structure **15** have been oxidized by two electrons relative to the core in cubane **2**. These 58 VSE clusters have been modeled by oxidizing **2*** with halogens and disulfides (**2*** = cluster **2** with C_5Me_4Et as the Cp' ligand). It was also found that benzene thiol oxidized **2*** to give the same product as obtained from the reaction of **2*** with diphenyldisulfide (Scheme V). The thiolato cluster, **16**, is paramagnetic with S = 1. The iodide, **19**, is a spin-ladder with an S = 1 ground state and thermally accessible S = 2 and S = 3 excited states. The Co...Co distances reflect these different magnetic properties. In **16**, the Co...Co distance is 2.74 Å, whereas in the halides, **17** and **19**, the corresponding lengths are 2.95 and 2.93 Å, respectively (*68*).

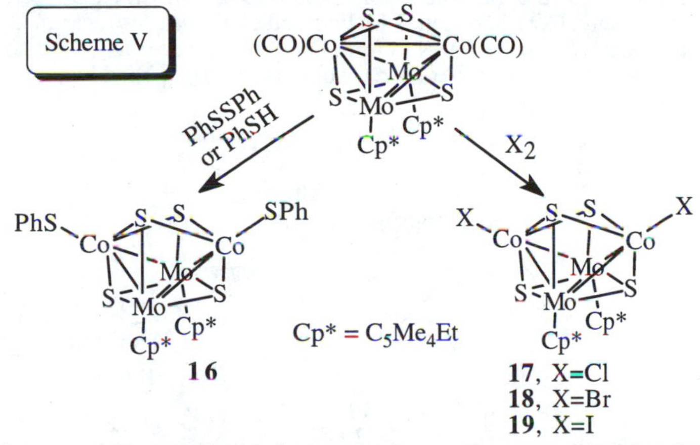

These modeling studies lend some support to the supposition that the insoluble sulfides obtained in the reaction depicted by eq. 5 are indeed cubane clusters linked by sulfide bridges. Some additional evidence was sought by attempting to convert the insoluble material back to a characterizable species. The thiolato cluster, **16**, was selected as a model for such reactions. Treatment of **16** with CO (400-1000 psi) regenerates **2** and produces PhSSPh (eq. 6).

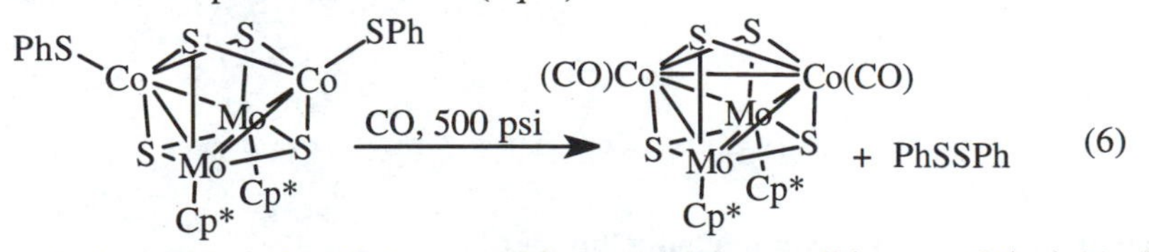

When the insoluble polysulfide was subjected to these conditions, a 40% yield of **2*** was obtained and COS was detected in the head gases by GC/MS. This transformation can be summarized by eq. 7. Yields higher than ca. 40% in eq. 7 have not been realized, possibly because some of the material has a structure that blocks access to the reagents.

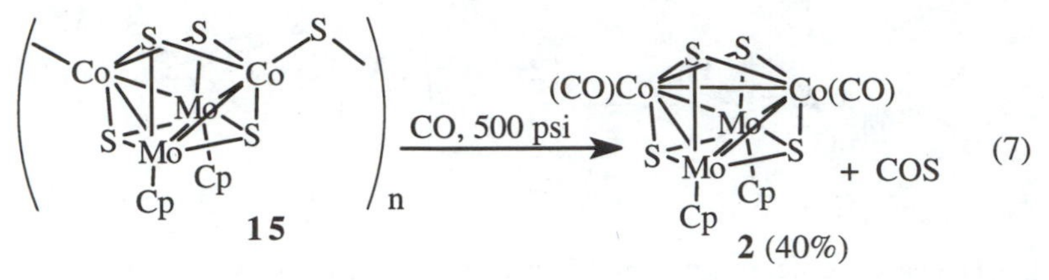

The conversion of **16** to **2*** (eq. 6), combined with the formation of **16** from PhSH (Scheme V), suggests the possibility of a catalytic cycle for the conversion of PhSH into PhSSPh and H_2. This conversion, shown in Scheme VI, was attempted with a 162-fold excess of PhSH under 1000 psi of CO at 150 °C. The products were PhSSPh (171%, based on **2***) and PhC(O)SPh (161%). A 56% yield of **16** was also recovered along with some of the black, insoluble polymer **16**. The PhSSPh was the expected product. The phenyl thiobenzoate was an unexpected product that must arise from a *catalytic* desulfurization of PhSH, and its formation also suggests that metal-carbon bonds are formed during the desulfurization reaction. We propose that under the high CO pressure, CO inserts into a M-Ph bond to give an acyl intermediate. This intermediate reacts with excess PhSH to form the thiobenzoate ester (Scheme VI). The mechanism of the initial desulfurization step shown in the right-hand loop in Scheme VI is at present unknown.

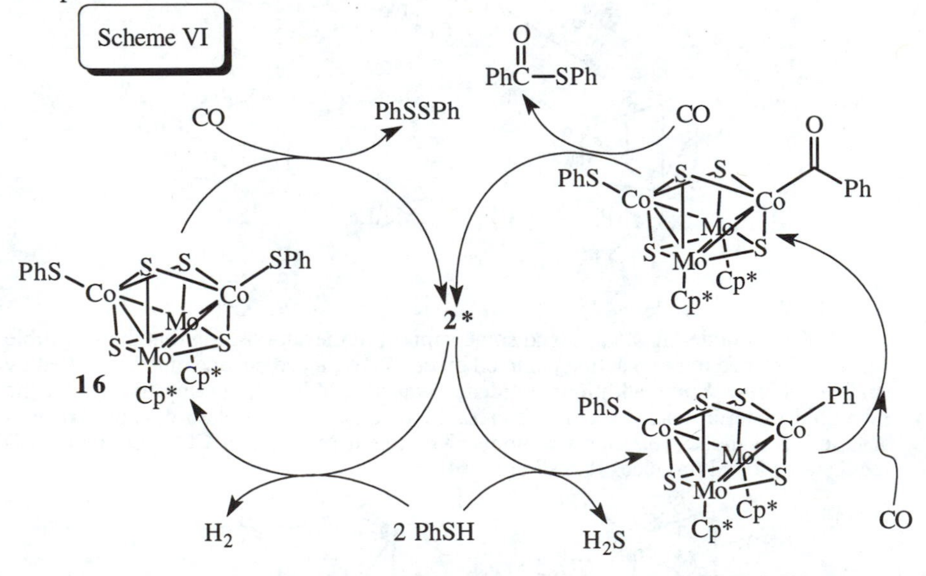

HDS Catalysts from Supported Clusters

Clusters **1** and **2** have been supported on γ-Al_2O_3 and used as CO hydrogenation and HDS catalysts (*39*). EXAFS spectra clearly showed that the clusters reacted with the surface at temperatures ≤ 100 °C to form oxo-surface ensembles (*39,69*). It appeared that the surface decompositions of **1** or **2** in either H_2 or He atmospheres led to the same surface oxo-ensembles. At least, there were no significant differences in the EXAFS spectra of these different preparations. Curve fitting showed that, after reaction with the surface, the Co-atoms were surrounded by six oxygen atoms at a distance of 2.06 Å. The Mo-XAS exhibited a pre-edge feature characteristic of molybdate, and curve fitting gave a coordination number of ca. 4 oxygen atoms at 1.77 Å.

Upon sulfidation with 15% H_2S in H_2 at 400 °C, the Mo-EXAFS changed to resemble that of disordered MoS_2, i.e., the EXAFS showed that the Mo atoms were sulfided and that they had coalesced into small MoS_2 crystallites. The curve fitting was consistent with 6 sulfur atoms as nearest neighbors at 2.46 Å and 3Mo next nearest neighbors at 3.23 Å (for highly crystalline MoS_2, these numbers are, respectively, 6S at 2.42 Å and 6Mo at 3.16 Å). The Fourier transform of the EXAFS is shown in Figure 3.

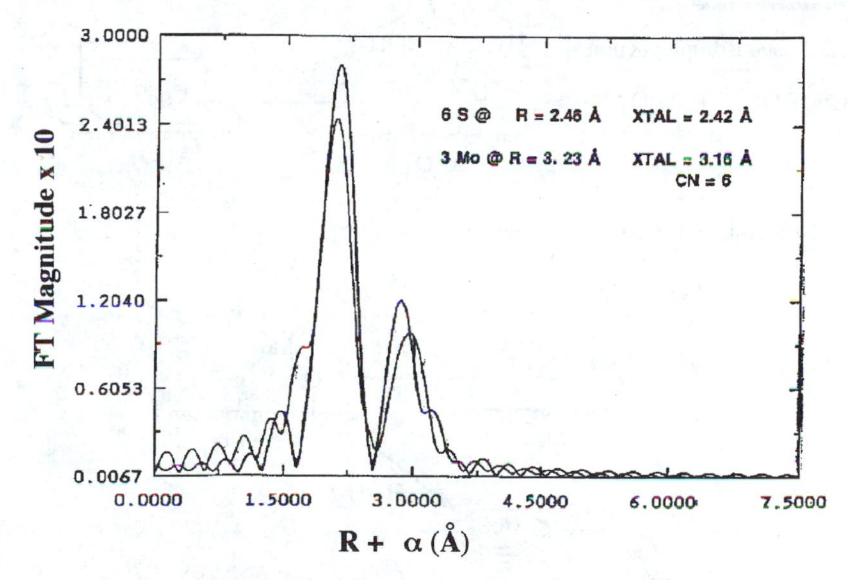

Figure 3. Phase shifted Fourier transform of the Mo EXAFS from sufided clusters supported on alumina.

The Co-EXAFS spectra show very little change during the sulfidation step (*69*). Curve fitting of the difference EXAFS for cluster **1** showed that, on average, the cobalt coordination changed by the addition of ≤1 S atom at R = 2.23 Å and a loss of ≤1 oxygen atom. Since EXAFS gives the average coordination environment, this result may mean that only about 1 in 6 Co-atoms are completely sulfided or that the Co is only partially sulfided and oxygen remains in the coordination sphere. A similar fitting procedure for the Co spectra from adsorbed cluster **2** showed that the cobalt picked up ca. 1.6 atoms of sulfur and lost 1.5 atoms of oxygen, i.e., slightly higher level of sulfidation was achieved with cluster **2** as the precursor, possibly reflecting the fact that the precursor cluster **2** has four S atoms as compared to the three S atoms in cluster **1**.

For comparison, a sample of 1% Co on Al_2O_3 was prepared by conventional impregnation with $Co(NO_3)_2$ solution, followed by calcination and sulfiding at 400 °C. This sample showed complete conversion of the Co to a sulfide, in contrast to the incomplete conversion with the cluster precursors. We attribute this difference to different Co species on the surface as a function of the deposition technique. The clusters are deposited onto the alumina surface from a non-aqueous medium (CH_2Cl_2). IR spectroscopy, as well as EXAFS data, show that the cluster is adsorbed intact initially. Upon heating, the cluster reacts with surface hydroxyl groups and is oxidized by them (even under H_2 atmosphere!). In this case, the Co is entrapped in the octahedral holes of the alumina lattice and, in this state, is not appreciably sulfided. The cobalt that is deposited from an aqueous solution, in contrast, apparently remains on the surface, first as an aquated complex and then as a surface oxide after calcination. This surface oxide is completely sulfided by the H_2S/H_2 mixture. These different behaviors are diagrammed in Scheme VII.

Scheme VII

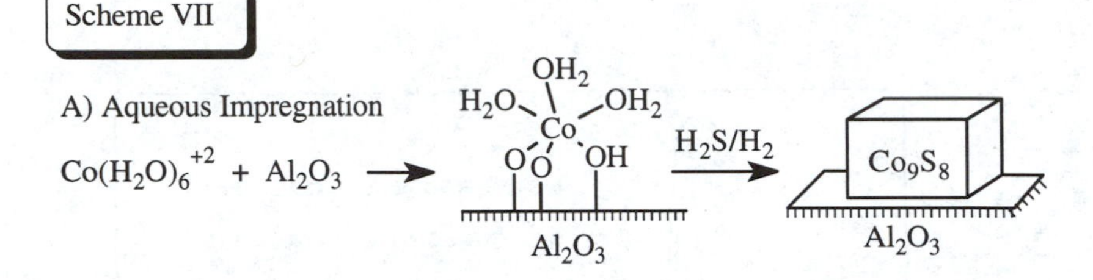

B) Non-Aqueous (Cluster) Impregnation

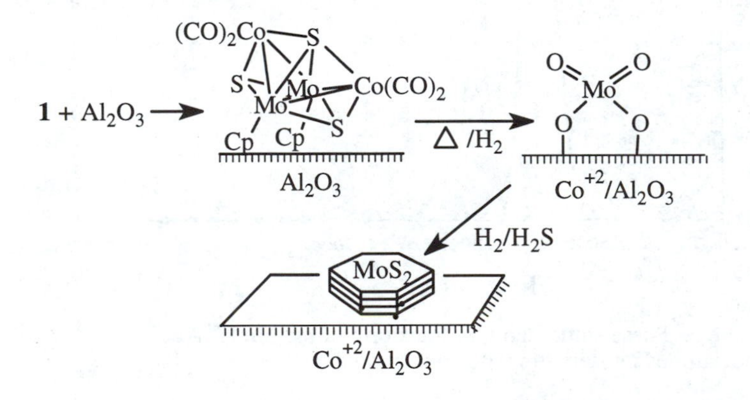

Table I shows the rates (turnover frequency or TOF) and product distributions for the HDS of thiophene at 300 °C catalyzed by 1% **1** on Al_2O_3, 1% **2** on Al_2O_3, and a commercial catalysts that contains ca. 14% MoO_3 and 4% CoO on Al_2O_3. The TOF is defined as moles of thiophene converted per second per mol of Mo. In addition to the butenes listed, small amounts of ethene, propane, propene, and butane were observed. The product distribution is essentially identical for all three catalysts. The equilibrium ratio of 1-butene: *cis*-2-butene: *trans*-2-butene at 300 °C is 1.0:1.9:2.8. The observed ratios are about 1.0:1.4:2.0. When extrapolated back to 0% conversion, the ratios are about 1.0:0.5:0.5. Therefore, the initially formed product slate is about equal parts 1- and 2-butenes, and there is no selectivity between the *cis*- and *trans*-2-

butenes. The initially formed butenes are then isomerized to a mixture that tends toward the equilibrium ratio with increasing conversion. Separate experiments showed that 1-butene was isomerized in a H_2-atmosphere over the HDS catalysts, but that under an atmosphere of He the rate of isomerization was very slow.

Table I. Product Distribution (% of Total) for HDS of Thiophene (2.7 mol% in 1 atm H_2) at 300°C

Catalyst[a]	TOF[b]	BuH	1-butene	c-2-butene	t-2-butene
1% **1**/Al_2O_3	26	4	22	30	43
1% **2**/Al_2O_3	3	4	22	28	38
Katalco[c]	14	3	22	27	37
1%CoMoS[d]	12	5	23	28	41

a) All catalysts calcined and presulfided @ 400°C. b) mol thiophene converted/s/mol Mo. c) 14% MoO_3, 4% CoO on Al_2O_3. d) Low loading conventional catalyst on Al_2O_3 with 1:1 Mo:Co.

The activity of the cluster-derived catalyst was about twice that of the commercial catalyst on a per mole of Mo basis. This may be due to a greater dispersion of the cluster-derived catalyst. Surprisingly, the activity of the catalyst derived from the cubane cluster, **2**, was nearly an order of magnitude less than that of the cluster **1** - derived catalyst. This decrease in activity correlates with the higher degree of Co-sulfidation, as observed by EXAFS spectroscopy, in the catalyst made from cluster **2**. Since the catalyst prepared from cluster **1** showed very little sulfidation of the cobalt by EXAFS, it may be that sulfidation of the cobalt is actually inimical to good catalyst performance. Such a conclusion is not in accord with the Topsøe "CoMoS" phase model, wherein all the active cobalt is presumably bound to sulfur at the edges of the basal planes of the MoS_2 crystallites.

MoS_2 Intercalation Catalysts

The results discussed above show that under HDS conditions, Mo/Co/S clusters adsorbed onto Al_2O_3 aggregate into MoS_2-like structures. The placement of the cobalt atoms remains problematic in that EXAFS showed that the Co atoms retained most of their oxygen coordination shell. We have begun a study of nanocomposites formed from the intercalation of Co species between the layers of MoS_2 in order to observe the behavior of the promoter/MoS_2 interface.

The layer structure of MoS_2 can be separated or exfoliated into a suspension of individual MoS_2 sheets. Typically, bulk MoS_2 is reacted with BuLi in hexane to form $LiMoS_2$. This compound, with Li^+ intercalated between the MoS_2 layers, is allowed to react with water under sonication. The $LiMoS_2$ reduces the water, giving H_2 and OH^- (equation 8). The hydrogen gas apparently blows the layers apart so that a suspension of single MoS_2 sheets, MoS_2(E), is obtained (*70*).

$$LiMoS_2 \xrightarrow[\text{sonicate}]{H_2O} LiOH + H_2 + MoS_2(E) \qquad (8)$$

Decreasing the pH of the suspension by the addition of strong mineral acid leads to flocculation and restacking of the MoS_2. Similarly, the addition of cations also leads to flocculation, but in this case the cations are sandwiched between the MoS_2 layers. The conditions (concentration and charge of the cation, etc.) determines the degree of intercalation. Only a single layer of metal ions, possibly with their hydration shells and hydroxide ions to impart electroneutrality, is incorporated between the MoS_2 sheets. The incorporation of molecules or ions between the MoS_2 sheets causes an expansion of the layer to layer distance, i.e., an expansion in the MoS_2 unit cell c-axis. This expansion, determined from the X-ray powder pattern, may be used to assess the presence and purity of the intercalated material.

In this way we have prepared the Co-intercalates shown listed in Table II (*71*). The sample labeled CoA/MoS_2 was prepared by adding 1M $Co(NO_3)_2$ solution (pH = 6.6) to the suspension of exfoliated MoS_2. CoC/MoS_2 was prepared from a solution of $Co(NH_3)_6Cl_3$, and Cp_2Co/MoS_2 was made by adding aqueous cobaltocenium chloride to the MoS_2 suspension. Table II lists the % Co in the isolated intercalate, the c-axis expansion, Δc, the crystallite size (l_c) in the c-direction as determined from the width of the 001 reflection, the BET surface area, and the turn-over frequency (mol thiophene/m^2-s) for the HDS of thiophene under 1 atm of H_2.

Table II. HDS Catalysts Prepared from MoS_2 Intercalates

Catalyst	% Co	Δc(Å)	l_c(Å)	BET (m^2/g)	TOF[a]x10^9
CoA/MoS_2	16.4	5.06	56	20	4.25
CoC/MoS_2	26.8	5.20	157	34	3.71
Cp_2Co/MoS_2	4.35	5.16	173	11	1.55
MoS_2	-	-	557	9	0.56
MoS_2(E)	-	-	200	52	1.19

a) mol thiophene converted/s/m^2(surface area).

The Co(III) hexammine complex ion is well known to be substitutionally inert, so we expected that CoC/MoS_2 would contain the $Co(NH_3)_6^{+3}$ ion between the MoS_2 layers. However, when subjected to sonication in mildly acidic HNO_3 solution, the cobalt was extracted from between the layers to give a solution of $Co(H_2O)_6^{+2}$, identified spectroscopically. From the concentration and volume of the resulting Co(II) solution, the amount of Co originally present in the intercalate was calculated, and this result agreed with the % Co determined by elemental analysis. Thus, it appears that the Co(III) was reduced by some residual charge in the MoS_2 sheets, and that CoA/MoS_2 and CoC/MoS_2 contain essentially the same Co species.

The magnetic susceptibilities of the samples were determined using a SQUID magnetometer (*71*). CoA/MoS_2 and CoC/MoS_2 are paramagnetic and the linear plots of $1/\chi$ vs. T were essentially superimposable with slopes corresponding to μ_{eff} = 4.89 B.M. per Co, a value in the range observed for high-spin Co(II) octahedral complexes. The Cp_2Co/MoS_2 was diamagnetic as expected if the intercalated species is Cp_2Co^+ and the charge is neutralized by hydroxide (the other possibility, a solid with negatively charged, reduced MoS_2 layers, Cp_2Co^+/MoS_2^-, should be paramagnetic).

TPDE (temperature-programmed decomposition) studies on the intercalates show loss of water in two peaks, one at ca. 210 ° and the second at 320 °C, for CoA/MoS_2. These results are similar to published observations on restacked MoS_2 that contained a bilayer of water (*72*). The Cp_2Co/MoS_2 showed two peaks (210 °, 380 °C), plus one near 100 °C corresponding to loss of surface water. In addition, the Cp_2Co/MoS_2 sample shows loss of CpH coincident with the water loss peak at 210 ° and a second Cp-loss peak near 320 °C.

The thiophene HDS activity of these intercalated and restacked MoS_2 nanocomposites was determined in a differential flow reactor under 1 atm of H_2. The results show that the restacked MoS_2 had an activity ca. 12 times that of the "native" MoS_2. However, the BET surface area of the restacked MoS_2 was almost six times that of the starting MoS_2, so that, when normalized for surface area, the exfoliated MoS_2 was about twice as active as the native MoS_2. This difference in reactivity is probably caused by a difference in the proportion of edge sites to rim sites as a function of crystallite size (*1,73*).

The Co-promoted catalysts derived from the intercalates all show slightly enhanced HDS activity, but no clear correlations are apparent. Activity seems to increase with Co-loading up to around 15-20% Co, but then stays constant or actually decreases somewhat. Increased disorder or decreased crystallinity, as judged by the l_c value, seems to promote activity. Part of the problem of attempting to make correlations with the structures or compositions of the intercalates is that these materials all decompose under HDS conditions, and the structures of the resultant materials are currently unknown. Very recently, EXAFS data have been collected for the intercalates before and after use as HDS catalysts. These results, when analyzed, should shed some new light on the structures of the intercalates and on their transformations under catalytic conditions (*74*).

Relevance of Homogeneous Cluster Reactivity to Heterogeneous HDS Catalysis

Our discovery that μ_3-coordination of thiols or thiolates to cluster **1** causes a dramatic decrease in the C-S BDE (80 → 20 kcal/mol) suggests that C-S bond homolysis should be seriously considered as a possible mechanism in heterogeneous HDS catalysis. This conclusion, reached on the basis of *homogeneous models*, was recently confirmed for the desulfurization of the two "radical clock" thiols (*64*), cyclopropylmethyl thiol and 6-mercaptohexene, over a *conventional* Co/Mo/S catalyst supported on Al_2O_3. The only organic products detected were those expected from rearrangements of the radicals arising from C-S bond homolysis, *viz.* 1-butene and methylcyclopentane, respectively (*75*). The ease with which the C-S bond is cleaved once the sulfur atom is coordinated to a μ_3-site is remarkable. This type of coordination is most likely to occur with thiolato or sulfido groups and the difficulty of desulfurizing thiophenes and benzothiophenes undoubtedly arises from their poor basicity and relatively low binding affinity to metals. Therefore, catalysts that are more effective at hydrogenating the double bonds in thiophenes and benzothiophenes will also be more active in HDS. This emphasizes the importance of the hydrogenation activity of HDS catalysts and points out a possible role of the promoter element, namely, to provide a higher hydrogenation activity. The latter may be achieved either by acting as a catalyst for the H_2-reduction of Mo(IV) (and thereby spilling over H^+ for removal of S^{-2} as H_2S) or by acting as a separate hydrogenation site for the thiophenic compounds. These proposals have been made many times previously, but the results obtained from model organometallic reactions now provide a factual basis for their importance. Another conclusion that follows from the drastic decrease in the C-S BDE upon coordination of the sulfur atom is that C-S bond breaking may not be the rate determining step in HDS catalysis. This conclusion is also suggested by some kinetic results (*10,76,77*) and by a recent theoretical paper (*78*).

The mechanism proposed in Scheme IV also introduces the question of whether the catalytic chemistry of sulfide catalysts is metal-centered or sulfur-centered. In sulfur-centered chemistry, the metal activates the sulfur as a leaving group, either by C-S bond homolysis, by an E_2- elimination to form a surface sulfhydryl group and an alkene (eq. 8), or by increasing the rate of displacement of the sulfur atom by an external nucleophile. In the metal-centered mode, oxidative addition reactions with C-S, C-H, and S-H bonds lead to M-H and M-R bonds that can then undergo reductive elimination to the alkane or β–elimination to the alkene (eq. 9). Both types of reactivity have now been observed in model organometallic systems and on surfaces under UHV conditions (*79-82*).

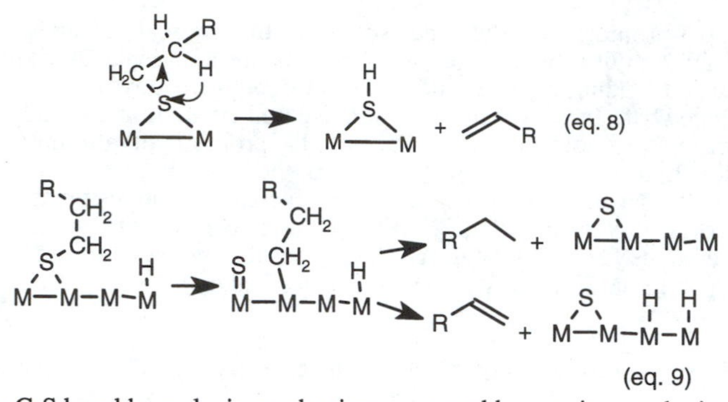

The C-S bond homolysis mechanism proposed here points to the importance of the "quality" of the S-vacancies as well as their number. The role of μ_3-S-coordination in activating the C-S bond now seems well established. Hence, any structural organization of the surface that promotes three-fold vacancies rather than two-fold or terminal vacancies will result in a catalyst more active toward C-S bond scission. However, such a catalyst may not be more active overall, since now the sulfur is bound more tightly to the metal surface, making its removal for the next turnover more difficult. Clearly, an optimized catalyst is one that can activate the C-S bond *and* whose surface can be easily cleaned by hydrogenation. As is often the case in catalysis, these two optimization goals are diametrically opposed to one another.

The mechanism shown in Scheme IV has other features that may be applied directly to heterogeneous catalysts. The fluxional "walking" of the bridging thiolate and sulfide ligands on the "surface" of the cluster has a direct analogy with the mobility of adsorbate molecules on the surfaces of heterogeneous catalysts. Also, we see in the displacement of the Co-(μ_4)S bond by an entering nucleophile (Schemes II and IV) another possible role for the Co promoter, namely, the creation of "latent vacancy" sites. A latent vacancy is defined as a structure that appears to have a complete coordination shell, but one that can accept a new bond to an adsorbate molecule by virtue of the latter displacing some other bond within the site structure. Since Co-S bonds are not as strong as Mo-S bonds, Co atoms located at the edges of the MoS_2 basal planes are attractive sites for latent vacancies and initial substrate binding. It is instructive to recall that an early, but now largely discarded, hypothesis for the catalytic action of Mo/Co/S HDS catalysts had MoS_2 merely acting as a support for the active Co-sulfide phase (*10,77,83*). The latent vacancy concept also explains the well-known difficulty of titrating the number of active sites on sulfide catalysts, since the number of sites determined by chemisorption will depend on the ability of the adsorbed species to displace a bond in the latent site. Both NO and O_2 have been used to determine the active site density of Co/Mo/S catalysts (*84*), and either of these species would be capable of displacing Co-S bonds. Hence, the apparent number of active sites will depend strongly on the conditions of the chemisorption experiment.

Summary

The structure and reactions of a class of Mo/Co/S clusters have revealed interesting clues to the mechanism of C-S bond cleavage over heterogeneous sulfide catalysts. In particular, a homolytic mechanism was suggested for the C-S bond breaking step in the cluster mediated desulfurization of radical clock reagents. It has now been shown that the same radical-derived products are obtained over the conventional heterogeneous catalyst. Hence, radical reactions must be seriously considered in any HDS mechanistic scheme. These homogeneous reactions have also suggested roles for the Co promoter element, namely as a hydrogenation enhancer and as an initial site for substrate bonding. High hydrogenation activity can convert aromatic derivatives of thiophenes and benzothiophenes to the more basic sulfides and thiols. The latter can then bond to three-fold site vacancies on the metal surface. It is the three-fold site vacancies that are expected to be most active in activating the C-S bond homolysis. Finally, these insights may aid the search for means to modify these catalysts for greater activity or selectivity.

Literature Cited

1) Chianelli, R. R.; Daage, M.; Ledoux, M. J. *Adv. Catal* **1994**, *40*, 177-232.
2) Hoffman, H. L. *Hydrocarbon Processing* **1991**, *70*, 37.
3) Massoth, F. E. *Adv. Catal.* **1978**, *27*, 265.
4) Chiu, N.-S.; Bauer, S. H.; Johnson, M. F. L. *J. Catal.* **1984**, *89*, 226.
5) Clausen, B. S.; Lengeler, B.; Topsøe, H. *Polyhedron* **1986**, *5*, 199.
6) Prins, R.; de Beer, V. H. J.; Somorjai, G. A. *Catal. Rev. - Sci. Eng.* **1989**, *31*, 1.
7) Parham, T. G.; Merrill, R. P. *J. Catal.* **1984**, *85*, 295.
8) Chiu, N. -S.; Bauer, S. H.; Johnson, M. F. L. *J. Catal.* **1986**, *98*, 32.
9) Clausen, B. S.; Topsøe, H.; Candia, R.; Villadsen, J.; Lengeler, B.; Als-Nielsen, J.; Christensen, F. *J. Phys. Chem.* **1981**, *85*, 3868.
10) Vissers, J. P. R.; DeBeer, V. H. J.; Prins, R. *J. Chem. Soc., Faraday Trans. 1* **1987**, *83*, 2145.
11) Parshall, G. W.; Ittel, S. D. *Homogeneous Catalysis*; Wiley, New York, NY, 1992.
12) Masters, C. *Homogeneous Transition Metal Catalysis: A Gentle Art*, Chapman and Hall, London, 1981.
13) Muetterties, E. L.; Stein, J. *Chem. Rev.* **1979**, *79*, 479.
14) Review: Angelici, R. J. *Accts. Chem. Res.* **1988**, *21*, 387.
15) Rauchfuss, T. B., *Prog. Inorg. Chem.* **1991**, *39*, 260.
16) Review: Rakowski Du Bois, M. *Chem. Rev.* **1989**, *89*, 1.
17) Lopez, L.; Godziela, G.; Rakowski Du Bois, M. *Organometallics* **1991**, *10*, 2660.
18) Gabay. J.; Dietz, S.; Bernatis, P.; Rakowski DuBois, M. *Organometallics* **1993**, *12*, 3630.
19) Giolando, D. M.; Rauchfuss, T. B.; Rheingold, A. L.; Wilson, S. R. *Organometallics* **1987**, *6*, 667.
20) Bolinger, C. M.; Rauchfuss, T. B.; Rheingold, A. L. *Organometallics* **1982**, 1, 1551; *idem. J. Am. Chem. Soc.* **1983**, *105*, 6321.
21) Ogilvy, A. E.; Draganjac, M.; Rauchfuss, T. M.; Wilson, S. R. *Organometallics*, **1988**, *7*, 1171.
22) Luo, S.; Ogilvy, A. E.; Rauchfuss, T. M.; Rheingold, A. L.; Wilson, S. R. *Organometallics* **1991**, *10*, 1002.
23) Luo, S.; Rauchfuss, T. B.; Gan, Z. *J. Am. Chem. Soc.*. **1993**, *115*, 4943.
24) Spies, G. H.; Angelici, R. J. *J. Am. Chem. Soc.* **1988**, *107*, 5569.
25) Lesch, D. A.; Richardson, Jr., J. W.; Jacobson, R. A.; Angelici, R. J. *J. Am. Chem. Soc.* **1984**, *106*, 2901.
26) Hachgenei, J. W.; Angelici, R. J. *J. Organometal. Chem.* **1988**, *355*, 359.
27) Chen, J.; Angelici, R. J. *Organometallics* **1990**, *9*, 879; and ibid., p. 879.
28) Benson, J. W.; Angelici, R. J. *Organometallics* **1993**, *12*, 680.
29) Benson, J. W.; Angelici, R. J. *Organometallics* **1992**, *11*, 922.
30) Huckett, S.C.; Angelici, R. J.; Ekman, M. E.; Schrader, G. L. *J. Catal.* **1988**, *113*, 36.

31) Kilanowski, D. R.; Tecueven, H.; de Beer, V. H. J.; Gates, B. C.; Schuit, G. C. A.; Kwart, H. *J. Catal.* **1978**, *55*, 129.
33) Jones, W. D.; Dong, L. *J. Am. Chem. Soc.* **1991**, *113*, 559.
39) Boorman, P. M.; Gao, X.; Fait, J. F.; Parvez, M. *Inorg. Chem.* **1991**, *30*, 3886.
34) Jones, W. D.; Chin, R. M. *Organometallics* **1992**, *11*, 2698.
35) Rosini, C. P.; Jones, W. D. *J. Am. Chem. Soc.* **1992**, *114,* 10767.
36) Dong, L.; Duckett, S. B.; Ohman, K. F.; Jones, W. D. *J. Am. Chem. Soc.* **1992**, *114,* 151.
37) Jones, W. D.; Chin, R. M. *J. Am. Chem. Soc.* **1992**, *114,* 9851.
38) Jones, W. D.; Chin, R. M. *J. Am. Chem. Soc.* **1994**, *116*, 198.
39) Curtis, M. D.; Penner-Hahn, J. E.; Schwank, J.; Baralt, O.; McCabe, D. J.; Thompson, L.; Waldo, G. *Polyhedron* **1988**, *7*, 2411.
40) Bianchini, C.; Meli, A.; Peruzzini, M.; Vizza, F.; Frediani, P.; Herrera, V.; Sanchez-Delgado, R. A. *J. Am. Chem. Soc.* **1993**, *115*, 2731.
41) Bianchini, C.; Meli, A.; Peruzzini, M.; Vizza, F.; Frediani, P. H.; Sanchez-Delgado, R. A. *J. Am. Chem. Soc.* **1993**, *115*, 7505.
42) Bianchini, C.; Meli, A.; Perizzini, M.; Vizza, F.; Herrera, V.; Sanchez-Delgado, R. A. *Organometallics* **1994**, *13*, 721.
43) Bianchini, C.; Meli, A.; Peruzzini, M.; Vizza, F.; Moneti, S.; Herrera, V.; Sanchez-Delgado, R. A. *J. Am. Chem. Soc.* **1994**,*116*, 4370.
44) Bianchini, C.; Jimenez, M. V.; Meli, A.; Moneti, S.; Vizza, F.; Herrera, V.; Sanchez-Delgado, R. A. *Organometallics* **1995**, *14,* 2342.
45) Adams, R. D. *Chem. Rev.* **1995**, *95*, 2587.
46) Adams, R. D.; Horvath, I. T. *Prog. Inorg. Chem.* **1985**, *33*, 127.
47) Adams, R. D.; Pompeo, M. P.; Wu, W.; Yamamoto, J. H. *J. Am. Chem Soc.* **1993**, *115*, 8207, and references therein.
48) Jaffey, D. M.; Madix, R. J. *J. Am. Chem. Soc.* **1994**,*116*, 3012.
49) Liu, A. C.; Friend, C. M. *J. Am. Chem. Soc.* **1991**, *113*, 820.
50) Stiefel, E. I.; Halbert, T. R.; Coyle, C. L.; Wei, L.; Pan, W. H.; Ho, T. C.; Chianelli, R. R.; Daage, M. *Polyhedron* **1989**, *8* , 1625-9.
51) Derouane, E. G.; Pedersen, E.; Clausen, B. S.; Gabelica, C.; Topsøe, H. *J. Catal.* **1987**, *107*, 587.
52) Harris, S. *Organometallics* **1994**, *13*, 2628.
53) Williams, P. D.; Curtis, M. D. *Inorg. Chem.* **1986**, *25*, 4562.
54) Curtis, M. D.; Williams, P. D.; Butler, W. M. *Inorg. Chem.* **1988**, *27*, 2853.
55) Li, P.; Curtis, M. D. *J. Am. Chem. Soc.* **1989**, *111*, 8279.
56) Li, P.; Curtis, M. D. *Inorg. Chem.* **1990**, *29*, 1242.
57) Curtis, M. D.; Riaz, U.; Curnow, O. J.; Kampf, J. W.; Rheingold, A. L.; Haggerty, B. J. *Organometallics* **1995**, *14*, 5337.
58) Li, P.; Curtis, M. D. unpublished results.
59) Riaz, U.; Curnow, O.; Curtis, M. D. *J. Am. Chem. Soc.*, **1991**, *113*, 1416.
60) Druker, S. H.; Mansour, M. A.; Sielemann, D.; Curtis, M. D. to be published.
61) Curtis, M. D.; Curnow, O. J. *Organometallics* **1994**, *13*, 2489.
62) Druker, S. H.; Curtis, M. D. *J. Am. Chem. Soc.* **1995**, *117*, 6366.
63) Dungey, K. E.; Curtis, M. D. to be published.
64) Beckwith, A. L. J.; Bowry, V. W. *J. Am. Chem. Soc.* **1994**, *116*, 2710.
65) Curnow, O. J.; Kampf, J. W.; Curtis, M. D.; Shen, J.-K.; Basolo, F. *J. Am. Chem. Soc.* **1994,** *116*, 224.
66) C-S bond dissociation energy calculated from thermochemical data in: Griller, D.; Kanabus-Kaminsks, J. M.; Maccoll, A. J. *J. Mol. Struct. (Theochem)* **1988**, *163*, 125.
67) Adams, R. D.; Pompeo, M. P. *Organometallics* **1992**, *11*, 103.

68) Mansour, M. A.; Curtis, M. D.; Kampf, J. W. *Organometallics* **1995**, *14*, 5460.
69) Curtis, M. D. *Appl. Organomet. Chem.* **1992**, *6*, 429.
70) Gee, M. A.; Frindt, R. F.; Joensen, P.; Morrison, S. R. *Mat. Res. Bull.* **1986**, *21*, 543.
71) Dungey, K. E.; Curtis, M. D. to be submitted for publication.
72) Joensen, P.; Crozier, E. D.; Alberding, N.; Frindt, R. F. *J. Phys. C: Solid State Phys.* **1987**, *20*, 4043.
73) Daage, M.; Chianelli, R. R. *J. Catal.* **1994**, *149*, 414.
74) Dungey, K. E.; Curtis, M. D.; Penner-Hahn, J. E. to be published.
75) Dungey, K. E.; Curtis, M. D. unpublished results.
76) Moser, W. R.; Rossetti, G. A.; Gleaves, J. T.; Ebner, J. R. *J. Catal.* **1991**, *127(1)*, 190-200.
77) DuChet, J. C.; Van Oers, E. M.; DeBeer, V. H. J.; Prins, R. *J. Catal.* **1983**, *80,*, 386.
78) Neurock, M.; van Santen, R. A. *J. Am. Chem Soc.* **1994**, *116*, 4427.
79) Serafin, J. G.; Friend, C. M. *J. Am. Chem. Soc.* **1989**, *111*, 8967.
80) Wiegand, B. C.; Uvdal, P. E.; Serafin, J. G.; Friend, C. M. *J. Am. Chem. Soc.* **1991**, *113*, 6686.
81) Wiegand, B. C.; Uvdal, P.; Friend, C. M. *J. Phys. Chem* **1992**, *96*, 4527.
82) Jaffey, D. M.; Madix, R. J. *J. Am. Chem. Soc.* **1994**, *116*, 3012.
83) DeBeer, V. H. J.; Duchet, J. C.; Prins, R. *J. Catal.* **1981**, *72*, 369.
84) Carvill, B. J.; Thompson, L. T. *Appl. Catal.* **1991**, *75*, 249, and references therein.

Chapter 9

Acid–Base Chemistry of Transition-Metal–π-Thiophene Complexes

Karen M. Koczaja-Dailey, Shifang Luo, and Thomas B. Rauchfuss[1]

School of Chemical Sciences, University of Illinois, Urbana, IL 61801–3080

Studies on the acid-base properties of (arene)Ru(C_4R_4S) complexes are summarized. The reactivity of π–thiophene complexes towards H^+ and OH^- depends on the oxidation state of the complex which in turn determines the hapticity of the thiophene ligand. Weak acids protonate (C_6Me_6)Ru(η^4-C_4R_4S) to give monocationic η^4-2-C_4R_4HS complexes. The S-C(sp^3) bond in the C_4H_5S complex spontaneously cleaves in the solid state as well as in solution to give a 1-thiapentadienyl derivative. For other R groups the protonated thiophene complexes do not undergo C-S bond scission. The thiophene ring in other (C_6Me_6)Ru(η^4-C_4R_4S) complexes can be cleaved using sources of ($C_5R'_5$)Ru^+ (R' = H, Me) to give products containing the Ru_2-(μ-η^3:η^4-SC_4R_4) group. This demonstrates the superior ability of two metals to effect C-S bond activation. In the case of Ru(II) complexes, such as (arene)Ru(η^5-$C_4R'_4S$)$^{2+}$, their C-S bonds are easily cleaved using aqueous base to give acylthiolato complexes of the type (arene)Ru(η^4-$SC_3R'_3COR'$). One of the curiosities of this chemistry is that the nucleophile, OH^-, initially attacks at sulfur. Amines and ammonia react with the dicationic complexes to give the iminium salts (arene)Ru(η^4-$SC_3R'_3C(NHR)R'$)$^+$.

The study of the transition metal complexes of thiophenes has uncovered a rich chemistry of a common class of heterocycles whose coordination chemistry had been examined ony lightly prior to 1990 (*1*). Research in this area has also provided insights into the hydrodesulfurization (HDS) reaction, an important industrial process with significant implications for the environment (*2*, *3*, see Chapter 1). While the underlying connection of organometallic chemistry to thiophene desulfurization was recognized in the 1960's (*4*), efforts over the last several years have dramatically advanced our understanding of HDS mechanisms.

The striking feature of thiophene as a ligand is the extremely low nucleophilicity of the sulfur. This is manifested in the reluctance of thiophene to form S-bound metal complexes and the difficulty in effecting S-alkylation and S-oxidation. Sulfur in thiophene has sulfonium character. Compensating for the low sigma basicity of the

[1]Corresponding author

0097–6156/96/0653–0176$15.00/0

thiophene sulfur, the pi-system is relatively nucleophilic as reflected by the fact that electrophilic substitution for thiophene is > 10^3 faster than for benzene (*5*). We have capitalized on this reactivity to prepare families of η^5-thiophene complexes starting from organometallic electrophiles (ring)M$(OTf)_2$, where M = Rh, Ir; ring = C_5Me_5, or where M = Ru, Os; ring = arenes such as cymene (4-isopropyltoluene) and C_6Me_6 (*6*).

Protonation and Metalation of Reduced Thiophene Complexes

The dicationic thiophene complex $(C_6Me_6)Ru(\eta^5\text{-}C_4Me_4S)^{2+}$, prepared from $(C_6Me_6)Ru(OTf)_2$ and C_4Me_4S, is easily reduced chemically as well as electrochemically to give the zero-valent species $(C_6Me_6)Ru(\eta^4\text{-}C_4Me_4S)$. With less substituted thiophenes, one can isolate the reduced products, although they are thermally unstable (*7*). It is even possible to prepare $(\eta^5\text{-}C_4Me_4S)Ru(\eta^4\text{-}C_4Me_4S)$ where the two thiophenes adopt different and rapidly interconverting coordination modes (*8*). These reduced thiophene complexes allow us to examine the path by which heterolytically activated hydrogen leads to the hydrogenolysis of thiophene. In heterolytic hydrogen activation, protons and reducing equivalents (electrons) are delivered to the substrate separately. Because of the high electron mobility in semiconductors such as MoS_2 and RuS_2 (*9*), these reducing equivalents can be readily transferred to substrates bound to the same particle of catalyst. If this is a valid description of thiophene HDS, then it should be possible to cleave thiophene C–S bonds by reduction followed by protonation.

Protonation of Reduced Thiophene Complexes. The reduced thiophene complexes $(C_6Me_6)Ru(\eta^4\text{-}C_4R_4S)$ (R = Me or H) are readily protonated by the weak acid NH_4^+ (*7*). In contrast, the protonation of tetramethylthiophene (*10*), one of the more basic thiophenes, requires strong acids (pK_a <-10). Thus, complexation as an η^4-ligand increases the basicity of thiophene by about 20 orders of magnitude. Protonation of $(C_6Me_6)Ru(\eta^4\text{-}C_4R_4S)$ causes a change in the coordination of the heterocycle such that the sulfur becomes bound and the tetrahedral carbon detaches from the metal (equation 1). The stereochemistry of the protonation is endo which

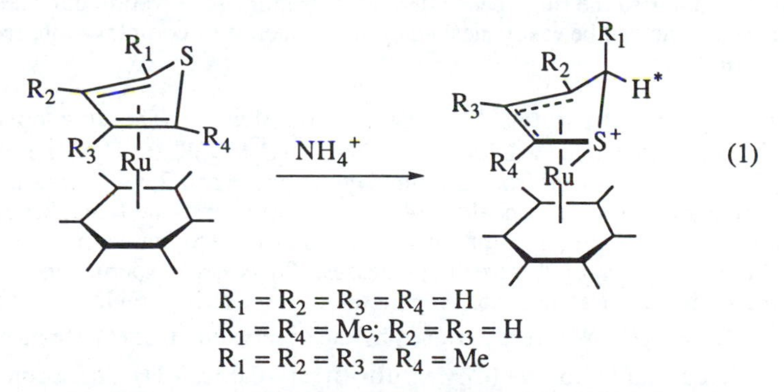

suggests that it may occur initially at Ru followed by migration to carbon. The $C(sp^3)$-S distance in $[(C_6Me_6)Ru(2,5\text{-}C_4Me_2H_3S)]PF_6$ is quite long at 1.91 Å, vs. 1.71 Å in free thiophene. The unusual pK_a of $(C_6Me_6)Ru(\eta^4\text{-}2\text{-}HC_4R_4S)^+$ vs. 2-$HC_4Me_4HC_4Me_4S^+$ is not due to a change in the site of protonation since tetramethylthiophene itself is also protonated at the α-carbon (*11*).

The protonated complex of thiophene $(C_6Me_6)Ru(\eta^4\text{-}C_4H_5S)^+$ slowly and reversibly undergoes C-S scission to give a thiapentadienyl derivative (equation 2) (*12*). It is possible to separate the ring opened isomer from the starting material by fractional crystallization (ring opened thiophenic ligands are referred to as SC_4R_n and ring closed forms as C_4R_nS). With these purified samples we were able to examine the approach to equilibrium from both directions. These measurements give K_{eq} (300 K) = 4.38. The rates of isomerization (k_1, k_{-1}) were both first order, as expected for simple unimolecular processes. Related thiapentadienyl complexes such as (C_5R_5)-$Ru(SC_4H_5)$ are formed in the reaction of $(C_5R_5)Ru(C_4H_4S)^+$ with hydride reagents (*13*).

The ring opening/closing processes are stereospecific as determined through studies on $(C_6Me_6)Ru(C_4H_4S\text{-}2\text{-}D)^+$. Decoupling and nOe measurements established the position of the deuterium shown in equation 2 (labeled H*). This transformation does not involve simple oxidative addition of the S-C bond since the

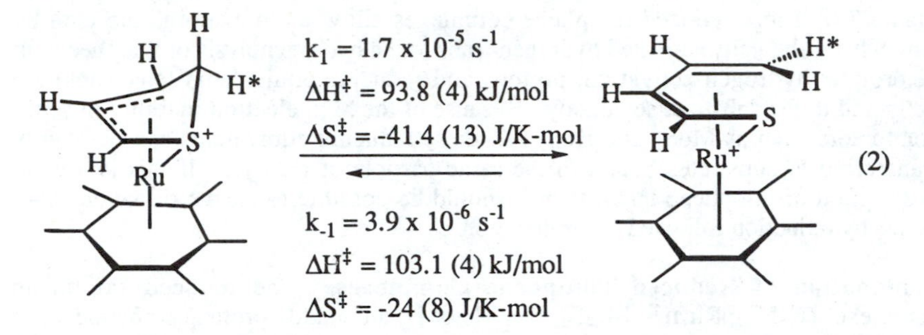

ruthenium remains divalent. Microcrystalline samples of $[(C_6Me_6)Ru(C_4H_5S)]BF_4$ were found to give the ring-opened product when heated for a few hours at 55 °C. The fact that the C-S scission also occurs in the solid state highlights the relevance of these results to heterogeneous catalysis. The reaction was conveniently monitored by solid state ^{13}C NMR spectroscopy, where we observed the steady conversion of the ring-closed complex to the ring opened derivative (Figure 1). Overall, our findings show that thiophene can be easily cleaved by the sequence of complexation, reduction, and protonation.

Ring Opening of Substituted Thiophenes Complexes. The ring opening discussed above is limited to the thiophene derivative $(C_6Me_6)Ru(C_4H_5S)^+$. Related protonated complexes derived from 2,5-dimethylthiophene and 2,3,4,5-tetramethylthiophene do not undergo ring opening (*14*). Since these substituted thiophenes are more representative of the kinds of substrates found in petroleum (*15*), we were interested in more general C-S scission processes. Successful experiments on this theme evolved from our attempts to prepare the triple decker complex $(C_6Me_6)_2$-$Ru_2(\eta^5\text{:}\eta^5\text{-}C_4Me_4S)^{2+}$. We were intrigued by the observation that $(C_6Me_6)Ru(\eta^4\text{-}C_4Me_4S)$ reacted with H_2 to give $(C_6Me_6)_2Ru_2(\eta^4\text{:}\eta^1\text{-}C_4Me_4S)H_2$ (*16*), a complex that is close in stoichiometry and structure to the desired triple decker (equation 3).

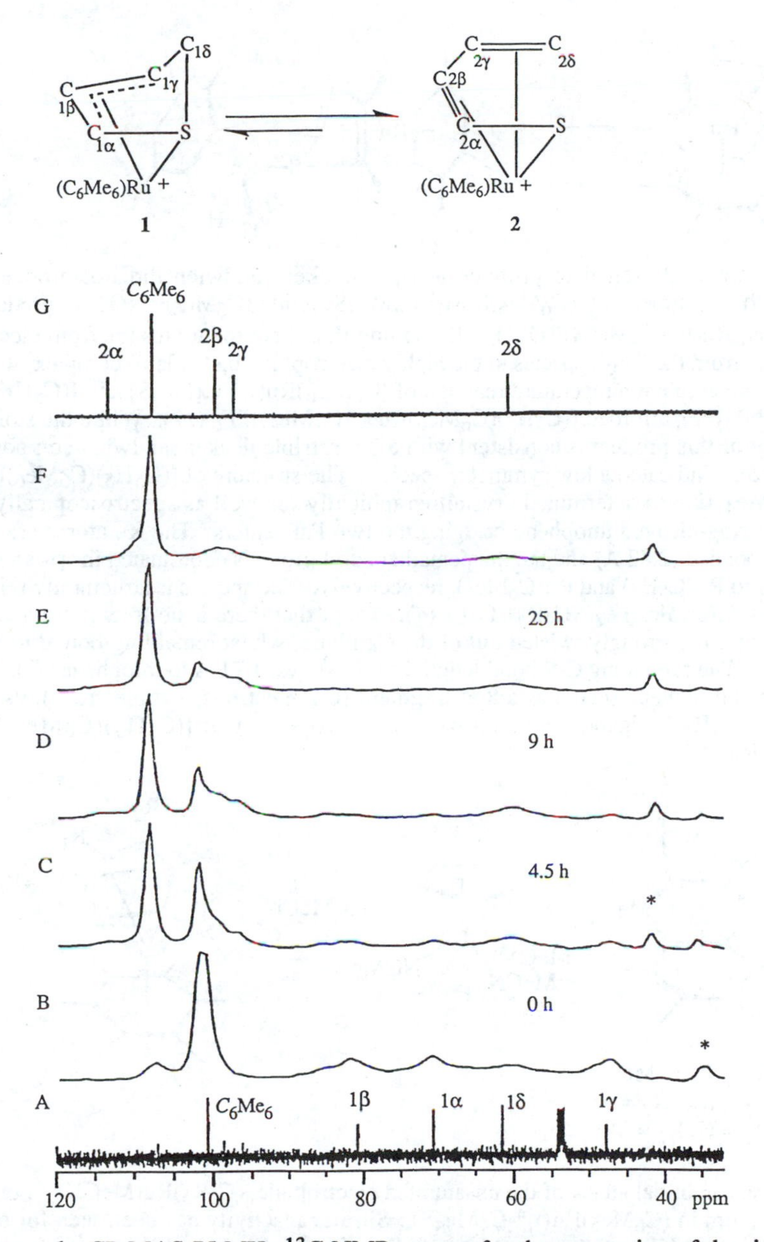

Figure 1. CP-MAS 75 MHz ^{13}C NMR spectra for the conversion of the ring closed isomer of $(C_6Me_6)Ru(C_4H_5S)^+$ to the ring opened isomer $(C_6Me_6)Ru(SC_4H_5)^+$. The sample was maintained at 55 °C, the spectra were recorded at ambient temperature. Spectra A and G are solution ^{13}C NMR spectra (100 MHz, acetone-d_6 soln) for the ring-closed and ring-opened isomers, respectively. Spectra B through E were recorded after the indicated reaction times. Spectrum F is for a purified sample of the ring-opened isomer. The asterisks mark spinning side bands.

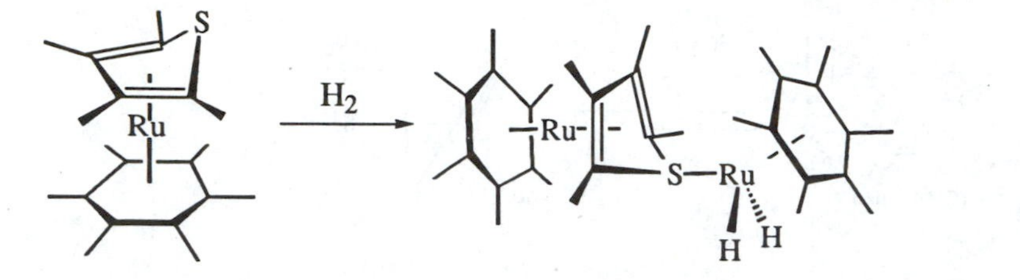

Experiments designed to produce a triple decker sandwich did not proceed as planned: treatment of $(C_6Me_6)Ru(\eta^4\text{-}C_4Me_4S)$ with $(C_6Me_6)Ru(OTf)_2$ produced $(C_6Me_6)Ru(\eta^5\text{-}C_4Me_4S)](OTf)_2$. Reasoning that this product results from electron transfer from the Ru(0) species to the highly electrophilic bis(triflato) complex, we

The ambient temperature reaction of $(C_6Me_6)Ru(\eta^4\text{-}C_4Me_4S)$ and $[(C_5H_5)Ru\text{-}(MeCN)_3]PF_6$ afforded $[(C_5H_5)(C_6Me_6)Ru_2(SC_4Me_4)]PF_6$ (*17*). While the stoichiometry of this product is consistent with a 32 e^- triple decker sandwich compound, NMR data indicated a low symmetry species. The structure of $[(C_5H_5)(C_6Me_6)Ru_2\text{-}(SC_4Me_4)]^+$ was determined crystallographically (as well as spectroscopically) to have a ring-cleaved thiophene bridging the two Ru centers. The Ru atoms are mutually bonded (2.82 Å) and the thiapentadienediyl group is coordinated in η^4- and η^3-modes to $Ru(C_5H_5)$ and $Ru(C_6Me_6)$, respectively. The species is structurally related to the thiaferroles $Fe_2(SC_4R_4)(CO)_6$ (*4*), except that there is no free olefin and the sulfur atom is strongly twisted out of the C_4 plane, whilst remaining bound to both metals. The remaining C-S bond length is 1.77 Å (vs. 1.71 Å in thiophene (*1*)). The ruptured thiophene serves as a 8 e^- fragment (6 π electrons, 2 σ electrons), vs. the 6e^- $\eta^5\text{-}C_4R_4S$ ligand (equation 4). We propose that $[(C_5H_5)(C_6Me_6)Ru_2(SC_4Me_4)]^+$

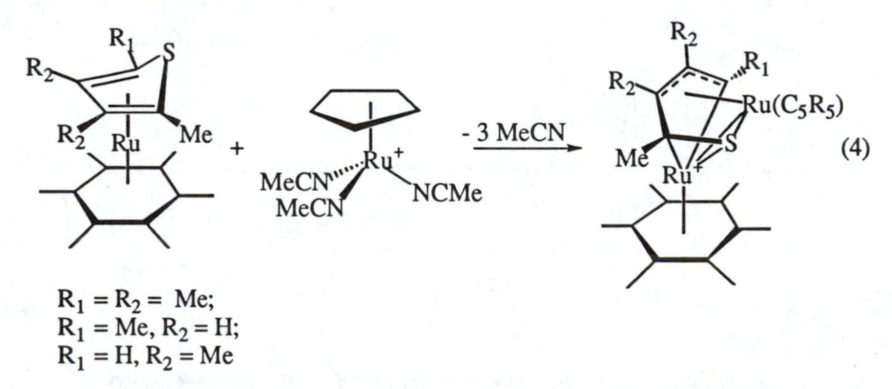

arises via an initial attack of the unsaturated electrophile, $(C_5R_5)Ru(MeCN)_2^+$, on the sulfur atom in $(C_6Me_6)Ru(\eta^4\text{-}C_4Me_4S)$. Similar reactivity has been seen for other metal electrophiles. Subsequent loss of two other MeCN ligands is compensated by C-S cleavage and Ru-Ru bond formation. Conceivably, the reaction proceeds via a triple decker species, although we detected no intermediates. η^4-Thiophene complexes are known to react with unsaturated *low valent* metal complexes ($Fe(CO)_4$, $Mo(CO)_5$) resulting in eventual desulfurization, but in these cases the added metal center serves as an S acceptor (*18,19*).

The generality of the new C–S cleavage reaction was tested by variations in the metal electrophile and the thiophene. For example, in place of $[(C_5H_5)Ru(MeCN)_3]^+$

one can use $[(C_5Me_5)Ru(MeCN)_3]^+$. We also showed that 2,3,4-trimethylthiophene and 2,5-dimethylthiophene complexes underwent C-S scission. The trimethylthiophene case is interesting because it afforded only a single regioisomer. The ^{13}C NMR and crystallographic analyses showed that the CH-S bond was cleaved (Figure 2).

It is noteworthy that the C-S oxidative addition is promoted by the attachment of a redox *inactive* electrophile. The ability of electrophiles to control the redox state of sulfur ligands is well known (e.g., the induced redox reaction (*20*)) and may be related to the catalytic activity of metal-sulfido ensembles. These results parallel and extend the proton-induced C-S cleavage (see equation 2) (*12*), with the advantage that the metal electrophile activates a more encumbered substrate.

On the basis of the aforementioned results, we propose that thiophenes can be cleaved by the combination of thiophene π-complexation, reduction, and electrophilic attack. An unusual aspect of this reaction is the implied coexistence of the electrophile and a reductant. We tested the feasibility of a three component reaction using a solution of both $(C_5H_5)Ru(MeCN)_3^+$, the *oxidized* thiophene complex $(C_6Me_6)Ru(C_4Me_4S)^{2+}$, and cobaltocene. This affords a 60% yield of $(C_6Me_6)(C_5H_5)Ru_2(SC_4Me_4)^+$, irrespective of the order in which these species were mixed. The cobaltocene employed in these experiments simulates the effect of the reducing equivalents provided by H_2 in HDS catalysis.

Relevant to our $Ru_2(\eta^3\eta^4\text{-}SC_4R_4)^+$ complex are the cations $[(triphos)M]_2(\eta^3{:}\eta^4\text{–}SC_4H_4)^{2+}$, where M = Rh, Ir and triphos = $CH_3C(CH_2PPh_2)_3$. These species are formed by the reaction of thiophene with $(triphos)Rh(C_2H_4)Cl/TlPF_6$ and $(triphos)Ir(\eta^4\text{–}C_6H_6)^+$, respectively. The pathway by which these species arise was not established (*21*).

Base Hydrolysis and Ammonolysis of Thiophene Complexes

Early work on reduced thiophene complexes led to the finding that the reduced thiophene complex $(C_5Me_5)Rh(\eta^4\text{-}C_4Me_4S)$ reacted with O_2 to give the sulfoxide $(C_5Me_5)Rh(\eta^4\text{-}C_4Me_4SO)$ (*22*). In view of the stoichiometry of the product, it appeared likely that the addition of hydroxide to the dication $(C_5Me_5)Rh(\eta^5\text{-}C_4Me_4S)^{2+}$ would produce the same species. This investigation led us to expand our study of the acid-base chemistry of coordinated thiophene ligands.

Base Hydrolysis. The base hydrolysis of dicationic thiophene complexes is best developed for derivatives of $(arene)Ru(C_4R_4S)^{2+}$, which are easily prepared for a wide range of substituents, R (*6*). At high $[OH^-]$, these dications convert to S-oxides (Scheme 1), whose structures are analogous to those initially obtained by oxygenation of reduced thiophene complexes. This transformation provides rare cases where divalent sulfur undergoes nucleophilic attack. On the basis of in situ NMR measurements we hypothesize that the S-oxides arise via $\eta^4\text{-}C_4Me_4S\text{-}OH$ ligands (*23*). In all cases S-oxides isomerize to the acylthiolates, often upon standing in solution, although chromatography on silica gel accelerates the process. For very stable S-oxides, such as $(C_5Me_5)Ir(\eta^4\text{-}C_4Me_4SO)$ and $(cymene)Os(\eta^4\text{-}C_4Me_4SO)$, the isomerization requires a two step procedure beginning with protonation by NH_4^+. This generates the 2-hydroxy derivatives such as $(C_5Me_5)Rh(\eta^4\text{-}2\text{-}HOC_4Me_4S)^+$. The structure of one such derivative was confirmed crystallographically (*24*); it is very similar to the aforementioned complexes prepared by protonation of $\eta^4\text{-}C_4R_4S$ complexes. It is not, however, clear if other "2-hydroxy" compounds adopt similar structures with two intact C-S bonds or, like the ring-opened form of $(C_6Me_6)Ru(SC_4H_5)^+$, they have only one C-S bond (see below). Deprotonation of the 2-hy-

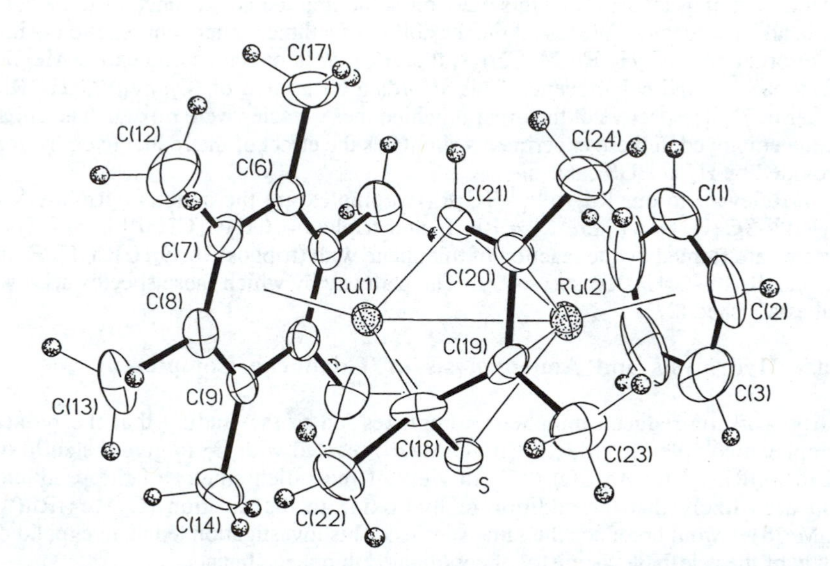

Figure 2. Structure of $[(C_6Me_6)(C_5H_5)Ru_2(SC_4Me_3H)]^+$ (BF_4^- salt) derived from the reaction of $[(C_5H_5)Ru(MeCN)_3]^+$ with the 2,3,4-trimethylthiophene complex $(C_6Me_6)Ru(\eta^4\text{-}2,3,4\text{-}C_4Me_3HS)$.

droxy compound cleanly gives the acylthiolates. It is noteworthy that virtually all base hydrolyses can be cleanly reversed with HOTf.

The acylthiolates exist in two diastereomeric forms, depending on the orientation of the acyl side. For $(C_5Me_5)Rh(SC_4Me_4O)$, the rate of isomerization is fast on the NMR time scale down to -60 °C where we observed two isomers. The rate of isomerization is slower for other acyl thiolates, and the equilibrium favors the second-formed, or thermodynamic, isomers. These thermodynamic isomers do not revert to the dicationic thiophene complexes upon treatment with acids, protonation occurs at carbon (*21*). Acylthiolato complexes derived from tetramethylthiophene thermally decompose to give tetramethylfuran and organometallic sulfido clusters (*25*).

Scheme 1

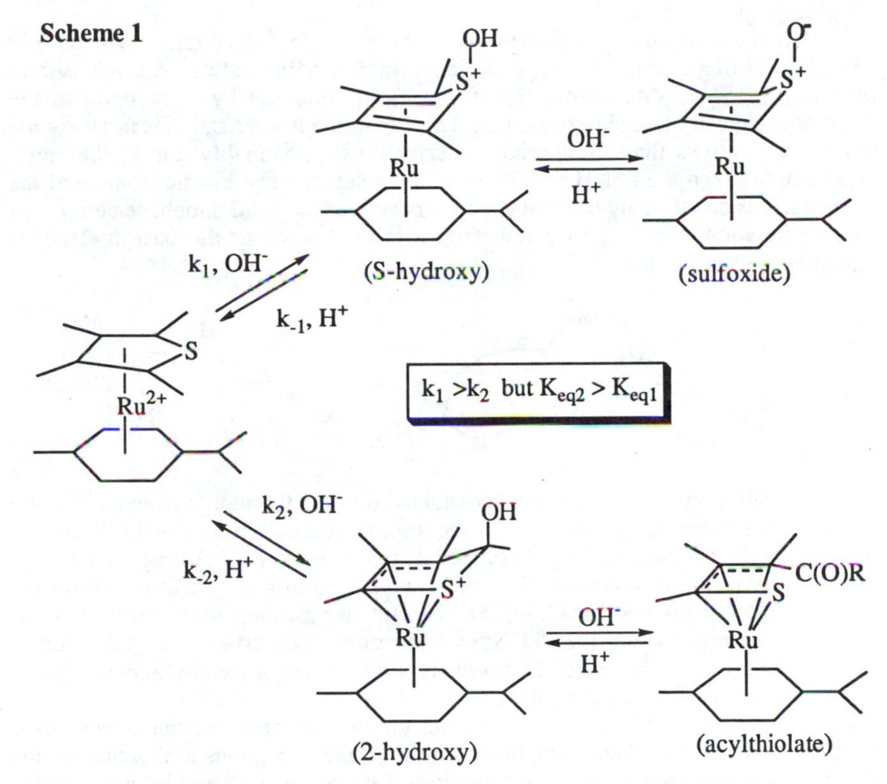

Amination of Thiophene Complexes. Dicationic π-thiophene complexes react efficiently with ammonia and amines to give monocations resulting from the net addition of NRH^- (*26*). The tetramethylthiophene complexes $(C_5Me_5)Rh(C_4Me_4S)^{2+}$ and $(arene)Ru(C_4Me_4S)^{2+}$ are the exception: they give ring adducts resulting from the addition of NHR^- without C-S scission, i.e., the products are $(ring)M(2\text{-}H_2NC_4Me_4S)^+$. Spectroscopic measurements confirm that these products adopt structures analogous to that of $(C_5Me_5)Rh(2\text{-}HOC_4Me_4S)^+$ (equation 5).

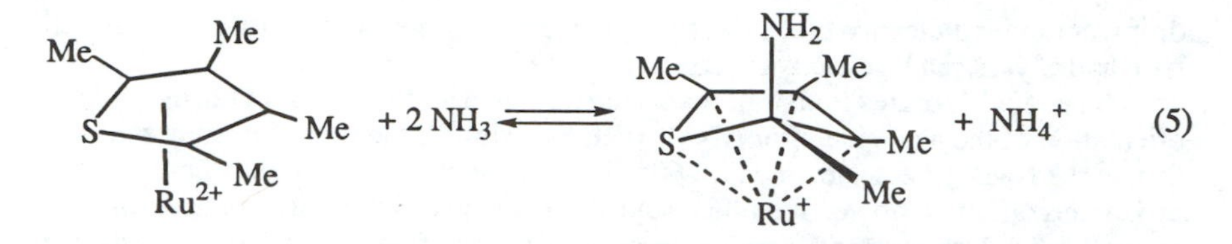

In contrast, ammonia reacts with complexes of thiophene, 2-methylthiophene, and 2,5-dimethylthiophene give deep red compounds that exist in the ring-opened form. The differing structures of the NH_2-derivatives of tetramethylthiophene and other thiophenes suggests that some species of the type (ring)M(2-HOC_4R_4S)$^+$ might also adopt ring-opened geometries.

Amination of other thiophene complexes results in C-S scission. The case of (ring)Ru(2,5-$Me_2C_4H_2S$)$^{2+}$ (ring = C_6Me_6, cymene) is illustrative. A single isomer of (ring)Ru($SC_3H_2MeCMeNH_2$)$^+$ forms first, followed by conversion to the thermodynamically favored exo isomer. The rate of the isomerization is faster for the aniline derivatives than the ammonia derivatives, presumably due to the steric repulsion between the PhNH substituent and the sulfur. The kinetic isomers of the NH_2^- adducts could easily be obtained by exposure of the solid thiophene complexes to gaseous ammonia. The overall pattern is analogous to the base hydrolysis reactions (equation 6).

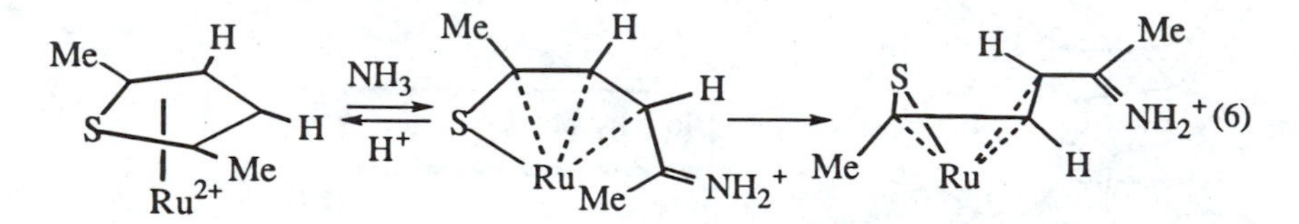

Crystallographic analyses were conducted on both thermodynamic and kinetic isomers. The distance between Ru and the iminium carbon in the kinetic isomer of (C_6Me_6)Ru($SC_3H_2MeCMeNH_2$)$^+$ is 2.627 Å, which is only 0.3 Å longer than other Ru-C distances in this complex. Two thermodynamic isomers are observed for the $PhNH^-$ adduct of (cymene)Ru(2,5-$C_4Me_2H_2S$)$^{2+}$; we attribute these isomers to cis-trans arrangements about the $PhHN^+$=CRMe unit. The crystallographic studies indicate conjugation via a π-bonding network with the rest of the allylthiolato ligand; this may explain their deep red colors.

Like the base hydrolysis reactions, most amination products can be converted back to the parent thiophene complexes with acid; the exceptions to this rule are the thermodynamic isomers. In these ring-closure processes it is highly unlikely that protonation occurs at the iminium nitrogen. We therefore suggest that acid-induced ring closure proceeds via an initial nucleophilic attack of the thiolate sulfur on the iminium carbon followed by protonation of the exocyclic amino group, as seen in (C_5Me_5)Rh(2-$NH_2C_4Me_4S$)$^+$.

The iminium centers in the ring-opened species are also reactive toward nucleophiles. Ammonia converts the $PhNH_2$-derived iminium compounds to the amino analogs. Aqueous base converts the kinetic iminium derivative of 2,5-dimethylthiophene to the kinetic isomer of the acyl thiolato complex.

We have applied the amination reaction to the optical resolution of π-thiophene complexes. Addition of (-)-S-2-aminoethylbenzene to (cymene)Ru(2-MeC_4H_3S)$^{2+}$ gives a 1:1 mixture of two diastereomeric iminium salts that can be separated by chromatography on silica gel. Treatment of the individual diastereomers with HOTf gives a single enantiomer of (cymene)Ru(2-MeC$_4$H$_3$S)$^{2+}$. The optically active thiophene complexes are configurationally stable in solution for many days (*27*).

Summary

Coordinated thiophenes exhibit rich reactivity towards simple reagents. The protonation of reduced thiophene ligands provides a plausible model for thiophene HDS. Our studies suggest that C–S cleavage is facilitated by reduction of a metal-thiophene ensemble followed by the addition of an electrophile. The base hydrolysis of cationic thiophene complexes is a particularly rich theme that has been fruitfully extended to the use of amines. The nucleophilic additions uncovered in this work merit further attention since the requisite (arene)Ru(C$_4$R$_4$S)$^{2+}$ salts are easily prepared at moderate cost and the transformations are efficient.

Acknowledgment

This research was funded by the U. S. Department of Energy through DEFG02-90-ER14146.

Literature Cited

(1) Rauchfuss, T. B. *Prog. Inorg. Chem.* **1991**, *39*, 259.
(2) TopØse, H.; Clausen, B. S.; Massoth, F. E. in *Hydrotreating Catalysis, Science and Technology;* Springer-Verlag: Berlin, 1996
(3) Angelici, R. J. In *Encyclopedia of Inorganic Chemistry*; King, R. B., Ed.; J. Wiley and Sons: New York, 1994, vol. 3; p 1433.
(4) Ogilvy, A. E.; Draganjac, M.; Rauchfuss, T. B.; Wilson, S. R. *Organometallics* **1988**, *7*, 1171.
(5) Taylor, R. In *Chemistry of Heterocyclic Compounds*; Gronowitz, S., Ed.; Wiley: New York, 1982; Vol. 4.
(6) Ganja, E. A.; Rauchfuss, T. B.; Stern, C. L. *Organometallics* **1991**, *10*, 270.
(7) Luo, S.; Rauchfuss, T. B.; Wilson, S. R. *J. Am. Chem. Soc.* **1992**, *114*, 8515.
(8) Luo, S.; Rauchfuss, T. B.; Rheingold, A. L. *Organometallics* **1992**, *11*, 3497.
(9) Tributch, H. In *Sulfur, Its Significance for Chemistry for the Geo-, Bio-, and Cosmosphere and for Technology*; Müller, A., Krebs, B., Eds.; Elsevier: Amsterdam, 1984, 277.
(10) Hogeveen, H. *Recueil* **1966**, *85*, 1072.
(11) Luo, S.; Feng, Q.; Olmstead, M.; Rauchfuss, T. B.; Stafford, P. R. unpublished data.
(12) Luo, S.; Rauchfuss, T. B.; Gan, Z. *J. Am. Chem. Soc.* **1993**, *115*, 4943.
(13) Hachgenei, J. W.; Angelici, R. J. *Angew. Chem., Int. Ed. Engl.* **1987**, *26*, 909.
(14) For the influence of the degree of ring methylation on C-S scission, see Jones, W. D.; Dong, L. *J. Am. Chem. Soc.* **1991**, *113*, 559.
(15) *Geochemistry of Sulfur in Fossil Fuels*; Orr, W. L.; White, C. M., Eds; American Chemical Society: Washington, DC, 1990.
(16) Luo, S.; Rauchfuss, T. B.; Rheingold, A. L. *J. Organometal. Chem.* **1994**, *472*, 295.

(17) Koczaja Dailey, K. M.; Rauchfuss, T. B.; Yap, G.; Rheingold, A. L. *J. Am. Chem. Soc.* **1995**, *117*, 6396.
(18) Luo, S.; Ogilvy, A. E.; Rauchfuss, T. B.; Rheingold, A. L.; Wilson, S. R. *Organometallics* **1991**, *10*, 1003.
(19) Chen, J.; Daniels, L. M.; Angelici, R. J. *J. Am. Chem. Soc.* **1991**, *113*, 2544.
(20) Young, C. J.; Kocaba, T. O.; Yan, X. F.; Tiekink, E. R. T.; Wei, L.; Murray, H. H., III; Coyle, C. L.; Stiefel, E. I. *Inorg. Chem.* **1994**, *33*, 6252 and references therein.
(21) Bacchi, A; Branchini, C.; Herrera, U.; Jimenez, M. V.; Mealli, C.; Meli, A.; Moneti, S.; Peruzzini, M.; Sanchez-Delgado, R. A.; Vizza, F. *J. Chem. Soc., Chem. Commun.* **1995**, 921.
(22) Skaugset, A. E.; Rauchfuss, T. B.; Stern, C. L. *J. Am. Chem. Soc.* **1990**, *112*, 2432.
(23) Krautscheid, H.; Feng, Q.; Rauchfuss, T. B. *Organometallics* **1993**, *12*, 3273.
(24) Skaugset, A. E.; Rauchfuss, T. B.; Wilson, S. R. *J. Am. Chem. Soc.* **1992**, *114*, 8521.
(25) Feng, Z.; Krautscheid, H.; Rauchfuss, T. B.; Skaugset, A. E.; Venturelli, A. *Organometallics* **1995**, *14*, 297.
(26) Feng, Q.; Rauchfuss, T. B.; Wilson, S. R. *Organometallics* **1995**, *34*, 5220.
(27) Koczaja Dailey, K. M.; Rauchfuss, T. B. unpublished results.

Chapter 10

Catalytic Hydrogenolysis of Thiophenic Molecules to Thiols by Soluble Metal Complexes

C. Bianchini[1], A. Meli[1], and R. A. Sánchez-Delgado[2]

[1]Istituto per lo Studio della Stereochimica ed Energetica dei Composti di Coordinazione, CNR, Via J. Nardi 39, 50132 Firenze, Italy
[2]Instituto Venezolano de Investigaciones Cientificas, Caracas 1020–A, Venezuela

The thermally generated 16-electron fragments [(triphos)RhH] and [(triphos)IrH] react with benzo[*b*]thiophene (**BT**) and dibenzo[*b,d*]thiophene (**DBT**) by C-S bond scission to give (triphos)Rh[η^3-S(C_6H_4)CH=CH_2] and (triphos)IrH(η^2-*C,S*-DBT), respectively [triphos = MeC(CH_2PPh_2)$_3$]. The Rh complex is an efficient catalyst precursor for the homogeneous hydrogenation of **BT**, forming 2-ethylthiophenol and, to a lesser extent, dihydrobenzo[*b*]thiophene. The Ir complex is a catalyst precursor for the homogeneous hydrogenation and hydrodesulfurization of **DBT** to 2-phenylthiophenol, biphenyl and H_2S. The mechanisms of these catalytic transformations have been elucidated by the isolation and characterization of key species related to catalysis combined with high pressure NMR spectroscopic studies.

The development of efficient catalysts for the simple hydrogenolysis of thiophenic molecules to thiols remains an attractive goal in hydrodesulfurization (HDS) catalysis. The thiol products can then be desulfurized over solid catalysts under milder reaction conditions than those required to accomplish the overall HDS of the thiophene precursors. This aspect is particularly important for the dibenzothiophenes since the conventional catalysts can desulfurize the corresponding aromatic thiols without affecting the benzene rings, necessary to preserve a high octane rating.

Some homogeneous modeling studies devoted to understanding the mechanisms through which thiophenes are degraded to thiols by transition metal complexes have recently been reported (*1-10*). Among these, only two examples have been described in which the hydrogenolysis reactions occur in catalytic fashion (*9,10*).

The present article is concerned with these two homogeneous hydrogenolysis reactions, which involve benzo[*b*]thiophene (**BT**) and dibenzo[*b,d*]thiophene (**DBT**) as model compounds.

Rh-Catalyzed Conversion of Benzo[*b*]thiophene into 2-Ethylthiophenol

The most widely accepted mechanisms for HDS of **BT** over solid catalysts are shown in Scheme 1. Path **a** begins with hydrogenation to form dihydrobenzo[*b*]thiophene

0097–6156/96/0653–0187$15.00/0

(**DHBT**) prior to C-S bond scission and desulfurization. The second pathway (**b**) involves initial C-S bond scission, followed by desulfurization of the 2-vinylthiophenol product, and hydrogenation of the vinyl group.

The results obtained with the homogeneous metal system described below show that the hydrogenolysis of **BT** to 2-ethylthiophenol (**ETSH**) can occur only after insertion has occurred into the C-S bond and that path **b** may be redirected (dotted line in Scheme 1) so as to contain the hydrogenation of 2-vinylthiophenol to **ETSH** prior to the desulfurization step.

The thermally generated 16-electron fragment [(triphos)RhH] reacts with **BT** by C-S bond scission to give (triphos)Rh[η^3-S(C_6H_4)CH=CH_2], (**1**) [triphos = MeC(CH_2PPh_2)$_3$] (*11*). The 2-vinylthiophenolate complex, **1**, forms by reductive coupling of a terminal hydride with the vinyl moiety of a metallabenzothiabenzene intermediate (Scheme 2) (*7*).

The 2-vinylthiophenolate complex, **1**, is an active catalyst precursor for the homogeneous transformation of **BT** into **ETSH**. At 160°C and 30 atm H_2 in either THF or acetone, **BT** is converted to **ETSH** with an average rate of 13 (mol per mol of catalyst per hour) in the first two hours. The reaction is not fully selective as **DHBT** is also formed (relative rate of 0.6) in an independent catalysis cycle (*vide infra*). Catalytic runs performed under different conditions show that the rate of formation of **ETSH** increases significantly with the concentration of **BT**, while it is only slightly affected by the H_2 pressure (Table I). Below 15 atm H_2 and 100°C, no appreciable transformation of **BT** is observed. The catalytic system is truly homogeneous up to 180°C. Above 200°C, appreciable decomposition of the catalyst occurs with formation of Rh metal particles which are responsible for the observed heterogeneous HDS of **BT** to ethylbenzene and H_2S (entries 6,14 of Table I).

The catalytic mechanism has been probed by high-pressure NMR (HPNMR) spectroscopy combined with the isolation and characterization of key species of the catalytic cycle. Under catalytic conditions, $^{31}P\{^1H\}$ NMR spectroscopy shows that all rhodium is incorporated into (triphos)Rh(H)$_2$[*o*-S(C_6H_4)C_2H_5], (**2**) and [(η^2-triphos)Rh{μ-*o*-S(C_6H_4)C_2H_5}]$_2$, (**3**). Below 100°C, only the dihydride complex, **2**, is present in solution. After quenching the catalytic reactions with dinitrogen, all rhodium is recovered as the bis-thiolate complex (triphos)RhH[*o*-S(C_6H_4)C_2H_5]$_2$ (**4**).

The nature of the chemical processes that connect compounds **1, 2, 3** and **4** has been elucidated by independently carrying out a variety of reactions using isolated compounds, some of which are summarized in Scheme 3. The 2-vinylthiophenolate complex, **1**, reacts in THF with H_2 (>15 atm) at 60°C, quantitatively converting to the dihydride, **2**. This reaction has been mimicked by the sequential addition to **1** of H^+, H^-, and H_2. From this experiment, it has been concluded that the conversion of **1** to **2** is a stepwise process in which the higher activation energy step is the first H_2 uptake to give the (alkyl)hydride (triphos)RhH[η^2-S(C_6H_4)CH(CH_3)], (**5**). This step may involve a heterolytic splitting of H_2. The dihydride, **2**, and the dimer, **3**, are in equilibrium in THF by reductive elimination/oxidative addition of H_2, while the dimer **3** reacts with **ETSH** yielding the bis-thiolate complex, **4**.

Incorporation of all of the experimental evidence leads to the mechanism shown in Scheme 4 for the reaction between **BT** and H_2 (15-60 atm) catalyzed by **1** in the temperature range from 120 to 180°C where the system is homogeneous.

Initially, the 2-vinylthiophenolate ligand in **1** is hydrogenated to 2-ethylthiophenolate (steps **a-b**). The unsaturated 16-electron fragment [(triphos)Rh{*o*-S(C_6H_4)C_2H_5}] either picks up further H_2 to give the dihydride **2** (step **d**) or dimerizes to **3** (step **c**), the latter path being favored at high temperature and low H_2 pressure. It is the dihydride complex that, upon interaction with **BT**, eliminates **ETSH** and forms an η^1-*S*-**BT** adduct (step **e**). In the η^1-*S* bonding mode, **BT** is

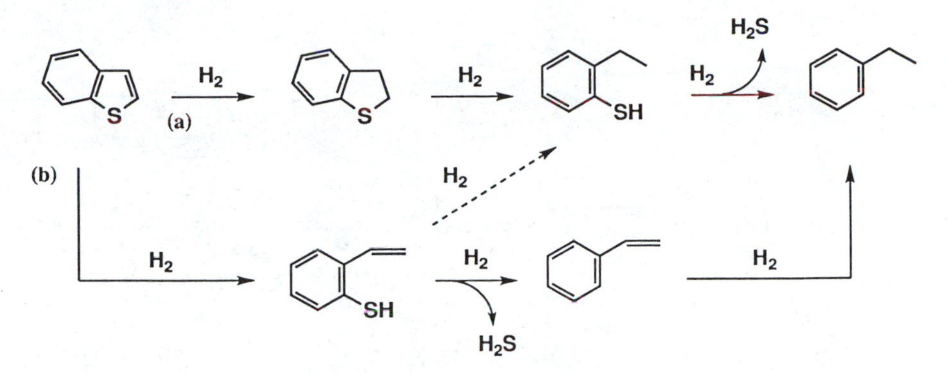

Scheme 1. Proposed heterogeneous mechanisms for HDS of **BT**.

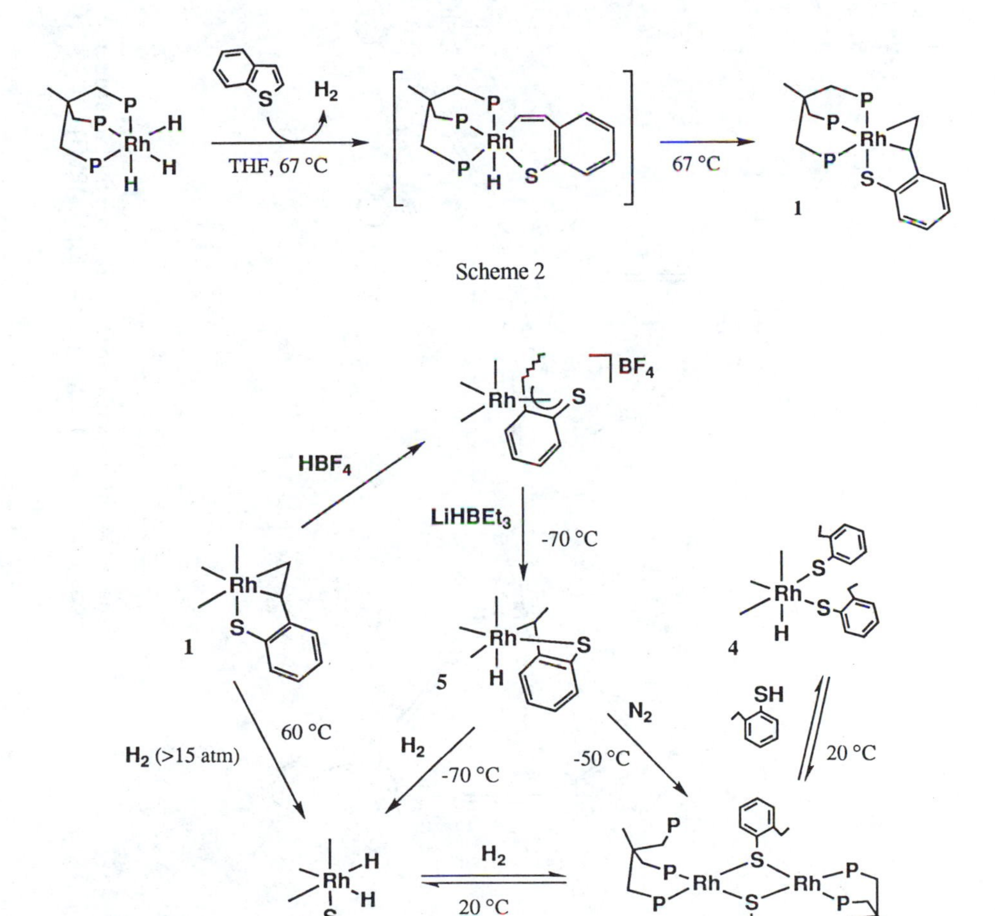

Scheme 2

Scheme 3

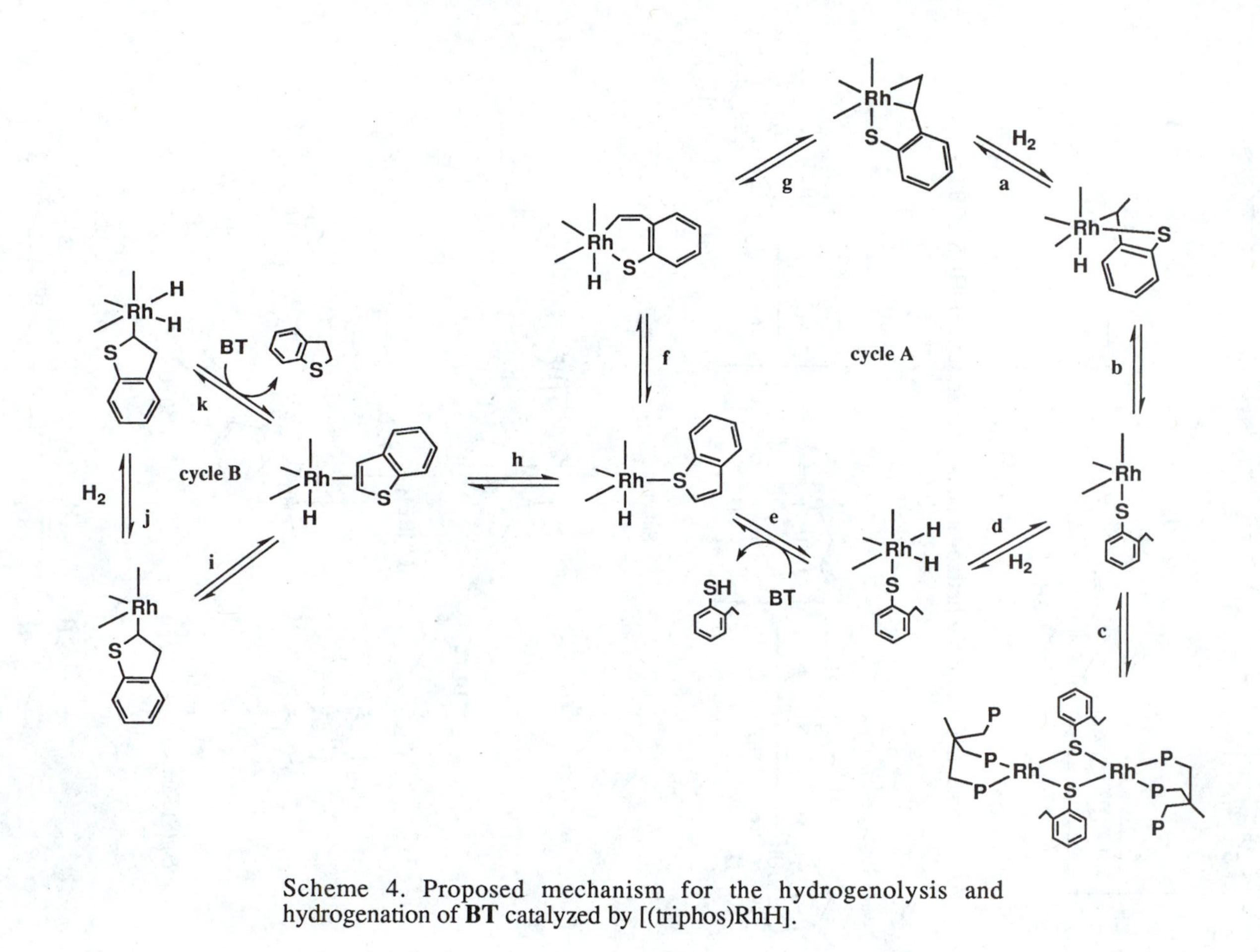

Scheme 4. Proposed mechanism for the hydrogenolysis and hydrogenation of **BT** catalyzed by [(triphos)RhH].

activated in such a way that C-S insertion is followed by attack by the electron-rich Rh(I) metal on the adjacent carbon atom (*via* electron donation into the C-S antibonding orbital). As a result of the C-S bond scission, the rhodabenzothiabenzene hydride intermediate is formed (step **f**), which regenerates the 2-vinylthiophenolate precursor **1** (step **g**) *via* hydride migration, thus closing the catalytic cycle **A**.

The observed catalytic production of **DHBT** is explained by taking into account a parallel catalysis cycle, quite similar to those previously described by Fish (*12-14*) and Sánchez-Delgado (*15-17*) for the chemoselective hydrogenation of **BT** to **DHBT** (**DHBT** is stable under the actual reaction conditions).

The occurrence of cycle **B** requires that the η^1-*S*-**BT** intermediate is in equilibrium with its η^2-2,3-**BT** isomer (step **h**), in which the C_2-C_3 double bond is activated for accepting a migrating hydrogen. As a result, an alkyl intermediate (step **i**) is formed, which can oxidatively add H_2 and later eliminates **DHBT** (steps **j**-**k**).

In the proposed mechanistic picture, the real catalyst for both transformations of **BT** is the 16-electron fragment [(triphos)RhH] generated from the dihydride **2** by reductive elimination of **ETSH**. This process is apparently the rate determining step in light of the HPNMR evidence as well as the dependence on both hydrogen pressure and substrate concentration (Table I).

The prevalence of hydrogenolysis of **BT** to **ETSH** over hydrogenation to **DHBT** is most likely driven by steric effects: although the Rh center is sufficiently electron-rich to bind the C_2-C_3 double bond of **BT**, the large steric hindrance provided by the six phenyl substituents of triphos favors the η^1-*S* coordination mode of **BT**, and ultimately controls the chemoselectivity of the reaction with H_2.

Iridium-Catalyzed Conversion of Dibenzo[*b*,*d*]thiophene to 2-Phenylthiophenol, Biphenyl and H_2S

There are two principal reaction pathways that have been proposed to account for the HDS of **DBT** over conventional heterogeneous catalysts (Scheme 5). Path **a** involves the hydrogenation of one of the arene rings of **DBT** to give tetrahydrodibenzothiophene, which, after C-S bond scission, is hydrogenated to 2-cyclohexylthiophenol. Path **b** is quite similar to one of the mechanisms suggested for HDS of **BT** (see Scheme 1) as it involves the opening and hydrogenation of the substrate to give 2-phenylthiophenol prior to desulfurization.

The homogeneous modeling study described below shows that the HDS of **DBT** can proceed *via* ring opening to 2-phenylthiophenol prior to desulfurization and hydrogenation and thus provides evidence that the desulfurization does not necessarily require the preliminary hydrogenation of one benzene ring of **DBT**.

The 16-electron fragment [(triphos)IrH], generated *in situ* by thermolysis of the (ethyl)dihydride complex (triphos)Ir$(H)_2(C_2H_5)$ is capable of selectively cleaving **DBT** in THF at 160°C to give the C-S insertion product (triphos)IrH(η^2-*C,S*-DBT) (**6**) (*10*) (Scheme 6). At lower temperature, kinetic C-H insertion compounds are also (< 160°C) or exclusively (< 120°C) produced (*10*).

Complex **6** in THF is hydrogenated (100°C, 5 atm of H_2) to the 2-phenylthiophenolate dihydride (triphos)Ir$(H)_2(SC_{12}H_8)$ (**7**), which gives biphenyl, H_2S, and 2-phenylthiophenol upon treatment with 30 atm of H_2 at 170°C (Scheme 7). In the presence of an excess of **DBT** this reaction is catalytic, although the rate of transformation of the thiophenic molecule is quite slow as only 10 mol of **DBT** per mol of catalyst precursor are converted in 24 h to either open (rate 0.25) or desulfurized (rate 0.16) products. Most importantly, the reaction is homogenous (mercury test).

Based on the results of several independent reactions with isolated compounds, some of which are summarized in Scheme 8, as well as the fact that, at the end of the

Table I. Catalytic Hydrogenation Experiments[a]

run	solvent	T (°C)	P_{H2} (atm)	t (h)	reaction mixture composition (%)[b] ETB	ETSH	BT	DHBT	other	ETSH rate[c]
1	acetone	160	30	2	0.2	25.5	73.2	1.1	--	12.7
2	acetone	160	30	4	0.2	39.8	57.4	2.6	--	9.9
3	acetone	160	30	8	0.3	45.1	51.8	2.8	--	5.6
4	acetone	160	30	12	0.4	51.0	44.6	4.0	--	4.2
5	acetone	160	30	16	0.4	57.4	37.6	4.6	--	3.6
6[d]	acetone	160	30	16	--	57.6	37.9	4.5	--	3.6
7	THF	160	30	16	0.2	52.8	41.3	5.7	--	3.3
8	acetone	160	15	16	0.3	55.3	40.2	4.2	--	3.5
9	acetone	160	60	16	0.4	60.2	34.3	5.1	--	3.8
10	THF	120	30	4	--	2.0	97.6	0.4	--	0.5
11	acetone	100	30	16	--	1.4	98.1	0.5	--	<0.1
12	acetone	180	30	16	0.9	64.2	29.0	5.9	--	4.0
13	THF	220	30	16	3.5	43.3	45.8	6.9	0.5	2.7
14[d]	THF	220	30	16	--	42.6	51.1	6.0	0.3	2.7
15[e]	acetone	160	30	16	0.3	61.2	34.0	4.5	--	1.9

[a]Reaction conditions: Parr reactor, **1** (0.12 g, 0.139 mmol), BT (1.86 g, 13.9 mmol), solvent (30 mL). [b]Key: ethylbenzene (ETB), 2-ethylthiophenol (ETSH), benzo[*b*]thiophene (BT), dihydrobenzo[*b*]thiophene (DHBT). [c]Rate expressed as mol of ETSH per mol of catalyst per hour. [d]Reactions carried out in the presence of excess elemental Hg. [e]Reaction conditions: Parr reactor, **1** (0.12 g, 0.139 mmol), BT (0.93 g, 6.95 mmol), solvent (30 mL).

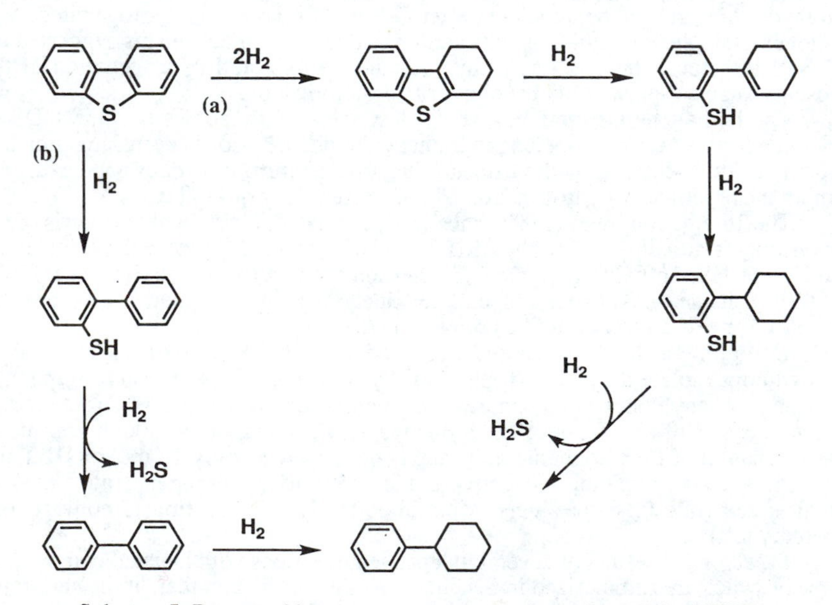

Scheme 5. Proposed heterogeneous mechanisms for HDS of **DBT**.

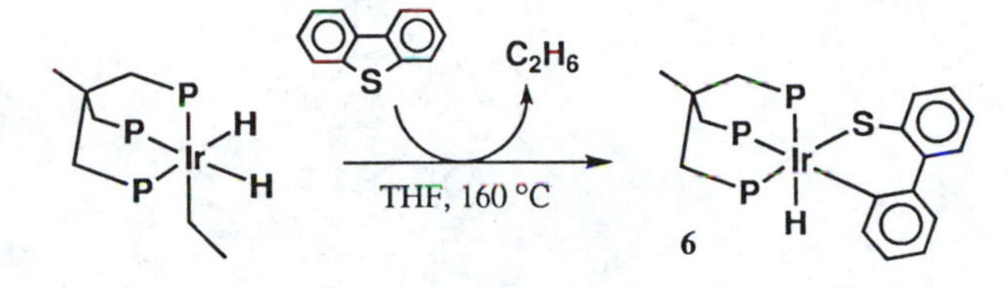

Scheme 6

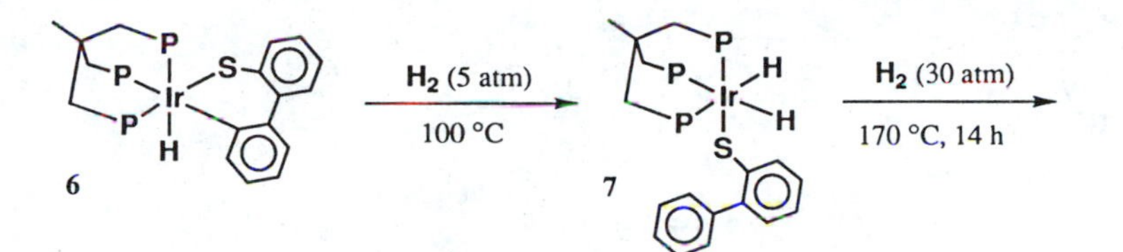

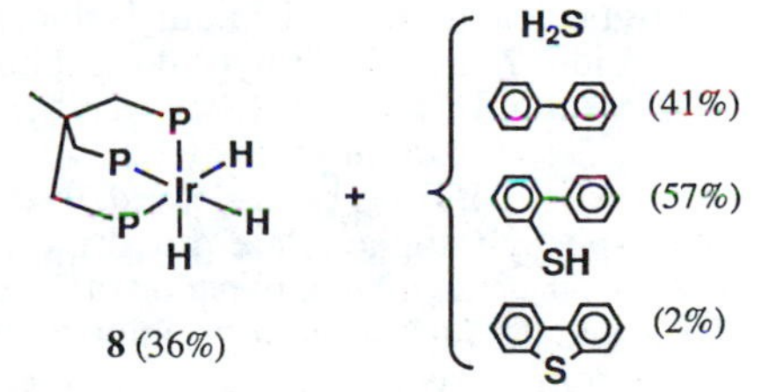

Scheme 7

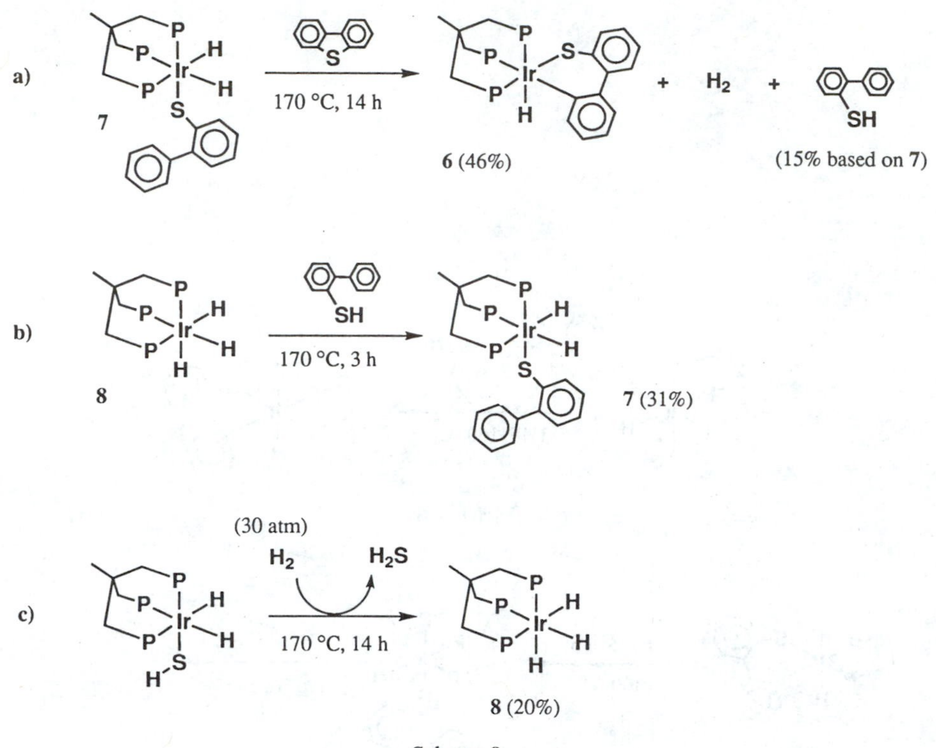

Scheme 8

catalytic reaction, all iridium is incorporated into two complexes, namely the dihydride, **7**, and the trihydride (triphos)IrH_3 (**8**) in a 12:88 ratio, a mechanism is proposed in Scheme 9 for both the hydrogenolysis of **DBT** to 2-phenylthiophenol and its desulfurization to biphenyl.

After **DBT** has been cleaved, the C-S insertion product **6** reacts with H_2 to give the 2-phenylthiophenolate dihydride complex **7**. This complex has two reaction options: the reductive elimination of 2-phenylthiophenol promoted by interaction with **DBT** and the further hydrogenation to give 2-phenylthiophenol and biphenyl + H_2S. In the former case, the precursor **6** is regenerated *via* an η^1-*S*-**DBT** intermediate (the existence of an equilibrium between (triphos)IrH(η^1-*S*-DBT) and **6** in the catalysis conditions is experimentally proven). In the second case, the trihydride **8** forms and the reaction would stop since the trihydride does not react with **DBT**. This, however, does not occur as the 2-phenylthiophenol is capable of converting **8** to **7**, which thus can reenter the catalysis cycle. The low tof of the catalytic reaction is attributed to the slowness of the latter reaction which is disfavored at the experimental high pressure of H_2.

The (thiolate)dihydride complex **7** is the key compound also in the desulfurization step as it independently reacts with H_2 (30 atm) at 170°C to produce H_2S and biphenyl, and the trihydride **8** (Scheme 7).

The mechanism by which the C-S bond of the *S*-bonded thiolate is cleaved by H_2 still needs to be addressed. The relevant literature contains convincing evidence that the C-S bond cleavage in metal thiolates occurs *via* migration of a hydride to the sulfur-bound carbon atom. If this occurs in the case of **7**, the M-S moiety, which forms after the elimination of biphenyl, might convert to an M(SH)(H) species upon reaction with H_2. Although not detected along the transformation of the thiolate **7** into

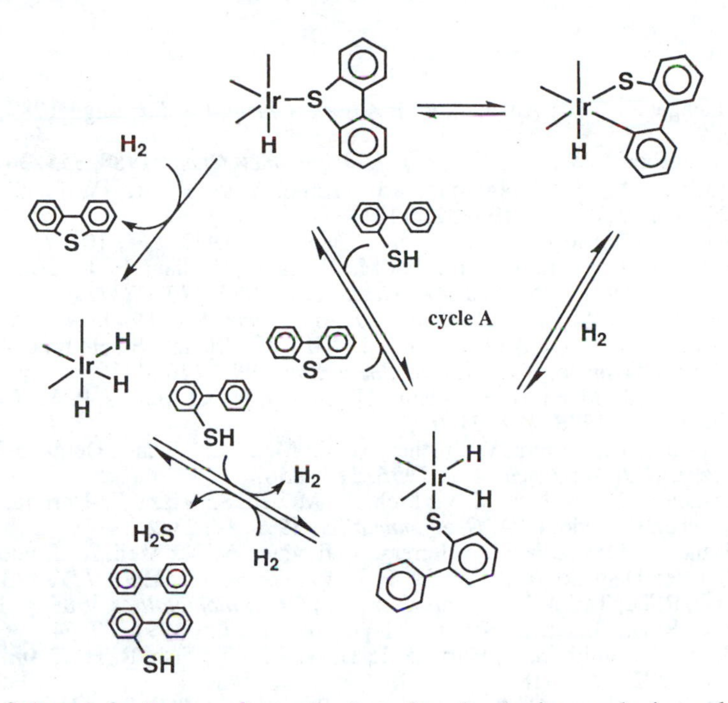

Scheme 9. Proposed mechanism for the hydrogenolysis and hydrodesulfurization of **DBT** catalyzed by [(triphos)IrH].

the trihydride **8**, biphenyl and H_2S, a hydrosulfide complex of the formula (triphos)Ir(H)$_2$(SH), independently synthesized, does react in THF with 30 atm of H_2 at 170°C to form the trihydride **8** and H_2S (Scheme 8c). On the other hand, a hydride(hydrosulfide) complex has been recently intercepted in the stoichiometric desulfurization of **DBT** to biphenyl + H_2S promoted by Pt(PEt$_3$)$_3$ (*8*).

Conclusions

The results presented here demonstrate that homogeneous metal catalysts are viable models for several steps in the heterogeneously catalyzed HDS reaction. However, although homogeneous modeling studies may continue to provide valuable mechanistic information, it is unrealistic to think of the use of a homogeneous catalyst in an HDS petrochemical plant. In contrast, liquid biphase catalysis could be used for the purification of distillates from residual sulfur contaminants up to the limit of commercial fuels (where international regulations will soon require reducing the sulfur content to less than 100 ppm).

The activity shown by the Rh and Ir complexes stabilized by the tripodal ligand triphos in the homogeneous hydrogenolysis of thiophenic molecules suggests the design and synthesis of triphos-like ligands capable of coordinating HDS-active metals with formation of complexes which, being soluble in water but not in hydrocarbons, may be used in liquid biphase catalysis.

Literature Cited

(*1*) Hachgenei, J. W.; Angelici, R. J. *Angew. Chem., Int. Ed. Engl.* **1987**, *26*, 909.

(*2*) Hachgenei, J. W.; Angelici, R. J. *J. Organomet. Chem.* **1988**, *355*, 359.

(*3*) Ogilvy, A. E.; Draganjac, M.; Rauchfuss, T. B.; Wilson, S. R. *Organometallics* **1988**, *7*, 1171.

(*4*) Rosini, G. P.; Jones, W. D. *J. Am. Chem. Soc.* **1992**, *114*, 10767.

(*5*) Bianchini, C.; Meli, A.; Peruzzini, M.; Vizza, F.; Frediani, P.; Herrera, V.; Sánchez-Delgado, R. A. *J. Am. Chem. Soc.* **1993**, *115*, 2731.

(*6*) Luo, S.; Rauchfuss, T. B.; Gan, Z. *J. Am. Chem. Soc.* **1993**, *115*, 4943.

(*7*) Bianchini, C.; Meli, A.; Peruzzini, M.; Vizza, F.; Moneti, S.; Herrera, V; Sánchez-Delgado, R. A. *J. Am. Chem. Soc.* **1994**, *116*, 4370.

(*8*) Garcia, J. J.; Mann, B. E.; Adams, H.; Bailey, N. A.; Maitlis, P. M. *J. Am. Chem. Soc.* **1995**, *117*, 2179.

(*9*) Bianchini, C.; Herrera, V; Jiménez, M. V.; Meli, A.; Sánchez-Delgado, R. A.; Vizza, F. *J. Am. Chem. Soc.* **1995**, *117*, 8567.

(*10*) Bianchini, C.; Jiménez, M. V.; Meli, A.; Moneti, S.; Vizza, F.; Herrera, V; Sánchez-Delgado, R. A. *Organometallics* **1995**, *14*, 2342.

(*11*) Bianchini, C.; Frediani, P.; Herrera, V; Jiménez, M. V.; Meli, A.; Rincón, L.; Sánchez-Delgado, R. A.; Vizza, F. *J. Am. Chem. Soc.* **1995**, *117*, 4333.

(*12*) Fish, R. H.; Tan, J. L.; Thormodsen, A. D. *Organometallics* **1985**, *4*, 1743.

(*13*) Fish, R. H.; Baralt, E.; Smith, S. J. *Organometallics* **1991**, *10*, 54.

(*14*) Baralt, E.; Smith, S. J.; Hurwitz, I.; Horváth, I. T.; Fish, R. H. *J. Am. Chem. Soc.* **1992**, *114*, 5187.

(*15*) Sánchez-Delgado, R. A.; González, E. *Polyhedron* **1989**, *8*, 1431.

(*16*) Sánchez-Delgado, R. A. In *Advances in Catalyst Design*; Graziani, M.; Rao, C. N. R., Eds.; World Scientific Publishing Co.: Singapore, 1991; pp 214-231.

(*17*) Sánchez-Delgado, R. A.; Herrera, V.; Rincón, L.; Andriollo, A.; Martín, G. *Organometallics* **1994**, *13*, 553.

Chapter 11

C–S Bond-Breaking and Bond-Making Reactions in Molybdenum and Tungsten Complexes

P. Michael Boorman, Xiaoliang Gao, Heinz-Bernhard Kraatz, Vivian Mozol, and Meiping Wang

Department of Chemistry, University of Calgary, Calgary, Alberta T2N 1N4, Canada

Selected C-S bond-breaking and bond-making reactions from our work are discussed that have implications to the modeling of both industrial and biological processes. The complexes $WCl_5(SR)$ were found to be subject to elimination of RCl, with concomitant formation of WCl_4S via a carbonium ion mechanism. Attempts to emulate the C-S bond cleavage step of hydrodesulfurization by introduction of hydride into the coordination sphere, or by reaction with a second transition metal hydride, resulted in the synthesis of a variety of homonuclear and heteronuclear complexes, with elimination of the appropriate alkane. The synthesis of complexes of the type $Cl_3W(\mu\text{-}R_2S)_3WCl_3$ led to the discovery of the reactivity of the C-S bonds toward nucleophiles, including hydride. A computational study provides an explanation for this lability. Reactions of the $[MoS_4]^{2-}$ and $[WS_4]^{2-}$ ions with organic halides, have been reinvestigated. $[PPh_4][WS_3(SR)]$ (R = Et, ^{i}Pr, ^{t}Bu) and $[PPh_4][MoS_3(S^tBu)]$ have been characterized crystallographically. These complexes undergo reductive elimination reactions forming polynuclear sulfidometalates such as $W_3S_9^{2-}$. The synthetic potential and possible biological significance of these reactions is discussed.

The rich chemistry of molybdenum and tungsten with sulfide and organosulfur ligands has been recognized for more than two decades. It is punctuated by complicated redox behavior of both metal and ligands, and also by unexpected reactions of the coordinated ligands. In this regard the metal-assisted cleavage of C-S bonds has been of particular interest, since this process is believed to be involved in the catalytic hydrodesulfurization of fossil fuels. The reverse process, namely the formation of C-S bonds has attracted less attention, but it is interesting to speculate as to the possible importance of this process in the biosynthesis of metalloenzymes.

0097–6156/96/0653–0197$15.00/0

Our interest in this area arose from the observation that when reacted with WCl_6, dimethyl sulfide yielded $[Me_3S]_2[WCl_6]$ thereby implying both cleavage and formation of C-S bonds (*1*). A proposed mechanism is shown in Scheme 1, and a number of experiments were performed that supported such a sequence.

Scheme 1

$$WCl_6 + \underset{\text{(excess)}}{Me_2S} \rightarrow WCl_6(Me_2S)_2 \xrightarrow[\text{(S-dealkylation)}]{} WCl_4(SMe)_2 + 2MeCl$$

$$WCl_4(SMe)_2 \rightarrow [WCl_4] + Me_2S_2 \xrightarrow[\text{(excess } Me_2S)]{} WCl_4(Me_2S)_2$$

$$WCl_4(Me_2S)_2 + 2MeCl \xrightarrow[\text{(S-alkylation)}]{} [Me_3S]_2[WCl_6]$$

Part I: A Series of C-S Bond-Breaking Studies

S-Dealkylation of W(VI) Thiolates: Evidence for Carbocation Formation

One of the key steps in Scheme 1 is the reductive elimination of Me_2S_2 from the proposed W(VI) bis(thiolate) intermediate, and we thus embarked on a study of the chemistry of W(VI) thiolates. By systematically replacing chloride by thiolate in WCl_6 it was established that the products were, in general, unstable (*2*). For 1:1 reactions the following pattern was observed:

$$WCl_6 + Me_3Si(SR) \rightarrow \underset{\mathbf{1}}{\{ WCl_5(SR) \}} \rightarrow WCl_4S + RCl$$

Only for the cases of R = Me or Ph could the complexes **1** be isolated as pure materials. This observation made us suspect that the mechanism of the decomposition involved carbocation formation, a suspicion that was confirmed by utilizing R groups known to rearrange when converted to carbonium ions. The following reaction illustrates this:

$$WCl_6 + Me_3Si(S^iBu) \rightarrow WCl_4S + {}^tBuCl$$

Similarly, when R = cyclohexyl, the predicted rearrangement product, 1,1`-chloro(methyl)cyclopentane was formed. The reversibility of this reaction was shown by reacting ^{i}BuCl with WCl_4S under reflux. Although no tungsten thiolate product was observed, the presence of ^{t}BuCl in the products suggested that the reaction is reversible, but thermodynamically the system strongly favors WCl_4S + RCl. Even in 1:1 reactions as described, there is always some dialkyldisulfide present in the decomposition products formed from $WCl_5(SR)$, and reductive elimination becomes the preferred pathway for 1:2 or higher ratios of $WCl_6:Me_3Si(SR)$. Hence, by replacing Cl by RS ligands in a high oxidation state

Mo or W complex, reductive elimination of R_2S_2 will leave coordinatively unsaturated metal centers in intermediate oxidation states, which are ideal precursors for cluster formation. This procedure proved to be a fruitful method for synthesizing binuclear complexes of W(IV), and, in some cases, mixed valent W(IV)/W(III) products were isolated (*3-6*). Even in these reactions the dealkylation of thiolate ligands to give sulfido complexes is frequently observed, and the products from such reactions are extremely unpredictable. In general, however, aryl thiolates, in which the C-S bonds are stronger than in their alkyl analogs, are less prone to C-S bond cleavage, and more likely to undergo reductive elimination reactions.

Possible S-Dealkylation with Hydride as the Leaving Ligand

The tungsten oxidation state was found to be critical in determining the stability of thiolates toward S-C bond cleavage via the carbocation mechanism. Only in the case of W(VI) described above was there any evidence for this pathway. However, in synthetic reactions starting with W(IV) or W(V), we saw sufficient examples of dealkylation of thiolate ligands to prompt us to consider if hydride could be introduced into the coordination sphere, and induced to participate as the departing ligand as shown in Scheme 2.

Scheme 2

Possible Extension to S-dealkylation with Hydride Ligands

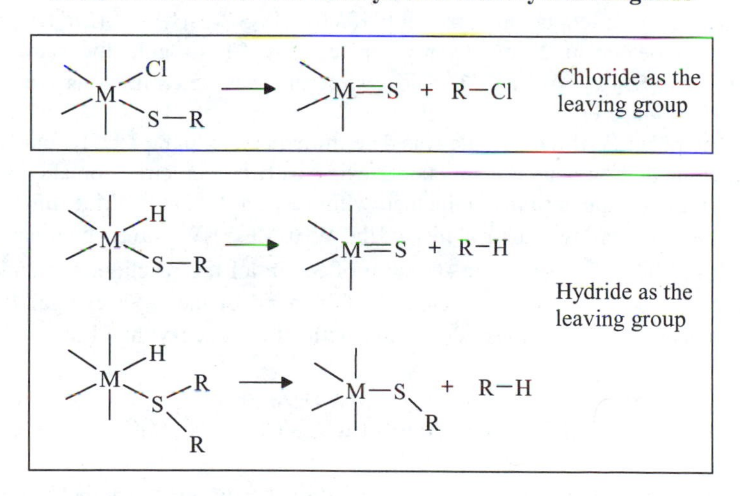

The most obvious strategy to test this hypothesis was to attempt the reaction:

$$WCl_6 + Me_3Si(SR) \rightarrow WCl_5(SR) \xrightarrow[Et_3SiH]{} WCl_4H(SR) \rightarrow RH + WSCl_4$$

This type of reaction is not feasible, and leads to a variety of reductive elimination products. Indeed, stable hydrido-thiolates of Mo and W were only isolated many years later, as a result of oxidative addition of thiols to zerovalent complexes, by Richards and Morris and their co-workers (*7,8*). The use of sterically hindered thiols was a major factor in these successful preparations, and it is of interest to note that some of these complexes are five coordinate, (e.g. $MoH(SC_6H_2R_3)_3(PR'Ph_2)$, (R = Me or ^iPr; R'= Me or Et), and highly reactive. Of particular interest was the observation that two molecules of these complexes can undergo C-S bond cleavage under mild conditions in THF/MeOH solution, eliminating the appropriate arene, and yielding $[\{Mo(SC_6H_2R_3\text{-}2,4,6)(OMe)(PR'Ph_2)\}_2(\mu\text{-}S)_2]$. The possible mechanism for this reaction will be referred to later. Since we were interested primarily in systems without stabilizing phosphine ligands, we probed the possibility of hydride transfer to a coordinated thioether molecule. The reactions attempted were as follows:

$$\text{(i)}\quad MoCl_4(Me_2S)_2 + Et_3SiH\ (\text{excess}) \rightarrow \text{products}$$

$$\text{(ii)}\quad WCl_4(Me_2S)_2 + Et_3SiH\ (\text{excess}) \rightarrow \text{products}$$

Reaction (i) gave rise to reduction of Mo, and two isomers (C_s and C_{2v} symmetry) of the binuclear complexes $Mo_2Cl_6(Me_2S)_3$, but no evidence for C-S bond cleavage (*9*). Reaction (ii), however, yielded a detectable amount of methane, and the major tungsten-containing product $[Cl_3W(\mu\text{-}Me_2S)_2(\mu\text{-}H)WCl_2(Me_2S)]$ **2**. The terminal thioether in **2** was readily replaced by Cl^- to give the stable anion $[Cl_3W(\mu\text{-}Me_2S)_2(\mu\text{-}H)WCl_3]^-$ **3**, whose structure was determined as the $[Ph_4P]^+$ salt (*10*) (Figure 1).

The position of the hydride was determined crystallographically, but it could not be refined. The neutral compound **2** exists as two geometric isomers, based on the position of the terminal thioether with respect to the hydride ligand. Of particular note here was the stability of the $W(\mu\text{-}Me_2S)_2W$ core, which has a very short W-W bond, based on the structure of **3**. In all the reactions we undertook (*11*), it showed no tendency to cleave a C-S bond of the $\mu\text{-}SMe_2$ ligands. The hydridic nature of the μ-H was shown in reactions with benzyl halides:

$$[Cl_3W(\mu\text{-}Me_2S)_2(\mu\text{-}H)WCl_2(Me_2S)] + C_6H_5CH_2X \rightarrow$$
$$[(C_6H_5CH_2)(CH_3)_2S]^+\,[Cl_3W(\mu\text{-}Me_2S)_2(\mu\text{-}X)WCl_3]^- + C_6H_5CH_3 \quad (X = Cl, Br)$$

The metal-assisted alkylation of a terminal thioether provides another example of C-S bond-making.

Intermolecular Reactions between Hydride and Thiolate Ligands

The transfer of hydride to carbon need not be restricted to one metal center. Indeed, as a model for catalytic reactions the use of homobimetallic, or even heterobimetallic complexes, in light of the role of promoters in hydrotreating catalysts, would seem to be a reasonable strategy. Two approaches were considered:

a) Attempt to make binuclear complexes in which H and the organosulfur ligands are on adjacent atoms:

$$M(H)\text{———}M'(SR)$$

Thermal elimination of R-H would likely result in M(μ-S)M′. The system could be synthesized with or without other supporting bridging ligands.

b) React separate complexes in an intermolecular reaction:

$$M(H) + M'(SR) \rightarrow M(\mu\text{-}S)M' + R\text{-}H$$

After a number of unsuccessful attempts to execute the first type of reaction, we focussed on (b) with some significant success. The requirements for such a reaction would be that the M-H moiety must be as hydridic as possible, which suggested the use of a metallocene hydride, or possibly a Group 13 hydrido complex. The choice of a suitable M′-SR system was based on the need to have sufficient activation of the C-S bond by the metal but not to the extent that reductive elimination of a disulfide would be the preferred pathway. This suggested an intermediate oxidation state of Mo or W, which of course would also be consistent with the +4 oxidation in an MS_2 phase. This limited the choice somewhat, and led to successful reactions between Cp_2ZrH_2 and $Mo(S^tBu)_4$, and Cp_2NbH_3 and $Mo(S^tBu)_4$ respectively. The initial reaction was carried out with Cp_2ZrH_2, and $Mo(S^tBu)_4$ with benzene as solvent (*12*). The insoluble hydride reacted rapidly to yield isobutane and an insoluble metal-containing product. The source of **H** in the Me_3C**H** was confirmed by using Cp_2ZrD_2 and identifying the labeled product using ^2H NMR. A cleaner reaction was observed with Cp_2NbH_3, which again gave isobutane, and an eventual metal-containing product that proved to be $Cp_2Nb(\mu\text{-}S)_2Mo(\mu\text{-}S)_2NbCp_2$ **4**. The reaction was followed using ^1H NMR and the intermediacy of an unstable compound believed to be $Cp_2Nb(\mu\text{-}S)_{\text{-}3}MoS^tBu$. The structure of **4** is shown in Figure 2 (*12*).

The mechanism proposed for the initial reaction, but for which there is only circumstantial support, involved a four-center interaction between -S-R and M′-H, from which the Mo-S-Nb bridge arises. It seems reasonable to assume an associative pathway, since the analogous reaction between $Mo(SCy)_4$ (Cy = cyclohexyl) and Cp_2ZrH_2 gave rise to cyclohexane, and no trace of isomerized product (methylcyclopentane) that would arise if a completely dissociative reaction occurs (*13*). The earlier observations of Boorman and O'Dell (*2*) set a precedent for this conclusion. However, it will be noted that, as predicted above, the polarization of both the C-S and M-H bonds will be required to facilitate this reaction. The involvement of the second, hydride-bearing, metal (M′) in the

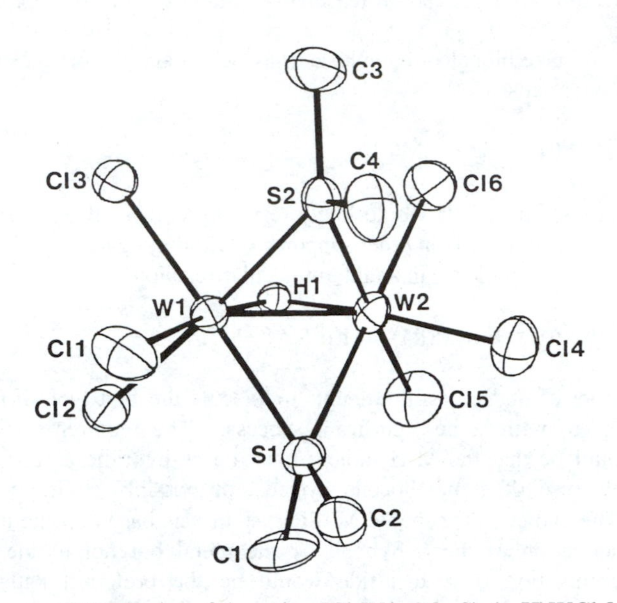

Figure 1 ORTEP plot of the anion $[Cl_3W(\mu\text{-}Me_2S)_2(\mu\text{-}H)WCl_3]^-$ **3** (Reproduced with permission from ref. 10. Copyright 1985 American Chemical Society.)

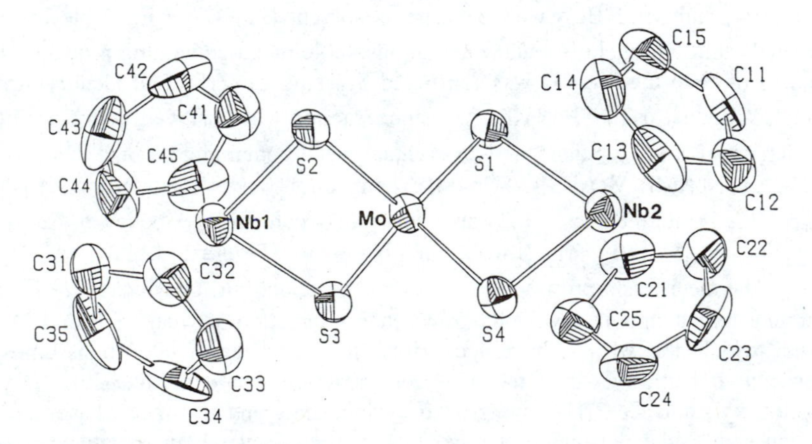

Figure 2 The structure of $Cp_2Nb(\mu\text{-}S)_2Mo(\mu\text{-}S)_2NbCp_2$ (Reproduced with permission from ref. 12. Copyright 1989 Pergamon Press.)

formation of a new M′-S bond appears to be important. A similar suggestion, namely that an associative reaction is likely involved, was recently made by Morris and co-workers to decribe the elimination of arene from the hydrido-thiolate complexes mentioned earlier (*7*). Our studies indicated that late transition metal (electron-rich) hydrides were ineffective in analogous reactions with $Mo(S^tBu)_4$ (*13*), adding further support for the suggested mechanism.

As noted earlier, it is known that S-C bonds in aromatic thiolates are stronger, and therefore more difficult to cleave, than their aliphatic counterparts. It was of interest to probe this reactivity in a system analogous to the above. A suitable thiolate was found to be $Mo(S\text{-tipt})_4(MeCN)$ (tipt = 2,4,6-triisopropylphenyl). This compound had been reported by Dilworth and co-workers (*14*). Although there was no reaction with Cp_2ZrH_2, Cp_2NbH_3 was shown to bring about S-C bond cleavage, as evidenced by the detection of 1,3,5-triisopropylbenzene in the organic reaction products (*13*).

The reactions between a metallocene hydride and a late transition metal thiolate gave completely different results from those with early transition metal thiolates. No C-S bond cleavage was observed but reduction, with the elimination of H_2, dominated these reactions. An example is the reaction between Cp_2ZrH_2 and $[\{Pd(SCH_2CH_2S)(PPh_3)\}_2]$ (*15*). The product, $[(Cp_2Zr)_2((\mu\text{-}SCH_2CH_2S))_2Pd]$ **5** is of particular interest, containing a Pd(0) center, (presumed on the basis of T_d symmetry) coordinated exclusively by thiolates (Figure 3).

This complex is structurally similar to the cationic complex $[\{Cp_2Zr(\mu\text{-}SCH_2CH_2S)_2ZrCp_2\}Ag]^+$ **6** reported by Stephan (*16*), but the synthetic methodologies employed were completely different. To prepare **6**, Stephan first synthesized the metalloligand $[Cp_2Zr(\mu\text{-}SCH_2CH_2S)_2ZrCp_2]$ then encapsulated a silver cation within it. These $d^0d^{10}d^0$ systems display relatively short metal-metal interactions,which was ascribed to dative $d^{10}\rightarrow d^0$ bonding (*17*). The Pd-Zr distances and bond angle information suggest that the metal-metal interaction is stronger in **5** than in **6**, as might have been anticipated based on the formal oxidation states of Ag(+1) and Pd(0). Indeed this adds further support to this assignment for Pd.

The results described so far illustrate that there is a potentially rich chemistry in the reactions between thiolate complexes and hydride sources. The possible involvement of reactions of this type in the heterogeneous catalytic HDS process is, of course, a matter of conjecture, but the process has at least been shown to be feasible in model systems. The variety of sulfur-containing compounds established to be present in Alberta heavy oils and bitumen suggests that although the thiophenic species are undoubtedly the most important, thiols and thioethers are also candidates for HDS (*18*). We will now consider some systems which illustrate the reactivity of simple thioethers.

Synthesis and Reactivity of Thioether-Bridged Mo(III) and W(III) Face-Sharing Bioctahedral Complexes

Thioether-bridged complexes of Mo(III), W(III) and also Nb(III),Ta(III), have some unusual structural features, the most notable of which is the shortness of the

M-(μ-S) bonds. In attempting to extend the series of known complexes that contain M-(μ-SR_2)-M entities, we carried out reductions of MCl_4 in the presence of excess thioether. Three isomeric forms of $M_2Cl_6(R_2S)_3$, were obtained, as shown in Figure 4.
The examples of these structures are:
C_{2v} geometry, found only for Mo (S = Me_2S **7**; S = THT **10**) (*17, 19, 20*)
C_s geometry, found only for Mo (S = Me_2S **8**; S = THT **11**) (*17, 19, 20*)
D_{3h} geometry found only for W (S = Et_2S **9**; S = THT **13**) (*16*)

Structural parameters for three representative complexes are given in Table I. The Et_2S complex of W is given rather that its THT analog, since the latter complex exhibited disorder in the carbon positions of the THT rings. Surprisingly, we were unable to obtain the Me_2S analog of **9** but could only isolate Na[Cl_3W(μ-$Me_2S)_2$(μ-Cl)WCl_3] **12**.

Of particular note is the effect of exchanging bridging ligands on the Mo-Mo bond length. Analogous complexes of Nb and Ta with C_{2v} geometry also display strong metal-metal interactions. In terms of magnetic properties, whereas the C_{2v} and D_{3h} complexes are diamagnetic, the C_s analogs display antiferromagnetic behavior. The unexplained structural features of these compounds were addressed by a computational study using Density Functional Theory (*21*).

Table I Structural Data for Three Isomers of $M_2Cl_6(SR_2)_3$

Complex → / **Parameter (Å) ↓**	Cl_3W(μ-$Et_2S)_3WCl_3$ **9** (*16*)	Cl_2(THT)Mo(μ-Cl)$_2$ (μ-THT)$MoCl_2$(THT) **10** (*17*)	Cl(THT)$_2$Mo(μ-Cl)$_3$ $MoCl_2$(THT) **11** (*17*)
	D_{3h}	**C_{2v}**	**C_s**
M—M	2.499	2.470	2.690
M—S(bridging) (average value)	2.40	2.409	—
M—S(terminal) (average value)	—	2.563	2.518
M—Cl (*trans* to μ-S) (average value)	2.388	—	

Density Functional Theory Studies Using the Hoffmann fragment approach the partial orbital manifold for M_2L_6 was computed, then the bridging fragment L_3 was allowed to interact with this enabling a sensitive analysis of the bonding of the bridging ligands to be accomplished. The qualitative results of this study were as follows. The interaction between filled orbitals of π symmetry on the part of bridging chloride with the occupied metal-metal bonding orbitals of the M_2L_6 fragment is repulsive in nature and will have the effect of destabilizing the metal-

metal bonding orbitals. Substitution of Cl bridges by thioether in the simplified models $[Mo_2Cl_{9-n}(SH_2)_n]^{(3-n)-}$ is shown in the DFT calculations to result in stronger Mo-Mo bonding. This is due, in part, to a reduction in the unfavorable chlorine p-π interactions with metal-metal bonding orbitals, since the thioether has one less lone pair. In addition, however, there is an interaction between a filled metal-metal bonding orbital of b_2 symmetry with a σ^* (H-S) orbital from the ligand (*21*). This is shown in Figure 5.

The σ^* orbitals in phosphines have, of course, become recognized as the likely acceptors of π back-donated electrons from metals, and so this result might have been anticipated (*22*). The DFT study was extended to both bridging and terminal complexes of R_2E (E = chalcogen; R = CH_3, H, F) and the bonding model was found to apply here too. Hence the π-acceptor orbitals of thioethers are better described in terms of σ-antibonding orbitals, rather than 3d orbitals. These results are of broad significance, but in the context of thioether-bridged face-sharing bioctahedral complexes the immediate result is to explain the shortness of the M-S bonds. The results also provide a reason for the lability of C-S bonds in these complexes, some examples of which will now be described.

Having prepared the binuclear tungsten complexes with strongly bonded $W(\mu\text{-}SR_2)_3W$ cores, it was of interest to attempt to do substitution reactions on these neutral molecules, with a view to building completely sulfur-ligated systems. We therefore reacted solutions of **9** with thiolate anion, anticipating that stepwise replacement of chloride might be possible (*24*). Instead, the nucleophile was found to attack at an α-carbon atom of thioether e.g.

$$Cl_3W(\mu\text{-}Et_2S)_3WCl_3 + (CH_3)C_6H_4S^- \rightarrow (CH_3)C_6H_4SC_2H_5 + [Cl_3W(\mu\text{-}Et_2S)_2(\mu\text{-}SEt)WCl_3]^-$$

This unexpected result led us to examine a wide range of other nucleophiles, including hydride. In addition to the S-deakylation reactions of Et_2S, we have been able to demonstrate ring opening reactions of THT in **13**. (Scheme 3.)

Scheme 3: Ring opening reaction of anionic nucleophiles (Nu^-) with **13**

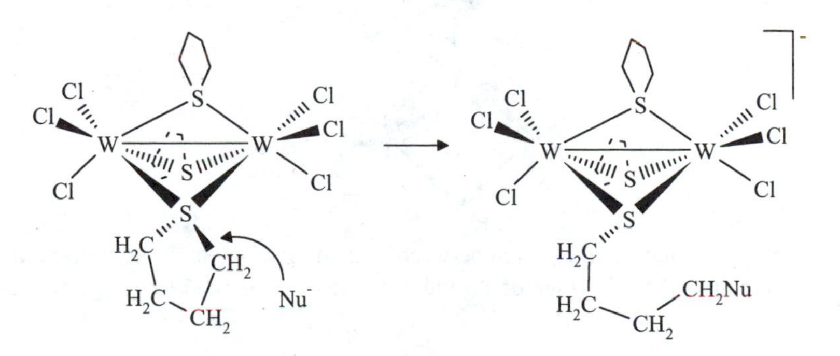

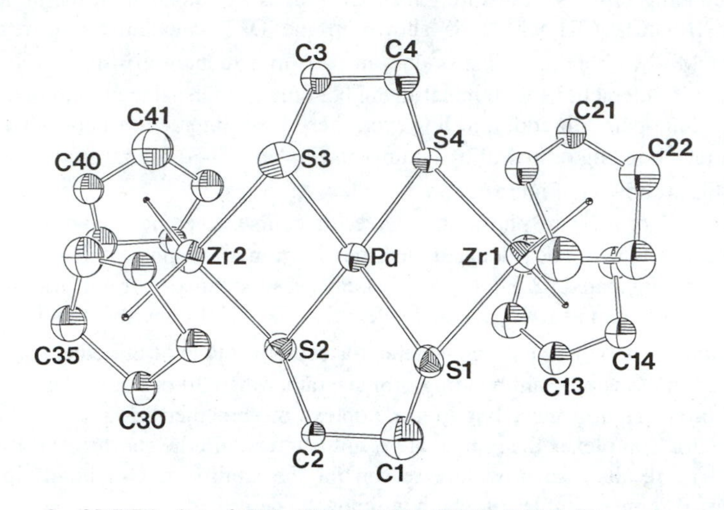

Figure 3 ORTEP plot of $[(Cp_2Zr)_2(\mu\text{-}SCH_2CH_2S)_2Pd]$ **5** (Reproduced with permission from ref. 15. Copyright 1992 Pergamon Press.)

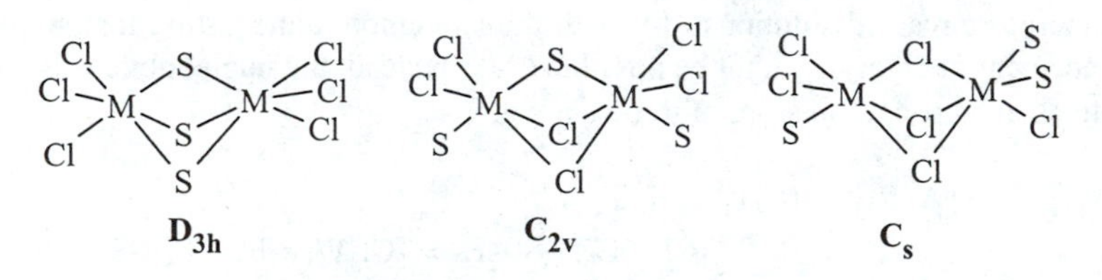

Figure 4 Three isomers of $M_2Cl_6(R_2S)_3$ (R_2S is abbreviated as S)

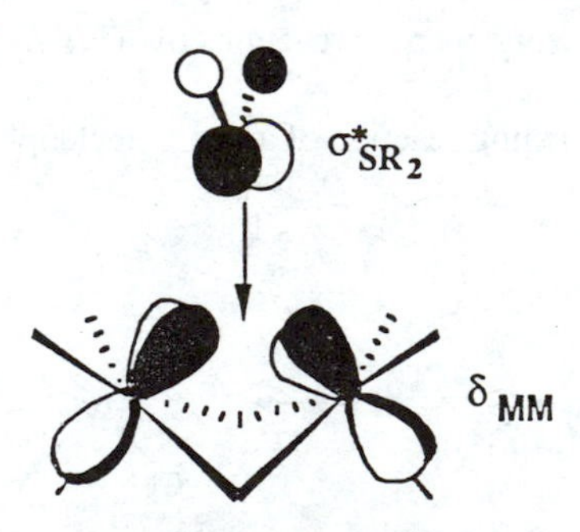

Figure 5 π-bonding interaction between a filled metal-metal bonding orbital and an empty H-S σ^* orbital of a bridging thioether (adapted from ref. 21)

Spectroscopic evidence supports the ring-opened structures of several such derivatives, including that in which Nu^- is hydride. Confirmation was provided by the successful structure determination of the $[PPh_4]^+$ salt of the anion **14** derived from attack by Cl^- (Figure 6).

It can be seen that the core of the binuclear structure is retained and that the 4-chlorobutylthiolate ligand still occupies a bridging position. The W-S bond lengths (2.444(4) and 2.434(4) Å) for the thiolate ligand are longer than those of the thioether ligands (average = 2.373(4) Å). The characterization of this complex has enabled extensive studies of related compounds to be carried out using ^{183}W NMR (*25*). The application of ^{183}W NMR is of some interest, since it can probe the effect of changes in the bridging region of the face-sharing bioctahedral ditungsten(III) complexes. It was observed that ^{183}W shielding increased in the order μ-Cl^- < μ-SR^- < μ-SR_2. For complexes of the structural form of **14**, in which only the terminal nucleophile on the butyl chain is varied (SePh, SPh, SH) the chemical shift was unchanged at δ 3046 as compared to 2850 ppm for the parent compound **13**.

The derivatization of a stable binuclear complex using the reaction of nucleophiles with **13**, provides a possible route to the preparation of potential metalloligands. The redox chemistry of the anion **14** suggested that it might be a useful probe for the presence of a metal ion coordinated by the functionalized butylthiolate chain, hence we have studied a variety of anionic nucleophiles with this aim in mind. The simplest of these is the thiolate anion, and a very recent determination of the crystal structure of the complex $[PPN][Cl_3W(\mu\text{-THT})_2(\mu\text{-}S(CH_2)_4(S\text{-ptol})WCl_3]^-$ **15** (ptol = *para*-tolyl) has confirmed its identity. More complex anionic nucleophiles, which would result in a chelating thiolate side chain (e.g. $Nu^- = [S(CH_2)_2S(CH_2)_2S]^{2-}$) are also under investigation (*26*).

The possibility of multiple S-C bond cleavages in the complexes **9** and **13** is an obvious question to be answered, especially in the context of the modeling of HDS reactions, in which the ultimate goal is the conversion of any organosulfur ligands to sulfide. This has been studied largely in the context of THT ring opening reactions of **13**. Stepwise reaction with different stoichiometries of nucleophile have shown that in the case of Nu^- = thiolate or selenolate, a second ring opening can be achieved (*26*). This happens in preference to cleaving the second C-S bond from the same THT ring. To this point we have been unable to achieve a clean second C-S bond cleavage with hydride as the nucleophile (using any of the usual Group 13 hydride reagents). Third and subsequent C-S cleavage has not been unambiguously achieved with any nucleophiles.

Realkylation of the $[Cl_3W(\mu\text{-}Et_2S)_2(\mu\text{-SEt})WCl_3]^-$ anion, and related systems, has not been achieved, suggesting that the thiolate sulfur carries very little charge in these species. Another example of the synthetic potential of S-dealkylation reactions is the synthesis of a unique face-sharing bioctahedral complex with three different bridging ligands. This was achieved by reacting the

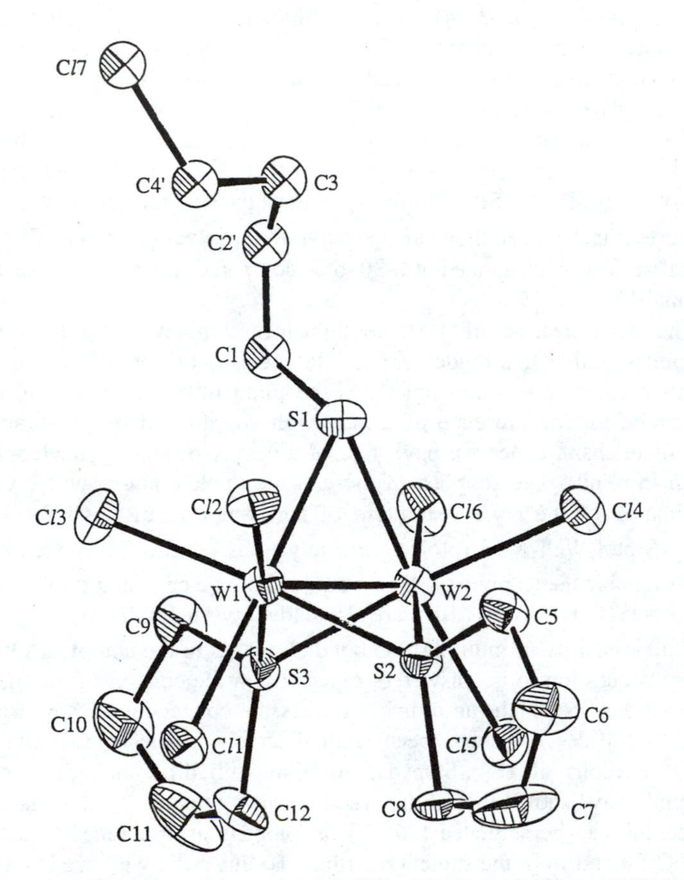

Figure 6 ORTEP plot of the anion $[Cl_3W(\mu\text{-THT})_2(\mu\text{-}S(CH_2)_4Cl)WCl_3]^-$ **14** (Reproduced with permission from ref. 23. Copyright 1991 American Chemical Society.)

anion $[Cl_3W(\mu\text{-}Et_2S)_2(\mu\text{-}Cl)WCl_3]^-$ with a thiolate anion, resulting in the formation of $[Cl_3W(\mu\text{-}Et_2S)(\mu\text{-}SEt)(\mu\text{-}Cl)WCl_3]^{2-}$ **16** (*27*).

The results described above need to be put into a more general context. The first example of dealkylation of a coordinated thioether by a nucleophile, of which we are aware, was reported by Roundhill and co-workers (*28*). Displacement of alkyl groups from the ligand *o*-$Ph_2PC_6H_4SR$ coordinated to Pd(II) or Pt(II) can be accomplished with an amine. It was proposed that the mechanism involved a nucleophilic substitution at the electrophilic α-carbon atom of the thioether. The alternative situation, that the metal is as electrophilic as the carbon atom, provides a competing reaction in many systems. We found this to be the case for the Mo(III) complex **10**, which shows reactivity of the C-S bonds of the THT bridge, but in addition it undergoes ligand substitution. Finally in this section we refer to the elegant studies of the Rakowski DuBois group, who have observed some remarkable examples of both C-S bond-making and bond-breaking in $[(CpMo)_2(S_2CH_2)(\mu\text{-}SMe_2)(\mu\text{-}SMe)]^+$ and related systems (*29*). In addition, and uniquely different from our systems, is the lability of the bridging thioether, and its easy replacement by other ligands.

Part II: Recent Studies Involving C-S Bond-Making Reactions

During our early investigations of the reactions of nucleophiles with the tris(thioether)-bridged ditungsten(III) complexes, **9** and **13**, we undertook reactions between $[PPh_4][MS_4]$ (M = Mo, W) and these species, anticipating that they could conceivably become incorporated into the binuclear structures to yield extended clusters. After finding that the major organosulfur product from the reaction with **9** was Et_2S_2 it became clear that **9** was essentially behaving as an exotic alkylating agent. The isolation of $[PPh_4]_2[W_3S_9(DMF)]$ from a DMF extract of the solid residues from this reaction (proven by an X-ray structure determination (*30*)) alerted us to the earlier literature reports of the use of MS_4^{2-} to convert organic halides into disulfides (*31*). It was claimed that the major metal-containing by-product, with minimal published substantiation, was $M_3S_9^{2-}$. The parallel of this chemistry to protonation reactions, which have been the focus of innumerable investigations, suggested that a mechanistic study of the condensation of MS_4^{2-} into higher aggregates might be feasible with simple alkyl labels. There is no universally accepted mechanism for the acid catalyzed reaction, although there seems to be general agreement that it is initiated by protonation, and Müller et al. prepared and characterized the $[WS_3(SH)]^-$ ion (*32*). Earlier work by Stiefel and co-workers had examined reactions of WS_4^{2-} with disulfides, alkynes and also provided a theoretical depiction of the tetrathiometalate ions (*33,34*). We found that by reacting organic bromides with WS_4^{2-} at 0°C in MeCN we could slow down the subsequent decomposition of the product and isolate and characterize the compounds $[PPh_4][WS_3(SR)]$ (R = Et, ^{i}Pr, ^{t}Bu) (*30*). The structure of the anion $[WS_3(SEt)]$ **17** is shown in Figure 7.

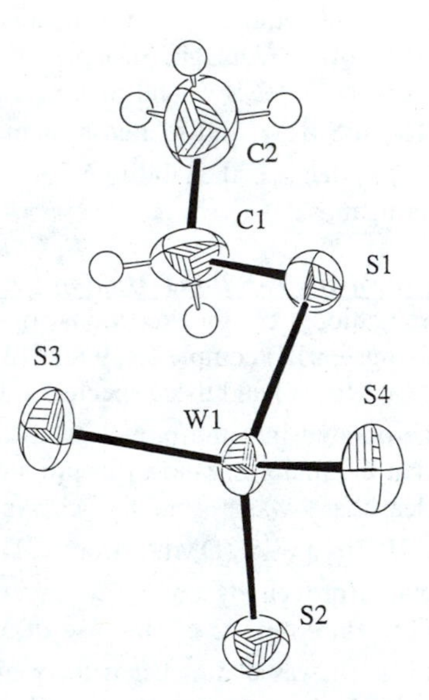

Figure 7 Structure of the anions $[WS_3(SEt)]^-$ **17** (Reproduced with permission from ref. 35. Copyright 1995 Royal Society of Chemistry.)

The reaction takes a different course when the more oxidizable iodide anion is present. Thus by reacting EtI with WS_4^{2-} under the same conditions, I_3^- was found to be a major product. A kinetic study of the EtBr / WS_4^{2-} reaction was undertaken, and it proved that it was cleanly first order in EtBr, with no sign of a second alkylation occurring. Solutions of $[PPh_4][WS_3(SEt)]$ in MeCN decompose at room temperauture to yield Et_2S_2, without addition of further EtBr, if the concentration is > 0.01 *M*. The production of Et_2S_2 and $W_3S_9^{2-}$ has been confirmed, but as is usually observed, an insoluble WS_x phase is slowly precipitated as well. This of course complicates any attempts to follow the kinetics of the reaction at this stage. To date, attempts to demonstrate the intermolecular nature of the decomposition of $[WS_3(SR)]^-$ by mixing $[WS_3(SEt)]^-$ and $[WS_3(S^tBu)]^-$ have been unsuccessful, since the ^tBu species is considerably more stable than its ethyl analog. Hence only Et_2S_2, and no $EtSS^tBu$ or $^tBu_2S_2$, is observed. It is interesting to note that Dhar and Chandrasekaran (*31*) had suggested that tertiary alkyl halides were unreactive toward $[WS_4]^{2-}$. The greater stability of $[WS_3(S^tBu)]^-$ prompted us to attempt to isolate and characterize its molybdenum analog, a target that we had previously considered unlikely based on the lability of the $[MoS_3(SEt)]^-$ ion. $[PPh_4][MoS_3(S^tBu)]$ is isostructural with its W analog (*30*). The possibility of using the two compounds as precursors to mixed Mo/W sulfido clusters is under investigation. Although the action of a second mole of RBr fails to induce a second alkylation, more potent alkylating agents are being studied to attempt this. A potentially significant observation is the rapid reaction of $[WS_4]^{2-}$ with Me_3SiCl in MeCN (36).

We draw the reader's attention to the ongoing work of Tatsumi and co-workers, who have recently reported on a number of reactions of $CpWS_3^-$ with alkylating agents, and also with alkynes (*37*). There are clearly a number of parallels with our work on the simple thiometallates. In both cases the highly unusual thiolate complexes of M(VI) result, species which have been regarded as very unstable.

The synthetic potential of the $[WS_3(SR)]^-$ salts has yet to be realized, but our early studies suggest that these ions react with other metal species to give completely different products from those obtained with $[WS_4]^{2-}$. Transient alkylated mixed metal species are certainly formed, leading us to speculate on the possibility of biosynthetic pathways to sulfido clusters adopting this kind of chemistry.

Part III: An Example of C-S Bond-Breaking and Bond-Making Without Metal Mediation

There have been many reports in the literature on the use of ethanedithiolate (edt) as a chelating ligand. In addition to its expected behavior as a chelating dianion it has been shown to undergo C-S bond cleavage quite readily (*37, 38*), and in at least one case a new C-S bond has been reported (*39*). This latter paper, by Tatsumi and co-workers (*39*), prompted us to conclude this review of our contributions to metal-promoted C-S bond reactivity by describing a serendipitous

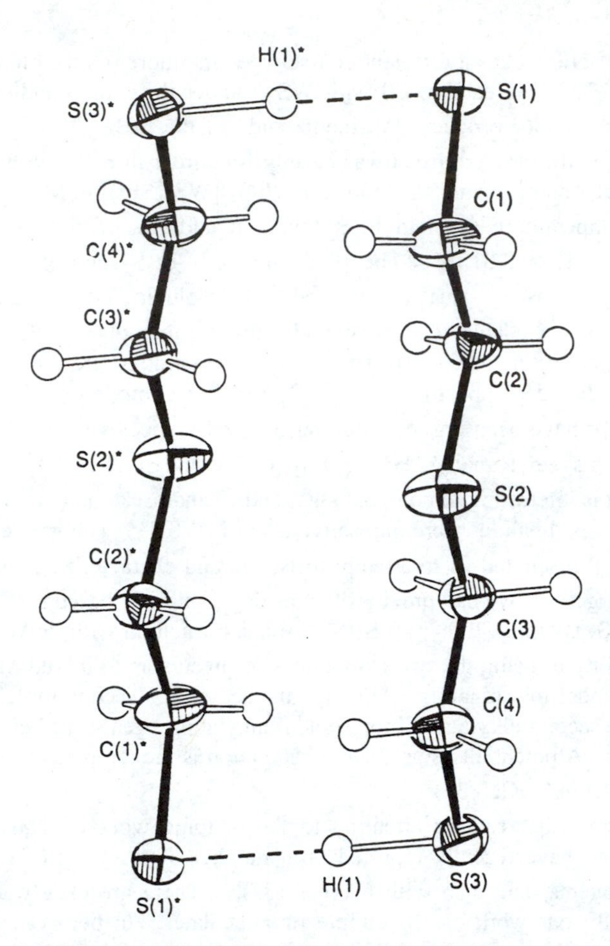

Figure 8 ORTEP plot of the S-H---S hydrogen bonded arrangement of anion pairs in the structure of $[Ph_4P][HSCH_2CH_2SCH_2CH_2S]$ (Reproduced with permission from ref. 40. Copyright 1992 Royal Society of Chemistry.)

discovery. In attempting to extend the range of phosphonium thiolates available for synthesis and mechanistic studies, we successfully prepared $[Ph_4P]_2[SCH_2CH_2S]$ and then attempted to prepare the salt of the monoanion, *viz* $[Ph_4P][SCH_2CH_2SH]$. The synthesis involves the deprotonation of edt by $Na^+[^iPrO]^-$ followed by exchange of the cations with $[PPh_4]Cl$. An orange crystalline salt was obtained on which an X-ray structure was determined. The identity of the anion proved to be $[HSCH_2CH_2SCH_2CH_2S]^-$ which packs as hydrogen bonded dimers in the crystal (Figure 8).

It is, incidentally, a uniquely well defined example of a H-S--H hydrogen bond (*40*). Of more interest in the present context is the origin of this ion. Evidence was obtained to support the following mechanism:

$$HSCH_2CH_2SH + {}^{i}PrO^- \rightarrow [SCH_2CH_2SH]^-$$

$$[SCH_2CH_2SH]^- + [SCH_2CH_2SH]^- \rightarrow [SCH_2CH_2SCH_2CH_2SH]^- + HS^-$$

$$[PPh_4]^+ + HS^- \rightarrow Ph_3PS + C_6H_6$$

The unexpected displacement of HS^-, generally considered a poor leaving group, is apparently facilitated by the immediate reaction of the hydrosulfide anion with the PPh_4^+ cation to form triphenylphosphine sulfide and benzene, both of which were conclusively identified. This example clearly illustrates the dangers of assuming that all reactions of organosulfur ligands, in which C-S bond are formed or broken, require the presence of a metal ion.

Acknowledgments

We thank the Natural Sciences and Engineering Research Council of Canada for its continued support of this work, and Dr Masood Parvez for many of the recent structure determinations.

Literature Cited

1. Boorman, P.M.; Chivers, T.; Mahadev, K.N. *Can. J. Chem.* **1977**, *55*, 869.
2. Boorman, P.M.; O'Dell, B.D. *J.Chem. Soc., Dalton Trans.* **1980**, 257.
3. Boorman, P.M.; Patel, V.D.; Kerr, K.A.; Codding, P.W.; Van Roey, P. *Inorg. Chem.,* **1980**, *19,* 3508.
4. Boorman, P.M.; Kerr, K.A.; Patel, V.D. *J. Chem. Soc., Dalton Trans.,* **1981**, 506.
5. Ball, J.M.; Boorman, P.M.; Moynihan, K.J.; Patel, V.D.; Richardson, J.F.; Collison, D.; Mabbs, F.E. *J. Chem. Soc., Dalton Trans.* **1983**, 2479.
6. Patel, V.D.; Boorman, P.M; Kerr, K.A.; Moynihan, K.J. *Inorg. Chem.*, **1982**, *21,* 1383.
7. a) Hughes, D.L.; Lazarowych, N.J.; Maguire, M.J.; Morris, R.H.; Richards, R.L. *J. Chem. Soc., Dalton Trans.,* **1995**, 5. b) Henderson, R.A.; Hughes, D.L.; Richards, R.L.; Shortman, C. *J. Chem. Soc., Dalton Trans.,* **1987.**
8. Burrow, T.E.; Hills, A.; Hughes, D.L.; Lane, J.D.; Lazarowych, N.J.; Maguire, M.J.; Morris, R.H.; Richards, R.L. *J. Chem. Soc., Chem. Commun.,* **1990**, 1757.
9. Boorman, P.M.; Moynihan, K.J.; Oakley, R.T. *J. Chem. Soc., Chem. Commun.,* **1982**, 899.
10. Boorman, P.M.; Moynihan, K.J.; Patel, V.D.; Richardson, J.F. *Inorg. Chem.,* **1985**, *24*, 2989.
11. Boorman, P.M.; Moynihan, K.J.; Richardson, J.F. *Inorg. Chem.,* **1988**, *27*, 3207.
12. Boorman, P.M.; Fait, J.F.; Freeman, G.K.W. *Polyhedron* **1989**, *8,* 1762.
13. Freeman, G.K.W., Ph.D. Thesis, University of Calgary, Calgary Alberta, Canada, 1990.
14. Bishop, P.T.; Dilworth, J.R.; Hughes, D.L. *J. Chem. Soc., Dalton Trans.,* **1988**, 2535.

15. Boorman, P.M.; Freeman, G.K.W.; Parvez, M. *Polyhedron* **1992**, *11,* 765.
16. Stephan, D.W. *Organometallics,* **1991**, *10*, 2037.
17. Wark, T.A.; Stephan, D.W. *Organometallics,* **1989**, *8*, 2836.
18. Strausz, O.P., in *AOSTRA Technical Handbook on Oil Sands, Bitumens and Heavy Oils*, Hepler, L.G. and Chiu, H., (Eds.), Alberta Oil Sands Technology and Research Authority, **1989**.
19. Moynihan, K.J.; Gao, X.; Boorman, P.M.; Fait, J.F.; Freeman, G.K.W.; Thornton, P.; Ironmonger, D.J. *Inorg. Chem.,* **1990**, *29,* 1648.
20. Boorman, P.M.; Gao, X.; Freeman, G.K.W.; Fait, J.F. *J. Chem. Soc., Dalton Trans.,* **1991**, 115.
21. a) Jacobsen, H.; Kraatz, H.-B.; Ziegler, T.; Boorman, P.M. *J. Am. Chem. Soc.*, **1992,** *114,* 7851. b) Boorman, P.M.; Kraatz, H.-B. *Coord. Chem. Rev.,* **1995**, *143*, 35.
22. Pacchioni, G.; Bagus, P.S. *Inorg. Chem.,* **1991**, *31*, 4391.
23. Boorman, P.M.; Gao, X.; Fait, J.F.; Parvez, M. *Inorg. Chem.*, **1991**, *30*, 3886.
24. Boorman, P.M.; Gao, X.; Freeman, G.K.W.; Fait, J.F. *J. Chem. Soc., Dalton Trans.,* **1991**, 115.
25. Kraatz, H.-B.; Aramini, J.M.; Gao, X.; Boorman, P.M.; Vogel, H.J. *Inorg. Chem*. **1993**, *32*, 3976.
26. Boorman, P.M.; Mozol, V.J., unpublished observations.
27. Boorman, P.M.; Gao, X.; Freeman, G.K.W.; Fait, J.F. *J. Chem. Soc., Dalton Trans.,* **1991**, 115.
28. Benefiel, A.; Roundhill, D.M. *Inorg. Chem.,* **1986**, *25*, 4027.
29. Tucker, D.S.; Dietz, S.; Parker, K.G.; Carperos, V.; Gabay, J.; Noll, B.; Rakowski DuBois, M. *Organometallics* **1995**, *14*, 4325
30. Boorman, P.M.; Wang, M.; Parvez, M. unpublished observations.
31. Dhar, P.; Chandrasekaran, S. *J. Org. Chem.*, **1989**, *54*, 2998.
32. Koniger-Ahlborn, E.; Schulze, H.; Muller, A. *Z. anorg. allg. Chem.,* **1977**, *428,* 5.
33. a) Harmer, M.A.; Halbert, T.R.; Pan, W.-H.; Coyle, C.L.; Cohen, S.A.; Stiefel, E.I. *Polyhedron,* **1986**, *5*, 341. b) Gea, Y.; Greaney, M.A.; Coyle, C.L.; Stiefel, E.I. *J. Chem. Soc., Chem. Commun.,* **1992**, 160.
34. a) Coyle, C.L.; Harmer, M.A.; George, G.N.; Daage, M.; Stiefel, E.I. *Inorg. Chem.,* **1990**, *29,* 14 b) Bernholc, J.; Stiefel, E.I. *Inorg. Chem.,***1985**, *24*, 1323.
35. Boorman, P.M.; Wang, M.; Parvez, M. *J. Chem. Soc., Chem. Commun*. **1995**, 999.
36. Boorman, P.M.; Downard, A.M.; unpublished observations.
37. Kawaguchi, H.; Tatsumi, K. *J. Am. Chem. Soc.,* **1995**, *117*, 3885.
38. Jiang, F.; Huang, Z.; Wu, D.; Kang, B.; Hong, M.; Liu, H. *Inorg. Chem.,* **1993**, *32*, 4971.
39. Tatsumi, K.; Sekiguchi, Y.; Nakamura, A.; Cramer, R.E.; Rupp, J.J. *J. Am. Chem. Soc.,* **1986**, *108*, 1358.
40. Boorman, P.M.; Gao, X.; Parvez, M. *J. Chem. Soc., Chem. Commun.,* **1992,** 1656.

Novel Structures

Chapter 12

Heterometallic Cuboidal Complexes as Derivatives of $[Mo_3S_4(H_2O)_9]^{4+}$

Interconversion of Single and Double Cubes and Related Studies

D. M. Saysell[1], M. N. Sokolov, and A. G. Sykes[2]

Department of Chemistry, University of Newcastle, Newcastle upon Tyne, NE1 7RU, United Kingdom

Heterometals can be incorporated into the Mo^{IV}_3 trinuclear ion $[Mo_3S_4(H_2O)_9]^{4+}$ to give cuboidal complexes of the kind Mo_3MS_4, or related edge-linked species $\{Mo_3MS_4\}_2$ or corner-shared $Mo_3S_4MS_4Mo_3$ double cubes, depending on the heteroatom and conditions used. All the products are present as aqua ions in acidic solutions. Specific topics included in this overview are the interconversion of single and double cubes with, e.g,. M = Pd and Sn. The tendency of cubes with M = Tl and In to react with H^+ with the formation of H_2 is discussed. The formation and properties of chalcogenide-rich clusters $Mo_3X_7^{4+}$ (X = S, Se, Te) is also considered.

The relevance of Fe_4S_4 clusters in biology (*1*), and the synthesis of an extensive series of analogue complexes (*2*), has helped target the area of metal-sulfur clusters as one meriting increased attention. Whereas the Fe atoms in Fe_4S_4 are tetrahedral and labile/highly reactive in aqueous solution, the octahedral Mo atoms in clusters $[Mo_4S_4(H_2O)_{12}]^{n+}$ n = 4,5,6, (*3*) are much more stable. It has also been noted that whereas the Fe_3S_4 incomplete cube occurs in numerous proteins, it has until recently (*4*) proved difficult to synthesize, due to the readiness with which the linear $Fe(\mu\text{-}S)_2Fe(\mu\text{-}S)_2Fe$ structure forms. The $[Mo_3S_4(H_2O)_9]^{4+}$ incomplete cube on the other hand is unusually stable (years!) if stored in acidic solution, and it has been possible to explore its rich and varied solution chemistry (*5*).

[1]Current address: Department of Chemistry, University of La Laguna, Tenerife, Canary Islands, Spain
[2]Corresponding author

0097–6156/96/0653–0216$15.00/0

Of particular interest in the studies described are the clusters formed by incorporating a heterometal atom into the trinuclear ion $[Mo_3S_4(H_2O)_9]^{4+}$ to give cuboidal complexes Mo_3MS_4 containing a heterometal M subsite, or related edge-linked $\{Mo_3MS_4\}_2$, or corner-shared $Mo_3S_4MS_4Mo_3$ double cubes (*6*). A subject not previously addressed in an overview of this kind is the ease of interconverting single and double cubes. More widely, double cubes are relevant with the occurrence of Fe_7MoS_9 as well as Fe_7VS_9 (composition to be confirmed) clusters in nitrogenase (*7-10*).

There are currently 16 different heterometal atoms which can be incorporated into the $[Mo_3S_4(H_2O)_9]^{4+}$ cluster, where examples range

Cr	Fe	Co	Ni	Cu				
Mo		Rh	Pd			In	Sn	Sb
			Pt		Hg	Tl	Pb	Bi

from Cr in Group 6 to Bi in Group 15. Preparative methods and procedures for characterization are also considered. The preparation of chalcogenide rich Mo_3X_7 clusters X = S, Se, Te is also seen as a route into this chemistry to be further explored.

Preparation of $[Mo_3S_4(H_2O)_9]^{4+}$

The most widely used procedure to date is to reduce with BH_4^- the cysteinato-Mo^V_2 complex $[(cys)OMo(\mu\text{-}S)_2MoO(cys)]^{2-}$, whereupon the cube $[Mo_4S_4(H_2O)_{12}]^{5+}$ (~20% yields) can be separated by Dowex 50W-X2 cation-exchange chromatography. Subsequently, this tetranuclear complex is air-oxidised to the trinuclear $[Mo_3S_4(H_2O)_9]^{4+}$ product by heating for up to 10h on a steam bath (~90°C). The latter is characterized by its UV-Vis peaks λ/nm ($\epsilon/M^{-1}cm^{-1}$ per Mo_3) at 366(5550) and 603(362). An alternative method makes use of the polymeric compound $\{Mo_3S_7Br_4\}_x$ obtained by heating Mo, S and Br_2 together in a sealed vessel (*11*). This compound can be solubilized as $[Mo_3S_7Br_6]^{2-}$. On treating with 4M Hpts (Hpts = p-toluenesulfonic acid), and then a phosphine PR_3, first $[Mo_3S_7(H_2O)_6]^{4+}$ and then $[Mo_3S_4(H_2O)_9]^{4+}$ can be obtained (*12*), in what is a relatively simple and rapid conversion (see below).

Preparation of Heterometal Clusters

Four methods have been used to date:

Reaction with Metals: The exposed triangular face on $[Mo_3S_4(H_2O)_9]^{4+}$ consisting of three μ_2-S core ligands, can interact directly with metals. Shibahara and colleagues (*13*) were the first to identify this method in their preparation of $[Mo_3FeS_4(H_2O)_{10}]^{4+}$, (1).

$$Mo_3S_4^{4+} + Fe \rightarrow Mo_3FeS_4^{4+} \quad (1)$$

So far it has been used for M = Fe, Co, Pd, Cu, Hg, In, Sn, Pb, Sb; see also Table I (*13,19,14,20,17,21*). Some of these reactions occur rapidly on addition of $[Mo_3S_4(H_2O)_9]^{4+}$ to the metal e.g. Hg (instant), Pb (~10 min), Sn (~1h). In the case of Ni, a number of days are required, and other metals (e.g., Cr, Mo) are inert. With Pd it is necessary to use freshly prepared Pd black (*14*). Heterometallic incorporation is also observed on electrochemical reduction of solutions containing $[Mo_3S_4(H_2O)_9]^{4+}$.

Reaction with Salts/Complexes + BH_4^-. The method was used first for Ni^{II} incorporation (*15*). It has also been applied in the case of $[Mo(H_2O)_6]^{3+}$, $[PtCl_4]^{2-}$, $In(ClO_4)_3$, TlCl, and bismuth citrate. Solutions of the latter are made up with $[Mo_3S_4(H_2O)_9]^{4+}$ and then syphoned onto a solution of BH_4^- (typically 100-fold excess), under air-free conditions, (2).

$$Mo_3S_4^{4+} + Ni^{2+} + 2e^- \rightarrow Mo_3NiS_4^{4+} \quad (2)$$

Dowex cation-exchange chromatography is required for purification. This is in contrast to direct reaction with the metals, which are cleaner, so that it is often possible to proceed directly without chromatographic purification. It is not always clear whether BH_4^- reduces $Mo_3S_4^{4+}$ to transient $Mo_3S_4^{2+}$ or the heterometal ion (or both) *in situ* thereby making reaction possible.

Direct Addition of Ionic Salts. There are two well established examples with Cu^+ (a slurry of insoluble CuCl can be used) (*16*), and Sn^{2+} (*17*), as in (3) and (4).

$$Mo_3S_4^{4+} + Cu^+ \rightarrow Mo_3CuS_4^{5+} \quad (3)$$

$$Mo_3S_4^{4+} + Sn^{2+} \rightarrow Mo_3SnS_4^{6+} \quad (4)$$

Use of Strong Reductants. The only example at present is with Cr^{2+} (-0.41V) (*18*), which presumably is able to serve as a reductant instead of BH_4^- (5).

Table I. X-ray Crystallographic Structures

A. Single cubes

Cluster Compound	References
$[Mo_3FeS_4(NH_3)_9(H_2O)]Cl_4$	13
$[Mo_3NiS_4(H_2O)_{10}](pts)_4.7H_2O$	19
$[Mo_3(PdCl)S_4(9\text{-aneN}_3)_3]Cl_3.4H_2O$ [a]	14
$[Mo_3(Inpts_2)S_4(H_2O)_{10}](pts)_3.13H_2O$	20
$(Me_2NH_2)_6[Mo_3(SnCl_3)S_4(NCS)_9]10.5H_2O$	17
$[Mo_4S_4(NH_3)_{12}]Cl_4.7H_2O$	21

[a] 9-aneN_3 is the 1,4,7-triazacyclonononane ligand.

B. Double cubes (edge-linked)

Cluster Compound	References
$[\{Mo_3S_4Cu(H_2O)_9\}_2](pts)_8.20H_2O$	22
$[\{Mo_3S_4Co(H_2O)_9\}_2](pts)_8.18H_2O$	23
$[\{Mo_3S_4Pd(H_2O)_9\}_2](pts)_8.24H_2O$	14

C. Double cubes (corner-shared i.e. sandwich type)

Cluster Compound	References
$[(H_2O)_9Mo_3S_4HgS_4Mo_3(H_2O)_9](pts)_8.20H_2O$	23
$[(H_2O)_9Mo_3S_3OInOS_3Mo_3(H_2O)_9](pts)_8.30H_2O$	24
$[(H_2O)_9Mo_3S_4SnS_4Mo_3(H_2O)_9](pts)_8.26H_2O$	25
$[(H_2O)_9MoS_4SbS_4Mo_3(H_2O)_9]^{8+}$	26
$[(H_2O)_9Mo_3S_4MoS_4Mo_3(H_2O)_9](pts)_8.18H_2O$	27

$$Mo_3S_4^{4+} + 3Cr^{2+} \rightarrow Mo_3CrS_4^{4+} + 2Cr^{3+} \quad (5)$$

The same procedure has been attempted with Eu^{2+} (-0.36V), so far without success (*6*).

Characterization of Heterometal Containing Products

X-ray crystallography has been used widely to establish the different structure types. The use of p-toluenesulfonic acid (H^+pts^-) at 4M levels to elute 8+ double cubes, as well as to obtain crystalline samples, has proved to be a major break-through. An important factor as far as crystallisation of the aqua species is concerned appears to be the ability of pts^- to participate in H-bonding. Many of the crystal structures summarized in Table I are from Shibahara's laboratory. Some of the reported structures have proved difficult to refine.

Chromatography. Elution behavior, often with $[Mo_3S_4(H_2O)_9]^{4+}$ as a marker, enables a fairly accurate estimate of the charge to be made. While 8+ double cube structures will elute in 2 – 3M HCl, when the net charge is reduced by complexing of the Cl^-, 4M levels of Hpts are required. Little or no elution of 8+ cations is observed with perchloric acid as high as 4M.

Metal Analysis. Inductively Coupled Plasma – Atomic Emission Spectroscopy (ICP-AES) is widely used, and can also provide information regarding the S (and Se) content, although not in the presence of pts^-.

Redox Titrations. Stoichiometries observed for redox processes with $[Fe(H_2O)_6]^{3+}$ or $[Co(dipic)_2]^-$ (dipic = 2,6-dicarboxylatopyridine) provide essential information. Spectrophotometric monitoring of the decay of the heterometallic cluster on titration with Fe^{III} or Co^{III} (small volumes from a micro-syringe), or colorimetric determination of the Fe^{II} produced, are procedures that can be used (*15-18*). Examples are as follows (6)-(9).

$$Mo_3NiS_4^{4+} + 2Fe^{III} \rightarrow Mo_3S_4^{4+} + Ni^{II} + 2Fe^{II} \quad (6)$$

$$Mo_3CuS_4^{5+} + Fe^{III} \rightarrow Mo_3S_4^{4+} + Cu^{II} + Fe^{II} \quad (7)$$

$$Mo_3SnS_4^{6+} + 2Fe^{III} \rightarrow Mo_3S_4^{4+} + Sn^{IV} + 2Fe^{II} \quad (8)$$

$$Mo_3CrS_4^{4+} + 3Fe^{III} \rightarrow Mo_3S_4^{4+} + Cr^{III} + 3Fe^{II} \quad (9)$$

The amount of $[Mo_3S_4(H_2O)_9]^{4+}$ can also be determined spectrophotometrically, hence the charge on the heterometal cube deduced.

An unusual feature in the case of the double cubes is that all the examples to date have 8+ charges. The only single cubes so far identified with more than one oxidation level are $Mo_3CuS_4^{n+}$ (n = 4, 5), and $Mo_4S_4^{n+}$ (n = 4, 5 and 6). The clusters are in many cases air-sensitive (N_2 used), exceptions being M = Ni, Pd, Rh, Pt. Even with the latter, the clusters are best stored air-free as a precaution.

Interconversion of Single and Double Cubes Type A

The kinetics of conversion of the edge-linked (Type A) double cube $[\{Mo_3PdS_4(H_2O)_9\}_2]^{8+}$ (Pd−S on one cube linked to S−Pd on the other), to the corresponding single cube $[Mo_3(PdX)S_4(H_2O)_9]^{4+}$ does not require air-free conditions (*28*). This system has therefore been the subject of recent attention. The reaction is induced by any of the six reagents X = CO, two water soluble phosphines $[P(C_6H_4SO_3)_3]^{3-}$ and 1,3,5-triaza-7-phosphaadamantane (PTA), Cl^-, Br^- or NCS^-, (10).

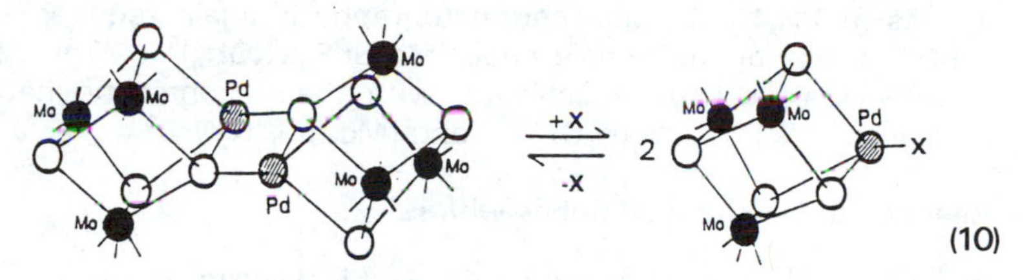

(10)

The first stage of reaction is fast and accompanied by color changes e.g. purple to dark blue in the case of Cl^-, which is assigned to the conversion of double to single cube. With X = CO or phosphine, absorbance changes are intense enough for stopped-flow monitoring with reactants at less than mM levels, when rate constants (25°C) in the range $10^5 - 10^6$ $M^{-1}s^{-1}$ are obtained. The reactions are independent of $[H^+]$, 0.30−2.00M, and no substitution at Mo is observed. The first stages with reagents X = Cl^-, Br^-, NCS^- are too fast to follow and only equilibrium constants K/M^{-1} (25°C) were obtained Cl^- (490), Br^- (8040), and NCS^- (630). Kinetic studies on the slower substitution processes occurring at Mo could be monitored in these cases. All the reagents show a high affinity for and react rapidly at the Pd to give single cube structures, whereas H_2O coordinated to the Pd is not able to retain the single-cube structure. Conversion of the single cube $[Mo_3(PdX)S_4(H_2O)_9]^{4+}$ back to the double cube can be achieved by chromatographic or chemical removal of X.

Interconversion of Single and Double Cubes Type B

This reaction involves the Type B corner-shared double cube. Air-free (N_2) conditions are required for these studies. Two-electron oxidations of the red-purple double cube $[Mo_6SnS_8(H_2O)_{18}]^{8+}$ with $[Fe(H_2O)_6]^{3+}$ or $[Co(dipic)_2]^-$ yields first the yellow single cube $[Mo_3SnS_4(H_2O)_{12}]^{6+}$, and finally (with excess oxidant) green $[Mo_3S_4(H_2O)_6]^{4+}$ (and Sn^{IV}), (*17*). The overall stoichiometry of the conversion is 4:1. With 2:1 amounts it is possible to recover by cation-exchange chromatography ~70% of the single cube $[Mo_3SnS_4(H_2O)_{12}]^{6+}$, (11).

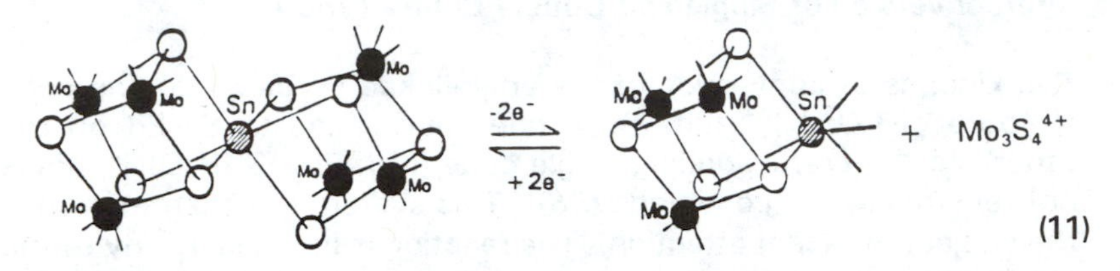

(11)

The reverse reaction can be achieved by mixing 1:1 amounts of $[Mo_3SnS_4(H_2O)_{12}]^{6+}$ and $[Mo_3S_4(H_2O)_9]^{4+}$ in the presence of 100-fold excess of BH_4^-. Column chromatography is again required with ~50% yields of the double cube $[Mo_6SnS_8(H_2O)_{18}]^{8+}$. The same interconversions can be achieved with the In-containing heterometallic cubes $[Mo_3InS_4(H_2O)_{12}]^{5+}$ and $[Mo_6InS_8(H_2O)_8]^{8+}$.

Reaction of Heterometal Cubes with H⁺.

The first example to be identified was with $[Mo_6TlS_8(H_2O)_{18}]^{8+}$ (*21*). With HCl in the range 0.5 – 2.0M, small bubbles of gas were observed consistent with (12).

$$2Mo_6TlS_8^{8+} + 2H^+ \rightarrow 4Mo_3S_4^{4+} + 2Tl^+ + H_2 \quad (12)$$

The rate law (13) has been confirmed,

$$\frac{-d[Mo_6TlS_8]^{8+}}{dt} = k_H[Mo_6TlS_8^{8+}][H^+] \quad (13)$$

with k_H (25°C) = $2.5 \times 10^{-4} M^{-1}s^{-1}$. The In single cube also reacts with H^+ but with a smaller rate constant k_H (25°C) ~ 10^{-5} $M^{-1}s^{-1}$. The behavior of the In cluster is of particular interest because it is known that the aqua In^+ ion reacts with H^+, (14).

$$In^+ + 2H^+ \rightarrow In^{3+} + H_2 \quad (14)$$

For this reaction k_H (25°C) = $11.2 \times 10^{-3} M^{-1}s^{-1}$, I = 0.70M ($LiClO_4$), (*30*).

Chalcogenide Rich Clusters

On melting the polymeric material $\{Mo_3S_7Br_4\}_x$ (*11*) with Ph_4PBr, $[Mo_3S_7Br_6]^{2-}$ is obtained in 70-80% yields. On treating with 4M Hpts lemon-yellow colored $[Mo_3S_7(H_2O)_6]^{4+}$ is produced, and can be purified by column chromatography (*12*). A similar procedure holds for the Se analogue, which yields orange-brown $[Mo_3Se_7(H_2O)_6]^{4+}$. On reacting with a water soluble phosphine or a slurry of triphenylphosphine, $[Mo_3S_4(H_2O)_9]^{4+}$ and $[Mo_3Se_4(H_2O)_9]^{4+}$, are obtained respectively. The rate law in the former case is (15)

$$\text{Rate} = k[Mo_3S_7^{4+}][PR_3] \tag{15}$$

where $k = 3.9 \times 10^4\ M^{-1}s^{-1}$ in 2.0M Hpts.

A tellurium analogue has been prepared and the crystal structure of $Cs_3[Mo_3Te_7(CN)_6]I.3H_2O$ reported (*31*). Here, no route through to $Mo_3Te_4^{4+}$ has yet been identified, and no reaction is observed with the phosphines. The Se clusters, $[Mo_3Se_4(H_2O)_9]^{4+}$ and $[Mo_4Se_4(H_2O)_{12}]^{5+}$, have been fully characterised (*32*). Heterometallic complexes are obtained from $[Mo_3Se_4(H_2O)_9]^{4+}$, but remain to be fully characterized.

Acknowledgments. We are grateful to the UK Engineering and Physical Sciences Research Council (MNS), the University of Newcastle (DMS), and the European Union HCM Program for financial support.

Literature Cited

(*1*) For recent reviews on Iron-Sulfur Proteins, see *Adv.Inorg. Chem.*, Cammack, R. and Sykes, A.G., ed. Vol. 38, **1992**.

(*2*) Holm, R.H., *Adv.Inorg.Chem.*, **1992**, *38*, 1-71.

(*3*) Hong, M.-C.; Li, Y.-J.; Lu, J.; Nasreldin, M.; Sykes, A.G., *J.Chem.Soc. Dalton Trans.*, **1993**, 2613.

(*4*) Zhou, J.; Holm, R.H., *J.Am.Chem.Soc.*, **1995**, *117*, 11353.

(*5*) Ooi, B.-L.; Sykes, A.G., *Inorg.Chem.*, **1989**, *28*, 3799.

(*6*) Saysell, D.M.; Sykes, A.G., *J.Clus.Sci.*, **1995**, *6*, 449.

(*7*) Coucouvanis, D., *Acc.Chem.Res.*, **1991**, *24*, 1.

(*8*) Kim, J.; Rees, D.C., *Nature (London)*, **1993**, *360*, 553.

(*9*) Rees, D.C.; Chan, M.K.; Kim, J., *Adv.Inorg.Chem.*, **1993**, *40*, 89-119.

(*10*) For recent reviews see *'Molybdenum Enzymes'* Stiefel, E.I.; Coucouvanis, D.; Newton, W.E., eds, American Chemical Society Publication, **1993**.

(*11*) Fedin, V.P.; Sokolov, M.N.; Geras'ko, O.A.; Virovets, A.V.; Podberezskaya, N.V.; Fedorov, V.Y., *Inorg.Chim.Acta*, **1991**, *187*, 81.

(*12*) Fedin, V.P.; Lamprecht, G.J.; Sykes, A.G., *J.Chem.Soc.Chem. Comm.*, **1994**, 2685.
(*13*) Shibahara, T.; Akashi, H.; Kuroya, H., *J.Am.Chem.Soc.*, **1986**, *108*, 1342.
(*14*) Murata, T.; Mizobe, Y.; Gao, H.; Ishii, Y.; Wakabayashi, T.; Nakano, F.; Tarase, T.; Yano, S.; Hidai, M.; Echizen, I.; Nanikawa, H.; Motomura, S., *J.Am.Chem.Soc.*, **1994**, *116*, 3389.
(*15*) Dimmock, P.W.; Lamprecht, G.J.; Sykes, A.G., *J.Chem.Soc. Dalton Trans.*, **1991**, 955.
(*16*) Nasreldin, M.; Li, Y.J.; Mabbs, F.E.; Sykes, A.G., *Inorg. Chem.*, **1994**, *33*, 4283.
(*17*) Varey, J.E.; Lamprecht, G.J.; Fedin, V.P.; Holder, A.; Clegg, W.; Elsewood, M.R.J.; Sykes, A.G., submitted.
(*18*) Routledge, C.A.; Humanes, M.; Li, Y.-Y.; Sykes, A.G., *J. Chem.Soc.*, **1994**, 1275.
(*19*) Shibahara, T.; Yamasaki, M.; Akashi, H.; Katayama, T., *Inorg. Chem.*, **1991**, *30*, 2693.
(*20*) Sakane, G.; Shibahara, T., *Inorg.Chem.*, **1993**, *32*, 777.
(*21*) Shibahara, T.; Kawano, E.D.; Okano, M.; Nishi, M.; Kuroya, H., *Chem.Lett.*, **1986**, 827.
(*22*) Shibahara, T.; Akashi, H.; Kuroya, H., *J.Am.Chem.Soc.*, **1988**, *110*, 3313.
(*23*) Shibahara, T.; Akashi, H.; Yamasaki, M.; Hashimoto, K., *Chem. Lett.*, **1991**, 689.
(*24*) Sakane, G.; Yao, Y.G.; Shibahara, T., *Inorg.Chim.Acta.*, **1994**, *216*, 13.
(*25*) Akashi, H.; Shibahara, T., *Inorg.Chem.*, **1989**, *28*, 2906.
(*26*) Shibahara, T.; Hashimoto, K.; Sakane, G., *J.Inorg.Biochem.*, **1991**, *43*, 280.
(*27*) Shibahara, T.; Yamamoto, T.; Kanadan, H.; Kuroya, H., *J.Am. Chem.Soc.*, **1987**, *109*, 3496.
(*28*) Saysell, D.M.; Lamprecht, G.J.; Darkwa, J.; Sykes, A.G., submitted.
(*29*) Varey, J.E.; Sykes, A.G., *Polyhedron*, **1996**, *15*, 1887.
(*30*) Taylor, R.S.; Sykes, A.G., *J.Chem.Soc.(A).*, **1969**, 2419.
(*31*) Fedin, V.P.; Imoto, H.; Saito, T.; McFarlane, W.; Sykes, A.G., *Inorg.Chem.*, **1995**, *34*, 5097.
(*32*) Nasreldin, M.; Henkel, G.; Kampman, G.; Krebs, B.; Lamprecht, G.J.; Routledge, C.A.; Sykes, A.G., *J.Chem.Soc. Dalton Trans.*, **1993**, 737.

Chapter 13

Characterization of Incomplete Cubane-Type and Cubane-Type Sulfur-Bridged Clusters

Genta Sakane and Takashi Shibahara

Department of Chemistry, Okayama University of Science, 1–1 Ridai-cho, Okayama 700, Japan

Incomplete cubane-type clusters with M_3S_4 cores and cubane-type clusters with $M_3M'S_4$ cores are characterized (M_3 = Mo_3, Mo_2W, MoW_2, W_3; M′= metal). The clusters $[M_3NiS_4(H_2O)_{10}]^{4+}$ take up ethylene to give $[M_3NiS_4(C_2H_4)(H_2O)_9]^{4+}$: an increase in the number of tungsten atoms results in the upfield chemical shift of ethylene-signal in the ^1H-NMR spectra. The molybdenum-iron cluster $[Mo_3FeS_4(H_2O)_{10}]^{4+}$ (**Mo3Fe**) does not react with ethylene. The formal oxidation states of metals in **Mo3Fe** and $[Mo_3NiS_4(H_2O)_{10}]^{4+}$ (**Mo3Ni**) are assigned as $Mo^{IV}Mo^{III}{}_2M'^{II}$ (M′ = Fe, Ni). The oxygen/sulfur-bridged clusters $[Mo_3S_4(H_2O)_9]^{4+}$ (**Mo3**) and $[Mo_3OS_3(H_2O)_9]^{4+}$ (**Mo3OS**) react with acetylene to give clusters with carbon-sulfur bonds. The reactivities of **Mo3Ni** and **Mo3** are interpreted by the electronic structures calculated by the Discrete Variational (DV)-Xα method. EPR studies of the mixed-metal cluster with $Mo_3CuS_4{}^{4+}$ core are also described.

Sulfide (S^{2-}), disulfide ($S_2{}^{2-}$), and thiolate (SR^-) ligands combine metal ions to give varied types of metal clusters. Much attention has been paid not only to metal-centered chemistry but also to ligand-centered chemistry, which is interesting inherently and of potential use in catalysis. Especially, molybdenum sulfur compounds have attracted much attention, and a large number of sulfur-bridged molybdenum compounds have appeared (*1*).

In this article, some of our recent results on incomplete cubane-type clusters with M_3S_4 cores and cubane-type clusters with $M_3M'S_4$ cores (M_3 = Mo_3, Mo_2W, MoW_2, W_3; M′ = metal) will be summarized.

Comparison of Molybdenum-Iron and Molybdenum-Nickel Clusters, $[Mo_3FeS_4(H_2O)_{10}]^{4+}$ (Mo3Fe) and $[Mo_3NiS_4(H_2O)_{10}]^{4+}$ (Mo3Ni)

The report (*2*) that the incomplete cubane-type sulfur-bridged molybdenum aqua cluster $[Mo_3S_4(H_2O)_9]^{4+}$ (**Mo3**) reacts with iron metal to give the molybdenum-

0097–6156/96/0653–0225$15.00/0

iron mixed-metal cluster $[Mo_3FeS_4(H_2O)_{10}]^{4+}$ (**Mo₃Fe**; Scheme 1) engendered much research on this type of metal incorporation reaction. This reaction is the first example where the missing corner of the incomplete cubane-type core is filled with another metal atom. Many metal incorporation reactions of the aqua cluster have been reported by us (*3*) and by other groups (*4*) to give mixed-metal clusters with $Mo_3M'S_4$ cores (M′ = metal). We can think of two kinds of driving force for the formation of the cubane-type mixed-metal clusters from the incomplete cubane-type aqua cluster **Mo_3** and metals. One factor is the affinity of the metal for the bridging sulfur atoms, and another is the reducing ability of the metal (*2, 3b*). In addition to direct reaction of **Mo_3**, with metals or metal ions (e. g., Sn^{2+}), $NaBH_4$ reduction of **Mo_3** in the presence of M^{2+} has also been used (*4b, 4c*). The reaction of **Mo_3** with $[Cr(H_2O)_6]^{2+}$ to give $[Mo_3CrS_4(H_2O)_{12}]^{4+}$ (*4d*), and other routes to clusters with $Mo_3M'S_4$ cores (M′ = Cu (*5a, 5c, 5d*), M′ = Sb (*5b, 5c*), M′ = W (*6*), M′ = Co (*7*)) have also been reported. Metal atom replacement of the incorporated metal atom M in the cubane-type $Mo_3M'S_4$ core with Cu^{2+} to give the cluster with $Mo_3CuS_4{}^{4+}$ core has been reported (*8*), and the existence of a new oxidation state of $Mo_3CuS_4{}^{5+}$ is known (*9*). The strong coloration on the reaction of the cluster **Mo_3** with mercury can be used for the analysis of mercury (*10*). The reactivity of clusters with the Mo_3PdS_4 (*11*) or Mo_3NiS_4 (*12*) core toward small molecules such as CO, alkenes, and alkynes has been reported.

The clusters $[Mo_3FeS_4(H_2O)_{10}]^{4+}$ (**Mo_3Fe**) and $[Mo_3NiS_4(H_2O)_{10}]^{4+}$ (**Mo_3Ni**) are obtainable through the reaction of $[Mo_3S_4(H_2O)_9]^{4+}$ (**Mo_3**) with iron and nickel, respectively, and are crystallized from 4M Hpts solution to give $[Mo_3M'S_4(H_2O)_{10}](pts)_4{\cdot}7H_2O$ (M′ = Fe (**Mo_3Fepts**), Ni (**Mo_3Nipts**); Hpts = *p*-toluenesulfonic acid). X-ray structural analyses of **Mo_3Fepts** and **Mo_3Nipts** revealed that the clusters, **Mo_3Fe** and **Mo_3Ni**, have an approximate symmetry of C_{3v}, with the iron and nickel atoms having fairly regular tetrahedral geometry in both clusters (*2, 3h*).

The ^{57}Fe-Mössbauer spectroscopy showed that the oxidation state of the iron atom in the mixed metal cluster **Mo_3Fepts** was assignable as +2.39 (*13*), which indicates that the reaction is reductive addition of iron to the Mo_3S_4 core in the molybdenum aqua cluster **Mo_3**.

Figure 1 shows cyclic voltammograms of **Mo_3**, **Mo_3Fe**, and **Mo_3Ni**. The cyclic voltammogram of **Mo_3** shows three consecutive one-electron reduction processes (the cathodic peak potentials, E_{pc}, are -0.45, -1.01, and -1.74 V, respectively), similar to those observed in $[M_3S_4(Hnta)_3]^{2-}$ obtained from $[M_3S_4(H_2O)_9]^{4+}$ and H_3nta (M_3 = Mo_3, Mo_2W, MoW_2, W_3) (*14*) and $[Mo_3S_4(ida)_3]^{2-}$ (*15*). These processes correspond to the change of oxidation states of the three metal atoms in each cluster:

$$M^{IV}{}_3 \rightarrow M^{IV}{}_2M^{III} \rightarrow M^{IV}M^{III}{}_2 \rightarrow M^{III}{}_3.$$

The cyclic voltammogram of **Mo_3Fe** also shows three reduction peaks with E_{pc} = -0.91, -1.47, and -1.81 V. The cathodic peak currents are close to each other and almost the same as those of **Mo_3** after being normalized by concentration, indicating that **Mo_3Fe** undergoes consecutive one-electron reduction processes. Four combinations of the oxidation states of the metals in **Mo_3Fe** before reduction are possible: $Mo^{IV}{}_3Fe^0$, $Mo^{IV}{}_2Mo^{III}Fe^I$, $Mo^{IV}Mo^{III}{}_2Fe^{II}$, $Mo^{III}{}_3Fe^{III}$. Of these oxidation states, $Mo^{IV}Mo^{III}{}_2Fe^{II}$ is most appropriate, because the oxidation state of the iron atom in **Mo_3Fe** has been determined to be +2.39 by ^{57}Fe-Mössbauer

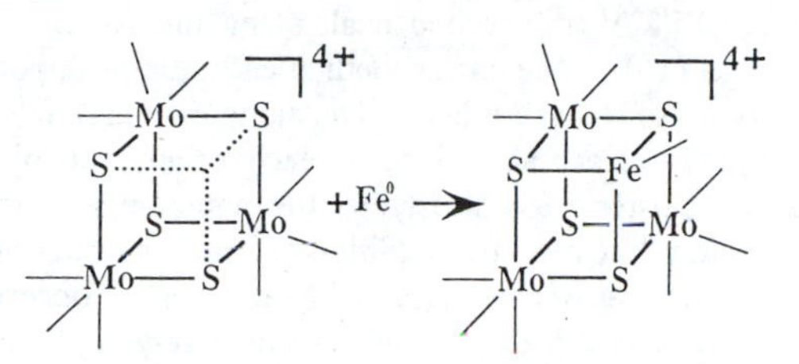

Scheme 1. Formation of mixed metal cluster $[Mo_3FeS_4(H_2O)_{10}]^{4+}$ (**Mo_3Fe**). Coordinated H_2O's are omitted for clarity.

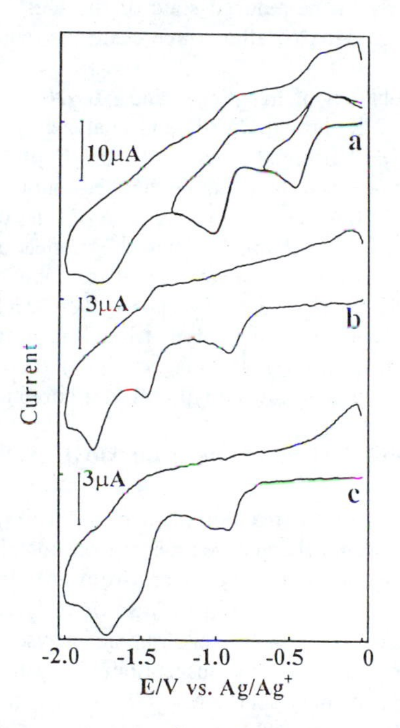

Figure 1. Cyclic voltammograms of a) $[Mo_3S_4(H_2O)_9]^{4+}$ (**Mo_3**, 1 mM), b)$[Mo_3FeS_4(H_2O)_{10}]^{4+}$ (**Mo_3Fe**, 0.3 mM), and c) $[Mo_3NiS_4(H_2O)_{10}]^{4+}$ (**Mo_3Ni**, 0.3 mM) at a glassy carbon electrode in 0.1 M tetrabuthylammonium hexafluorophosphate/acetonitrile with a scan rate of 100 mV/s. Reproduced with permission from reference 3h.

spectroscopy. After the complete reduction of **Mo_3Fe,** involving three electrons, the oxidation state of the metals is $Mo^{III}{}_3Fe^0$. The cyclic voltammogram of **Mo_3Ni** also shows three chemically irreversible consecutive one-electron reduction peaks at E_{pc} = -0.91, -1.48, -1.72 V; the second peak being unclear due to the overlap of second and third peaks (16). The corresponding cathodic peak potentials of **Mo_3Fe** and **Mo_3Ni** are very close to each other. The standard potentials of $Fe^{2+/0}$ (-0.44 V) and $Ni^{2+/0}$ (-0.257 V) are also close to each other, and nickel metal should therefore reduce the cluster **Mo_3** nearly to the same extent as iron metal. In conformity with the view "the reductive addition of iron- or nickel-metal to **Mo_3**", the orders of the binding energies of Mo $3d_{3/2}$ and Mo $3d_{5/2}$ are observed from the XPS measurements: E(**Mo_3**; 233.7 and 230.7 eV, respectively) > E(**Mo_3Fe**; 233.1 and 230.0 eV) and E(**Mo_3**; 233.7 and 230.7 eV) > E (**Mo_3Ni**; 233.3 and 230.3 eV). The formal oxidation states of metals in **Mo_3Fe** and **Mo_3Ni** are more adequately expressed as $Mo^{IV}Mo^{III}{}_2M'^{II}$ (M' = Fe, Ni) than $Mo^{IV}{}_3M'^0$, if we take the experimental results of Mössbauer spectroscopy, electrochemistry, and XPS into consideration. Similarly to the reduced state of the cluster **Mo_3Fe**, the oxidation state of the metals in **Mo_3Ni** after three-electron reduction is expressed as $Mo^{III}{}_3Ni^0$.

Magnetic susceptibilities of **Mo_3Fepts** and **Mo_3Nipts** have been measured at temperatures from ca. 2 K through 270 K and analyzed on the basis of the vector model formalism (*17*). Both the clusters **Mo_3Fepts** and **Mo_3Nipts** show antiferromagnetic behavior, and the effective magnetic moments are as follows: 2.78 B.M. at 2.16 K and 3.26 B.M. at 269.95 K for **Mo_3Fepts**; 0.11 B.M. at 2.00 K and 1.26 B.M. at 260.70 K for **Mo_3Nipts**. Using the models of $Mo^{IV}Mo^{III}{}_2Fe^{II}$ and $Mo^{IV}Mo^{III}{}_2Ni^{II}$, the following exchange integrals are obtained: $J_{1(Mo\text{-}Mo)}/k$ = -25 K, $J_{2(Mo\text{-}Fe)}/k$ = -75 K for **Mo_3Fepts**; $J_{1(Mo\text{-}Mo)}/k$ = -35 K, $J_{2(Mo\text{-}Ni)}/k$ = -60 K for **Mo_3Nipts**. The incomplete cubane-type trinuclear cluster of molybdenum (IV) $Ca_{1.5}[Mo_3S_4(Hnta)(nta)_2]\cdot 12H_2O$ is diamagnetic (*18*); therefore, the introduction of an iron atom into the incomplete cubane-type Mo_3S_4 core induced paramagnetism.

EPR Studies of the Mixed-Metal Cluster with $Mo_3CuS_4{}^{4+}$ Core

Shibahara et al. reported that the reaction of $[Mo_3S_4(H_2O)_9]^{4+}$ (**Mo_3**) with copper in diluted HCl gives a molybdenum-copper mixed-metal cluster with a single $Mo_3CuS_4{}^{4+}$ core (**Mo_3Cu**), which crystallizes from 4M Hpts solution as a double-cubane-type cluster, $[(H_2O)_9Mo_3S_4CuCuS_4Mo_3(H_2O)_9](pts)_8\cdot 20H_2O$ (*3c*). They examined the cluster by X-ray crystallographic analysis, XPS analysis, electronic absorption spectroscopy, and magnetic susceptibility measurement: the XPS analysis indicated the oxidation state of copper in **Mo_3Cu** as Cu(I) rather than Cu(II).

Sykes and coworkers measured EPR spectra for **Mo_3Cu** in 2M HCl or 2M Hpts, and reported that in the ground state, the unpaired electron is delocalized over the cube rather than localized on the copper, either as copper(I), which would be EPR silent, or as copper(II) (*9*).

Iwaizumi and his group, and our group carried out a joint research on the EPR behavior of the molybdenum-copper mixed-metal cluster **Mo_3Cu** in 1M HCl (*19*). Using ^{95}Mo *enriched* and unenriched (i.e., natural abundance) clusters, we arrived at the following conclusions:

1)The observation of the spectrum with g-value at about 2 indicates that the **Mo_3CuS_4** is a paramagnetic species of S = 1/2.

2)The four line splitting attributable to the hyperfine interaction with copper having a nuclear spin of I = 3/2 indicates that the interaction with copper nucleus is very small. 3)The spectral patterns at both high-field and low-field edges in the EPR spectrum of the ^{95}Mo enriched cluster are very similar to those of the satellite signals which arise from the hyperfine interaction with $^{95,97}Mo$ nuclei of natural abundance. This indicates that the molybdenum hyperfine splitting comes from hyperfine coupling with a single molybdenum nucleus; the unpaired electron is not distributed over the three molybdenum nuclei but mainly localized on a *single* molybdenum atom. The line broadening effect suggests that the spin center on the molybdenum ion is jumping among the three molybdenum ions. The following average spin densities were obtained: Mo(5s) = 0.061, Mo(4d) = 0.85, Cu(4s) = 0.011, Cu(3d) = 0.037. EHMO calculation indicates distortion of the $Mo_3CuS_4^{4+}$ core from C_{3v} to C_s symmetry

Sulfur-Bridged Incomplete Cubane-Type Molybdenum/Tungsten Clusters with M_3S_4 Cores and Cubane-Type Molybdenum/Tungsten-Nickel Clusters with M_3NiS_4 Cores (M_3 = Mo_3, Mo_2W, MoW_2, W_3)

A series of incomplete cubane-type sulfur-bridged molybdenum/tungsten clusters, $[M_3S_4(H_2O)_9]^{4+}$ (M_3 = Mo_3, Mo_2W, MoW_2, W_3) have been synthesized. From solutions of these aqua clusters, crystals of $[Mo_3S_4(H_2O)_9](pts)_4 \cdot 9H_2O$ (**Mo_3pts**), $[Mo_2WS_4(H_2O)_9](pts)_4 \cdot 9H_2O$ (**Mo_2Wpts**), $[MoW_2S_4(H_2O)_9](pts)_4 \cdot 9H_2O$ (**MoW_2pts**), and $[W_3S_4(H_2O)_9]$ $(pts)_4 \cdot 9H_2O$ (**W_3pts**) have been isolated, and their X-ray structures have been reported (*14, 15, 20*). Corresponding derivatives $[M_3S_4(Hnta)_3]^{2-}$ [M_3 = Mo_3 (**Mo_3nta**), Mo_2W (**Mo_2Wnta**), MoW_2 (**MoW_2nta**), W_3 (**M_3nta**); H_3nta = nitrilotriacetic acid] have also been obtained (*14,20*).

Binding energies of molybdenum ($3d_{3/2}$ and $3d_{5/2}$) and tungsten ($4f_{5/2}$ and $4f_{7/2}$) are obtained from XPS spectra of the clusters with M_3S_4 cores (M_3 = Mo_3, Mo_2W, MoW_2, W_3). The binding energies of Mo in **Mo_3** and **Mo_3nta** change little on the replacement of Mo with W, and those of W in **W_3** and **W_3nta** change little on the replacement of W with Mo, also: these phenomena can be explained by the softness of the bridging sulfurs which act as a buffer for the electron density changes on the Mo and W atoms.

Table I summarizes the data on current-sampled dc-polarograms and cyclic voltammograms of **Mo_3nta**, **Mo_2Wnta**, **MoW_2nta**, and **W_3nta**, each of which shows three consecutive one-electron reductive steps in alkaline solution (*14*). These steps correspond to the change of oxidation states of the three metals in each cluster: (IV, IV, IV) → (IV, IV, III) → (IV, III, III) → (III, III, III). Electronic spectra of one-electron reduction products, the oxidation state being (IV, IV, III), obtained by bulk electrolysis of **MoW_2nta** and **Mo_2Wnta** have been reported for the first time. The half-wave potentials, $E_{1/2}$, are significantly dependent on the cluster metals. In all the reduction processes the $[M_3S_4(Hnta)_3]^{2-}$ clusters (M_3 = Mo_3, Mo_2W, MoW_2, W_3) are more easily reduced as the numbers of Mo atoms in the cluster increase. The site of reduction of the molybdenum-tungsten mixed-metal clusters is mainly the Mo atom(s) rather than W atom(s).

The reaction of the incomplete cubane-type molybdenum/tungsten clusters $[M_3S_4(H_2O)_9]^{4+}$ (M_3 = Mo_3, Mo_2W, MoW_2, W_3) with nickel metal gave corresponding cubane-type molybdenum/tungsten-nickel clusters $[M_3NiS_4(H_2O)_9]^{4+}$, respectively (M_3 = Mo_3 (**Mo_3Ni**), Mo_2W (**Mo_2WNi**), MoW_2 (**MoW_2Ni**), W_3(**W_3Ni**)). X-ray analyses of the clusters revealed that if pts^- was

Table I. Comparison of $E_{1/2}$ for the Reduction of the Molybdenum-Tungsten Mixed-Metal Clusters, **Mo_3nta**, **Mo_2Wnta**, **MoW_2nta**, and **W_3nta**[a, b, c]

Compounds	$E_{1/2}$/V vs. Ag/AgCl		
	IV,IV,IV/ IV,IV,III	IV,IV,III/ IV,III,III	IV,III,III/ III,III,III
Mo_3nta	-0.64(70)	-1.08(63)	-1.39[d]
Mo_2Wnta	-0.73(66)	-1.22(69)	-1.66[d]
MoW_2nta	-0.84(59)	-1.40(63)	-1.78[d]
W_3nta	-1.12(59)	-1.41(55)	-1.88[d]

Note

[a] Obtained from sampled dc polarogram.

[b] Wave slopes (mV) are in parentheses.

[c] **Mo_3nta**, $K_2[Mo_3S_4(Hnta)_3]\cdot 9H_2O$;
Mo_2Wnta, $Na_2[Mo_2WS_4(Hnta)_3]\cdot 5H_2O$;
MoW_2nta, $Na_2[MoW_2S_4(Hnta)_3]\cdot 5H_2O$;
W_3nta, $K_2[W_3S_4(Hnta)_3]\cdot 10H_2O$.

[d] Accompanied by the catalytic hydrogen wave.

used as counter anions, the clusters, **Mo_3Ni** and **Mo_2WNi**, crystallized out as of single cubane-type, $[Mo_3NiS_4(H_2O)_{10}](pts)_4\cdot 7H_2O$ (**Mo_3Nipts**) (*3b*), and $[Mo_2WNiS_4(H_2O)_{10}](pts)_4\cdot 7H_2O$ (**Mo_2WNipts**) (*21*), respectively, and the clusters, **MoW_2Ni** and **W_3Ni** crystallized out as of double cubane-type, $[\{MoW_2NiS_4(H_2O)_9\}_2](pts)_8\cdot 20H_2O$ (**MoW_2Nipts**) (*22*) and $[\{W_3NiS_4(H_2O)_9\}_2](pts)_8\cdot 20H_2O$ (**W_3Nipts**) (*23*), respectively. The clusters $[M_3NiS_4(H_2O)_{10}]^{4+}$ (M_3 = Mo_3, Mo_2W, MoW_2, W_3) can be regarded as mononuclear nickel complexes with ligands "$M_3S_4(H_2O)_9$" and H_2O.

Contrary to the high reactivity of **Mo_3** towards metals, very little is known about the reactivity of the corresponding tungsten aqua cluster **W_3** (*20*). Only the reaction of the aqua ion **W_3** with Sn (or Sn^{2+}) has been reported so far (*24*). Other routes to the clusters with W_3MS_4 cores (M = metal) are also limited and only clusters with W_3CuS_4 cores have been reported (*25*).

Uptake of Ethylene by Sulfur-Bridged Cubane-Type Mixed-Metal Clusters with Molybdenum/Tungsten-Nickel Cores M_3NiS_4 (M_3 = Mo_3, Mo_2W, MoW_2, W_3): Syntheses, Structures, and ^1H NMR spectra

Despite the large number of investigations of nickel olefin π-complexes, there has not been any study of mixed-metal clusters containing nickel atom(s). The π-complexes so far reported were, to our knowledge, mononuclear or polynuclear ones containing only nickel atom(s) (*26*).

The sulfur-bridged molybdenum/tungsten-nickel mixed-metal cubane-type clusters $[M_3NiS_4(H_2O)_{10}]^{4+}$ (M_3 = Mo_3, Mo_2W, MoW_2, W_3) take up ethylene in aqueous or organic solutions to give olefin π-complexes $[M_3NiS_4(C_2H_4)(H_2O)_9]^{4+}$. Passing ethylene into the solution of **Mo_2WNi**, **MoW_2Ni**, and **W_3Ni** in 4M Hpts gave clusters $[Mo_2WNiS_4(C_2H_4)(H_2O)_9]^{4+}$ (**Mo_2WNiC$_2$**), $[MoW_2NiS_4(C_2H_4)\text{-}(H_2O)_9]^{4+}$ (**MoW_2NiC$_2$**), or

$[W_3NiS_4(C_2H_4)(H_2O)_9]^{4+}$ (**W_3NiC_2**), respectively. Cooling the solutions gave $[Mo_2WNiS_4(C_2H_4)(H_2O)_9](pts)_4(Hpts)_{1/2}{\cdot}6H_2O$ (**Mo_2WNiC_2pts**), $[MoW_2NiS_4(C_2H_4)(H_2O)_9](pts)_4(Hpts)_{1/2}{\cdot}6H_2O$ (**MoW_2NiC_2pts**), and $[W_3NiS_4(C_2H_4)(H_2O)_9](pts)_4{\cdot}(Hpts)_{1/2}{\cdot}6H_2O$ (**W_3NiC_2pts**), respectively (*27*). The reactivity of **Mo_3Ni** with ethylene is not as high as the tungsten containing clusters, **Mo_2WNi**, **MoW_2Ni**, and **W_3Ni**, and the reaction does not go to completion in aqueous solution. However, in organic solvents such as CH_2Cl_2, the reaction is complete.

X-ray structural analyses of **Mo_2WNiC_2pts**, **MoW_2NiC_2pts**, and **W_3NiC_2pts** revealed coordination of ethylene to the nickel site in each cluster. The three clusters are isomorphous to each other, and a schematic drawing of the complex cation **W_3NiC_2** is shown in Figure 2. The ethylene molecule on the three-fold axis is disordered, and molybdenum and tungsten atoms (in **Mo_2WNiC_2** and **MoW_2NiC_2**) are statistically disordered.

Peak positions of ^{1}H-NMR spectra of **Mo_2WNiC_2**, **MoW_2NiC_2**, and **W_3NiC_2** in D_2O containing 0.1 M methanesulfonic acid are shown in Figure 3 (standard: dimethylsilapentanesulphonic acid, Na salt). ^{1}H-NMR spectra of **Mo_3Ni** with C_2H_4 under pressure (ca. 2 atm; **Mo_3NiC_2**) is also included in the Figure 3.

The signals marked **b1**, **c1**, and **d1** are assigned to protons of free ethylene molecules dissociated from the clusters (the signal **a1** is also assignable to protons of free ethylene molecule), while those of **a2**, **b2**, **c2**, and **d2**, can be assigned to protons of ethylene molecules coordinated to the nickel atoms. An increase in the number of tungsten atoms in the clusters results in an increase in the upfield chemical shift of the ^{1}H-NMR signal due to the coordinated ethylene. This tendency can be explained by the fact that tungsten withdraws electron less than molybdenum, which causes the higher electron density at the hydrogen atoms of ethylene in the tungsten-containing clusters.

While the nickel complexes so far reported that are reactive toward ethylene contain zero-valent nickel, the formal oxidation states of the nickel atoms in the clusters **Mo_3Ni**, **Mo_2WNi**, **MoW_2Ni**, and **W_3Ni** are two as discussed above. Binding energies of Ni($2p_{3/2}$), Mo($3d_{3/2}$, $3d_{5/2}$), and W($4f_{5/2}$, $4f_{7/2}$) are measured by XPS of nickel metal and the molybdenum/tungsten-nickel clusters. These data also support Ni(II) rather than Ni(0). The binding energies of nickel in the clusters are similar to each other and distinctly larger than that of nickel metal (Ni $2p_{3/2}$: Ni metal, 852.7 eV; **Mo_3Ni**, 854.7; **Mo_2WNi**, 854.6 eV; **W_3Ni**, 854.5 eV). Binding energies of Mo and W decrease slightly but appreciably on the introduction of Ni into the incomplete cubane-type clusters, which indicates that Mo and W in the incomplete cubane-type clusters are reduced by Ni metal.

Carbon-Sulfur Bond Formation through the Reaction of Sulfur-Bridged Incomplete Cubane-Type Molybdenum Clusters with Acetylene.

Incomplete cubane-type molybdenum aqua clusters, $[Mo_3(\mu_3\text{-}S)(\mu\text{-}O)(\mu\text{-}S)_2\text{-}(H_2O)_9]^{4+}$ (**Mo_3OS**) and $[Mo_3(\mu_3\text{-}S)(\mu\text{-}S)_3(H_2O)_9]^{4+}$ (**Mo_3**), react with acetylene to produce clusters with alkenedithiolate ligands $[Mo_3(\mu_3\text{-}S)(\mu\text{-}O)(\mu_3\text{-}S_2C_2H_2)\text{-}(H_2O)_9]^{4+}$ (**Mo_3OSAc**) and $[Mo_3(\mu_3\text{-}S)(\mu\text{-}S)(\mu_3\text{-}S_2C_2H_2)(H_2O)_9]^{4+}$ (**Mo_3Ac**), respectively, with *carbon-sulfur bond formation* as shown in Scheme 2. (*28*). On the other hand, the oxygen-bridged cluster $[Mo_3O_4(H_2O)_9]^{4+}$ (**Mo_3O_4**) does not react either with metals or acetylenes. Acetylene addition was confirmed by the X-ray structure analysis of $[Mo_3(\mu_3\text{-}S)(\mu\text{-}O)(\mu_3\text{-}S_2C_2H_2)(H_2O)_9](pts)_4{\cdot}7H_2O$

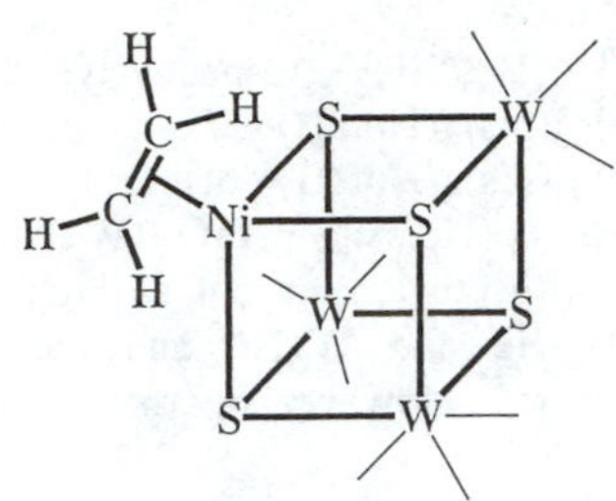

Figure 2. Schematic drawing of $[W_3NiS_4(C_2H_4)(H_2O)_9]^{4+}$(**$W_3NiC_2$**). Coordinated H_2O's are omitted for clarity.

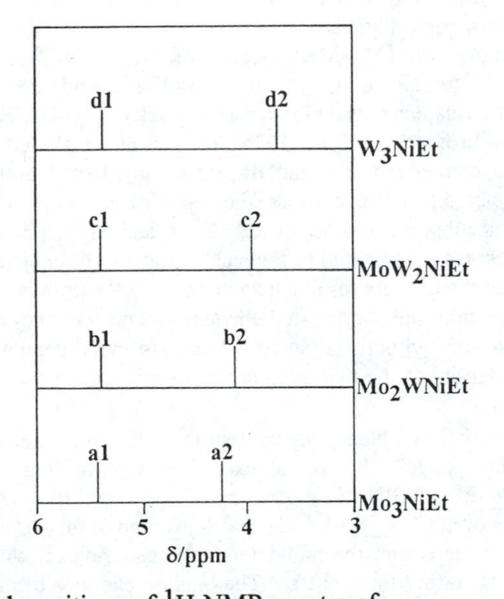

Figure 3. Peak positions of ^{1}H-NMR spectra of $[Mo_3NiS_4(C_2H_4)(H_2O)_9]^{4+}$ (**Mo_3NiC_2**), $[Mo_2WNiS_4(C_2H_4)(H_2O)_9]^{4+}$(**$Mo_2WNiC_2$**), $[MoW_2NiS_4(C_2H_4)(H_2O)_9]^{4+}$ (**MoW_2NiC_2**), and $[W_3NiS_4(C_2H_4)(H_2O)_9]^{4+}$ (**W_3NiC_2**) in D_2O containing 0.1 M methanesulfonic acid. (standard: dimethylsilapentanesulphonic acid, Na salt). See text.

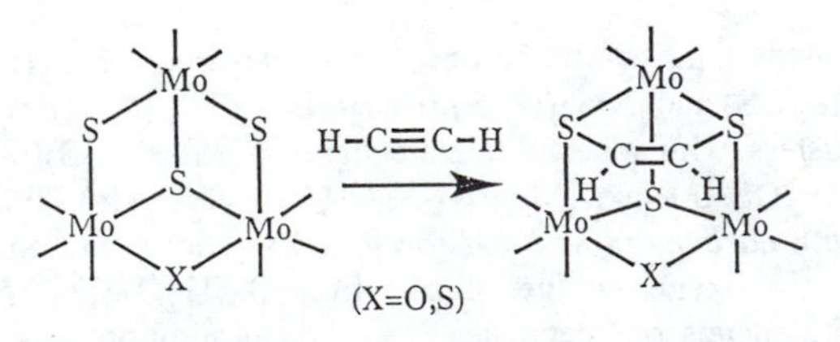

Scheme 2. Reaction of $[Mo_3(\mu_3\text{-S})(\mu\text{-X})(\mu\text{-S})_2(H_2O)_9]^{4+}$ (X = O, S) with acetylene. Aqua ligands and Mo-Mo bonds are omitted for clarity.

(**$Mo_3OSAcpts$**) (*28*). The carbon-carbon distance in the cluster is 1.321(11)Å, which is closer to that of ethylene (1.339 Å) than to that of acetylene (1.203 Å). The *cis* geometry of the two hydrogen atoms attached to the carbon atoms was demonstrated by the X-ray analysis. The bond angles about two carbon atoms of the acetylene are close to 120°, which also indicates the sp^2 character of the orbitals of the two carbon atoms.

Examples of *carbon-bridging-sulfur bond formation* through the reaction of sulfur-bridged metal compounds with acetylene and acetylene derivatives are rather limited. Rakowski DuBois and coworkers developed the chemistry of the reaction of dinuclear molybdenum compounds, for example $[\{(C_5H_5)Mo(\mu\text{-}S)(\mu\text{-}SH)\}_2]$, with acetylene (*29*). Similar chemistry of dinuclear sulfur-bridged iron (*30*) and vanadium (*31*) compounds has also been reported. No report has appeared on the reaction of trinuclear metal clusters with acetylene or acetylenederivatives to form carbon-bridging-sulfur bonds. Several other types of carbon-sulfur bond formation have been reported. Stiefel and coworkers (*32*), and Coucouvanis and coworkers (*33*) reported addition of activated acetylenes to coordinated polysulfide ligands of molybdenum compounds to form new C-S bonds. Similar reactions have been reported with iron (*34*) and tungsten (*35*) compounds.

The electronic spectra of **Mo_3OSAc** and **Mo_3Ac** have intense bands in the near infrared region, which are shown in Figure 4 together with those of **Mo_3OS** and **Mo_3**. The appearance of characteristic absorption in near infrared region indicates that derivatives of the aqua clusters **Mo_3OS** and **Mo_3** also react with acetylene and acetylene derivatives not only in aqueous solution but also in organic solvents.

The carbon-sulfur bond formation may lead to new synthetic routes to sulfur-containing organic compounds and the use of unsymmetrical acetylene derivatives could possibly give optically active clusters.

Elucidation of the Reactivity Differences between $[Mo_3NiS_4(H_2O)_{10}]^{4+}$ (Mo_3Ni) and $[Mo_3FeS_4(H_2O)_{10}]^{4+}$ (Mo_3Fe), and between $[Mo_3S_4(H_2O)_9]^{4+}$ (Mo_3) and $[Mo_3O_4(H_2O)_9]^{4+}$ (Mo_3O_4) by Discrete Variational (DV)-Xα Calculation

Differences between $[Mo_3NiS_4(H_2O)_{10}]^{4+}$ (Mo_3Ni) and $[Mo_3FeS_4(H_2O)_{10}]^{4+}$ (Mo_3Fe). Although the structures of **Mo_3Fe** and **Mo_3Ni** are very similar to each other (**Mo_3Fepts** is isomorphous with **Mo_3Nipts**), the reactivities toward small molecules such as carbon monoxide, ethylene and acetylene are very different from each other: **Mo_3Fe** does not react and **Mo_3Ni** does. This difference can be explained by Discrete Variational (DV)-Xα Calculation (*3h*).

Atomic orbital components constituting HOMO's (a, **Mo_3Fe** (93a′); b, **Mo_3Ni** (94a′)) are shown in Figure 5 (those of sulfur are not shown). Figure 6 shows LUMO's of the small molecules calculated by EHMO. The atomic orbital component of iron ($3d_{z^2}$, 27%) is smaller than that of nickel ($3d_{xz}$, 46%), and, moreover, the iron orbital does not match the LUMO of the small molecules, and can not back-donate electrons to the small molecules.

The calculated electronic structures near the HOMO and LUMO seems fairly reasonable as judged by the experimental results such as electronic spectra of **Mo_3Fe** and **Mo_3Ni** (*3h*).

Difference between $[Mo_3S_4(H_2O)_9]^{4+}$ (Mo_3) and $[Mo_3O_4(H_2O)_9]^{4+}$ (Mo_3O_4). While the sulfur-bridged cluster **Mo_3** reacts with acetylene, the oxygen-bridged

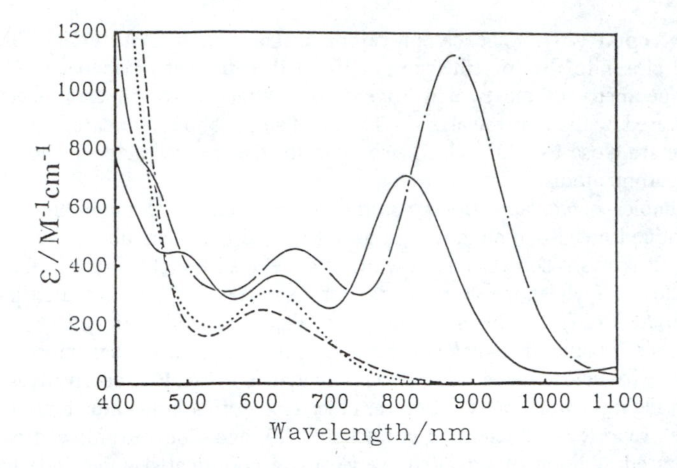

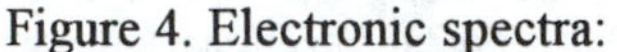

Figure 4. Electronic spectra:
----------------, $[Mo_3(\mu_3\text{-}S)(\mu\text{-}O)(\mu\text{-}S)_2(H_2O)_9]^{4+}$ (**Mo3OS**) in 1 M HCl;
.................., $[Mo_3(\mu_3\text{-}S)(\mu\text{-}S)_3(H_2O)_9]^{4+}$ (**Mo3**) in 1 M HCl;
————, $[Mo_3(\mu_3\text{-}S)(\mu\text{-}O)(\mu_3\text{-}S_2C_2H_2)(H_2O)_9]^{4+}$ (**Mo3OSAc**) obtained by passing acetylene through **Mo3OS** in 1 M HCl;
—·—·—·, $[Mo_3(\mu_3\text{-}S)(\mu\text{-}S)(\mu_3\text{-}S_2C_2H_2)(H_2O)_9]^{4+}$ (**Mo3Ac**) obtained by passing acetylene through **Mo3** in 1 M HCl.

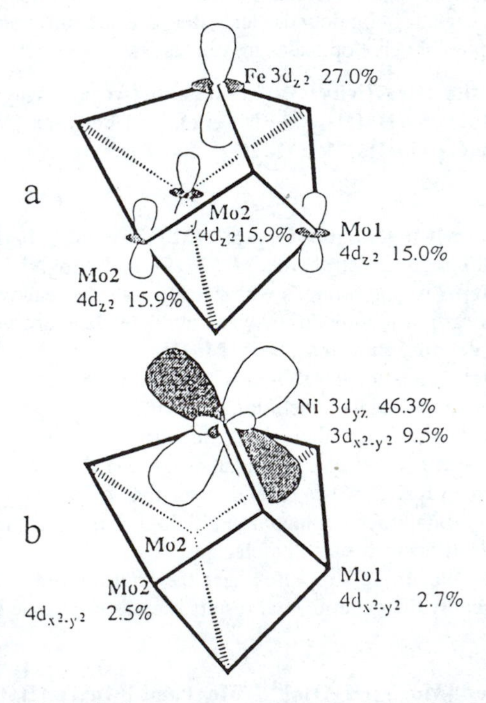

Figure 5. Atomic orbital components constituting HOMO's: a, $[Mo_3FeS_4(H_2O)_{10}]^{4+}$ (**Mo3Fe**); b, $[Mo_3NiS_4(H_2O)_{10}]^{4+}$ (**Mo3Ni**). Reproduced with permission from reference 3h.

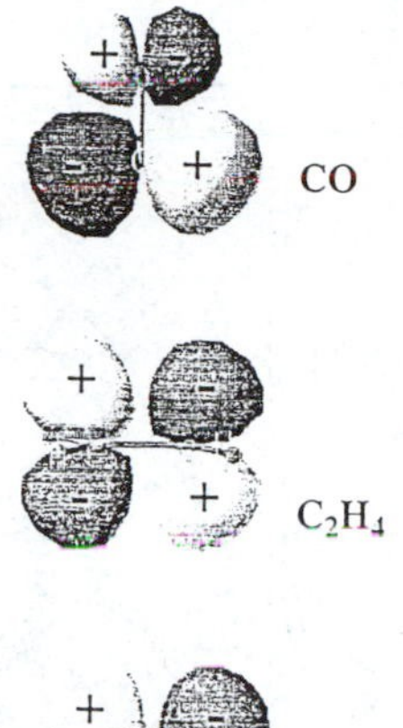

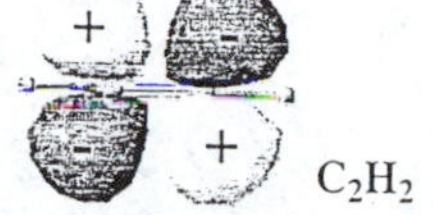

Figure 6. LUMO's for CO, C_2H_2, and C_2H_4. Reproduced with permission from reference 3h.

cluster $\mathbf{Mo_3O_4}$ does not react with acetylene. This difference can be understood in terms of the DV-Xα calculations of $\mathbf{Mo_3}$ and $\mathbf{Mo_3O_4}$ (*36*).

Reports have appeared on the calculation of clusters with $Mo_3S_4^{4+}$ or $Mo_3O_4^{4+}$ cores. Several calculation methods having been employed: 1) bare cores, $Mo_3S_4^{4+}$ (*37a, 37b, 37d, 37e*), $Mo_3O_4^{4+}$ (*37a, 37e, 37f*), and $Mo_3S_{4-n}O_n^{4+}$ (n=0-4) (*38*). 2) full clusters, $[Mo_3S_4Cl_6(PH_3)_3]^{2-}$ (*37c*), $[Mo_3O_4(OH)_6(H_2O)_3]^{2-}$ (*37c, 37f*) However, no reports on the full aqua clusters, $\mathbf{Mo_3}$ and $\mathbf{Mo_3O_4}$, have appeared to the best of our knowledge.

Fairly large mixing of Mo 4d and S 3p (or O 2p) atomic orbitals has been observed in some of the orbitals, regardless of the method of calculation. As for the HOMO of the bare cores of $Mo_3S_4^{4+}$ and $Mo_3O_4^{4+}$, however, the CNDO/2 calculation (*37a, 37b*) indicates no contribution of bridging-sulfur (or -oxygen) orbitals, while SCCC-EHMO (*37e*), Fenske-Hall (*37f*), and *ab initio* (*38*) methods indicate contribution of both molybdenum and sulfur (or oxygen). Our results on the bare cores, $Mo_3S_4^{4+}$ and $Mo_3O_4^{4+}$, as well as the full clusters, $\mathbf{Mo_3}$ and $\mathbf{Mo_3O_4}$, indicate the mixing of Mo 4d and S 3p (or O 2p) orbitals in each HOMO.

If the HOMO (45e, X-Z plane; Figure 7a) of $\mathbf{Mo_3}$ and the HOMO (40e, X-Z plane; Figure 7b) of $\mathbf{Mo_3O_4}$ are compared, it is found that the orbital lobe of μ-S expands toward the Z-axis, which is favorable to the overlapping of the μ-S orbital lobe with the π-orbital of acetylene. On the other hand, the orbital lobe of μ-O does not expand toward Z-axis; furthermore, the magnitude of expansion of the lobe is much less than that in $\mathbf{Mo_3}$. Therefore, the possibility of the orbital lobe of μ-O overlapping with the π-orbital of acetylene is much less than the case of $\mathbf{Mo_3}$. Another factor for the reactivity difference in $\mathbf{Mo_3}$ and $\mathbf{Mo_3O_4}$ clusters is the difference in energy levels. However, the energy differences in HOMO's (0.84 eV) and in LUMO's (0.67 eV) are not so large, and the different shapes seem to be the larger factor for the C-S bond formation.

Calculated transition energies are in fairly good agreement with experimental ones for both clusters.

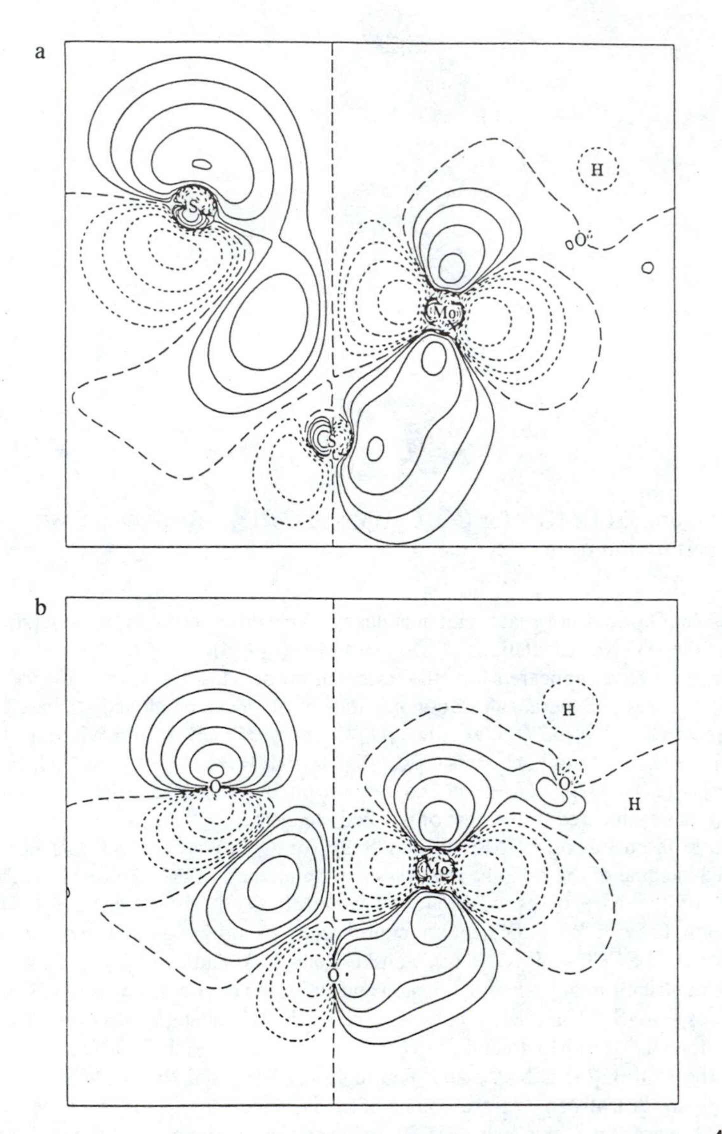

Figure 7. Contour maps: a) HOMO (x-z plane) for $[Mo_3S_4(H_2O)_9]^{4+}$ (**Mo_3**). The x-z plane contains one Mo, one μ-S, one μ_3-S, one O (water), and two H atoms (water). b) HOMO (x-z plane) for $[Mo_3O_4(H_2O)_9]^{4+}$ (**Mo_3O_4**). The x-z plane contains one Mo, one μ-O, one μ_3-O, one O (water), and two H atoms (water). Solid, dotted, and dashed lines indicate positive, negative, and zero contour lines, respectively. Reproduced with permission from reference 36.

Acknowledgments
This work was partly supported by a Grant-in-Aid for Scientific Research Nos. 59470039, 02453043, and 04241102 (on Priority Area of "Activation of Inactive Small Molecules") from the Ministry of Education, Science and Culture of Japan.

Literature Cited

(1) Some of recently published books and reviews: a) *The Chemistry of Metal Cluster Complexes*; Shriver, D. F.; Kaesz, H. D.; Adams, R. D., Ed; VCH: New York, 1990. b) Mingos, D. M. P.; Wales, D. J.; *Introduction to Cluster Chemistry*; Prentice Hall: New Jersey, 1990. c) Shibahara, T. *Coord. Chem. Rev.* **1993**, *123*, 73-147. d) Curtis, M. D. *Appl. Organomet. Chem.* **1992**, *6*, 429-436. e) Adams, R. D. *Polyhedron* **1985**, *4*, 2003-2025. f) Shibahara, T. *Adv. Inorg. Chem.* **1991**, *37*, 143-173. g) Lee, S. C.; Holm, R. H. *Angew. Chem. Int. Ed. Engl.* **1990**, *29*, 840-856. h) Chen, Z. *J. Cluster Science* **1995**, *6*, 357-377. i) Kang, B.-S.; Hong, M.-C.; Wen, T.-B.; Liu, H.-K.; Lu, J.-X. *J. Cluster Science* **1995**, *6*, 379-401. j) Saysell, D. M.; Sykes, A. G. *J. Cluster Science* **1995**, *6*, 449-461. k) Tsai, K. R.; Wan, H. L. *J. Cluster Science* **1995**, *6*, 485-501. l) Müller, A.; Krahn, E. *Angew. Chem. Int. Ed. Engl.* **1995**, *34*, 1071-1078.

(2) Shibahara, T.; Akashi, H.; Kuroya, H. *J. Am. Chem. Soc.* **1986**, *108*, 1342-1343.

(3) a) M=Ni: Shibahara, T.; Kuroya, H. *J. Coord. Chem.* **1988**, *18*, 233-236. b) M=Ni: Shibahara, T.; Yamasaki, M.; Akashi, H.; Katayama, T. *Inorg. Chem.* **1991**, *30*, 2693-2699. c) M=Cu: Shibahara, T.; Akashi, H.; Kuroya, H. *J. Am. Chem. Soc.* **1988**, *110*, 3313-3314. d) M=Sn: Akashi, H.; Shibahara, T. *Inorg. Chem.* **1989**, *28*, 2906-2907. e) M=Co, Hg: Shibahara, T.; Akashi, H.; Yamasaki, M.; Hashimoto, K. *Chem. Lett.* **1991**, 689-692. f) M=In: Sakane, G.; Shibahara, T. *Inorg. Chem.* **1993**, *32*, 777-778. g) M=Sb: Shibahara, T.; Hashimoto, K.; Sakane, G. *J. Inorg. Biochem.* **1991**, *43*, 280. h) M=Fe, Ni: Shibahara, T.; Sakane, G.; Naruse, Y.; Taya, K.; Akashi, H.; Ichimura, A.; Adachi, H. *Bull. Chem. Soc. Jpn.* **1995**, *68*, 2769-2782.

(4) a) M = Pd: Murata, T.; Gao, H.; Mizobe, Y.; Nakano, F.; Motomura, S.; Tanase, T.; Yano, S.; Hidai, M. *J. Am. Chem. Soc.* **1992**, *114*, 8287-8288. b) M = Fe: Dimmock, P.W.; Dickson, D. P. E.; Sykes, A. G. *Inorg. Chem.* **1990**, *29*, 5120-5125. c) M = Ni, Fe: Dimmock, P. W.; Lamprecht, G. J.; Sykes, A. G. *J. Chem. Soc., Dalton Trans.* **1991**, 955-961. d) M=Cr: Routledge, C. A.; Humanes, M.; Li, Y.-J.; Sykes, A. G. *J. Chem. Soc., Dalton Trans.* **1994**, 1275-1282.

(5) a) Wu, X.-T.; Lu, S.-F.; Zu, L.-Y.; Wu, Q.-I.; Lu, J.-X. *Inorg. Chim. Acta* **1987**, *133*, 39-42. b) Lu, S.-F.; Huang, J.-Q.; Lin, Y.-H.; Huang, J.-L. *Huaxue Xuebao* **1987**, *45*, 666-675. c) Huang, J.-Q.; Huang, J.-L.; Shang, M.-Y.; Lu, S.-F.; Lin, X.-T.; Lin, Y.-H.; Huang, M.-D.; Zhuang, H.-H.; Lu, J.-X. *Pure Appl. Chem.* **1988**, *60*, 1185-1192. d) Lu, S.-S.; Chen, H.-B.; Huang, J.-Q.; Wu, Q.-J.; Sun, Q.-L.; Li, J.; Lu, J.-X. *Inorg. Chim. Acta* **1995**, *232*, 43-50.

(6) Deeg, A.; Keck, H.; Kruse, A.; Kuchen, W.; Wunderlich, H. *Z. Naturforsch.* **1988**, *43b*, 1541-1546.

(7) Curtis M. D. *App. Organomet. Chem.* **1992**, *6*, 429-436.

(8) Shibahara, T.; Asano, T.; Sakane, G. *Polyhedron* **1991**, *10*, 2351-2352.

(9) Nasreldin, M.; Li, Y.-J.; Mabbs, F. E.; Sykes, A. G. *Inorg. Chem.* **1994**, *33*, 4283-4289.

(10) a) Aikoh, H.; Shibahara, T. *Physiol. Chem. Phys. & Med. NMR* **1990**, *22*, 187-192. b) Aikoh, H.; Shibahara, T. *Analyst* **1993**, *118*, 1329-1332.
(11) Murata, T.; Mizobe, Y.; Gao, H.; Ishii, Y.; Wakabayashi, T.; Nakano, F.; Tanase, T.; Yano, S.; Hidai, M.; Echizen, I.; Nanikawa, H.; Motomura, S. *J. Am. Chem. Soc.* **1994**, *116*, 3389-3398.
(12) Shibahara, T.; Mochida, S.; Sakane, G. *Chem. Lett.* **1993**, 89-92.
(13) Katada, M.; Akashi, H.; Shibahara, T.; Sano, H. *J. Radioanal. Nucl. Chem., Letters* **1990**, *145*, 143-149.
(14) Shibahara, T.; Yamasaki, M.; Watase, T.; Ichimura, A. *Inorg. Chem.* **1994**, *33*, 292-301.
(15) a) Shibahara, T.; Kuroya, H. *Polyhedron* **1986**, *5*, 357-361. b) Akashi, H.; Shibahara, T.; Kuroya, H. *Polyhedron* **1990**, *9*, 1671-1676.
(16) Although the first reduction wave seems to be splitting into two peaks, the magnitude of the second shoulder peak is less than 5% of that of the first main peak, and the second is ignored.
(17) Akashi, H.; Uryu, N.; Shibahara, T., to be published.
(18) Kobayashi, H.; Shibahara, T.; Uryu, N. *Bull. Chem. Soc. Jpn.* **1990**, *63*, 799-803.
(19) a) Miyamoto, R.; Kawata, S.; Iwaizumi, M.; Akashi, H.; Shibahara, T., *The 40th Symposium on Coordination Chemistry of Japan, Kanazawa*, October **1990**, Abstr. No. 2BP15. b)submitted for publication.
(20) a) Shibahara, T.; Yamasaki, M.; Sakane, G.; Minami, K.; Yabuki, T.; Ichimura, A. *Inorg. Chem.* **1992**, *31*, 640-647. b) Yamasaki, M.; Shibahara, T. *Anal. Sciences* **1992**, *8*, 727-729.
(21) Watase, T.; Yamasaki, M.; Ishigaki, Y.; Honbata, A.; Shibahara, T. *The 42nd Symposium on Coordination Chemistry of Japan, Nara*, October **1992**, Abstr., No. 1AP02.
(22) a) Sakane, G.; Watase, T.; Yamate, M.; Shibahara, T. *The 44th Symposium on Coordination Chemistry of Japan, Yokohama*, November **1994**, Abstr. No. 1B07. b) to be published.
(23) Shibahara, T.; Yamamoto, T.; Sakane, G. *Chem. Lett.*, **1994**, 1231-1234.
(24) W_3SnS_4: a) Ref. 19. b) Müller, A. ; Fedin, V. P.; Diemann, E.; Bogge, H.; Krickemeyer, E. Solter, D.; Giuliani, A. M.; Barbieri, R.; Adler, P. *Inorg. Chem.* **1994**, *33*, 2243-2247.
(25) W_3CuS_4: a) Zheng, Y.-F.; Zhan, H.-Q. ; Wu, X.-T.; Lu, J.-X. *Transition Met. Chem.* **1989**, *14*, 161-164. b) Zhan, H.-Q.; Zheng, Y.-F. ; Wu, X.-T.; Lu, J.-X. *Inorg. Chim. Acta* **1989**, *156*, 277-280.
(26) For example, a) P. W. Jolly, "Comprehensive Organometallic Chemistry", ed by G. Wilkinson, F. G. A. Stone, and E. W. Abel, Pergamon Press, Oxford, New York, Toronto, Sydney, Paris, Frankfurt (1982), Vol. 6, p. 1. b) "Dictionary of Organometallic Compounds", ed by J. Buckingham and J. Macintyre, Chapman and Hall, London, New York, Toronto(1984), Vol. 2, p. 1320.
(27) a) Sakane, G.; Yamamoto, T.; Yamada, T.; Shibahara, T. *41st Symposium on Organometallic Chemistry, Japan, Osaka*, October **1994**, Abstr., No. A208. b) Shibahara, T.; Sakane, G.; Maeyama, M.; Kobashi, H.; Yamamoto, T.; Watase, T., Inorg. Chim. Acta, in press.
(28) Shibahara, T.; Sakane, G.; Mochida, S. *J. Am. Chem. Soc.* **1993**, *115*, 10408-10409.
(29) a) Tanner, L. D.; Haltiwanger, R. C.; Rakowski DuBois, M. *Inorg. Chem.* **1988**, *27*, 1741-1746. b) Rakowski DuBois, M. ; VanDerveer, M. C.; DuBois, D.

L.; Haltiwanger, R. C.; Miller, W. K. *J. Am. Chem. Soc.* **1980**, *102*, 7456-7461. c) Rakowski DuBois, M. *Chem. Rev.* **1989**, *89*, 1-9. d) Ropez, L.; Godziela, G.; Rakowski DuBois, M. *Organometallics* **1991**, *10*, 2660-2664.

(30) Weberg, R.; Haltiwanger, R. C.; Rakowski DuBois, M. *Organometallics* **1985**, *4*, 1315-1318.

(31) Bolinger, C. M.; Rauchfuss, T. B.; Rheingold, A. L. *J. Am. Chem. Soc.* **1983**, *105*, 6321-6323.

(32) a) Halbert, T. R.; Pan, W.-H.; Stiefel, E. I. *J. Am. Chem. Soc.* **1983**, *105*, 5476-5477. b) Pilato, R. S.; Eriksen, K. A.; Greaney, M. A.; Stiefel, E. I. *J. Am. Chem. Soc.* **1991**, *113*, 9372-9374.

(33) a) Draganjac, M.; Coucouvanis, D. *J. Am. Chem. Soc.* **1983**, *105*, 139-140. b) Coucouvanis, D.; Hadjikyriacou, A.; Draganjac, M.; Kanatzidis, M. G.; Ileperuma, O. *Polyhedron* **1986**, *5*, 349-356. c) Coucouvanis, D.; Hadjikyriacou, A.; Toupadakis, A.; Koo, Sang-Man; Ileperuma, O.; Draganjac, M.; Salifoglou, A. *Inorg. Chem.* **1991**, *30*, 754-767.

(34) Kanatzidis, M. G.; Coucouvanis, D. *Inorg. Chem.* **1984**, *23*, 403-409.

(35) a) Ansari, M. A.; Chandrasekaran, J.; Sarkar, S. *Polyhedron* **1988**, *7*, 471-476. b) Ansari, M. A.; Chandrasekaran, J.; Sarkar, S. *Inorg. Chim. Acta* **1987**, *130*, 155-156.

(36) Sakane, G.; Shibahara, T.; Adachi, H. *J. Cluster Science* **1995**, *6*, 503-521.

(37) a) Chen, Z.-D.; Lu, J.-X.; Liu, C.-W.; Zhang, Q.-N. *Polyhedron* **1991**, *10*, 2799-2807. b) Chen. Z.-D.; Lu, J.-X.; Liu, C.-W.; Zhang, Q.-E. *J. Mol. Struct. (Theochem)* **1991**, *236*, 343-357. c) Cotton, F. A.; Feng, X.-J. *Inorg. Chem.* **1991**, *30*, 3666-3670. d) Wang, Y.; Wang, J.; Li, J. *J. Mol. Struct. (Theochem)* **1991**, *251*, 165-171. e) Chen, W.-D.; Zhang, Q.-N.; Huang, J.-S.; Lu, J.-X. *Polyhedron* **1990**, *9*, 1625-1631. f) Bursten, B. E.; Cotton, F. A.; Hall, M. B.; Najjar, R. C. *Inorg. Chem.* **1982**, *21*, 302-307.

(38) a) Li, J.; Liu, C.-W.; Lu, J.-X. *J. Chem. Soc. Faraday Trans.* **1994**, *90*, 39-45. b) Li, J.; Liu, C.-W.; Lu, J.-X. *J. Polyhedron* **1994**, *13*, 1841-1851.

$[Mo_3O_4(H_2O)_9]^{4+}$ (**Mo_3O_4**). The x-z plane contains one Mo, one μ-O, one μ_3-O, one O (water), and two H atoms (water). Solid, dotted, and dashed lines indicate positive, negative, and zero contour lines, respectively. Reproduced with permission from reference 36.

Chapter 14

Synthesis and Structure of Raft-Type Molybdenum Chalcogenide Clusters

T. Saito, H. Imoto, K. Tsuge, S. Mita, J. Mizutani, S. Yamada, and H. Fujita

Department of Chemistry, School of Science, University of Tokyo, Hongo, Tokyo 113, Japan

Tetranuclear molybdenum cluster complexes $[Mo_4S_6X_2(PMe_3)_6]$ (X = SH, Cl, Br, I, NCS) and $[Mo_4S_6(dtc)_2(PMe_3)_4]$ have been synthesized. They have rhombic metal cores of molybdenum atoms in +3.5 oxidation state with 10 cluster valence electrons. Hexanuclear molybdenum cluster complexes $[Mo_6E_8Cl_6(PEt_3)_6]$ (E = S, Se) with 14 cluster valence electrons form by reduction of the trinuclear cluster complexes with six cluster valence electrons. An intermediate seven-electron cluster complex $[Mo_3S_4Cl_3(dppe)_2(PEt_3)]$ has been isolated. The X-ray structures of these cluster complexes are described.

Chemistry of molybdenum chalcogenide cluster complexes has made a remarkable progress in the last decade with the discoveries of cluster compounds of various types (*1 - 7*). A number of compounds with triangular and cuboidal cluster cores are now known but relatively few cluster compounds with raft-type frameworks have been reported. The raft structures are built from triangular cores by fusion of the edges. The simplest one is a tetranuclear rhombus and a pentanuclear trapezoid and a hexanuclear parallelogram are also found. Although there have been several examples of such tetranuclear clusters of other metals, there are still a very limited number of molybdenum chalcogenide clusters. The present article is concerned with synthesis and structural characterization of a few tetranuclear and hexanuclear cluster complexes coordinated by trialkylphosphines. A part of the results have been published previously (*28, 29, 35*).

Tetranuclear Cluster Complexes

Synthesis. The reaction of $(NH_4)_2Mo_3S_{13}$ (*8*) dissolved in butylamine and a THF

0097–6156/96/0653–0240$15.00/0

solution of trimethylphosphine at room temperature for 2 days gave $[Mo_4S_6(SH)_2(PMe_3)_6]$ **1** in 69% yield. The mixture of **1** and $SnCl_2$ in THF was refluxed for 1.5 h to give $[Mo_4S_6Cl_2(PMe_3)_6]$ **2** in 74 % yield. Similar reactions with $SnBr_2$ or SnI_2 formed $[Mo_4S_6Br_2(PMe_3)_6]$ **3** and $[Mo_4S_6I_2(PMe_3)_6]$ **4**, and $[Mo_4S_6(NCS)_2(PMe_3)_6]$ **5** was prepared from **2**. The reaction of **3** with dtc (= diethyldithiocarbamate) afforded $[Mo_4S_6(dtc)_2(PMe_3)_4]$ **6**. All of these are crystalline compounds and fairly stable in the solid states but decompose gradually in solutions.

Structure. The structures of **1**, **2**, **3**, and **6** have been determined by single crystal X-ray crystallography at room temperature using Rigaku four-circle diffractometers with Mo Kα radiation. The interatomic distances between molybdenum atoms are compared in Table I. The molecular structures of **2** and **6** are shown in Figure 1 and Figure 2 respectively.

Table I. Interatomic Distances (av. Å)

Compound	Mo-Mo	Mo-μ_3-S	Mo-μ_2-S
$[Mo_4S_6(SH)_2(PMe_3)_6]$	2.829	2.369	2.435
$[Mo_4S_6Cl_2(PMe_3)_6]$	2.821	2.363	2.433
$[Mo_4S_6Br_2(PMe_3)_6]$	2.824	2.367	2.430
$[Mo_4S_6(dtc)_2(PMe_3)_4]$	2.819	2.374	2.441
$[W_4S_6(SH)_2(PMe_2Ph)_6]$ [a]	2.817	2.377	2.356

[a] Ref. (*9*)

The cluster cores of the tetranuclear cluster complexes consist of flat arrays of four molybdenum atoms capped by two sulfur atoms above and below the Mo_3 planes. Each Mo-Mo edge is bridged by a sulfur atom. The cluster structure is a rhombus of four molybdenum atoms but can also be regarded as a fusion of two incomplete cubanes when the Mo-Mo bonds are neglected and only Mo-S bonds are taken into account. The cluster can be viewed as a flat butterfly structure with a two-fold axis passing through the hinge molybdenum atoms and relating the two wing-tip atoms. Each hinge molybdenum atom is coordinated by two trimethyl-phosphine ligands. Each wing-tip molybdenum atom is coordinated by a trimethylphosphine ligand and an SH group in the cluster **1**, a trimethylphosphine and a halogen in the clusters **2** and **3**, and a dtc ligand in **6**. If Mo-Mo bonds are neglected, the hinge molybdenum atoms are hexacoordinated and the wing-tip molybdenum atoms pentacoordinated. The SH groups in **1** show an SH stretching frequency in the infrared spectrum at 2516 cm^{-1}.

The formal oxidation states of two molybenum atoms are Mo(III) and the other two Mo(IV). The average oxidation state is +3.5 and there are 10 cluster valence electrons.

The Mo-Mo interatomic distances range from 2.813 to 2.845 Å and a pair of cluster valence electrons can be assigned to five Mo-Mo single bonds. The differences in the Mo-Mo interatomic distances in **1**, **2**, **3**, and **6** are very small indicating that the change of the terminal ligands has little effect on the geometry of the cluster cores. The tungsten analogue of **1** has W-W distances that are ca. 0.01 Å shorter than those for **1** (*9*). The dtc ligands in **6** are coordinated in a bidentate fashion perpendicular to the Mo_4 plane. No trimethylphosphine is coordinated to the wing-tip molybdenum atoms.

Similar arrangements of four metals with two capping atoms are seen in molecular compounds; $[Ti_4(OEt)_{16}]$ (*10*), $(NH_4)_2[V_4S_2(SCH_2CH_2S)_6]$ (*11*), $[Nb_4Cl_{10}(PMe_3)_6]$ (*12*), $[Mo_4S_2O_4(H_2O)_{10}](pts)_4$ (*13*), $[W_4(OEt)_{16}]$ (*14*), $[Mn_4O_2(OAc)_6(bipy)_2]$ (*15*), and $[Mn_4(L)_2O_2(OAc)_2]$ (*16*). Rhombic tetranuclear metal cores also exist in some solid compounds; $CsNb_4Cl_{11}$ (*17*), $NaMo_2O_4$ (*18*), MMo_2S_4 (*19*), and ReS_2 (*20*). These solid state compounds contain discrete M_4 cluster units linked together by intercluster M-M bonds or by bridging atoms depending on the number of cluster valence electrons. For example, $CoMo_2S_4$ (= $Co_2Mo_4S_8$) has 12 cluster valence electrons and the rhombic Mo_4 cluster units are linked together by another Mo-Mo bond.

Molecular Orbitals. Molecular orbitals on model compounds $[Mo_4S_6X_2(PH_3)_6]$ (X = SH, Cl, Br) have been calculated by the DV-Xα method (*21 - 23*) and the electronic levels for the SH case are shown in Figure 3. Other derivatives give very similar energy levels. A characteristic feature of the energy levels is a larger HOMO-LUMO gap than those for other molybdenum sulfide cluster complexes such as $[Mo_4S_4\{(C_2H_5O)_2PS_2\}_6]$ (*24*), $[Mo_6S_8(PH_3)_6]$ (*25*), or $[Mo_3S_5(PH_3)_6]$ (*26*). The gaps are 1.74 eV for SH, 1.55 eV for Cl, and 1.52 eV for Br complexes. The large HOMO-LUMO gaps indicate that the overall energy schemes are not changed significantly by metal-ligand orbital mixing in the series $[Mo_4S_6X_2(PH_3)_6]$. The similarity of the level diagrams of the $[Mo_4S_6X_2(PH_3)_6]$ complexes is consistent with the fact that the change of the terminal ligands has little effect on the Mo-Mo distances seen in Table I. The tentative assignments of the peaks in the UV-vis spectrum of the real compound $[Mo_4S_6(SH)_2(PMe_3)_6]$ have been made by using the results of the calculations of the model compound (Figure 4). The nature of these transitions is from Mo-Mo bonding to anti-bonding orbitals.

Electrochemistry. The cyclic voltammetry of $[Mo_4S_6Br_2(PMe_3)_6]$ in CH_2Cl_2 has indicated two oxidation waves at -0.10 and 0.50 V *vs.* SCE. This suggests that nine-electron and eight-electron cluster complexes are attainable, but the reduction to form a cluster with more than ten electrons is difficult reflecting the fairly large energy gap between HOMO and LUMO.

Hexanuclear Cluster Complexes

Synthesis. $Mo_3S_7Cl_4$ (*27*) was treated with PEt_3 in THF at room temperature for 24 h. The homogeneous solution was stirred with magnesium at -20 °C for 3 h. When the color changed to purple, volatile materials were removed under reduced pressure. After washing with hexane, the residue was extracted by benzene to give $[Mo_6S_8Cl_6(PEt_3)_6]$

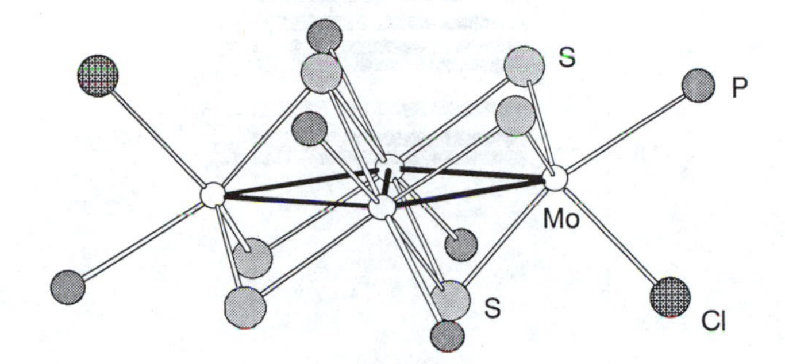

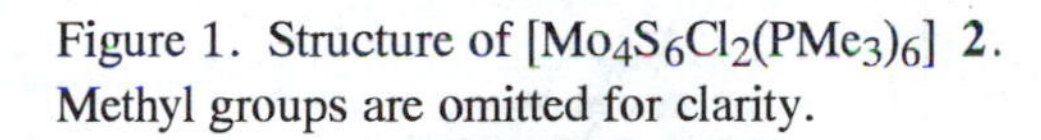
Figure 1. Structure of $[Mo_4S_6Cl_2(PMe_3)_6]$ **2**. Methyl groups are omitted for clarity.

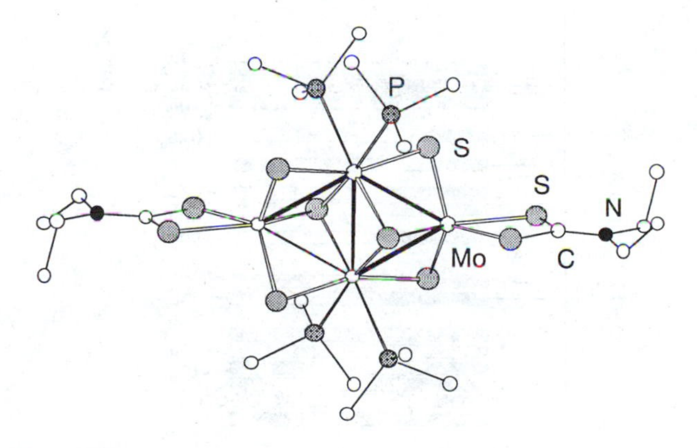

Figure 2. Structure of $[Mo_4S_6(dtc)_2(PMe_3)_4]$ **6**.

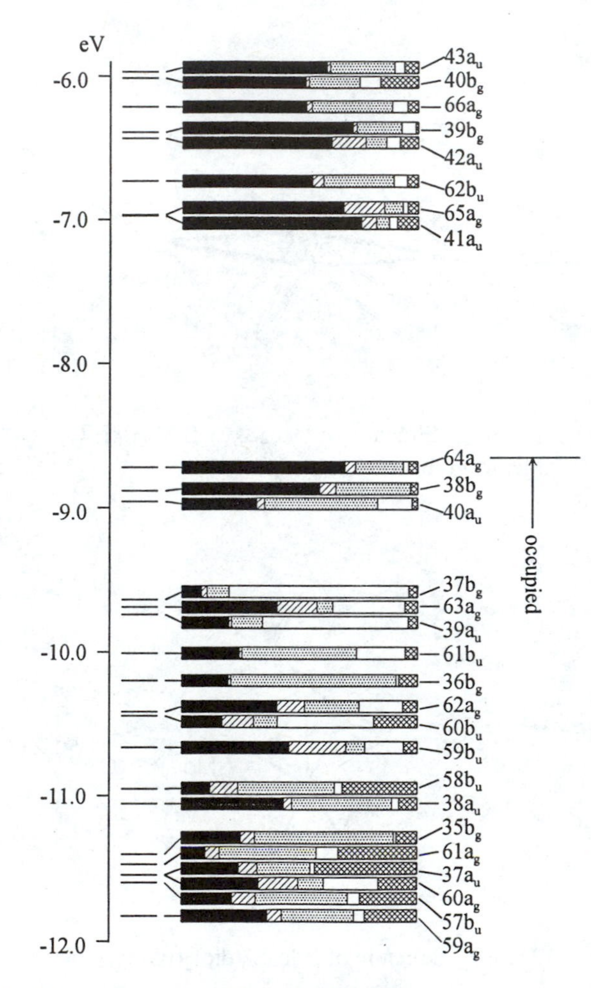

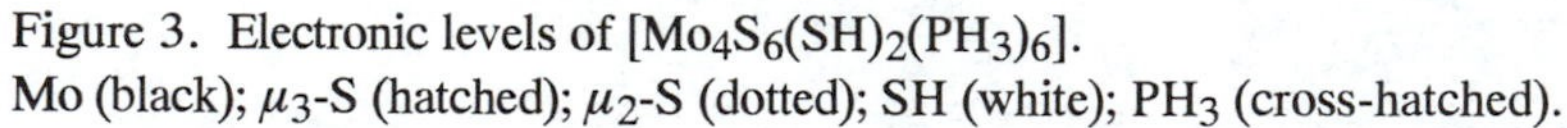

Figure 3. Electronic levels of $[Mo_4S_6(SH)_2(PH_3)_6]$.
Mo (black); μ_3-S (hatched); μ_2-S (dotted); SH (white); PH_3 (cross-hatched).

7 in 26% yield. The selenium analogue $[Mo_6Se_8Cl_6(PEt_3)_6]$ **8** was prepared similarly starting from $Mo_3Se_7Cl_4$.

Structure. The molecular structure of **7** is shown in Figure 5 and the Mo-Mo interatomic distances of **7** and **8** are compared in Table II. The cluster core is

Table II. Interatomic Distances (av. Å)

Compound	Mo-Mo	Mo-μ_3-E	Mo-μ_2-E	Mo-μ_2-Cl	Mo1-Mo2'
$[Mo_6S_8Cl_6(PEt_3)_6]$	2.809	2.373	2.282	2.497	3.548
$[Mo_6Se_8Cl_6(PEt_3)_6]$	2.887	2.501	2.409	2.504	3.683
$[Mo_6S_{10}(SH)_2(PEt_3)_6]$	2.838	2.393	2.339		3.057

composed of six molybdenum atoms in a parallelogram arrangement. The structure can be regarded as the linking of two incomplete cubanes Mo_3S_4. There is a crystallographic inversion center at the middle of Mo1-Mo1' and the molecule has 1 (C_i) point group symmetry. The sulfur atoms cap the Mo_3 faces from above and below the Mo_6 plane and bridge the Mo-Mo edges. The two Mo_3S_4 units are bridged by chlorine atoms and there are four terminal chlorine atoms. One triethylphosphine ligand coordinates to each molybdenum atom. The Mo1 and Mo2 atoms have pseudo-octa-hedral coordination and Mo3 has pseudo-trigonalbipyramidal coordination. The Mo-Mo interatomic distances range from 2.69 to 3.54Å and the seven distances shorter than 3 Å are assigned to Mo-Mo bonds. There are two Mo(III) and four Mo(IV) and the average oxidation state is +3.67. The number of the Mo-Mo bonds corresponds to 14 cluster valence electrons. The distance 3.55 Å between Mo1 and Mo2' is too long to invoke a Mo-Mo bond. This distance is longer than the corresponding 3.06 Å in the isoelectronic cluster $[Mo_6S_{10}(SH)_2(PEt_3)_6]$ (*28*) and has been explained by the difference of the coordination number of Mo2' (*29*).

The selenium analogue **8** has a very similar structure with somewhat longer Mo-Mo distances (Table II).

Dimerization of reduced clusters. The hexanuclear molybdenum cluster complexes **7** and **8** have been obtained by reduction of $[Mo_3E_4Cl_4(PEt_3)_n(thf)_{5-n}]$ (E = S, Se) with magnesium metal. These trinuclear complexes were generated by the treatment of the solid-state compounds $Mo_3(\mu_3\text{-}E)(\mu_2\text{-}E_2)_3Cl_2Cl_{4/2}$ with triethylphosphine in THF and have six cluster valence electrons. The reaction with magnesium at a low temperature abstracts one chlorine atom and the trinuclear complexes $[Mo_3E_4Cl_3(PEt_3)_n(thf)_{6-n}]$ with seven cluster valence electrons are likely to form. This oxidation state is not very stable and washing with a non-coordinating solvent removes thf ligands and dimerization occurs to form a 14-electron hexanuclear cluster complex. However, it has been found that further reduction of the trinuclear complexes at higher temperatures to remove four chlorine atoms leads to dimerization to form the 20-electron octahedral complexes $[Mo_6E_8(PEt_3)_6]$ (*30*). Therefore, the treatment with magnesium metal must be stopped at the one-electron reduction stage to obtain the raft-type clusters. The metastable seven-electron cluster has been stabilized by dppe (=

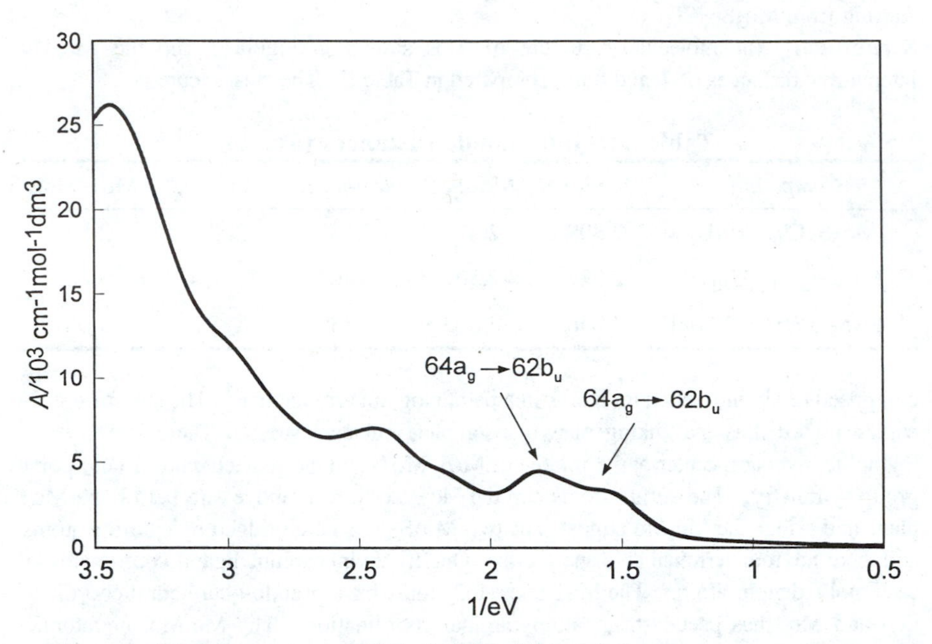

Figure 4. UV-Vis spectrum of $[Mo_4S_6(SH)_2(PMe_3)_6]$ **1**.
Tentative assignments of the peaks from the calculations on $[Mo_4S_6(SH)_2(PH_3)_6]$.

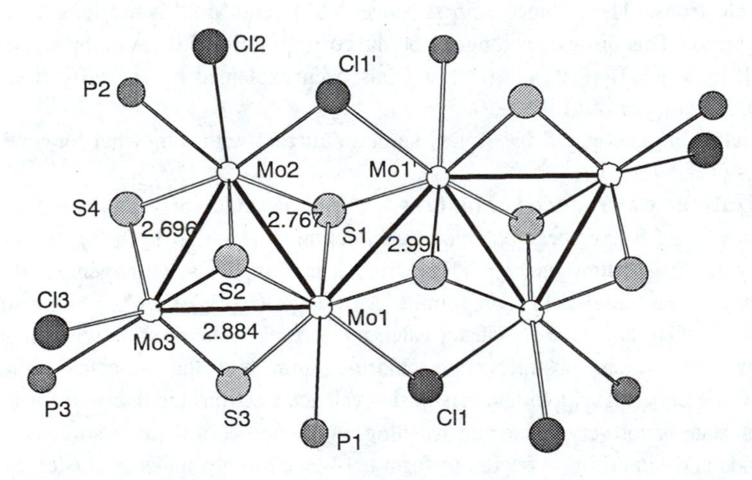

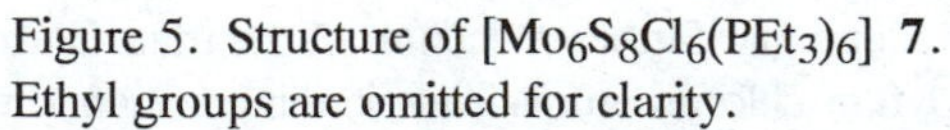
Figure 5. Structure of $[Mo_6S_8Cl_6(PEt_3)_6]$ **7**.
Ethyl groups are omitted for clarity.

1,2-bisdiphenylphosphinoethane) (*vide infra*). These results indicate that the dimerization of the trinuclear clusters by the conversion of μ_2-E into μ_3-E accompanied by formation of new M-M bonds can take place both in face-to-face or side-by-side modes depending on the reduction stage. Further reaction of **7** or **8** with magnesium at refluxing temperatures did not form the octahedral clusters suggesting that these raft-type dimers are not the intermediates to the octahedral cluster complexes.

Electrochemistry. The cyclic valtammetry of $[Mo_6S_8Cl_6(PEt_3)_6]$ in CH_2Cl_2 has indicated two oxidation steps at 0.33 and 0.85 V and two reduction steps at -0.75 and -1.16 V *vs*. SCE. The structures of the 12-electron stage corresponding to a dimer of six-electron clusters and the 16-electron stage corresponding to a cluster complex with one more Mo-Mo bond, would be interesting if these clusters can be isolated.

Trinuclear Seven-Electron Cluster

Synthesis. $Mo_3S_7Cl_4$ was treated with triethylphosphine in THF at room temperature for 24 h. Volatile materials were removed from the solution under reduced pressure and the residue was washed with hexane and redissolved in THF. The solution was stirred with magnesium at -20 °C for 5 h and the resulting purple solution was added to a THF solution of dppe and left standing at room temperature for two weeks to form the crystalline product $[Mo_3S_4Cl_3(dppe)_2(PEt_3)]$ **9**. The color of this cluster complex is red in contrast with the green of the six-electron clusters.

Structure. The molecular structure of **9** is shown in Figure 6. The interatomic Mo-Mo distances are compared with those of six-electron clusters in Table III. The

Table III. Interatomic Distances (av. Å)

Compound	Mo-Mo	Mo-μ_3-S	Mo-μ_2-S
$[Mo_3S_4Cl_3(dppe)_2(PEt_3)]$	2.806	2.367	2.342
$[Mo_3S_4Cl_3(dmpe)_2]Cl$	2.771	2.347	2.302
$[Mo_3S_4Cl_4(PEt_3)_3(MeOH)_2]$	2.766	2.355	2.288

cluster has a nearly equilateral triangle core with the Mo-Mo distances ranging from 2.804 to 2.809 Å. The cluster framework is capped by a sulfur atom and bridged by three sulfur atoms. A dppe coordinates to Mo1 and a PEt_3 to Mo2. Each molybdenum atom has a terminal chlorine ligand in the direction trans to the μ_2-S atom. The cluster **9** has seven cluster valence electrons and the interatomic Mo-Mo distances are *ca.* 0.04 Å longer than the six-electron cluster complex $[Mo_3S_4Cl_3(dmpe)_3]X$ (X = Cl or PF_6) (*31*).

The change of the triangular cluster cores of molybdenum by the addition of electrons has been a subject of considerable interest (*32*). Eight-electron (*33*) and nine-electron (*34*) cluster complexes have been isolated and structurally characterized, but the present cluster is the first example of the triangular cluster with seven electrons

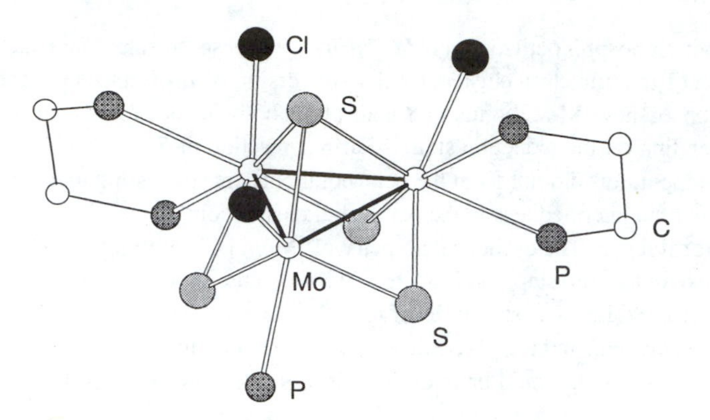

Figure 6. Structure of $[Mo_3S_4Cl_3(dppe)_2(PEt_3)]$ **9**.
Ethyl and phenyl groups are omitted for clarity.

(*35*), although it has been known for some time that electrochemical reduction leads to seven-electron clusters (*36* - *39*). In the triangular cluster complexes with C_{3v} symmetry, the seventh electron should occupy either $2a_1$ or 2e orbital after $1a_1$ and 1e orbitals are occupied by six electrons (*40*). The occupation of the nondegenerate $2a_1$ orbital by a single electron would not distort the cluster framework nor elongate the Mo-Mo bonds because this orbital is Mo-Mo non-bonding or weakly bonding. If the level order is $1a_1 < 1e < 2e < 2a_1$ (*41,42*), the seventh electron should occupy the degenerate 2e orbital resulting in Jahn-Teller distortion. The equilateral triangular geometry and slight elongation of cluster **9** can be explained by neither scheme. It has been suggested that the Mo-Mo bond distance of a cluster with seven electrons would be essentially unchanged (*32*) and the isolation of **9** has shown that this is the case.

The isolation of seven-electron cluster complex is interesting not only from the point of the structural argument but also as a key compound in the dimerization path to either octahedral or raft-type hexanuclear cluster complexes.

Electronic Relationship to Fused Ring Hydrocarbons

The concept of quasi-aromaticity of the Mo_3S_4 cluster compounds has been developed by the group of Lu (*43, 44*). In a recent article Zhang refers to our hexanuclear complex **7** as an equivalent of biphenylene $C_{12}H_8$ (*45*). However we have a different theory of the equivalence of the fused incomplete cubane complexes and aromatic hydrocarbons. We consider that these cluster compounds are equivalents of fused ring hydrocarbons and the number of CVE (cluster valence electron) is equal to the number of π electrons in the fused hydrocarbons. Thus, the tetranuclear cluster complexes **1** - **6** with 10 CVE are equivalents of naphthalene ($C_{10}H_8$) , the hexanuclear cluster complex **7** and **8** with 14 CVE are equivalents of tetrahydrotetracene ($C_{18}H_{16}$).

Literature Cited

(1) Saito, T. In *Early Transition Metal Clusters with π-Donor Ligands*; M.H. Chisholm, Ed.; VCH: New York, 1995; pp 63-164.
(2) Dance, I.; Fisher, K. *Prog. Inorg. Chem.* **1994**, *41*, 637-803.
(3) Kanatzidis, M. G.; Huang, S. P. *Coord. Chem. Rev.* **1994**, *130*, 509-621.
(4) Roof, L. C.; Kolis, J. W. *Chem. Rev.* **1993**, *93*, 1037-1080.
(5) Shibahara, T. *Coord. Chem. Rev.* **1993**, *123*, 73-147.
(6) Holm, R. H. *Adv. Inorg. Chem.* **1992**, *38*, 1-71.
(7) Shibahara, T. *Adv. Inorg. Chem.* **1991**, *37*, 143-173.
(8) Müller, A.; Krickemeyer, E. *Inorg. Synth.* **1990**, *27*, 47-51.
(9) Kuwata, S.; Mizobe, Y.; Hidai, M. *J. Chem. Soc., Chem. Commun.* **1995**, 1057-1058.
(10) Ibers, J. A. *Nature* **1963**, *197*, 686.
(11) Money, J. K.; Huffman, J. C.; Christou, G. *J. Am. Chem. Soc.* **1987**, *109*, 2210-2211.
(12) Cotton, F. A.; Shang, M. *J. Am. Chem. Soc.* **1988**, *110*, 7719-7722.
(13) Kobayashi, S.; Sasaki, M.; Sakane, G.; Shibahara, T., The 45th Symposium on Coordination Chemistry of Japan, Fukuoka, 1995, 1AP32.
(14) Chisholm, M. H.; Huffman, J. C.; Kirkpatrick, C. C.; Leonelli, J.; Folting, K. *J. Am. Chem. Soc.* **1981**, *103*, 6093-6099.
(15) Christmas, C.; Vincent, J. B.; Huffman, J. C.; Christou, G.; Chang, H. R.; Hendrickson, D. N. *J. Chem. Soc. Chem. Commun.* **1987**, 1303-1305.
(16) Mikuriya, M.; Nakadera, K.; Kotera, T.; Tokii, T.; Mori, W. *Bull. Chem. Soc. Jpn.* **1995**, *68*, 3077-3083.
(17) Broll, A.; Simon, A.; Schnering, H. G. v.; Schäfer, H. *Z. anorg. allg. Chem.* **1969**, *367*, 1-18.
(18) McCarley, R. E.; Lii, K. H.; Edwards, P. A.; Brough, L. F. *J. Solid State Chem.* **1985**, *57*, 17-24.
(19) Chevrel, R.; Sergent, M.; Meury, J. L.; Quan, D. T.; Colin, Y. *J. Solid State Chem.* **1974**, *10*, 260-269.
(20) Murray, H. H.; Kelty, S. P.; Chianelli, R. R.; Day, C. S. *Inorg. Chem.* **1994**, *33*, 4418-4420.
(21) Adachi, H.; Tsukada, M.; Satoko, C. *J. Phys. Soc. Jpn.* **1978**, *45*, 875-883.
(22) Satoko, C.; Tsukada, M.; Adachi, H. *J. Phys. Soc. Jpn.* **1978**, *45*, 1333-1340.
(23) Adachi, H.; Shiokawa, S.; Tsukada, M.; Satoko, C.; Sugano, S. *J. Phys. Soc. Jpn.* **1979**, *47*, 1528-1537.
(24) Coyle, C. L.; Eriksen, K. A.; Farina, S.; Francis, J.; Gea, Y.; Greaney, M. A.; Guzi, P. J.; Halbert, T. R.; Murray, H. H.; Stiefel, E. I. *Inorg. Chim. Acta* **1992**, *198-200*, 565-575.
(25) Imoto, H.; Saito, T.; Adachi, H. *Inorg. Chem.* **1995**, *34*, 2415-2422.
(26) Tsuge, K.; Imoto, H.; Saito, T. *Inorg. Chem.* **1995**, *34*, 3404-3409.

(27) Opalovskii, A. A.; Fedorov, V. E.; Khaldoyanidi, K. A. *Dokl. Akad. Nauk SSSR* **1968**, *182*, 1095-1097.
(28) Tsuge, K.; Imoto, H.; Saito, T. *Inorg. Chem.* **1992**, *31*, 4715-4716.
(29) Mizutani, J.; Yamada, S.; Imoto, H.; Saito, T. *Inorg. Chem.* **1996**, *35*, 244-247.
(30) Saito, T.; Yamamoto, N.; Nagase, T.; Tsuboi, T.; Kobayashi, K.; Yamagata, T.; Imoto, H.; Unoura, K. *Inorg. Chem.* **1990**, *29*, 764-770.
(31) Cotton, F. A.; Kibala, P. A.; Matusz, M.; McCaleb, C. S.; Sandor, R. B. W. *Inorg. Chem.* **1989**, *28*, 2623-2630.
(32) Cotton, F. A.; Feng, X. *Inorg. Chem.* **1991**, *30*, 3666-3670.
(33) Bino, A.; Cotton, F. A.; Dori, Z. *Inorg. Chim. Acta* **1979**, *33*, L133-L134.
(34) Cotton, F. A.; Shang, M.; Sun, Z. S. *J. Am. Chem. Soc.* **1991**, *113*, 6917-6922.
(35) Mizutani, J.; Imoto, H.; Saito, T. *J. Cluster Sci.* **1995**, *6*, 523-533.
(36) Shibahara, T.; Kuroya, H. *Polyhedron* **1986**, *5*, 357-361.
(37) Cotton, F. A.; Llusar, R. *Polyhedron* **1987**, *6*, 1741-1745.
(38) Shibahara, T.; Yamasaki, M.; Sakane, G.; Minami, K.; Yabuki, T.; Ichimura, A. *Inorg. Chem.* **1992**, *31*, 640-647.
(39) Shibahara, T.; Yamasaki, M.; Watase, T.; Ichimura, A. *Inorg. Chem.* **1994**, *33*, 292-301.
(40) Müller, A.; Jostes, R.; Cotton, F. A. *Angew. Chem. Int. Ed. Engl.* **1980**, *19*, 875-882.
(41) Li, J.; Liu, C.; Lu, J. *Polyhedron* **1994**, *13*, 1841-1851.
(42) Li, J.; Liu, C.; Lu, J. *J. Chem. Soc. Faraday Trans.* **1994**, *90*, 39-45.
(43) Huang, J.; Shang, M.; Huang, J.; Zhuang, H.; Lu, S.; Lu, J. *Pure & Appl. Chem.* **1988**, *60*, 1185-1192.
(44) Chen, Z. *J. Cluster Sci.* **1995**, *6*, 357-377.
(45) Zhang, Q. *J. Cluster Sci.* **1995**, *6*, 347-356.

Chapter 15

Synthesis and Reactions in the Void Created in Sulfur-Bridged Dinuclear Ruthenium Complexes

Kazuko Matsumoto, T. Koyama, and T. Furuhashi

Department of Chemistry, Waseda University, Tokyo 169, Japan

Several sulfur-bridged dinuclear ruthenium compounds have been synthesized and their reactions, especially those of their bridging sulfur-ligands, have been examined. Polysulfide complexes $[Ru_2(\mu\text{-}S_n)(\mu\text{-}S_2CNMe_2)(S_2CNMe_2)(CO)_2PPh_3)_2]$ (**1**, n = 5; **2**, n = 6) undergo cleavage of the chelating and bridging polysulfide ligand when reacted with NH_3. The disulfide-bridged dinuclear ruthenium complex $[\{Ru(CH_3CN)_3(P(OMe)_3)_2\}_2(\mu\text{-}S_2)]^{4+}$ (**7**) is highly reactive and its *trans*-RuSSRu core easily closes to sandwich small molecules between the two metals, while retaining its core structure. Compound **7** reacts with N_2H_4 to give $[\{Ru(CH_3CN)_2\text{-}(P(OMe)_3)_2\}_2(\mu\text{-}N_2H_4)_2(\mu\text{-}S_2)](CF_3SO_3)_3$ (**9**), while reaction with acetone leads to $[\{Ru(CH_3CN)_2(P(OMe)_3)_2\}(\mu\text{-}CH_3COCH_2S_2)\text{-}\{Ru(CH_3CN)_3(P(OMe)_3)_2\}]^{3+}$, in which a novel C-S bond is formed. Reaction of $[\{RuCl(P(OMe)_3)_2\}_2(\mu\text{-}Cl)(\mu\text{-}N_2H_4)(\mu\text{-}S_2)]$ (**13**) with O_2 in CH_2Cl_2 gives a diazene-coordinated compound $[\{RuCl(P(OMe)_3)_2\}_2(\mu\text{-}Cl)(\mu\text{-}N_2H_2)(\mu\text{-}S_2)]$ (**17**). The O_2 oxidation of **13** in CH_3CN gave a novel $S_2O_5^{2-}$ bridged complex.

Transition-metal complexes with S^{2-} ligand are widely distributed in nature in ores and in redox active centers of metalloproteins such as ferredoxins and nitrogenases (1-4). Polysulfides (S_x^{2-}, $x \geq 2$) are known to act as chelating or bridging ligands to metals (4), and above all, sulfide (S^{2-}) and disulfide (S_2^{2-}) are remarkably versatile ligands. Disulfide has also recently been proposed as a possible ligand in the P-cluster of the nitrogenase enzyme system (5, 6). We have attempted to synthesize sulfide-bridged dinuclear ruthenium compounds, whose RuS_xRu core structures are robust with respect to ligand substitution. We hoped to construct a stable but reactive space between two Ru atoms; the stability of the space is provided by the RuS_xRu core, while the reactivity is provided by other labile ligands on the Ru atoms. In our RuSSRu core system, small molecules are sandwiched between the two metal centers, thereby acting as another bridging ligands, and undergoing redox reactions. We have unexpectedly discovered that the disulfide ligand is highly

0097–6156/96/0653–0251$15.00/0

susceptible to nucleophilic attack, and even stable molecules such as acetone or O_2 reacts with the disulfide ligand to form a stable C-S or O-S covalent bond on the disulfide bridge. The RuSSRu core has also been found to stabilize hydrazine by coordination to the two Ru atoms, and this bridged hydrazine can be oxidized by O_2 while its bridging structure is retained. These reactions suggest that the void created between the two Ru atoms in the RuSSRu core is a suitable space, in which a very unstable molecule can be stabilized, and novel redox reaction can occur owing to the extensive electron delocalization in the core.

Syntheses and Reactions of Polysulfide-Bridged Diruthenium Complexes

There are several means by which polysulfide ligands (S_n^{2-} (n≥1)) coordinate to a metal. The source of the sulfide can be an alkali metal sulfide such as Li_2S_n and Na_2S_n, or simply elemental sulfur. In our effort to synthesize polysulfide complexes, the following reaction was attempted. Although we expected to prepare a disulfide-bridged complex, pentasulfide- and hexasulfide-bridged complexes were obtained instead (7).

$$RuH(MeNCS_2)(CO)(PPh_3)_2 + S_8 \xrightarrow[\text{benzene}]{}$$

$$[Ru_2(\mu\text{-}S_5)(\mu\text{-}S_2CNMe_2)(S_2CNMe_2)(CO)_2(PPh_3)_2]\ \mathbf{(1)}\ +$$

$$[Ru_2(\mu\text{-}S_6)(\mu\text{-}S_2CNMe_2)(S_2CNMe_2)(CO)_2(PPh_3)_2]\ \mathbf{(2)}$$

X-ray structural analysis of a single crystal of the product showed the presence of **1** and **2** at 78.5 % and 21.5 %, respectively. In the crystal lattice, the μ-S_5 and μ-S_6 ligands are disordered, whereas the remainder of the two molecules are not disordered. Both compounds can be obtained in pure form by recrystallization of the mixture from benzene. The ORTEP drawing of the twligandso molecules are shown separately in Figures 1 and 2. Both the S_5^{2-} and the S_6^{2-} ligands act as bridging as well as chelating ligands. The average S-S distance of the S_5 chain is 2.06 Å, which is slightly shorter than that in $[Os_2(\mu\text{-}S_5)(\mu\text{-}S_3CNEt_2)(S_2CNEt_2)_3]$ (8), in which the value is 2.09 Å. The average S-S distance of the S_6 chain in **2** is 2.03 Å, which is slightly shorter than that of the S_5 chain in **1**. Compound $(PPh_4)[Ru(NO)(NH_3)(S_4)_2]$ (9) is the first example reported for a polysulfide ligand coordinated to a ruthenium atom, but the S_4 ligands only chelate to a ruthenium atom. The present compounds **1** and **2** are the first examples having polysulfides that are both chelating and bridging.

Although **1** is orange, it turns to yellow when it is dissolved in CH_2Cl_2 and excess pyridine (py) is added. Quite interestingly, the yellow crystals obtained from the solution turns to orange again when they are dissolved in CH_2Cl_2. The ^{31}P NMR spectra show that the starting compound **1** is recovered on dissolution in CH_2Cl_2. In the CH_2Cl_2 solution of **1** with excess py, the chelation of the S_5 ligand is ruptured to give a py-coordinated compound **3**, in which the S_5 ligand acts only as a bridging ligand (10). The reaction is schematically shown in Figure 3. Although the structure of **3** could not be confirmed by X-ray analysis, similar reaction occurs to **2** with other bases such as NH_3 or NH_2NH_2 in place of py, and the X-ray structures of the products corresponding to **3** have been solved. The ORTEP drawings of the two products $[\{Ru(S_2CNMe_2)(CO)(PPh_3)(NH_3)\}_2(\mu\text{-}S_6)]$ (**4**) and $[\{Ru(S_2CNMe_2)\text{-}(CO)(PPh_3)\}_2(\mu\text{-}S_4)(\mu\text{-}NH_2NH_2)]$ (**5**) are shown in Figures 4 and 5, respectively. In

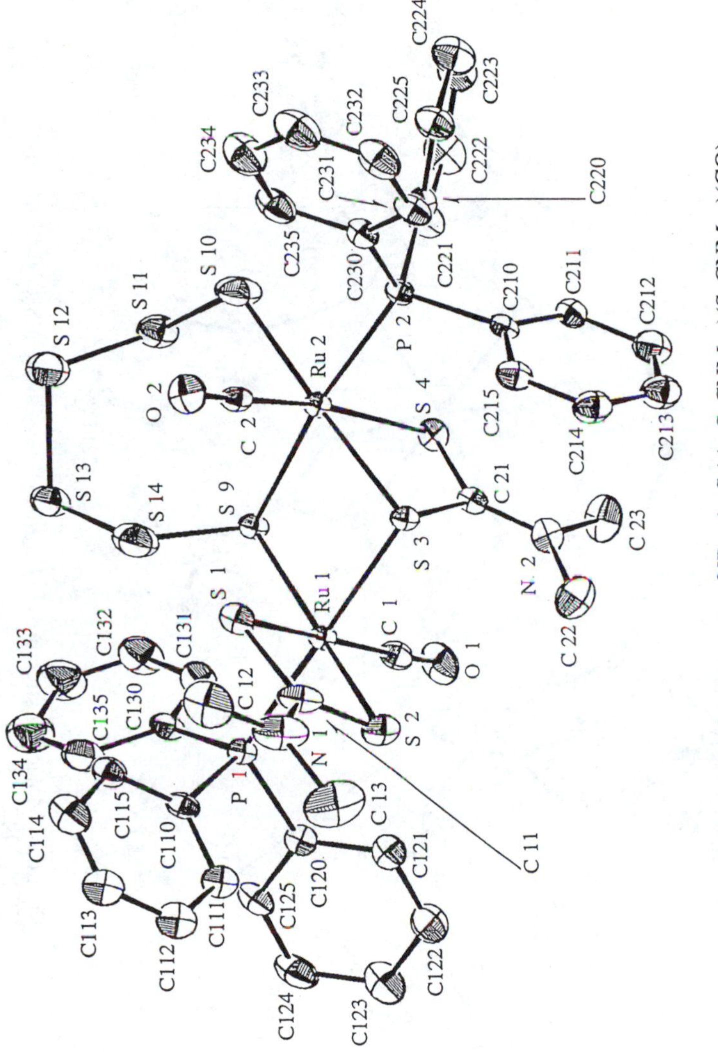

Figure 1. Molecular structure of $[Ru_2(\mu\text{-}S_5)(\mu\text{-}S_2CNMe_2)(CO)_2(PPh_3)_2]$ (**1**).

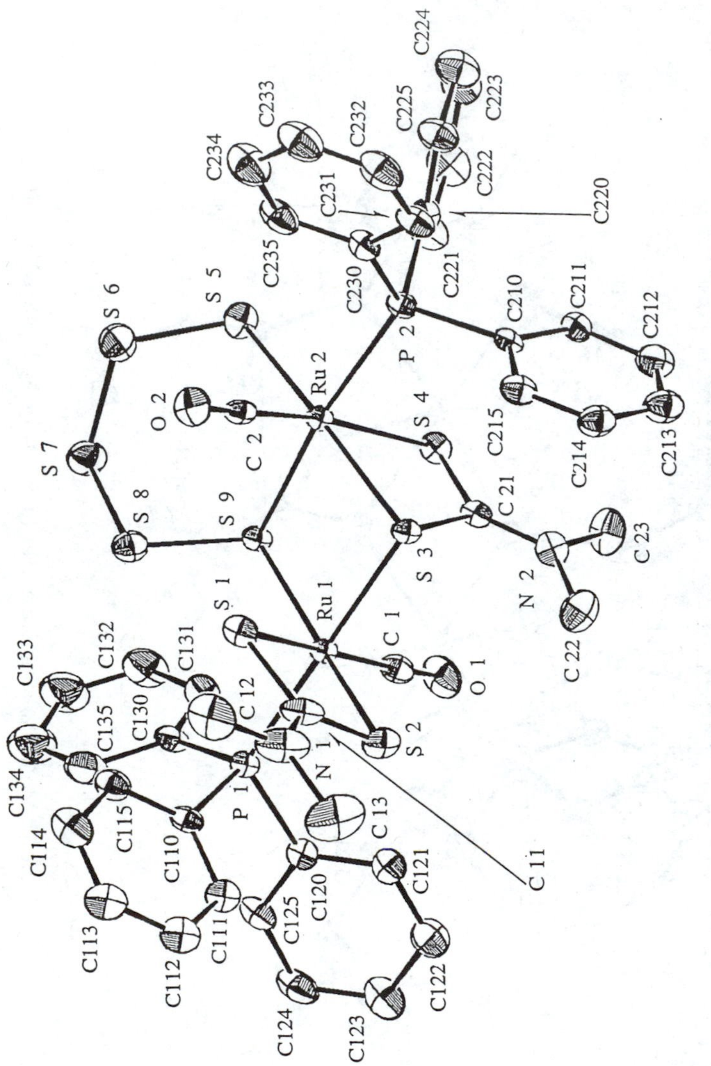

Figure 2. Molecular structure of $[Ru_2(\mu\text{-}S_6)(\mu\text{-}S_2CNMe_2)(CO)_2\text{-}(PPh_3)_2]$ (**2**).

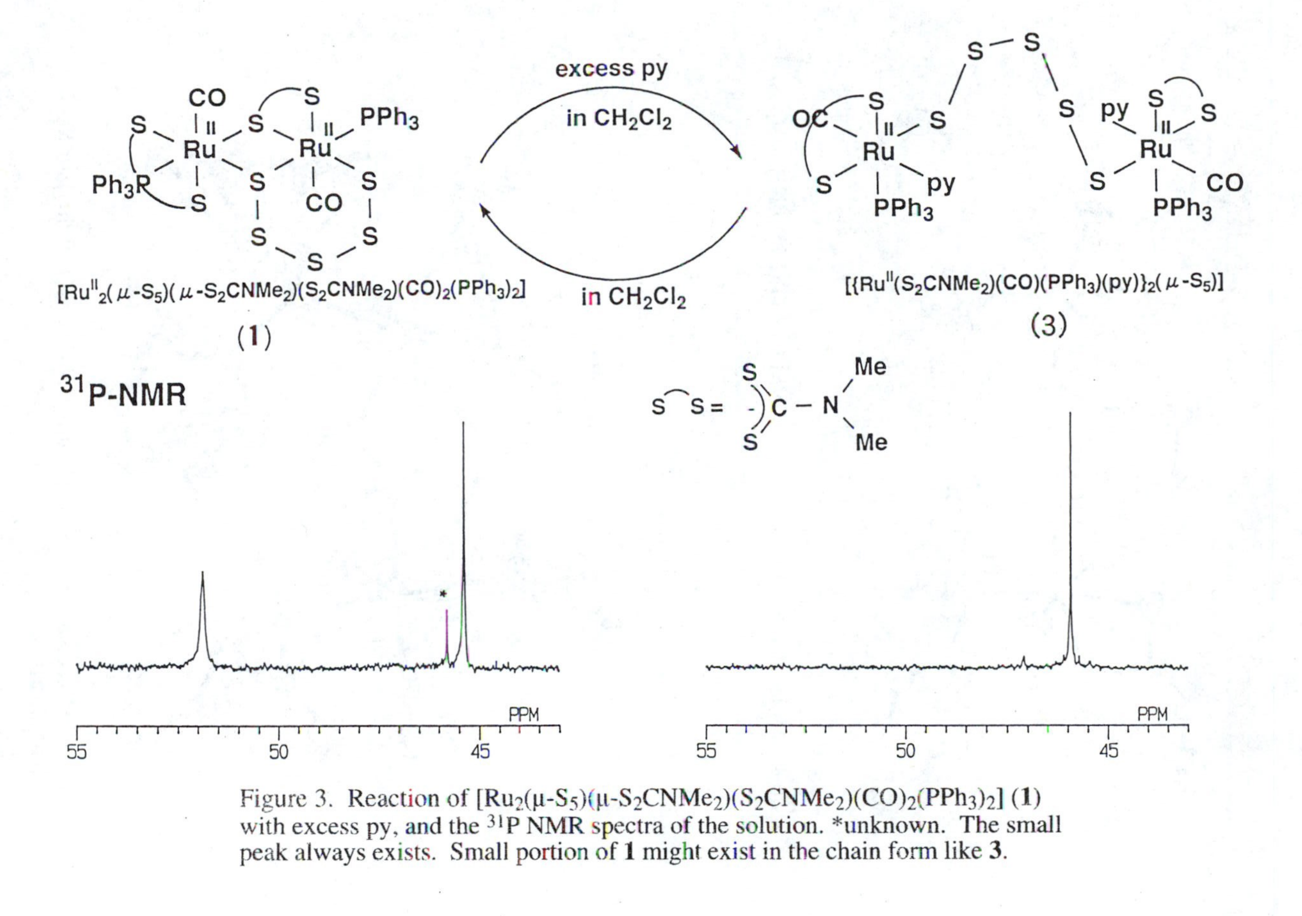

Figure 3. Reaction of $[Ru_2(\mu\text{-}S_5)(\mu\text{-}S_2CNMe_2)(S_2CNMe_2)(CO)_2(PPh_3)_2]$ (**1**) with excess py, and the ^{31}P NMR spectra of the solution. *unknown. The small peak always exists. Small portion of **1** might exist in the chain form like **3**.

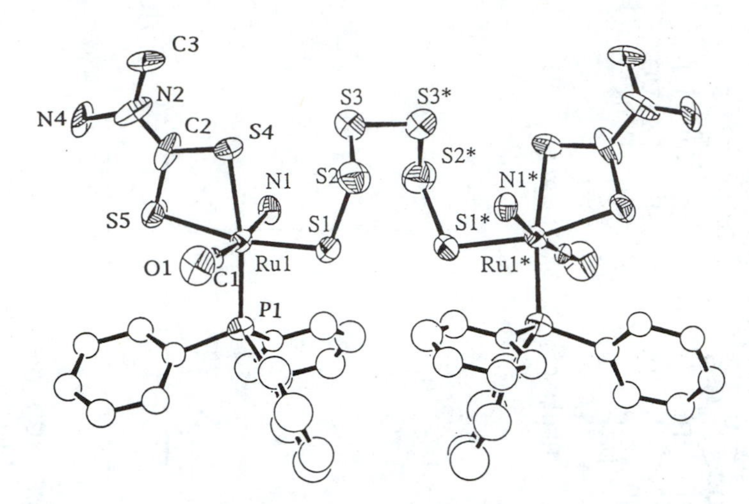

Figure 4. Molecular structure of $[\{Ru(S_2CNMe_2)(CO)(PPh_3)(NH_3)\}_2(\mu\text{-}S_6)]$ (**4**).

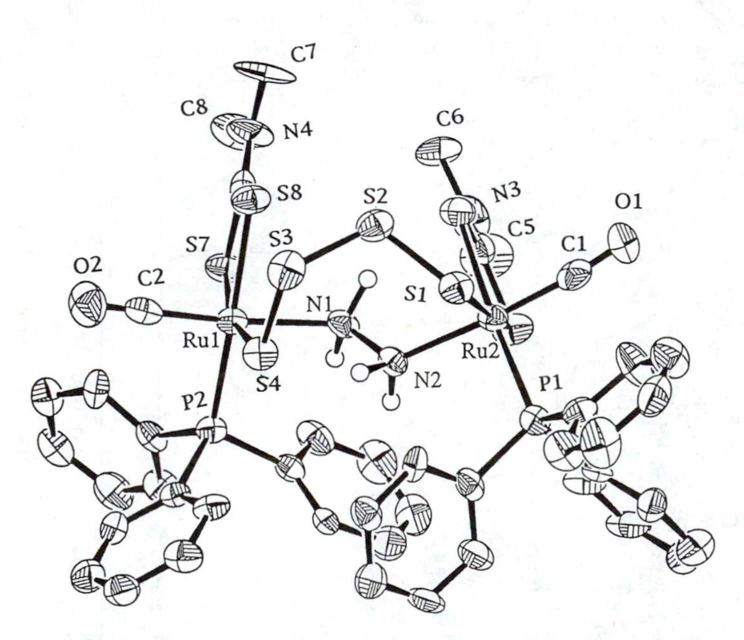

Figure 5. Molecular structure of $[\{Ru(S_2CNMe_2)(CO)(PPh_3)\}_2(\mu\text{-}S_4)(\mu\text{-}NH_2NH_2)]$ (**5**).

compound **5**, the S_6 ligand in the starting compound **2** has lost two sulfur atoms and is now a S_4 ligand, whereas in **4**, the S_6 ligand is retained. All the amine-coordinated complexes are stable in CH_2Cl_2 with excess amine, however without excess amine in the solution, the complexes revert to **2**. These reactions reveal that chelates of polysulfide ligands are thermodynamically not as stable as other common chelating ligands. This is perhaps caused by the strain due to the zigzag S-S bonds of the polysulfide ligands.

Syntheses and Reactions of Disulfide-Bridged Diruthenium Complexes

Formation of trans-Disulfide-Bridged Complexes and Their reactions. Complex $[\{RuCl(P(OMe)_3)_2\}_2(\mu\text{-}Cl)_2(\mu\text{-}S_2)]$ (**6**) with a disulfide bridging ligand shown in Figure 6 has been prepared by the following reaction (11).

$$trans\text{-}RuCl_2(P(OMe)_3)_4 + \text{excess } S_8 \xrightarrow[CH_2Cl_2]{} [\{RuCl(P(OMe)_3)_2(\mu\text{-}Cl)_2(\mu\text{-}S_2)] \ (\mathbf{6})$$

Although compound **6** is fairly stable in air, it becomes very reactive when it loses its bridging and terminal chloride ligands as shown in Figure 7. Compound **7** with a trans-disulfide bridge reacts with acetone to give $[\{Ru(CH_3CN)_2(P(OMe)_3)_2(\mu\text{-}S_2CH_2COCH_3)\{Ru(CH_3CN)_2(P(OMe)_3)_2\}]^{3+}$ (**8**), in which a novel $CH_3COCH_2S_2$ ligand bridges the two ruthenium atoms (12). This reaction shows the highly electrophilic nature of the bridging disulfide ligand in **7**. In another reaction in Figure 7, **7** easily closes its *trans*-RuSSRu core by forming a new bridge in addition to the disulfide. Specifically, **7** reacts with anhydrous NH_2NH_2 to give $[\{Ru(CH_3CN)(P(OMe)_3)_2\}_2(\mu\text{-}N_2H_4)_2(\mu\text{-}S_2)]$ (**9**) (13). The structures of **8** and **9** are shown in Figures 8 and 9, respectively. Compound **10** in Figure 7 corresponds to a one-electron reduced form of **7**, and is obtained from CH_3CN solution of **7** (14).

Reductive Coupling Reaction. Compound **6** and its derivatives undergo reductive coupling to form cluster compounds with higher nuclearity. For instance, $[\{Ru(CH_3CN)(P(OMe)_3)_2\}_2(\mu\text{-}Cl)_2(\mu\text{-}S_2)]^{2+}$ (**11**), which was prepared by addition of 2 equiv of $AgCF_3SO_3$ to **6** in CH_3CN, forms the tetranuclear compound $[Ru_4(\mu_2\text{-}H)_2(\mu_2\text{-}S)_2(\mu_4\text{-}S)_2(P(OMe)_3)_8]$ (**12**) in the reaction with Mg powder as shown in Figure 10 (15). The ORTEP drawing of **12** is shown in Figure 11. Compound **12** is a mixed-valent complex having two Ru^{II} and two Ru^{III}. This reaction occurs via initial reduction of Ru^{III} in **11** to a mixed-valent state accompanied by 90° rotation of the disulfide bridge to effect octahedral coordination around each Ru atom, which has lost the chloride bridges in **11** as $MgCl_2$ precipitates in the initial Ru reduction. The disulfide is rotated by 90° and is reduced to two sulfides which coordinate to the vacant sites formerly occupied by chloride ligands, providing octahedral environment to both Ru atoms. The two independent Ru-H distances are different, one being a normal Ru-H distance (1.65(6) Å), whereas the other is significantly longer (1.95(6) Å). The 18 dectron rule is satisfied for a Ru(III) atom, whereas the total number is more than 18 for a Ru(II) atom, which might be the cause of the unusually long Ru1-H distance in **12**.

Oxidation of a Hydrazine-Bridged Complex by O_2. Oxidation of a hydrazine-bridged complex by O_2 was attempted in the hope of obtaining a diazene-coordinated complex. Such an experiment is especially interesting, since hydrazine and diazene are the intermediates redox states between dinitrogen and ammonia. The recently

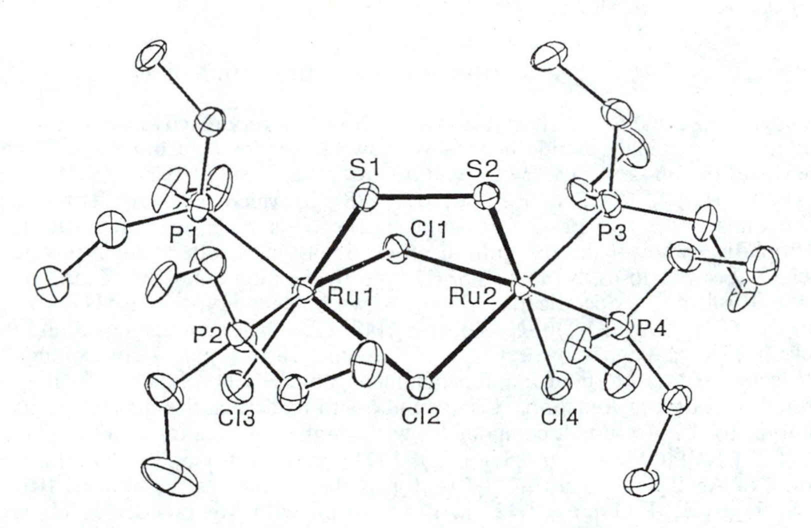

Figure 6. Molecular structure of $[\{RuCl(P(OMe)_3)_2\}_2(\mu\text{-}Cl)_2(\mu\text{-}S_2)]$ (**6**) (Reproduced with permission from reference 11. Copyright 1992 Japan Chemical Society).

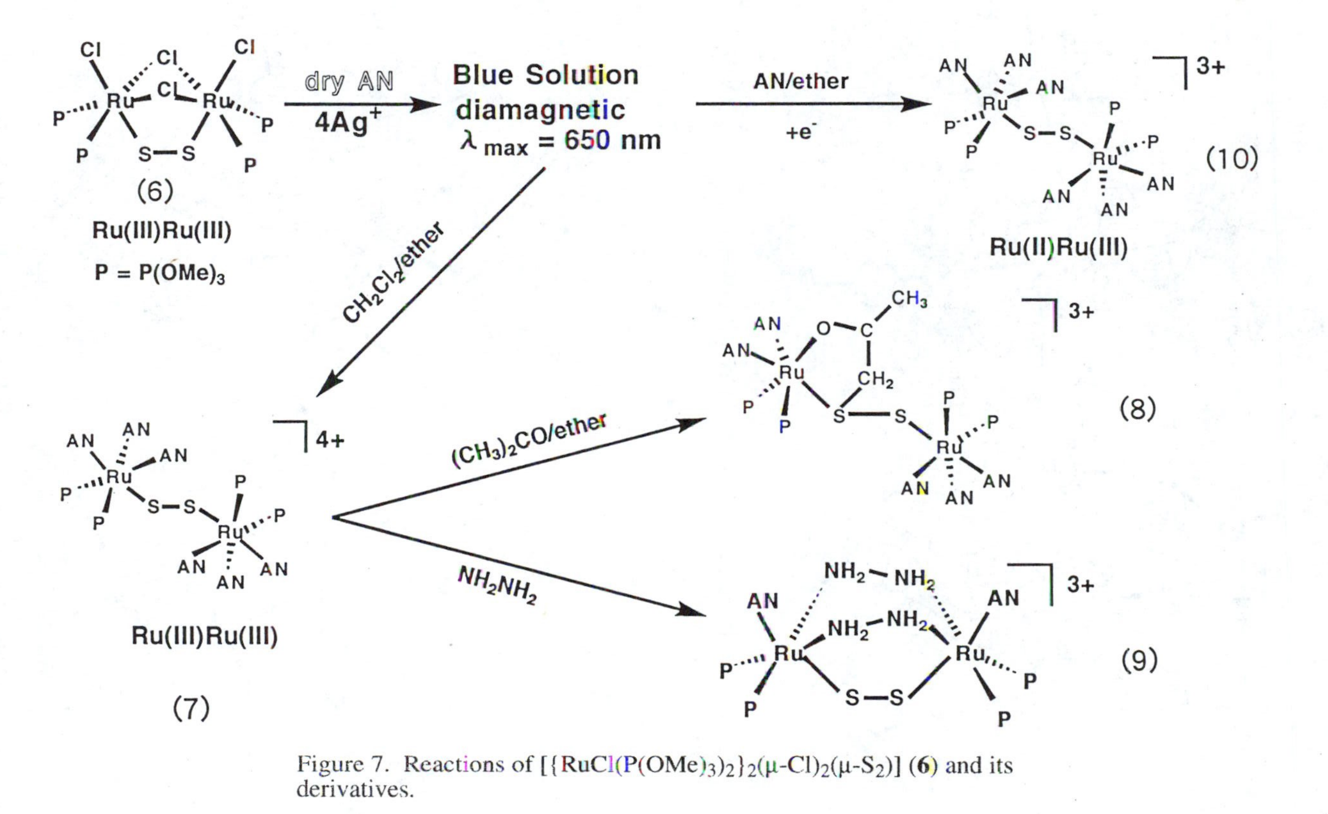

Figure 7. Reactions of $[\{RuCl(P(OMe)_3)_2\}_2(\mu\text{-}Cl)_2(\mu\text{-}S_2)]$ (**6**) and its derivatives.

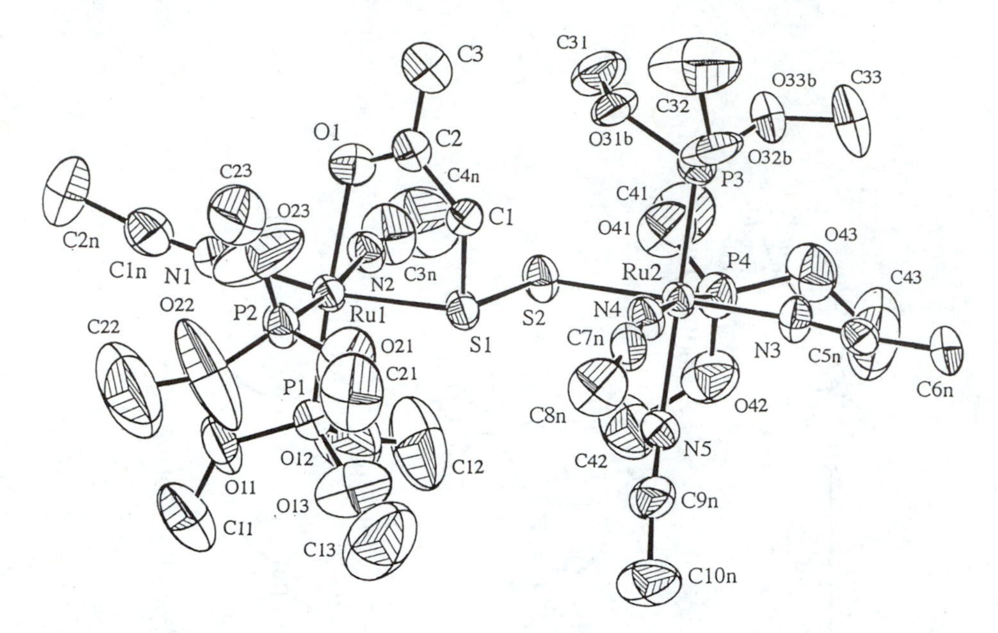

Figure 8. Molecular structure of $[\{Ru(CH_3CN)_2(P(OMe)_3)_2(\mu\text{-}S_2CH_2COCH_3)\text{-}\{Ru(CH_3CN)_2(P(OMe)_3)_2\}]^{3+}$ (**8**).

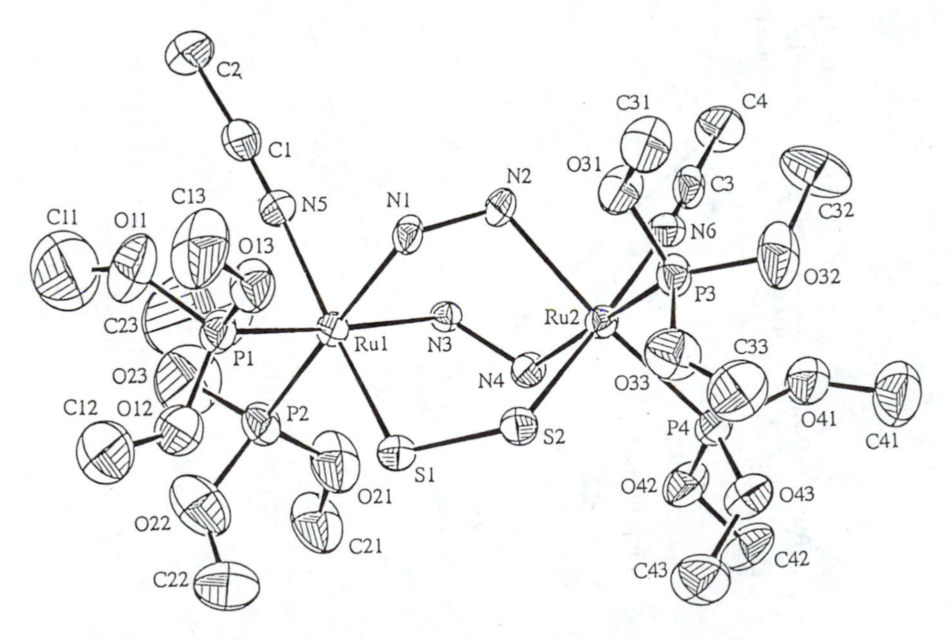

Figure 9. Molecular structure of $[\{Ru(CH_3CN)(P(OMe)_3)_2\}_2(\mu\text{-}N_2H_4)_2(\mu\text{-}S_2)]$ (**9**).

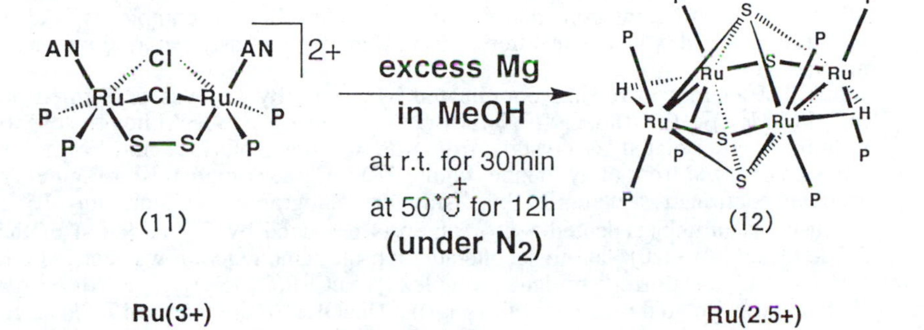

Figure 10. Reaction of $[\{Ru(CH_3CN)(P(OMe)_3)_2\}_2(\mu\text{-}Cl)_2(\mu\text{-}S_2)]^{2+}$ (**11**).

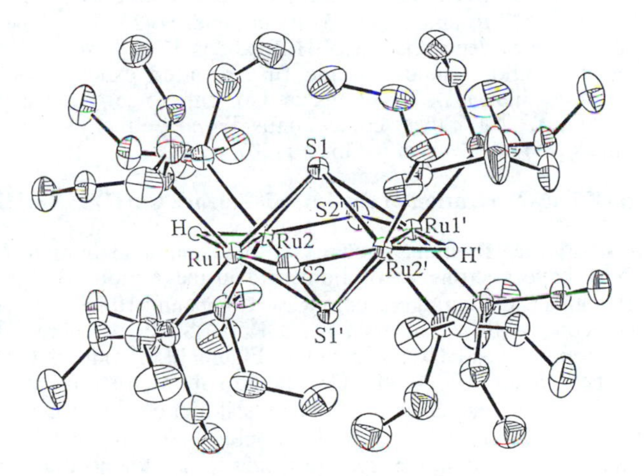

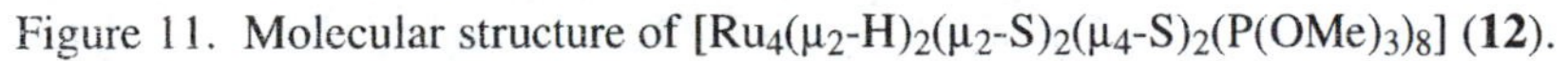
Figure 11. Molecular structure of $[Ru_4(\mu_2\text{-}H)_2(\mu_2\text{-}S)_2(\mu_4\text{-}S)_2(P(OMe)_3)_8]$ (**12**).

revealed X-ray structure of nitrogenase cofactor (16) shows a possible reaction site for dinitrogen reduction on the iron molybdenum cofactor, which contains a symmetrical cavity surrounded by six Fe atoms and three S atoms. This geometry suggests that N_2 molecule may be surrounded by the six Fe atoms in the cavity. If this is the actual situation for the enzyme, this interaction mode of N_2 molecule with metals is quite beyond what researchers had imagined. The X-ray structural work on the nitrogenase enzyme has highlighted the importance of a multi-metal center. Our diruthenium system, although much simpler than the natural system, should stabilize diazene between the two metals with cis-coordination. This mode of diazene bonding is totally different from that reported for other diazene complexes, and is expected to show redox behavior different from that of previously reported synthetic compounds (17, 18).

Our first attempt to oxidize coordinated hydrazine by O_2 was performed on $[\{RuCl(P(OMe)_3)_2\}_2(\mu\text{-}Cl)(\mu\text{-}S_2)(\mu\text{-}N_2H_4)]$ (**13**) (19) in CH_3CN. Although several compounds were isolated as crystals from the reaction solution, none of them contains an oxidized from of hydrazine (Figure 12). All the compounds in Figure 12 have been confirmed structurally by X-ray crystallography. Compound **16** is noteworthy, as an unprecedented $S_2O_5^{2-}$ anion is produced by O_2 oxidation of the disulfide ligand in **13** (20). In our second attempt, the same reaction was carried out in CH_2Cl_2 and the diazene-bridged complex $[\{RuCl(P(OMe)_3)_2\}_2(\mu\text{-}Cl)(\mu\text{-}S_2)(\mu\text{-}N_2H_2)]$ (**17**) was isolated together with **16** (20). The ORTEP drawing of **17** shown in Figure 13, clearly shows that Ru1-N1-N2-Ru2 and the two hydrogen atoms H1 and H2 are almost coplanar. The N-N distance of 1.33(2) Å is relatively long compared with that of trans-diazene (1.301 Å) in the previously reported diruthenium complex (21), but is distinctly shorter than those of hydrazine (1.43-1.46 Å) (19, 22, 23) or hydrazide $N_2H_2^{2-}$ (1.410(9) and 1.391(15) Å) complexes (23, 24). These structural features are definitive evidence that the N_2H_2 ligand in **17** is diazene. Compound **17** deserves special attention, since it is the first reported example of cis diazene coordination. Free diazene is cis, trans or iso, and so far only a few diazene complexes are known. All of these contain trans diazene, either as a bridge between two metal centers (25) or as a terminal ligand (20).

Spectroscopic Characterization of the Mixed-Valence Ru(II)SSRu(III) Core (27)

UV-Vis and Resonance Raman Spectra. All the compounds with a RuSSRu core described above have a strong absorption band in the region 600-800 nm, while compound **16** with a $S_2O_5^{2-}$ bridge is colorless. Compound **10** has a Ru(II)SSRu(III) mixed-valence core, while all others have a Ru(III)SSRu(III) core. The UV-vis spectra of compounds **6** and **10** are shown in Figure 14. None of the compounds absorbs in the near-IR up to 2000 nm. Compound **6** shows a strong absorption band at 737 nm, and exhibits resonance Raman bands at 385 cm^{-1} (ν(Ru-S), strong), 456 cm^{-1} (ν(S-S), very weak), and 769 cm^{-1} (double harmonic of 385 cm^{-1}). The assignment was made based on the reported values for analogous compounds (4, 28-31). The 385 cm^{-1} (ν(Ru-S)) and 456 cm^{-1} (ν(S-S)) bands of **6** can be compared to the reported values: 384 and 372 cm^{-1} (ν(Ru-S)) and 536 and 525 cm^{-1} (ν(S-S)) in the two isomers of $[(\mu\text{-}S_2)\{Ru(III)(PPh_3)'S_4'\}_2]$, where 'S$_4$' is 1,2-bis[(2-morcaptophenyl)thio]ethane(2-) (29), 415 cm^{-1} (ν(Ru-S)) and 519 cm^{-1} (ν(S-S)) in $[\{Ru(III)(NH_3)_5\}_2(\mu\text{-}S_2)]Br_4$ (30), and 409 cm^{-1} (ν(Ru-S)) and 530 cm^{-1} (ν(S-S)) in $[\{CpRu(III)(PPh_3)_2\}_2(\mu\text{-}S_2)](BF_4)_2$ (31). These compounds are the only *trans*-Ru(III)SS(III) cores, for which UV-vis and resonance Raman spectra have been reported; no Raman spectra data are available for *cis*-RuSSRu compounds other than

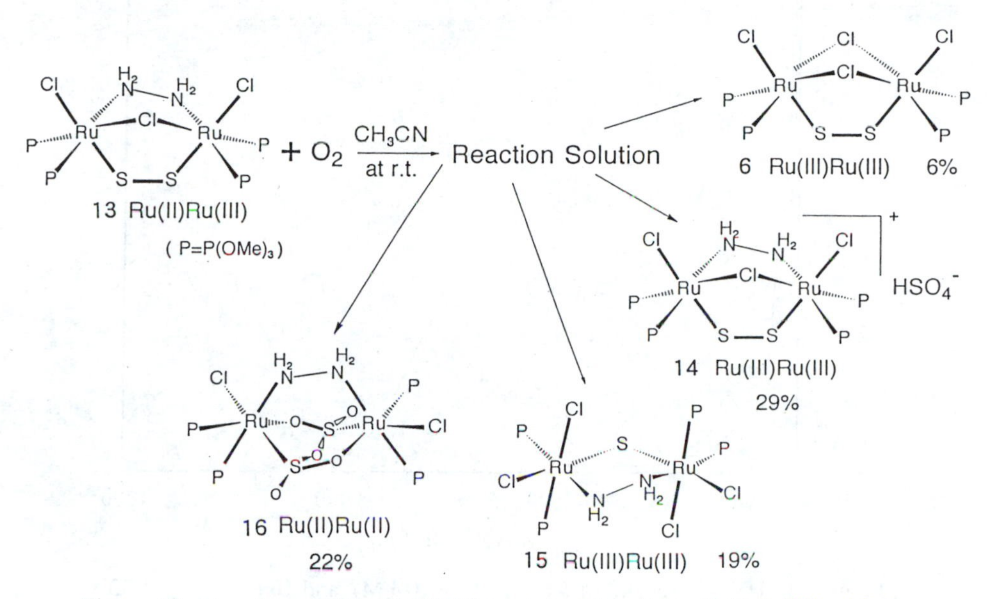

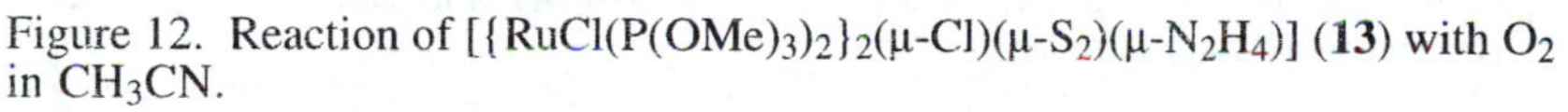
Figure 12. Reaction of [{RuCl(P(OMe)$_3$)$_2$}$_2$(μ-Cl)(μ-S_2)(μ-N_2H_4)] (**13**) with O_2 in CH_3CN.

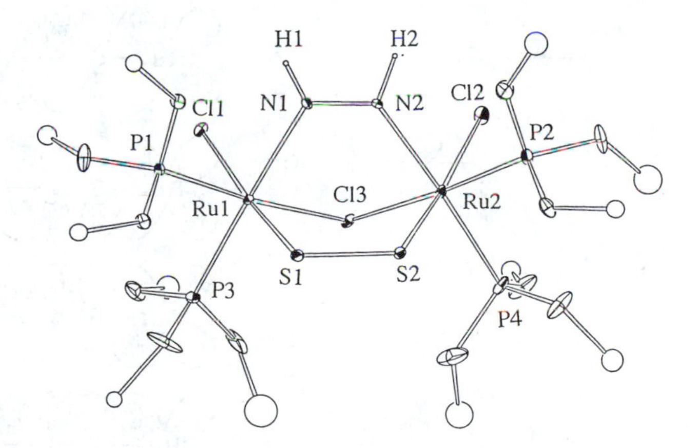

Figure 13. Molecular structure of [{RuCl(P(OMe)$_3$)$_2$}$_2$(μ-Cl)(μ-S_2)(μ-N_2H_2)] (**17**).

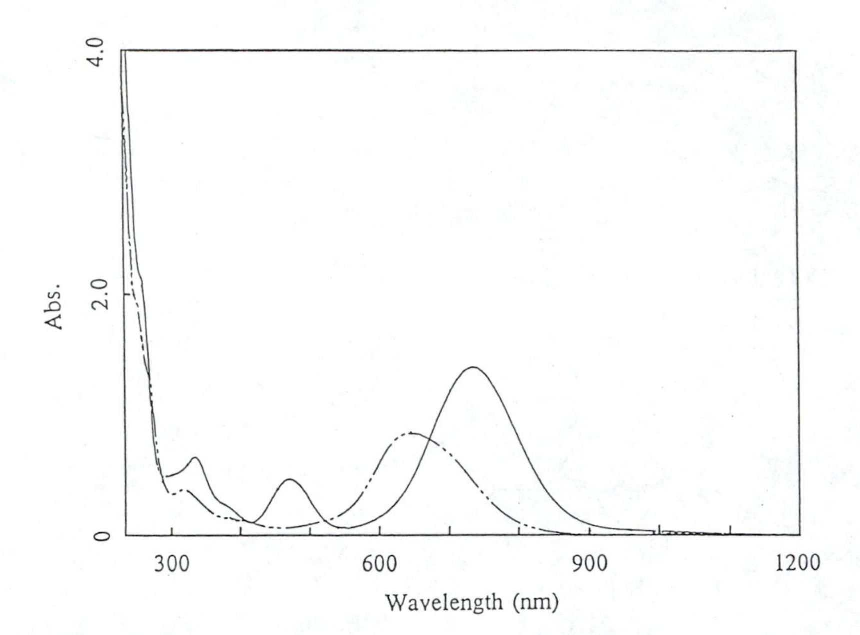

Figure 14. UV-Vis spectra of **6** (— 1.44 x 10^{-4} M), and **10** (— - - - — 1.33 x 10^{-4} M) in CH_3CN.

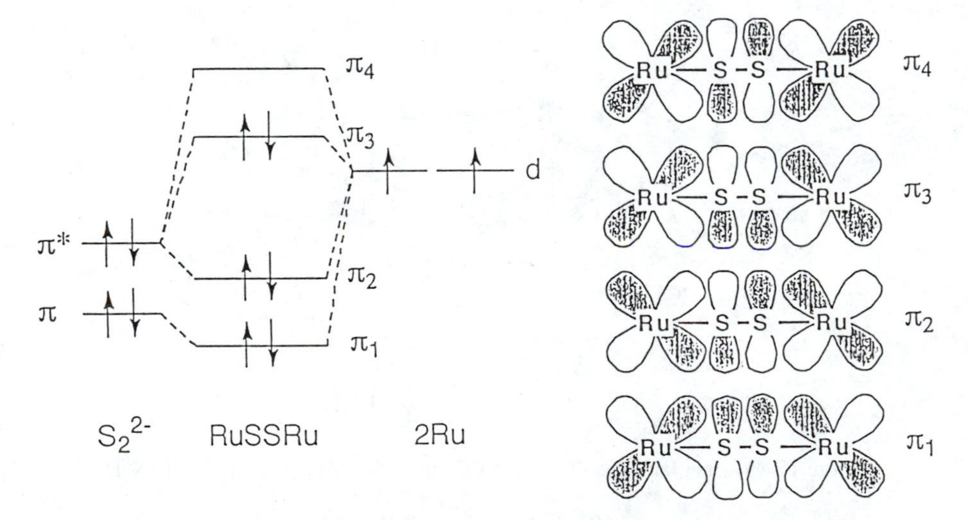

Figure 15. The π-MO scheme for a Ru(III)SSRu(III) core.

6. The ν(S-S) frequencies in disulfide complexes generally range from 480 to 600 cm^{-1} (4), which should be compared to free S_2 (725 cm^{-1}) (32), S_2^- (589 cm^{-1}) (33, 34) and S_2^{2-} (446 cm^{-1}) (4). The emission band of **6** at 385 cm^{-1} is more strongly enhanced by 647.1 nm radiation than by 568.2 nm. Therefore, the UV-vis absorption at 737 nm is assigned to an electronic transition within the Ru_2S_2 core. Compound **10** exhibits strong visible absorption at 646 nm, which is assigned analogously. The resonance Raman spectrum of **10** shows a strong ν(S-S) band at 561 cm^{-1}. A ν(Ru-S) band could not be observed for **10** in in CH_3CN, while a strong ν(S-S) band was observed at 561 cm^{-1}. The ν(Ru-S) band should be in the range 370-400 cm^{-1}, which is obscured by the Raman band of the CH_3CN solvent. Even if the ν(Ru-S) band is present in the area, its intensity may be very weak. Acetonitorile has to be used as the solvent, in order to avoid the release of the coordinated CH_3CN.

It should be noted that for complex **6** with a *cis*-RuSSRu core, ν(Ru-S) is strongly enhanced, whereas ν(S-S) is only very weakly observed. This is in remarkable contrast with complex **10** with a *trans*-RuSSRu core, which exhibits only a strong ν(S-S) in CH_3CN; ν(Ru-S) is very weak or is not enhanced. The resonance Raman spectrum of $[(H_3N)_5RuSSRu(NH_3)_5]^{4+}$ with a planar *trans*-Ru(III)SSRu(III) core (30) exhibits strong ν(Ru-S) but no ν(S-S) when excited at 647.1 nm. This excitation wavelength is close to the visible absorption maximum of the complex at 715 nm. The ν(S-S) at 514 cm^{-1} is observed only when it is excited by the shorter wavelength i.e., by 568.2 nm or less (30). The resonance Raman and electronic bands can be reasonably explained by a qualitative MO description of the Ru SSRu core (27). The electronic transitions in the visible region correspond to a LMCT (ligand to metal charge transfer) from S_2^{2-} to Ru(III). A resonance Raman band with significant magnitude should be observable only for symmetric ν(Ru-S) when the solution is irradiated in the visible band. The theory also predicts that the intensity of ν(S-S) band is zero (30). A basically similar but simpler explanation can be given for the electronic absorption bands of $[\{Ru(PPh_3)'S_4'\}_2(\mu\text{-}S_2)]\cdot CS_2$ (29). In order to obtain a clear image of the electronic states and to explain the Raman and ESR spectra (see later section) of the present compounds, the π-MO scheme for a RuSSRu core is given in Figure 15, which is basically similar to what is described in Ref. 29. The strong visible absorptions of compounds **6** and **10**, and all other compounds with RuSSRu cores, are the transitions from π_3 to π_4, which is LMCT. Compound **6** is diamagnetic, since the two unpaired electrons of the two low-spin Ru(III) ions are paired as shown in Figure 15. Compound **10** with a Ru(II)SSRu(III) core is paramagnetic, since its one unpaired electron is in the π_4 orbital. It is noteworthy that, for the three compounds with a *trans*-Ru(III)SSRu(III) core, $[\{CpRu(III)(PPh_3)_2\}_2(\mu\text{-}S_2)](BF_4)_2$ (31), $[\{Ru(III)(NH_3)_5\}_2(\mu\text{-}S_2)]Cl_4\cdot 2H_2O$ (30), and $[\{Ru(PPh_3)'S_4'\}_2(\mu\text{-}S_2)]\cdot CS_2$ (29), the ν(S-S) Raman bands are very weak and the ν(Ru-S) are strong. This relative intensity relation is completely reversed in compound **10** having a mixed-valent *trans*-Ru(II)SSRu(III) core. Only a strong ν(S-S) band is observed for **10**. Kim et al. explains that the strong ν(Ru-S) band in *trans*-Ru(III)SSRu(III) is enhanced by a transition associated with a symmetric Ru-S stretch, and thus only the ν(Ru-S) band is enhanced by the visible band to a significant extent (30). The stretching of the S-S bond does not contribute to the dipole moment responsible for the visible electronic absorption. If, however, the core is reduced to *trans* Ru(II)SSRu(III), a dipole moment, raised by the electronic transition between the two mixed-valent metals, would operate along the S-S bond (29). In addition, the compound would experience less intense LMCT, since the metal is reduced by one electron. Both of these two factors would in effect enhance the ν(S-S) strongly and weaken the ν(Ru-S). The intensity of ν(S-S) relative to ν(Ru-

S) is large for $[\{Ru(PPh_3)'S_4'\}_2(\mu\text{-}S_2)]\cdot CS_2$, compared to that of $[(H_3N)_5RuSSRu(NH_3)_5]^{4+}$ since the *trans* RuSSRu core of the former complex deviates significantly from a planar structure, caused by steric demands of the $'S_4'$ and PPh_3 ligands. Although the frequencies of ν (S-S) for **6** (456 cm^{-1}) are considerably lower than those of trans cores, the additional Cl^- bridges in **6** might cause significant differences in the electronic structures of the cores, lowering the ν(S-S) frequencies.

ESR Spectra of 10. Complex **10** is the first well-characterized mixed-valent complex with a *trans*-Ru(II)SSRu(III) core, which is ESR active due to the Ru(III) atom. All other Ru(III)SSRu(III) complexes in the present study and those previously reported are diamagnetic.

The powder spectrum of **10** at 288 K shows a rhombic signal with $g_1 = 2.12$, $g_2 = 2.05$, and $g_3 = 1.995$. Although several ESR spectra of mononuclear and dinuclear paramagnetic Ru(III) complexes have been analyzed using the matrix of a spin-orbit coupling Hamiltonian, most of them have axial symmetry with only $g_\perp$ and $g_{//}$ parameters (32). Rhombic spectral analysis of g_x, g_y, and g_z, using spin-orbit coupling constant λ, axial splitting parameter Δ, and rhombic splitting parameter V, is reported only for monomeric Ru(III) complexes (35, 36) with phosphine and other nitrogen-donor ligands. The anisotropy of **10** is remarkably small, compared to those of the reported monomeric and dimeric Ru(III) complexes with axial or rhombic symmetries (36, 37). Low spin Ru(III) species exhibit ESR spectra that are usually highly anisotropic with axial or rhombic symmetry (35,38-40).

The powder spectral pattern was analyzed according to the matrix calculation described in Refs 41-43. The final best fit was obtained when $\lambda = 100$ cm^{-1}, $V = 1680$ cm^{-1}, $\Delta = -2690$ cm^{-1}, $k = 0.92$, and the corresponding calculated g values were $g_x = 2.110$, $g_y = 2.050$, and $g_z = 1.994$. These g values are in excellent agreement with the experimental values. The best fit λ value of 100 cm^{-1} is extraordinarily low, compared to the previously reported values for monomeric Ru(III) complexes; λ values are 884 cm^{-1} for $RuCl_3(P(n\text{-}Bu)_2Ph)_3$ (35), 1007 cm^{-1} for $[Ru(H_2O)_6]^{3+}$ (44), and 1150 cm^{-1} for low -spin Ru(III) complexes (45). The low λ values of the present complex is, however, not extraordinary, since the extensive electron delocalization of the metal-ligand bond results in the transfer of some of the unpaired electron density onto the ligand. As a result, the orbital angular momentum is decreased, i.e., the orbital contribution is reduced, and thus the magnetic parameters became closer to the spin-only value of $g = 2$ (45), as observed in the present complex **10**. The orbital reduction factor k of 0.92 is normal, compared to the literature values; 0.865 for $[Ru(H_2O)_6]^{3+}$ (44), 0.959 for $[Ru(NH_3)_6]^{3+}$ (44), 0.932 for $[Ru(bipy)_3](PF_6)_3$ (bipy is 2, 2'-bipyridyl) (46), 0.912 for $[Ru(phen)_3](PF_6)_3$ (phen is phenanthroline) (46), 0.95 for $RuCl_3(PMe_2Ph)_3$ (35), and 0.99 for $RuCl_3(AsPr_3)_3$ (35). The reduced spin-orbital interaction in **10** does not contradict the π-MO scheme in Figure 15, since the distinct π-MO, composed of the π-orbitals of two Ru and two S atoms, corresponds to a delocalized electronic state between the metal and the sulfur atoms. The lack of near-IR absorption of **10** is not inconsistent with the π-MO scheme, but suggest that the electronic state of **10** is actually beyond what the ligand field theory covers, and that a MO treatment is a more realistic way to deal with the compound. The ESR spectrum of **10** with such small anisotropy is not common to Ru(III) complexes, and means that the parameters, Δ and V, are small. Actually the results of the present calculation shows relatively small splitting parameters, which should be compared to larger values of $\Delta = 5600$ cm^{-1} for $RuCl_3(P(n\text{-}Bu)_2Ph)_3$ (35).

Conclusion

The present study has revealed that a RuSSRu core is remarkably highly reactive, as indicated by the C-S bond formation of **7** in the reaction with acetone and by the formation of $S_2O_5^{2-}$ ligand by O_2 oxidation of **13**. These reactions suggest that the disulfide ligands undergo nucleophilic attack, and the resultant new bridging ligand, that is usually very unstable, can be stably coordinated in the RuSSRu core. These reactions together with the formation of the diazene complex **17** may give insight into the mechanism of the reaction chemistry of transition metal sulfide such as hydrodesulfurization (HDS) and the redox chemistry of hydrazine and diazene, if the reaction mechanism of the RuSSRu core system is more extensively and systematically investigated and understood.

Literature Cited

1. Müller, A.; Krebs, B. eds. *Sulfur: Its Significance for Chemistry, for the Geo-, Bio- and Cosmosphere and Technology,* Elsevier, New York, **1984**.
2. Holm, R. H. *Chem. Soc. Rev.* **1981**, *10*, 455.
3. Coucouvanis, D. *Acc. Chem. Rev.* **1981**, *14*, 201.
4. Müller, A.; Jaegermann, W.; Enemark, J. H. *Coord. Chem. Rev.* **1982**, *46*, 245.
5. Kim, J. C.; Rees, D. C. *Science* **1992**, *257*, 1677.
6. Christiansen, J.; Tittsworth, R. C.; Hales, B. J.; Cramer, S. P. *J. Am. Chem. Soc.* **1995**, *117*, 10017.
7. Uemura, H.; Kawano, M.; Watanabe, T.; Matsumoto, T.; Matsumoto, K. *Inorg. Chem.* **1992**, *31*, 5137.
8. Maheu, L. F.; Pignolet, L. H. *J. Am. Chem. Soc.* **1980**, *102*, 6346.
9. Müller, A.; Ishaque Khan, M.; Krickemmeyer, E.; Bogge, H. *Inorg. Chem. 1991*, *30*. 2040.
10. Matsumoto, K.;Furuhashi, T. manuscript in preparation.
11. Matsumoto, T.; Matsumoto, K. *Chem. Lett.* **1992**, 559.
12. Matsumoto, K.; Uemura, H.; Kawano, M. *Inorg. Chem.* **1995**, *34*, 658.
13. Matsumoto, K.; Uemura, H.; Kawano, M. *Chem. Lett.* **1994**, 1215.
14. Matsumoto, T.; Matsumoto, K. *Chem. Lett.* **1992**, 1539.
15. Matsumoto, K.; Ohnuki, H.; Kawano, M. *Inorg. Chem.* **1995**, *34*, 3838.
16. Kim, J.; Rees. D. C. *Nature* **1992**, *360*, 553.
17. Henderson, R. A.; Leigh, G. J.; Pickett, C. J. *Adv. Inorg. Chem. Radiochem.* **1983**, *27*, 197.
18. Pelikam, P.; Boca, R. *Coord. Chem. Rev.* **1984**, *55*, 55.
19. Kawano, M.; Hoshino, C.; Matsumoto, K. *Inorg. Chem.* **1992**, *31,* 5158.
20. Matsumoto, K.; Koyama, T. manuscript in preparation.
21. Sellmann, D.; Böhlen, E.; Waeber, M.; Huttner, G.; Zsolnai, L. *Angew. Chem. Int. Ed. Engl.* **1985**, *24,* 981.
22. Collin, R. L.; Lipscomb, W. *Acta Cryst.* **1951**, *4,* 10.
23. Blum, L.; Williams, I. D.; Schrock, R. R. *J.Am. Chem. Soc.* **1984,** *106,* 8316.
24. Churchill, M. R.; Li, Y.-J.; Blum, L.; Schrock, R. R. Organometallics **1984**, *3*, 109.
25. Sellmann, D.; Soglowek, W.; Knoch, F.; Moll, M. *Angew. Chem. Int. Ed. Engl.* **1989**, *28*, 1271.
26. Smith III, M. R.; Cheng, T.-Y.; Hillhouse, G. L. *J. Am. Chem. Soc.* **1993**, *115,* 8638.

27. Matsumoto, K.; Matsumoto, T.; Kawano, M.; Ohnuki, H.; Shichi, Y.; Nishide, T.; Sato, T. *J. Am. Chem. Soc.* **1996,** *118,* 3597.
28. Kato, M.; Kawano, M.; Taniguchi, H.; Funaki, M.; Moriyama, H.; Sato, T.; Matsumoto, K. *Inorg. Chem.* **1992,** *31,* 26.
29. Sellmann, D.; Lechner, P.; Knoch, F.; Moll, M. *J. Am. Chem. Soc.* **1992,** *114,* 922.
30. Kim, S.; Otterbein, E. S.; Rava, R. P.; Isied, S. S.; San Filippo, J. Jr.; Waszcyak, J. V. *J. Am. Chem. Soc.* **1983,** *105,* 336.
31. Amarasekera, J.; Rauchfuss, T. B.; Wilson, R. S. *Inorg. Chem.* **1987,** *26,* 3328.
32. Bunker, B. C.; Drago, R. S.; Hendrickson, D. N.; Richman, R. M.; Kessell, S. L. *J. Am. Chem. Soc.* **1978,** *100,* 3805.
33. Holzer, W.; Murphy, W. F.; Bernstein, H. J. *J. Mol. Spectrosc.* **1969,** *32,* 13.
34. Clark, R. J. H.; Cobbold, D. G. *Inorg. Chem.* **1978,** *17,* 3169.
35. Hudson, A.; Kennedy, M. J. *J. Chem. Soc. (A)* **1969,** 116.
36. Lahiri, G. K.; Bhattacharya, S.; Ghosh, B. K.; Chakravorty, A. *Inorg. Chem.* **1987,** *26,* 4324.
37. Cotton, F. A.; Torralba, R. C. *Inorg. Chem.* **1991,** *30,* 4392.
38. DeSimone, R. E. *J. Am. Chem. Soc.* **1973,** *95,* 6238.
39. Sakai, S.; Hagiwara, N.; Yanase, Y.; Ohyoshi, A. *J. Phys. Chem.* **1978,** *82,* 1917.
40. Raynor, J. B.; Jeliazkowa, B. G. *J. Chem. Soc. Dalton Trans.* **1982,** 1185.
41. Bleaney, B.; O'Brien, M. C. M. *Proc. Phys. Soc.* **1956,** *69,* 1216.
42. Stevens, K. *Proc. R. Soc. London, Ser. A,* **1953,** *219,* 542.
43. Kamimura, H. *J. Phys. Soc. Jpn.* **1956,** *11,* 1171.
44. Daul, C.; Goursot, A. *Inorg. Chem.* **1985,** *24,* 3554.
45. Goodman, B. A.; Raynor, J. B. *Adv. Inorg. Chem. Radiochem.* **1970,** *13,* 136.
46. DeSimone, R. E.; Drago, R. S. *J. Am. Chem. Soc.* **1970,** *92,* 2343.

Chapter 16

Syntheses, Structures, and Reactions of Cyclopentadienyl Metal Complexes with Bridging Sulfur Ligands

M. Rakowski DuBois, B. Jagirdar[1], B. Noll, and S. Dietz[2]

Department of Chemistry and Biochemistry, University of Colorado, Boulder, CO 80309–0215

Homogeneous cyclopentadienyl complexes of molybdenum and rhenium that contain sulfido and/or disulfido ligands have been synthesized. The structures and reactivities of the complexes are compared, and reactions which are relevant to those of heterogeneous metal sulfide catalysts are discussed.

An intriguing feature of the heterogeneous transition metal sufides is their ability to participate in organometallic chemistry by virtue of their reactivity with hydrogen and with organic substrates. The most prominent class of reactions catalyzed by metal sulfide surfaces is the commercially important hydrotreating process, in which the hydrogenolysis of C-S and C-N bonds in aromatic heterocycles is effected (*1-8*). While a molybdenum sufide based system is the catalyst used most widely in industry, other metal sulfides have been shown to be more active than the molybdenum system in this process (*9*). Additional reactions for metal sulfides have also been characterized, either as a part of the hydrotreating sequence or in independent studies (*1,2*). These include hydrogen activation and hydrogen/deuterium exchange, the hydrogenation and isomerization of olefins, and the dehydrogenation of alkanes. Reactions with carbon monoxide have been identified for molybdenum sulfide surfaces, including the reduction of CO to methane or to alcohols (*10-13*), and its conversion to CO_2 in the water gas shift reaction (*14*). Despite this extensive range of reactivity, examples of homogeneous metal complexes with sulfido ligands that react with dihydrogen and with organic molecules are still relatively rare (*15*). In this paper we focus on two of the metal sulfides, those of molybdenum and rhenium. We briefly compare characteristics of MoS_2 and ReS_2 that contribute to their catalytic chemistry, and we describe molecular derivatives of molybdenum and rhenium that permit us to probe reactions related to those of the heterogeneous systems.

Heterogeneous Molybdenum and Rhenium Sulfides

Both MoS_2 and ReS_2 crystallize as layered sulfides, but there are significant

[1]Current address: Indian Institute of Science, Bangalore 560 012, India
[2]Current address: TDA Research, 12345 West 52 Avenue, Wheatridge, CO 80033

0097–6156/96/0653–0269$15.00/0

differences between their structures. For example, the molybdenum ions are coordinated in a trigonal prismatic geometry, while the rhenium ions occupy octahedral holes. In MoS_2, the sulfide ions occupy distorted tetrahedral sites created by the metal lattice, while in ReS_2 the sulfides occur in a trigonal pyramidal geometry bridging three rhenium ions. More detailed descriptions of the structures are provided elsewhere (*16,17*).

Comparisons of the activity of molybdenum sulfide and rhenium sulfide catalysts have established that the rhenium system is more active in catalyzing hydrogenations (*18*) as well as the hydrodesulfurization and hydrodenitrogenation reactions (*19-21*). Relative activities of metal sulfides have been correlated with their electronic properties, including the metal d character of the HOMO and the degree of covalency and strength of the metal sulfur bond (*22,23*). For example, the weaker M-S bond and the resulting lower ΔH_f value for ReS_2 (42.7 ± 3Kcal/mol) compared to that of MoS_2 (65.8 ± 1 Kcal/mol) may contribute to the catalytic activity of the rhenium system by allowing easier formation of anion vacancies during the catalytic cycle. The study of homogeneous molybdenum and rhenium sulfide complexes may provide insights into the questions of how the structural and electronic differences between the two metal sulfides contribute to their different activities.

An important fundamental reaction of metal sulfides, which is poorly understood, is the dissociative addition of dihydrogen. Inelastic neutron scattering data suggest that the sulfide atoms are the site of hydrogen addition and SH groups on the surface have been identified. However, the involvement of the metal ion in this process has not been ruled out, and pathways have been suggested in the literature that do or do not involve metal hydride formation (*24-27*), e.g., equations 1-2. These have been referred to as homolytic and heterolytic cleavage of hydrogen, respectively. Metal-hydrides and/or sulfhydryls are proposed to be involved in hydrogen transfer reactions to the substrates during hydrogenation and hydrogenolysis reactions. Further reaction of the SH sites under thermal conditions leads to the elimination of H_2S and the generation of anion vacancies on the catalyst surface.

$$S\text{-}S^{2-} + H_2 \longrightarrow 2\,HS^- \quad (1)$$

$$M\text{-}S + H_2 \longrightarrow M(H)\text{-}SH \quad (2)$$

Reactions of Molecular Molybdenum Sulfide Complexes with Dihydrogen

In recent years we have studied the reactions of dihydrogen with selected dinuclear cyclopentadienyl-molybdenum complexes with bridging sulfido ligands. The reactions, which proceed under mild conditions (1-3 atm, 25-50°C), are characterized by formation of hydrosulfido products. For example, in some of our earliest work we found that the complex $(MeCpMo)_2(\mu\text{-}S)_2(\mu\text{-}S_2)$ reacted quantitatively with hydrogen to form the hydrosulfido derivative, equation 3 (*28*). The isomeric dimer with

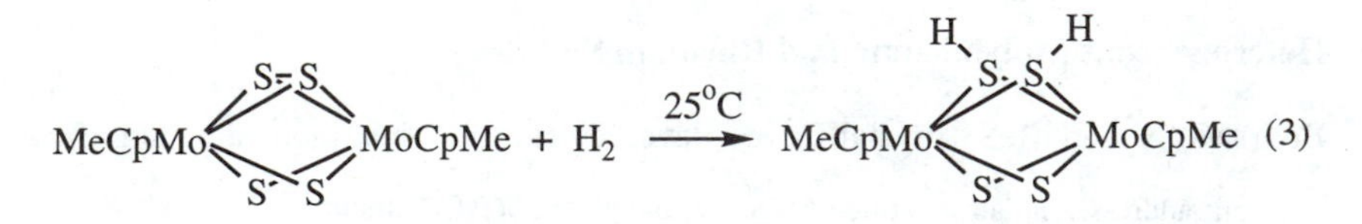

terminal sulfido ligands $[(MeCpMoS(\mu\text{-}S)]_2$ was also converted under hydrogen to the same product, but at a much slower rate (*29*). Photoisomerization of the analogous permethylcyclopentadienyl derivatives has been found to occur readily (*30*), and similar interconversions between isomers appear to take place in solution, although less cleanly, for the MeCp complexes. For this reason, the reaction rates of individual isomers with H_2 have not been quantitatively compared.

The dihydrogen addition reaction shown in equation 3 does not appear to be reversible; nor has the elimination of hydrogen sulfide been observed in this system at elevated temperatures. However, H_2 elimination can be promoted by reduction of the dimer to a formal Mo(III) derivative (*28*), and a rationale for this elimination reaction has been developed on the basis of molecular orbital calculations (*31*). The hydrosulfido complex has been found to function as a hydrogenation catalyst for certain unsaturated molecules (*32*), and this work has been reviewed previously(*15*).

Reactions of Cationic Molybednum Complexes with Dihydrogen. A second facile hydrogen addition reaction has been observed for the one-electron oxidation product of the Mo(IV)/(IV) methanedithiolate bridged complex, which exists as a dimer of dimers linked by a bond at the μ-sulfido ligands (Structure A, Cp' = MeCp) (*33*). Addition of hydrogen proceeds rapidly at room temperature to form the hydrosulfido cation, equation 4. The tetranuclear product has also been found to function as a

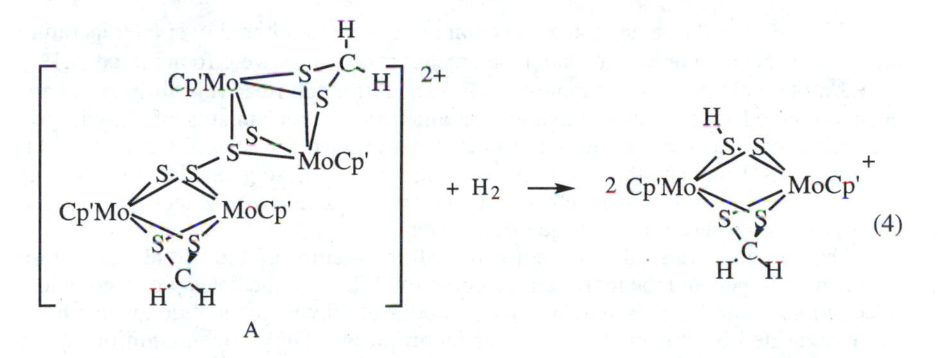

hydrogen atom acceptor in reactions with other hydrogen donors such as thiophenol. The ease of reactivity is attributed in part to the relatively weak and strained nature of the sulfur-sulfur bond between dimeric units. The S-S bond distance was found to be 2.147 (4) Å.

More recently we have studied the addition of hydrogen to cationic sulfido bridged derivatives of the formula $[(CpMo)_2(\mu\text{-}S_2CH_2)(\mu\text{-}S)(\mu\text{-}SR)]^+$. The reactions proceed in the presence of base to form the neutral hydrosulfido complex and one proton equivalent, equation 5 (*34*). The formation of a hydrosulfido product is a common feature of the reactions of the molybdenum dimers with hydrogen, but this hydrogen addition appears to differ mechanistically from those in equations 3 and 4 because of the heterolytic nature of the reaction. Although formal hydride addition to a sulfido ligand is an unusual reaction, the electrophilic nature of this ligand in the cations has been demonstrated previously in reactions with alkyl lithium reagents. The latter reactions result in nucleophilic additions to the sulfido ligand to form neutral bis(thiolate) complexes (*35*).

Mechanistic Studies. Kinetic studies of equation 5 have been carried out for the derivatives where R = thienyl or CH_2CO_2Me (*34*). The reactions have been found to be first order in hydrogen and first order in dimer with second order rate constants for both systems at 20°C of approx. 1.5 x 10^{-3} M^{-1} sec^{-1}. A deuterium isotope effect, k_H/k_D, of 2.5 was determined at 50°C for the reaction for the derivative where R =

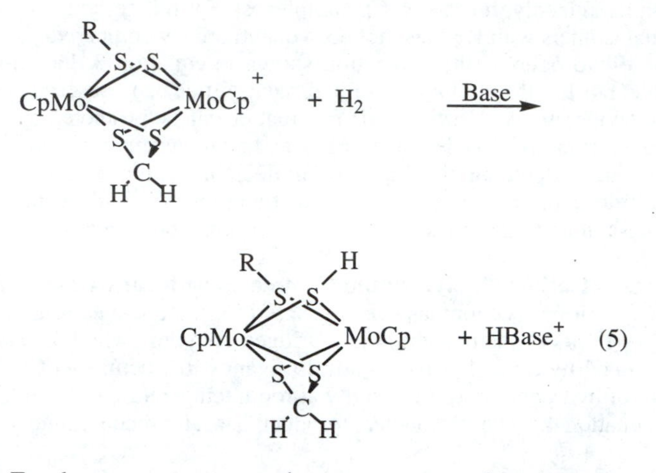

CH_2CO_2Me. For the same system, reaction rates were monitored over a temperature range of 20-60°C to obtain activation parameters; the values were found to be $\Delta H^{\pm}$ = 81 KJ/mole and $\Delta S^{\pm}$ = -23 J/mole-K (*34, 36*). Although relatively little data have been collected that relates activation parameters to mechanisms of dihydrogen activation, comparisons of our data to those determined for several Group 8 metal complexes (*37*) indicate that the values are higher than those generally observed for oxidative addition to a single metal center and are consistent with values obtained for other proposed heterolytic cleavage mechanisms.

The mechanistic studies do not establish the nature of the initial interaction between hydrogen and the molybdenum complex. The intermediates proposed below in equations 6 and 8 represent two possible modes of dihydrogen addition which have been suggested by model studies in our laboratories. Dihydrogen addition to the sulfur ligands, equation 6, is similar to the characterized interaction of alkenes with the same cationic structure, equation 7 (*38*). Alternatively, the addition of dihydrogen

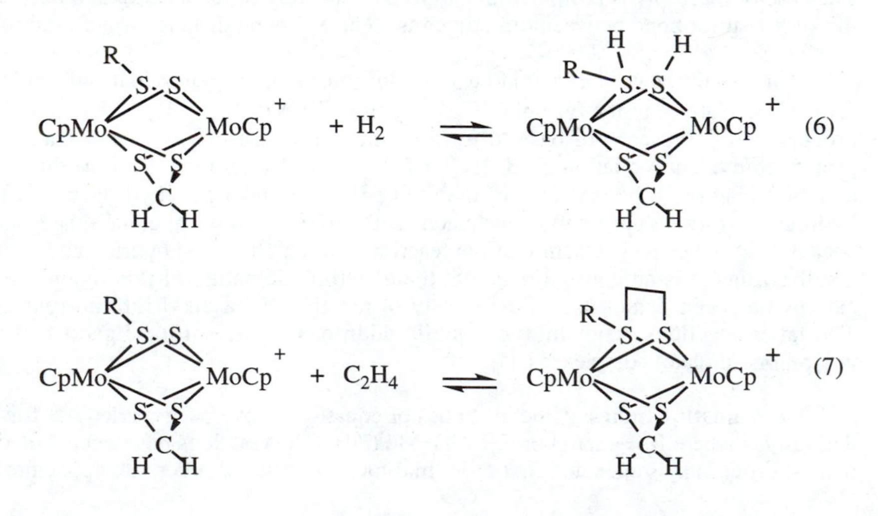

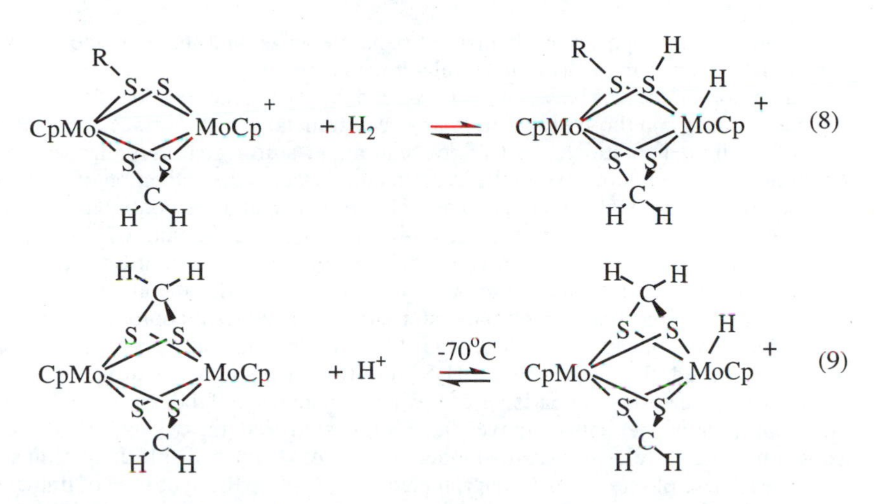

across a metal sulfide bond may be feasible, equation 8, because protonation of Mo(III)/(III) bis(thiolate) complexes at low temperature appears to form similar metal hydride structures of low stability, equation 9 (*39*).

Further Reactions under Hydrogen

The ability of the cationic complexes to activate hydrogen has led to the development of carbon-heteroatom bond cleavage reactions under a hydrogen atmosphere. For example, under 2-3 atm. of dihydrogen at room temperature, the reaction of the hydrosulfido cation with nitriles proceeds to form ammonia and a molybdenum structure which incorporates the hydrocarbon fragment of the nitrile (*40*). The first step in the reaction is proposed to involve the formation of an iminothiolate cation as shown in Scheme 1. This intermediate, although it is not detected spectroscopically, has been trapped in the presence of excess acid, and the resulting N-protonated dication (not shown) has been characterized spectroscopically. The reaction of the iminothiolate intermediate with hydrogen is proposed to form a neutral hydrosulfido intermediate and an equivalent of protic acid by analogy to reactions of other thiolate cations. Intramolecular insertion of the C=N bond into the adjacent S-H bond leads to an amino-substituted 1,1-dithiolate complex. Analogous intramolecular insertions of vinylthiolate ligands into the hydrosulfido ligand have been observed in related systems (*41*). Protonolysis of the C-N single bond in the 1,1-dithiolate ligand by the protic acid results in the release of NH_3 and the formation of the final observed product. Similar steps have been proposed in the reductive cleavage of isonitriles (*40*) and of the carbon oxygen multiple bonds of acyl derivatives under a hydrogen atmosphere (*42*).

Comparisons to Metal Sulfide Surfaces. This mechanism demonstrates several elementary reaction steps which may be relevant to reactions of metal sulfide surfaces with hydrogen. For example, the formation of hydrosulfido sites by hydrogen addition has been well established for the molecular systems, and the above mechanism demonstrates a specific role of hydrosulfido ligands in reducing the bond order of the unsaturated substrate. In contrast, insertion reactions into metal-hydride bonds are not identified in these systems. Protonolysis of the carbon-heteroatom single bond is the proposed mechanism of bond cleavage, and this is made possible by proton production in the heterolytic dihydrogen activation.

Important differences between the C-X bond cleavage reactions identified for the molecular dimers and those proposed to proceed on metal sulfide surfaces are that the former are limited to stoichiometric conversions at sulfur centers under the mild conditions studied. In contrast, in the hydrotreating process the activation of substrate by interaction with metal centers is commonly proposed, and the continuation of the catalytic cycle depends on the generation of anion vacancies by ligand hydrogenolysis. For example, in HDS, the conversion of the coordinated sulfide atoms to labile hydrogen sulfide creates reactive sites at the metal centers (*3-8*). Methods have been explored in our laboratory to generate vacant coordination sites in the dinuclear molybdenum complexes (*43-44*). While some success in this project has been achieved, the elimination of H_2S from the dimers under hydrogen pressure has not been observed. Comparisons of the heterogeneous molybdenum and rhenium sulfide properties, described above, have suggested that the weaker Re-S bond strength may be correlated with the higher activity of rhenium disulfide in reductive processes. These observations have prompted us to explore the synthesis of dinuclear sulfido bridged complexes containing rhenium, and to compare their reactivity with the related molybdenum derivatives.

Syntheses of Dinuclear Rhenium Complexes

The reaction of $Cp'ReCl_4$ (Cp' = Me_4EtCp) (*45*) with one equivalent of bis(trimethylsilyl)sulfide results in a dinuclear complex that has been isolated and found to have the formulation $[Cp'ReS(Cl)_2]_2$, confirmed by mass spectrometry and elemental analyses. An X-ray diffraction study of this product has not been completed, but the dimer is presumed to be bridged by two sulfido or one disulfido ligand. $Cp'ReCl_4$ reacted with two equivalents of bis(trimethylsilyl)sulfide to form a new complex of formulation $[Cp'ReS_2Cl]_2$, **1**, in 80% yield, equation 10. In the infrared spectrum of the product a relatively weak band at 460 cm^{-1} suggested the presence of a sulfur-sulfur bond (*46*), but no strong absorptions were observed in the

$$Cp'ReCl_4 + 2\,(Me_3Si)_2S \longrightarrow \underset{\mathbf{1}}{[Cp'ReS_2Cl]_2} \qquad (10)$$

region 480-530 cm^{-1}, associated with terminal Re=S bonds (*47*). The mass spectrum showed only a very weak parent ion and the base peak corresponded to the fragment P-Cl at m/e 835. An X-ray diffraction study was carried out in order to obtain more information about the coordination geometry about each rhenium ion.

Structural Characterization of 1. Complex **1** crystallized from dichloromethane-pentane in space group $P2_1/n$ with cell dimensions a = 8.596(2) Å, b = 9.804(2) Å, c = 15.726(3) Å, and $\beta = 90.64^{\circ}$. A perspective drawing and numbering scheme for the complex are shown in Figure 1. The structure shows discrete dinuclear dications of Re(IV) with two μ-η^2-S_2 ligands. The chloride counterions do not interact with the Re centers, but the anions lie between disulfide ligands and show close contacts of 2.65 Å and 2.78 Å with S1' and S2, respectively. A crystallographically imposed inversion center midway along the Re-Re vector relates the disulfide, chloride, and parallel Cp' ligands. During refinement of the structure, it became apparent that a disorder of the sulfur ligands was present. A model that used two pairs of bridging

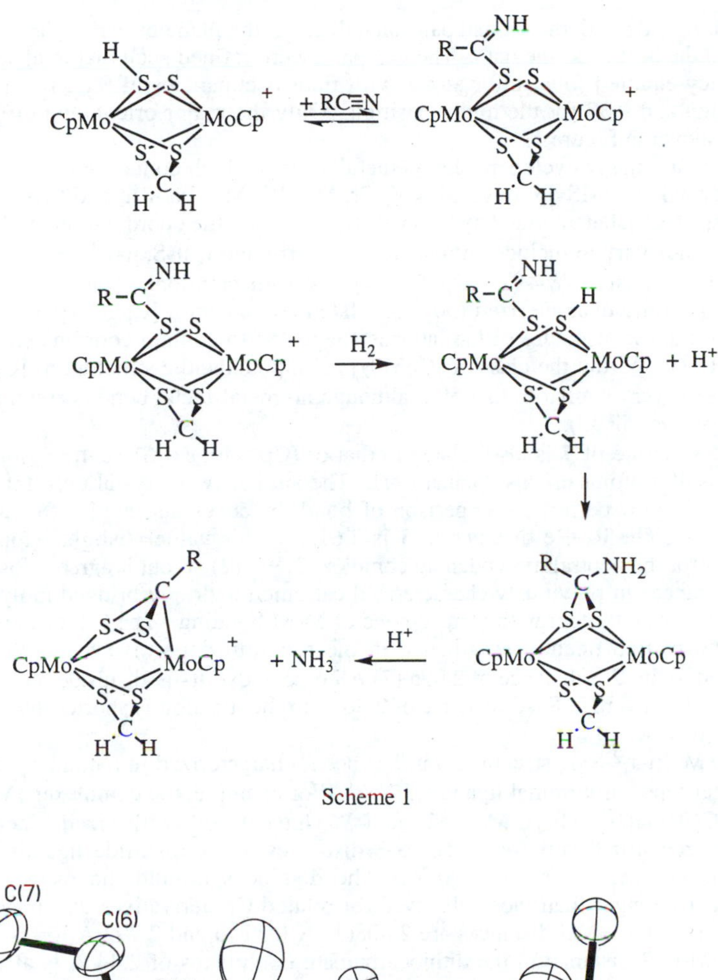

Scheme 1

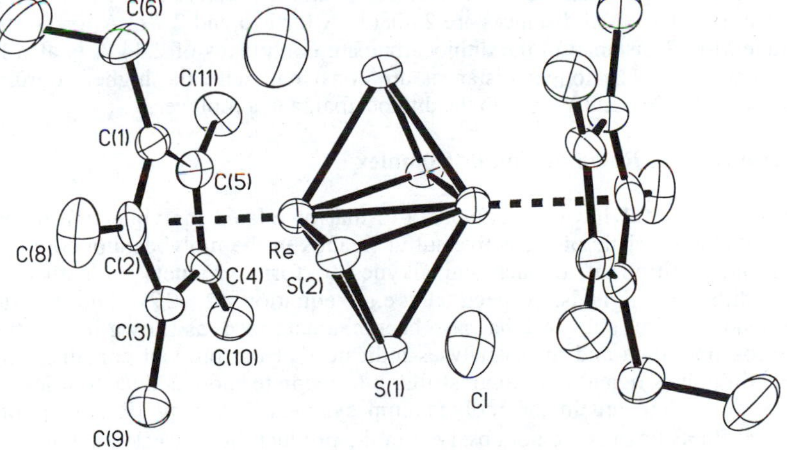

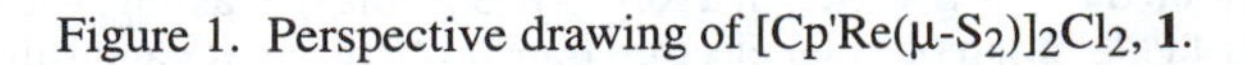
Figure 1. Perspective drawing of $[Cp'Re(\mu\text{-}S_2)]_2Cl_2$, **1**.

disulfide ligands with the second pair rotated 22° in the plane bisecting the metal ions provided the best fit to the data. The two pairs were refined such that total sulfur site occupancy equaled four sulfur atoms with final occupancies of 0.62 for the major orientation and 0.38 for the minor position. Only the major orientation of disulfide pairs is shown in Figure 1.

There are many cyclopentadienyl metal dimers which contain four sulfur atoms of the formula $[CpMS_2]_2$, (e.g., M = V, Cr, Mo, W, Mn, Fe, Ru, and Co). Many of these have been characterized by X-ray diffraction, and the coordination modes of the sulfur ligands vary to include combinations of terminal S, μ-S, μ-η^1-S_2, μ-η^2-S_2 and μ-η^1, η^2-S_2 ligands (*48-49*). $[Cp^*ReS_2]_2$ has been mentioned briefly, but has not been structurally characterized (*50*). A related composition $[Cp'RuS_2Cl]_2$ has been reported, but the structure of the latter is suggested to contain coordinated chloride ligands (*51*). Within the class of $[CpMS_2]_2$ compounds, the structure of $[Cp^*Fe(\mu$-η^2-$S_2]_2^{2+}$ is very similar to that of **1**, although no metal-metal bond is assigned to the iron derivative (*52,53*).

The structure of **1** is also related to that of $(Cp^*Mo)_2(\mu$-$S)_2(\mu$-η^2-$S_2)$, in which the four sulfur atoms are also planar (*54*). The similarity in crystal radii for Mo(IV) and Re(IV) (*55*) permit a comparison of bond distances and angles for these two derivatives. The Re-Re distance in **1** is 2.613 Å; this value is slightly longer that observed for the neutral molybdenum complex, 2.599 (2) Å, but it agrees closely with M-M distances in previously characterized cationic quadruply bridged molybdenum dimers and is consistent with some degree of M-M bonding. The S-S distance in the S_2 ligand is significantly lengthened in the rhenium derivative to 2.228 (10) Å, compared to the S-S distance of 2.095 (7) Å in the above μ-η^2-disulfide molybdenum complex. The average Re-S distance of 2.36 Å in the dication is shorter than the Mo-S (μ-S_2) distance of 2.45 Å.

The $M_2(\mu$-η^2-$S_2)_2$ structural unit has been characterized in a number of dimers with other types of terminal ligands (*56-60*). For example, the complexes $[M_2(\mu$-η^2-$S_2)_2(S_2CNR_2)_4]^{+2}$, where M = Mo and W, have been synthesized recently and characterized structurally (*60*). The S-S distances in the disulfide ligands in these dications are very close to 2.00 Å. The distances around the metal ions are significantly longer than those observed for related Cp derivatives discussed above. For example, the M-M distances are 2.808(1) Å for Mo and 2.792 Å for W, and the average M-μ-S distance in the dithiocarbamate complexes of 2.44 Å is also longer than that in **1**. The longer distances are consistent with the higher coordination numbers of the Mo and W ions in the dithiocarbamate complexes.

Reactions of the Rhenium Sulfide Complex

The reactivity of **1** has been compared to that of related molybdenum dimers. A reaction characteristic of the active sulfur ligands in the molybdenum complexes is the rapid addition of alkenes and alkynes to form products with alkane- and alkenedithiolate ligands, respectively, e.g., equation 11 (*29*). The addition of acetylene to sulfur sites in **1** has also been characterized, as shown in equation 12. Spectroscopic data and microanalyses indicate that the product contains only one alkenedithiolate ligand. In contrast this intermediate monoadduct has never been successfully detected in the molybdenum systems. Strong IR absorptions for terminal Re=S ligands are not observed in the product, but a weak band at 522 cm^{-1} might be attributed to a S-S stretch in a $Re^{IV}{}_2(\mu$-$S_2)$ linkage, as suggested in equation 12. However facile intramolecular electron transfer between Re and sulfur ligands

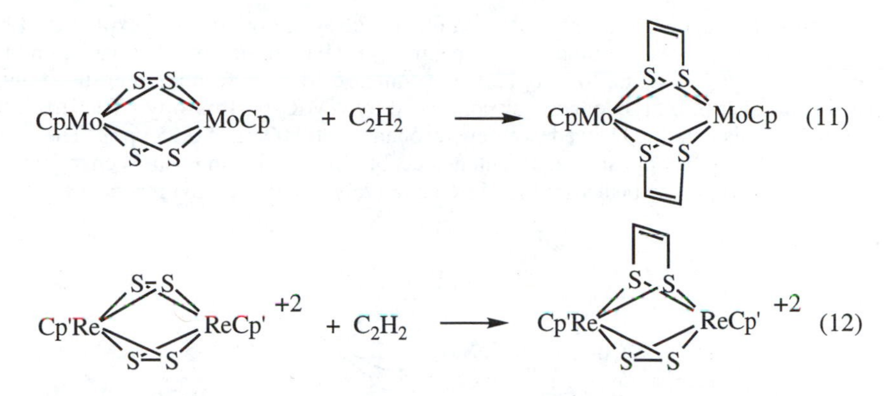

has been well established in other systems (*61*), and the formation of a $Re^{V}_2(\mu\text{-}S)_2$ structure cannot be ruled out at this stage. More definitive characterization of the structure and the resulting electron distribution in the product will be important in understanding the extent of the reaction.

Reactions of 1 with Dihydrogen and Other Hydrogen Donors. The reactions of the molybdenum sulfide dimers with dihydrogen formed the basis for much of the chemistry that related these systems to the heterogeneous metal sulfides. The rhenium disulfide derivative **1** also reacted with dihydrogen under exceptionally mild conditions (1 atm and 25°C). The products of this reaction include hydrogen sulfide and a trinuclear rhenium cluster with the tentative formulation shown in equation 13.

$$[Cp'Re(\mu\text{-}S)_2]_2Cl_2 + H_2 \longrightarrow [(Cp'Re)_3S_4H_x]^{n+} + H_2S \qquad (13)$$

The hydrogenolysis of bridging sulfur ligands to liberate hydrogen sulfide is an unusual reaction for dinuclear complexes (*29*). The reaction represents an important difference in the chemistry of the Re_2S_4 and Mo_2S_4 structures, and the difference appears to reflect the trend in metal sulfur bond strengths observed for the molybdenum and rhenium sulfides. These preliminary studies suggest that the rhenium complexes may permit us to explore mechanistic models for the formation of anion vacancies and their role in substrate activation in simple homogeneous systems.

The nature of the rhenium cluster formed in equation 13 is not yet completely defined. Mass spectral data confirm the trinuclear nature of the product. The presence of hydrosulfido ligands is suggested by the infrared spectrum which shows a weak band at 2375 cm^{-1}. The ^1H NMR spectrum of the major product shows the characteristic pair of methyl singlets for the Cp' ligands at 2.32 and 2.30 ppm and the Cp' triplet at 1.24 ppm. Resonances for SH ligands have not been definitively assigned. The cluster shows limited stability under nitrogen and slowly rearranges to a product with different spectroscopic features. Attempts to crystallize a derivative of the trinuclear cluster product are in progress.

The addition of dihydrogen to the molybdenum complex $(MeCpMo)(\mu\text{-}S)_2(\mu\text{-}S_2)$ proceeded to from a bis (hydrosulfido) product (see equation 3 above), and we wished to determine whether a similar product was formed in the initial interaction of **1** with dihydrogen. If an intermediate dihydrogen activation product could be detected, it might be possible to obtain mechanistic information on its further reaction to form H_2S. Because an intermediate was not detected in the reaction of **1** with excess molecular hydrogen, reactions of **1** with other hydrogen donors were explored. For example, one equivalent of the bis(hydrosulfido) molybdenum complex was found to serve as a hydrogen donor to **1**, as shown in equation 14. The formation of the known (μ-S_2) product of molybdenum was confirmed by NMR spectroscopy. The rhenium product of this reaction showed two new resonances at 0.074 and 0.15 ppm. The two resonances initially appear in equal intensities, but the relative intensities change with time to a 2:1 ratio; the peaks are therefore tentatively assigned to SH resonances of

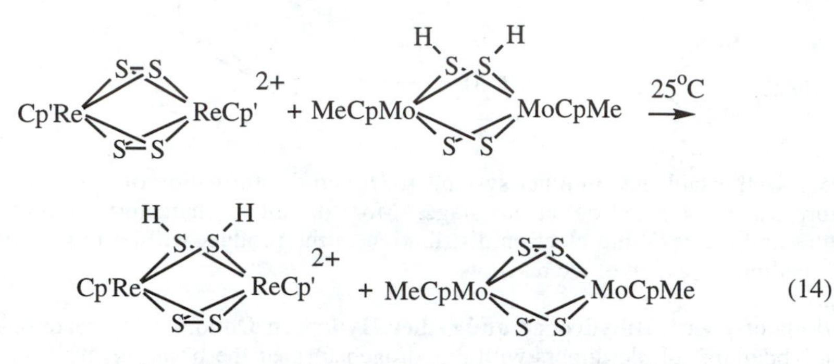

two isomers of a rhenium-bis(hydrosulfido) complex. Individual S-H and S-S bond dissociation energies in these metal complexes are not known, but we believe a driving force for the hydrogen transfer reaction may be the formation of a stronger S-S bond in the molybdenum dimer relative to the one broken in the rhenium complex. This is suggested by the significantly longer S-S distances observed for the η^2-S_2 ligands in the X-ray structure of **1**.

Summary and Conclusions

Dinuclear cyclopentadienyl complexes of molybdenum and of rhenium with four bridging sulfur atoms have been prepared and characterized. Extensive reactions of the molybdenum systems have been identified under hydrogen pressure, including the cleavage of C-S, C-N and C-O bonds. Features of these reactions include either homolytic or heterolytic hydrogen activation and the formation of SH ligands, and the reactions of S-H ligands either by hydrogen atom transfer or by S-H addition to unsaturated substrates. Cleavage of C-heteroatom single bonds which occur within the substituents of the bridging sulfur ligands in the dimer has been found to be induced by the protic acid which is generated in the heterolytic hydrogen cleavage. Throughout these reactions under hydrogen the Mo_2S_4 core of the complexes remains intact.

The recently synthesized rhenium derivative $[Cp'Re(\mu\text{-}S_2)]_2^{2+}$ is structurally similar to the quadruply bridged molybdenum systems, and it undergoes facile reactions with molecular hydrogen and with hydrogen atom donors. Although further work will be necessary to develop the chemistry of the rhenium complexes, some important differences have been observed in our preliminary studies of the rhenium reactions. These include the apparently weaker nature of the M-S bonds and of S-S bonds relative to those in the molybdenum structures which result in a tendency for dimer fragmentation and reformation into larger clusters. The lower bond energies open the way for synthesizing new clusters as well as for exploring new types of reactions that may be relevant to those involving substrate activation on heterogeneous metal sulfide surfaces.

Acknowledgments. We thank the National Science Foundation for support of recent work, and the Division of Chemical Sciences, Office of Basic Energy Sciences, Office of Energy Research, U.S. Department of Energy for support of earlier aspects of this work.

Literature Cited

1. Weisser, O.; Landa, S. *Sulphide Catalysts, Their Properties and Applications;* Pergamon Press: New York, 1973.
2. Gates, B. C.; Katzer, J. R.; Schuit, G. C. A. *Chemistry of Catalytic Processes*; McGraw-Hill: New York, 1979; p. 390.
3. Prins, R.; de Beer, V. H. G.; Somorjai, G. A. *Catal. Rev.-Sci. Eng.* **1989**, *31*, 1-41.
4. Ho, T. C. *Catal. Rev.-Sci. Eng.* **1988**, *30*, 117.
5. Chianelli, R. R. *Catal. Rev.-Sci. Eng.* **1984**, *26*, 361-393.
6. Topsoe, H.; Clausen, B. S. *Catal. Rev.-Sci. Eng.* **1984**, *26*, 395-420.
7. Grange, P. *Catal. Rev.-Sci. Eng.* **1980**, *21*, 135.
8. Massoth, F. E. *Adv. Catal.* **1978**, *27*, 265.
9. Pecararo, T. A.; Chianelli, R. R. *J. Catal.* **1981**, *67*, 430.
10. Hou, P.; Wise, H. *J. Catal.* **1985**, *93*, 409.
11. Woo, H. C.; Nam, I. S.; Lee, J. S.; Chung, J. S.; Lee, K. H.; Kim, Y. G. *J. Catal.* **1992**, *138*, 525.
12. Frank, A. J.; Dick, H. A.; Goral, J.; Nelson, A. J.; Graetzel, M. *J. Catal.* **1990**, *126*, 674.
13. Mirimadi, B. K.; Morrison, S. R. *J. Catal.* **1983**, *80*, 280.
14. Hou, P.; Meeker, D.; Wise, H. *J. Catal.* **1983**, *80*, 280.
15. Rakowski DuBois, M. *Chem. Rev.* **1989**,*89*, 1-9.
16. Tsigdinos, G. "The Properties and Structures of Inorganic Molybdenum-Sulfur Compounds;" Bulletin Cdb-17, Climax Molybdenum Company: Greenwich, CT, 1973; and references within.
17. Murray, H. H.; Kelty, S. P.; Chianelli, R. R.; Day, C. S. *Inorg. Chem.* **1994**, *33*, 4418.
18. Broadbent, H. S.; Slaugh, L. H.; Jarvis, N. L. *J. Am. Chem. Soc.* **1954**, *76*, 1519.
19. Eijsbouts, S.; de Beer, V. H. J.; Prins, R. *J. Catal.* **1988**, *109*, 217.
20. Stern, E. W. *J. Catal.* **1979**, *57*, 390.
21. Ledoux, M.; Michaux, O.; Agostina, G.; Panissod, P. *J. Catal.* **1986**, *102*, 275.
22. Harris, S.; Chianelli, R. R. *J. Catal.* **1984**, *86*, 400-412.
23. Norskov, J. K., Clausen, B. S.; Topsoe, H. *Catal. Lett.* **1992**, *13*, 1-8.
24. Vasudevan, S.; Thomas, J. M.; Wright, C. J.; Sampson, C. *J. Chem. Soc. Chem. Commun.* **1982**, 418-419.
25. Polz, J.; Zeilinger, H.; Muller, B.; Knozinger, H. *J. Catal.* **1989**, *120*, 22-28.
26. Topsoe, N. Y.; Topsoe, H. *J. Catal.* **1993**, *139*, 641-651.
27. LaCroix, M.; Yuan, S.; Breysse, M.; Doremieux-Morin, C.; Fraissard, J. *J. Catal.* **1992**, *138*, 409-412.
28. Rakowski DuBois, M.; VanDerveer, M. C.; DuBois, D. L.; Haltiwanger, R. C.; Miller, W. K. *J. Am. Chem. Soc*, **1980**, *102*, 7456.
29. Rakowski DuBois, M.; DuBois, D. L.; VanDerveer, M. C.; Haltiwanger, R. C. *Inorg. Chem.* **1981**, *20*, 3064, and unpublished results.

30. Bruce, A. E.; Tyler, D. R. *Inorg. Chem.* **1984**, *23*, 3433.
31. DuBois, D. L.; Miller, W. K.; Rakowski DuBois, M. *J. Am. Chem. Soc.* **1981**, *103*, 3429.
32. Casewit, C. J.; Coons, D. E.; Wright, L. L.; Miller, W. K.; Rakowski DuBois, M. *Organometallics* **1986**, *5*, 951.
33. Birnbaum, J.; Godziela, G.; Maciejewski, M.; Tonker, T. L.; Haltiwanger, R. C.; Rakowski DuBois, M. *Organometallics*, **1990**, *9*, 394-401.
34. Lopez, L. L.; Bernatis, P.; Birnbaum, J.; Haltiwanger, R. C.; Rakowski DuBois, M. *Organometallics* **1992**, *11*, 2424-2435.
35. Casewit, C. J.; Haltiwanger, R. C.; Noordik, J.; Rakowski DuBois, M. *Organometallics* **1985**, *4*, 119-129.
36. Lopez, L. L.; Bernatis, P.; Birnbaum, J.; Haltiwanger, R. C.; Rakowski DuBois, M. *Organometallics* (Correction) **1992**, *11*, 3928.
37. James, B. R. in *Comprehensive Organometallic Chemistry*; Wilkinson, G.; Stone, F. G. A.; Abel, E. W., Eds.; Pergamon: New York, 1982; Vol. 8, Chap. 51, pp. 285-302.
38. Birnbaum, J.; Haltiwanger, R. C.; Bernatis, P.; Teachout, C.; Parker, K.; Rakowski DuBois, M. *Organometallics* **1991**, *10*, 1779-1786.
39. Bernatis, P.; Haltiwanger, R. C.; Rakowski DuBois, M. *Organometallics* **1992**, *11*, 2435-2443.
40. Bernatis, P.; Laurie, J. C. V.; Rakowski DuBois, M. *Organometallics* **1990**, *9*, 1607-1617.
41. Laurie, J. C. V.; Duncan, L.; Haltiwanger, R. C.; Weberg, R. T.; Rakowski DuBois, M. *J. Am. Chem. Soc.* **1986**, *108*, 6234-6241.
42. Coons, D. E.; Haltiwanger, R. C.; Rakowski DuBois, M. *Organometallics* **1987**, *6*, 2417-2425.
43. Gabay, J.; Dietz, S.; Bernatis, P.; Rakowski DuBois, M. *Organometallics* **1993**, *12*, 3630-3635.
44. Tucker, D. S.; Dietz, S.; Parker, K. G.; Carperos, V.; Gabay, J.; Noll, B.; Rakowski DuBois, M. *Organometallics* **1995**, *14*, 4325-4333.
45. Herrmann, W. A.; Herdtweck, E.; Floel, M.; Kulpe, J.; Kusthardt, N.; Okuda, J. *Polyhedron* **1987**, *6*, 1165.
46. Weldon, M. K.; Napier, M. E.; Wiegand, B. C.; Friend, C. M.; Uvdal, P. *J. Am. Chem. Soc.* **1994**, *116*, 8328.
47. Tahmassebi, S. K.; Mayer, J. M. *Organometallics* **1995**, *14*, 1039-1043, and references within.
48. Wachter, J. *J. Coord. Chem.* **1987**, *15*, 219-236, and references within.
49. Rauchfuss, T. B.; Rodgers, D. P. S.; Wilson, S. R. *J. Am. Chem. Soc.* **1986**, *108*, 3114-3115.
50. Herrmann, W. A. *Angew. Chem, Int. Ed. Engl.* **1988**, *27*, 1297.
51. Houser, E. J.; Dev, S.; Ogilvy, A. E.; Rauchfuss, T. B.; Wilson, S. R. *Organometallics* **1993**, *12*, 4678-4681.
52. Ogino, H.; Tobita, H.; Inomota, S.; Shimoi, M. *J. Chem. Soc., Chem. Commun.* **1988**, 586.
53. Brunner, H.; Merz, A.; Pfauntsch, J.; Serhadli, O.; Wachter, J.; Ziegler, M. L. *Inorg. Chem.* **1988**, *27*, 2055-2058.
54. Brunner, H.; Meier, W.; Wachter, J.; Guggolz, E.; Zahn, T.; Ziegler, M. L. *Organometallics* **1982**, *1*, 1107.
55. Douglas, B.; McDaniel, D.; Alexander, J. *Concepts and Models of Inorganic Chemistry*, 3rd Ed.; John Wiley: New York, N.Y.,1994; pp. 225-227.

56. Halbert, T. R.; Hutchings, L. L.; Rhodes, R.; Stiefel, E. I. *J. Am. Chem. Soc.* **1986**, *108*, 6437-6438.
57. Muller, A.; Nolte, W. O.; Krebs, B. *Inorg. Chem*. **1980**, *19*, 2835.
58. Fenske, D.; Czeska, B.; Schumacher, C.; Schmidt, R. E.; Dehnicke, K. *Z. Anorg. Allg. Chem*. **1985**, *520*, 7.
59. Coucouvanis, D.; Hadjikyriacou, A. *Inorg. Chem*. **1987**, *26*, 1.
60. Young, C. G.; Kocaba, T. O.; Yan, X. F.; Tiekink, E. R. T.; Wei, L.; Murray III, H. H.; Coyle, C. L.; Stiefel, E. I. *Inorg. Chem.* **1994**, *33*, 6252-6260.
61. Murray, J. J.; Wei, L.; Sherman, S. E.; Greaney, M. A.; Eriksen, K. A.; Carstensen, B.; Halbert, T. R.; Stiefel, E. I. *Inorg. Chem*. **1995**, *34*, 841.

Chapter 17

New Aspects of Heterometallic Copper (Silver) Cluster Compounds Involving Sulfido Ligands

Xiantao Wu, Qun Huang, Quanming Wang, Tianlu Sheng, and Jiaxi Lu

State Key Laboratory of Structural Chemistry, Fujian Institute of Research on the Structure of Matter, Chinese Academy of Sciences, Fuzhou, Fujian 350002, China

The reactivity of sulfido ligands with different coordination configurations are systematically introduced leading to "unit construction", as a convenient method for designed synthesis of some heterometallic copper(silver)-sulfur cluster compounds. Heterometallic copper(silver)-sulfur cluster crystals with acentric space groups as potential precursors of crystal materials with non-linear physical properties are discussed. Discussions of synthetic methods and crystal structures for larger M-Cu-S cluster derivatives containing a square-type unit $\{M_4Cu_4S_{12}\}$ (M = Mo, W) are also provided. Additionally, polymeric M-Ag-S cluster complexes and the coexistence of polymeric cluster anions and hydrogen-bond supramolecular structure cations are rudimentarily explored.

In recent years, the synthetic and structural chemistry of heterometallic copper (silver)-sulfur cluster compounds has developed rapidly (*1*). There are at least three factors stimulating this rapid development; i) the biological Mo-Cu antagonism (*2-3*) and the desire to produce analogues of active sites of metal enzymes (*4*); ii) the rich structural variety of complexes of this kind (*1-4*); iii) the nonlinear optical properties of transition metal sulfido cluster compounds (*5-8*).

"Unit Construction": Designed Synthesis of Some Heterometallic Copper(Silver)-Sulfur Cluster Compounds (*9*)

The Coordination Configuration of the Sulfido Ligand and Constructive Reactivity. "Unit construction" is a convenient method for the rational synthesis of transition metal clusters using reactive fragments as building blocks. It has been shown that the "unit", which contains sulfido ligand(s) with lone-pair electrons and unsaturated coordination, may add to another metal complex that has unsaturated coordination positions or easily displaceable ligand(s). In table I, the relationship between the six coordination configurations for the sulfido ligands is concisely described. Their constructive reactivity is summarized below.

0097–6156/96/0653–0282$15.00/0

1). Terminal sulfido ligand (term-S). A typical example is MoS_4^{2-}. The bond between the Mo atom and the sulfido ligand is a double bond. The terminal sulfido ligand has potent constructive value, because it is only coordinated by one metal atom and it contains two lone-pairs of electrons.

2). μ_2-Bridging sulfido ligand (μ_2-S). This type of coordination configuration can be classified into three subtypes [(a), (b), (c) below].

Table I. The relationship between coordination configuration of the sulfido ligands and constructive reactivity. (Reproduced with permission from reference 9. Copyright 1994 Plenum.)

No.	Type of coordination	Coordination configuration	No. of lone-pair electrons	C. N. of S atoms	Assembly activity	Example	Ref.
1	term-S	M=S	4	1		MoS_4^{2-}	20
2	μ_2-S	(a) 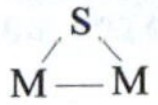 S M—M	≤ 4	2	↑	$Pt_2S(PPh_3)_4$	21
		(b) 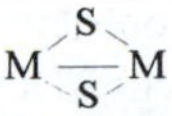 M S—S M	≤ 4	2		$[Mo_2O_2S_2(S_2)\cdot(S_4)]^{2-}$	22
		(c) 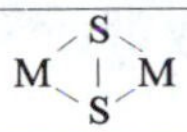 M S | S M	2-3	3		$Fe_2S_2(CO)_6$	12
3	μ_3-S	(a)  M S M M	2	3		$[Mo_3S_4]^{4+}$	13
		(b)  S M M M S	2	3		$Co_3S_2(C_5H_5)_3$	15
		(c) 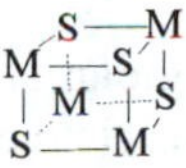 S—M M S S M S S—M	2	3		$[Mo_3CuS_4]^{5+}$	16
		(d) 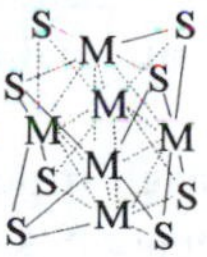 S M S S M S M M S M S S M S	2	3	increase	$Mo_6S_8(PEt_3)_6$	17
4	μ_4-S	(a) 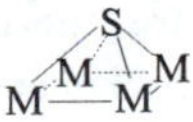 S M M M—M	2	4		$Co_4S_2(CO)_{10}$	23
		(b) 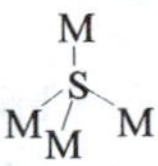 M S M M M	0	4	No	$Fe_4S(CO)_{12}\cdot(SR)_2$	24
5	μ_5-S	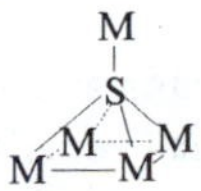 M S M M M—M	0	5	activity	$Os_5(CO)_{15}(\mu_5\text{-}S)[W(CO)_4PPh_3]$	18
6	μ_6-S	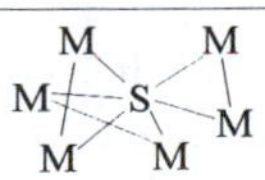 M M M S M M M	0	6		$[(\mu_6\text{-}S)Cu_6S_6\cdot(S_2)_6Mo_6O_6]^{2-}$	19

2a). μ_2-S(a). The sulfido ligand bridges two metal atoms that have metal-metal bonding to form M_2S triangular configuration (the so-called M_2S island), which is the fundamental structural fragment of a cluster. Because of d-p bonding involving the metal atoms and the sulfido atom, the number of lone-pair electrons on the sulfido atom is equal to or less than four. Several complexes have this type of M_2S island fragment, but there is no bond between the metal atoms, such as $[Mo_2O_2S(S_2)_4]^{2-}$ (*10*).

2b). μ_2-S(b). Two sulfido ligands bridge two metal atoms to form a M_2S_2 configuration. This configuration is common, as in Roussins Red salt (*11*). It can also be regarded as two M_2S triangles sharing a M-M edge. Because the d-p conjugation may exist, the number of lone pair electrons is equal to or less than four.

2c). μ-S_2(disulfide). In this ligand there is a bond between the two sulfur atoms (*12*). Each sulfur atom is similar to a μ_3-bridging sulfido ligand. It can have two or three lone-pair electrons. The S-S bond is easily broken and its reactivity is very similar to the μ_2-S ligand(b).

3). μ_3-Bridging sulfido ligand. This sulfido ligand contributes four electrons to the three Mo-μ_3-S covalent bonds, and reduces its reactivity. The remaining one pair of lone-pair electrons completes the tetrahedral geometry of the sulfido-ligand. Because of the stability of this structural unit, it exists widely in nature. The μ_3-sulfido atom can be classified into four subtypes [(a), (b), (c), (d) below].

3a). μ_3-S(a): Typical examples are the two well-known clusters $Mo_3S_4^{4+}$ (*13*) and $W_3S_4^{4+}$ (*14*), both of which contain three μ_2-bridging sulfido atoms and one μ_3 -bridging sulfido atom.

3b). μ_3-S(b): In the cluster core Co_3S_2 (*15*), the two μ_3-S atoms cap the Co_3 triangle which has strong interaction among the three Co atoms. This type of cluster system is very stable and easily obtained.

3c). μ_3-S(c): This type of ligand caps a triangular group of four metal atoms and is commonly found in cubane-like clusters. In these clusters the four metal atoms and the four sulfido ligands arrange in an advantageous cubane-like geometry. Each sulfido ligand simultaneously bonds to three metal atoms, and each metal atom also simultaneously bonds to three sulfido ligands.

3d). μ_3-S(d): The eight sulfido ligands cap the triangular plane of the octahedral skeleton formed by six metal atoms. This configuration permits each sulfido ligand to bond to three metal atoms simultaneously. Furthermore, there exists strong metal-metal interactions. It is apparent that this type of clusters is very stable. For example, the recently reported cluster $[Mo_6S_8(PEt_3)_6]$ (*17*), contains the core structural unit of superconductive Chevrel phases.

4). μ_4-bridging sulfido ligands. This ligand serves as a bridge across four metal atoms. It can be divided into two different subtypes according to the coordination geometry.

4a). μ_4-S(a). This ligand acts as a four-electron donor and is located above the plane of four metal atoms. This type of sulfido atom has some reactivity because it contains one lone-pair electrons.

4b). μ_4-S(b). This type of sulfido ligand has a tetrahedral coordination by four metal atoms. All six electrons are used in the metal-sulfur bonding.

5). μ_5-Bridging sulfido ligand: This type of sulfido ligand serves as a six electron donor. The cluster contains the M_5S structural unit. Up to now only two examples (*18*), $Ru_4(CO)_7(\mu\text{-}CO)_2(PMe_2Ph)_2(\mu_4\text{-}S)(\mu_5\text{-}S)[W(CO)_4PMe_2Ph]$ and $Os_5(CO)_{15}$ -$(\mu_5\text{-}S)[W(CO)_4PPh_3]$, have been reported. The μ_5-sulfido ligand has a tetragonal-pyramidal coordination.

6). μ_6-Bridging sulfido ligand: This type of sulfido ligand also acts as a six electron donor. The cluster contains a M_6S structural unit. The unique example of this cluster is $[(\mu_6\text{-}S)Cu_6S_6(S_2)_6Mo_6O_6]^{2-}$ (*19*) which was recently synthesized by our group. The cluster has the core $[(\mu_6\text{-}S)Cu_6]$, in which the sulfido ligand is located in the center of the six copper atoms. Obviously, all six electrons of the sulfido ligand are used in forming the six metal-sulfur bonds.

In summary, sulfido ligands that contain lone pair of electrons are coordinatively "unsaturated" and have the potential to react with other complexes to form new and larger clusters. This constructive reactivity decreases with the decreasing number of lone pair electrons and increases with decreasing coordination number. That is, the reactivity tends to decrease in the order, term-S>μ_2-S>μ_3-S>μ_4-S.

Application of "Unit Construction" in the Designed Synthesis of Some Heterometallic Copper(Silver)-Sulfur Cluster Compounds. A transition metal complex that contains an unsaturated sulfido ligands can be expected to have constructive reactivity. It may react with other metal complexes that have the following three conditions: i) sulfidophilicity, ii) appropriate steric configuration, and iii) appropriate oxidation state, usually a low oxidation state.

In the past several years, using this methodology, our group have successfully synthesized a series of heterometallic copper(silver)-sulfur cluster compounds.

1. Cubane-like cores $\{M_3CuS_4\}^{5+}$ (M=Mo, W) (*16, 25-27*):

$$\{M_3S_4\}(dtp)_3(RCOO)(L) + CuI \longrightarrow \{M_3CuS_4\}(I)(dtp)_3(RCOO)(L)$$

where dtp = $S_2P(OEt)_2^-$; R = CH_3, C_2H_5 and CCl_3; L = H_2O, CH_3CN, PhCN, C_5H_5N, DMF and DMSO.

2. [2+1] and [3+1] additions:

$$[\{M_2S_4\}(edt)_2]^{2-} + CuPPh_3{}^{1+} \longrightarrow \{M_2CuS_4\}(edt)_2(PPh_3)^{1-} \quad (28)$$

$$\{M_2CuS_4\}(edt)_2(PPh_3)^{1-} + CuPPh_3{}^{1+} \longrightarrow \{M_2Cu_2S_4\}(edt)_2(PPh_3)_2 \quad (29)$$

where edt = $(SCH_2CH_2S)^{2-}$.

3. Linear heterotrimetallic clusters:

$[Et_4N][(CN)\{CuS_2MS_2Ag\}(PPh_3)]$ (M = Mo, W) (*30*):

$$[MS_4Cu(CN)]^{2-} + AgPPh_3{}^+ \longrightarrow [(CN)\{CuS_2MS_2Ag\}(PPh_3)]^+$$

$[Et_4N][Cl_2\{FeS_2MS_2M'\}(PPh_3)]$ (M = Mo, W; M' = Ag, Cu) (*31-32*):

$$[MS_4FeCl_2]^{2-} + M'PPh_3{}^+ \longrightarrow [Cl_2\{FeS_2MS_2M'\}(PPh_3)]^+$$

4. Chiral heterotrimetallic clusters: Two butterfly type heterotrimetallic clusters $[Et_4N][\{OMS_3AgCu\}(PPh_3)_2(CN)]$ (M=Mo, W) (*33-34*) were obtained by reacting the "active unit" $[OMS_3CuCN]^{2-}$ (M=Mo, W) with $AgPPh_3^+$.

Observing the anion cluster in Figure 1 from the μ_2-S site, we find that the four atoms μ_3-S, Mo, Cu and Ag form a distorted tetrahedron (see Figure 2).

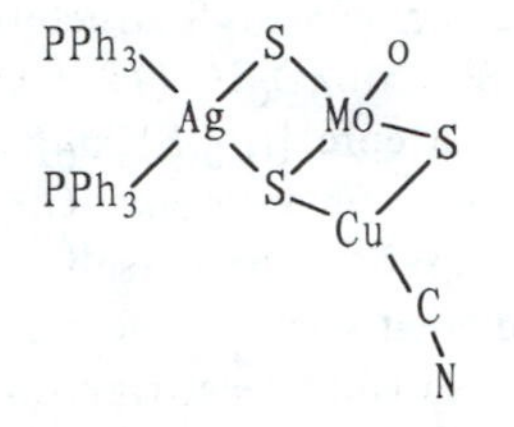

Figure 1. Chemical diagram of $[\{OMS_3AgCu\}(PPh_3)_2(CN)]$

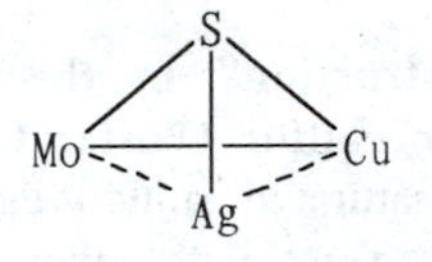

Figure 2. The asymmetry MoCuAg(μ_3-S) tetrahedron framework

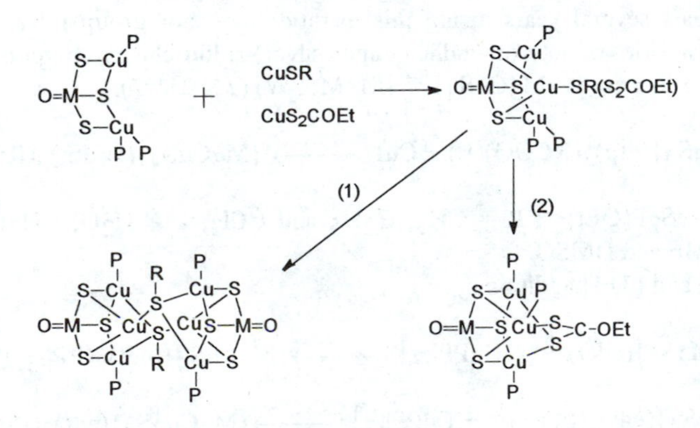

Figure 3. Possible pathways of $[\{MCu_2S_3\}(O)(PPh_3)_3]$ reacting with CuSR or CuS_2COEt. (1) with monothiolato ligand; (2) with 1,1-dithiolato ligand.

5. Double-cubane-like complexes $\{M_2Cu_6S_6(SCME_3)_2\}(O)_2(PPh_3)_4$ (M=Mo, W) (*35-36*) and single-cubane-like complexes $[\{MCu_3S_3(S_2COEt)\}(O)(PPh_3)_3$ (M=Mo, W). Two possible routes to formation are shown in Figure 3.

6. A novel dodecanuclear cluster $[Et_4N]_2[(\mu_6\text{-}S)Cu_6S_6(S_2)_6Mo_6O_6]$ has been obtained (*19*) containing a rare μ_6-S ligand. Its configuration is shown in Figure 4.

Most of the "unit construction" examples given above start from stable units with sulfido ligands, which are important in the unit construction of transition metal clusters with bridging-sulfido ligands.

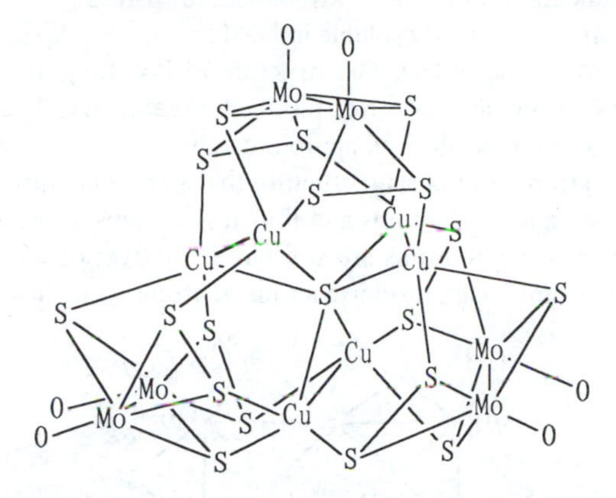

Figure 4. Diagram of anion $[(\mu_6\text{-}S)Cu_6S_6(S_2)_6Mo_6O_6]^{2-}$.

Larger Heterometallic M-Cu-S Clusters Containing a Square-Type Unit $\{M_4Cu_4S_{12}\}$ (M = Mo, W)

It is known that M-Cu-S complexes can show rich structural variety. So far, many M-Cu-S complexes with various structures have been obtained, such as: $[(PPh_3)_2CuS_2MS_2Cu(PPh_3)]$ (linear) (*37*), $[Et_4N][(PPh_3)_2AgS_3MOCu(CN)]$ (butterfly-type) (*33-34*), $[Et_4N][M_2CuS_4(PPh_3)(SCH_2CH_2S)_2]$ (incomplete cubane-like) (*28*), $[MCu_3S_4Cl]$ (cubane-like) (*38*), $\{M_3CuS_4\}$ (cubane-like) (*16, 25-27*), $[M_2Cu_2S_4(PPh_3)_2(SCH_2CH_2S)_2]$(cubane-like) (*29*), $[M_2Cu_4S_6O_2]^{4-}$ (cage) (*39*), $[MS_4(CuL)_4]^{2-}$ (chain) (L = Br, SCN) (*40-41*), $[(H_2O)_9Mo_3S_4CuCuS_4Mo_3\text{-}(H_2O)_9]^{8+}$ (double-cubane-like) (*42*), $[(CuCl)_5Cl_2MS_4]^{4-}$ (double-cubane-like) (*43*), $[MCu_3S_4Cl_3]^{2-}$ (*44*), $[Cu_6S_6M_2(SCMe_3)_2(O)_2(PPh_3)_4]$(double-cubane-like) (*35-36*), $[Et_4N]_2[(\mu_6\text{-}S)Cu_6S_6(S_2)_6Mo_6O_6]$ (*19*), and $[M_2Cu_5S_8(S_2CNMe_2)_3]^{2-}$ (*45*) (M = Mo, W). These complexes are interesting from the structural point of view. Of these complexes, only some contain Cu atoms only coordinated by inorganic sulfur atoms from $MO_nS_{4-n}{}^{2-}$ (n = 0, 1, 2), e.g., $[Bu_4N]_4[Cu_{12}Mo_8S_{32}]$ (*46*). Below, we concisely discuss six recently obtained complexes of this kind with a square-type unit $\{M_4Cu_4S_{12}\}$.

Octanuclear Clusters $[Et_4N]_4[M_4Cu_4S_{12}O_4]$ (M = Mo, 1; W, 2). These two novel compounds were synthesized by reaction of $[Et_4N]_2[MS_4]$, $[Et_4N]_2[MS_2O_2]$, $Cu(NO_3)_2{\cdot}3H_2O$, and KBH_4 (or Et_4NBH_4) (*47*). Cu^+ ions were introduced into the reaction in a special way using KBH_4 (or Et_4NBH_4) to reduce Cu^{2+}; thus, without using other copper-philic ligands, new novel M-Cu-S structures may be obtained. According to the sequence of addition of reaction materials, the syntheses of **1** and **2**

possibly derive from an initially formed $[MS_4]^{2-}/Cu^{1+}$ cluster and the $[MO_2S_2]^{2-}$ anion; however, this initial M-Cu cluster compound has not yet been characterized in our laboratory. The cluster anion can be considered to consist of two butterfly-type $[CuOMS_3Cu]$ fragments bridged by two bidentate $[MOS_3]^{2-}$ groups. The two cluster compounds are only slightly soluble in DMF.

Both clusters are isomorphous. The structure of $[W_4Cu_4S_{12}O_4]^{4-}$ is shown in Figure 5. Each metal atom has a distorted tetrahedral geometry. The configuration of eight metal atoms can be described as approximately square, and these metal atoms together with sulfur atoms form a ring structure that is rare in cluster chemistry. It is noted that there exist eight μ_2-S atoms and four μ_3-S atoms in the anion. According to our past experiments, μ_2-S atoms are still quite reactive; that is to say, the eight μ_2-S atoms can in principle react further with metal atoms to form a larger cluster.

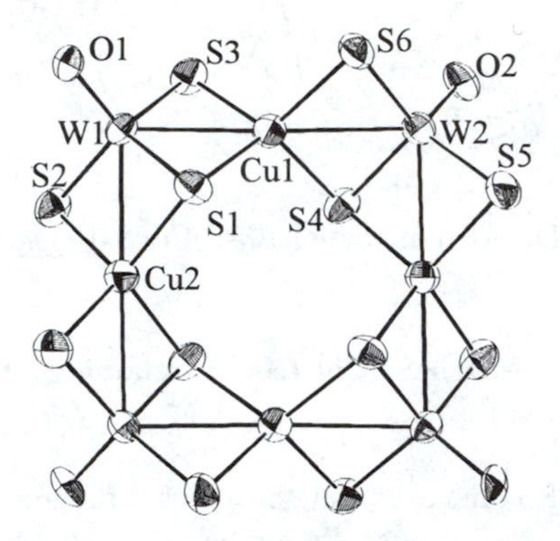

Figure 5. ORTEP drawing of $[W_4Cu_4S_{12}O_4]^{4-}$

Dodecanuclear Clusters $[M_4Cu_4S_{12}O_4(CuTMEN)_4]$ (M = Mo, 3; W, 4; TMEN = N, N, N', N'-Tetramethylethylenediamine). These two dodecanuclear compounds were produced by reaction of $[Et_4N]_2[WOS_3]$, $Cu(NO_3)_2{\cdot}3H_2O$ and TMEN, and by reaction of $[Me_4N]_2[Mo_2O_2S_8]$, $Cu(NO_3)_2{\cdot}3H_2O$, and TMEN, respectively (*47*). Both compounds crystallize in the trigonal space group $P3_221$. It is unexpected that the structure of the octanuclear clusters (**1** and **2**) and that of the dodecanuclear clusters (**3** and **4**) are closely related. It seems that Compounds **3** and **4** might, in principle, be synthesized from Compounds **1** and **2** by using the unit-construction method. However, because of poor solubility of **1** and **2**, this route was not successful. Compounds **3** and **4** are slightly soluble in CH_2Cl_2 and DMF.

Clusters **3** and **4** are isomorphous. The structure of **3** is shown in Figure 6. Each W atom has slightly distorted tetrahedral coordination, Cu1 and Cu2 atoms are each coordinated by four μ_3-S atoms to give a distorted tetrahedral geometry. Cu3 and Cu4 atoms both display highly distorted tetrahedral coordination, with two nitrogen atoms of TMEN and two μ_3-S atoms. The molecule can be considered to consist of one $[W_4Cu_4S_{12}O_4]^{4-}$ (the anion of **1**) unit and four $Cu(TMEN)^{1+}$ groups. These $Cu(TMEN)^{1+}$ groups are bound to eight μ_2-S atoms of $[W_4Cu_4S_{12}O_4]^{4-}$ as capping units on W atoms.

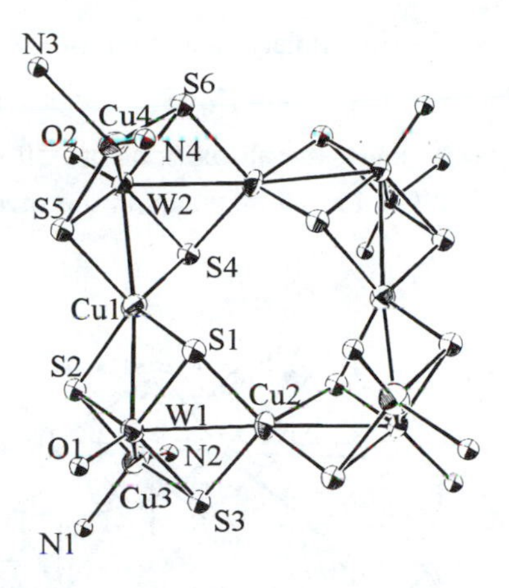

Figure 6. ORTEP drawing of $[W_4Cu_4S_{12}O_4(CuTMEN)_4]$. C atoms are omitted for clarity.

Eicosanuclear Clusters $[Et_4N]_4[M_8Cu_{12}S_{28}O_4]\cdot$DMF (M = Mo, 5; W, 6). The clusters were obtained from reaction of $Cu(NO_3)_2\cdot 3H_2O$, KBH_4, $(NH_4)_2MoS_4$, $(NH_4)_2MoO_2S_2$, and Et_4NBr in DMF solvent (Huang, Q.; Wu, X.-T.; Sheng, T.-L.; Wang, Q.-M. *Polyhedron*, in press). MOS_3^{2-} in the molecule can be thought to derive from the interaction of $MO_2S_2^{2-}$ and MS_4^{2-}. In this reaction system, different sequences of addition and different ratios of MS_4^{2-} to $MO_2S_2^{2-}$ can result in different products; for example, the 1:2 molar ratio of MS_4^{2-} to $MO_2S_2^{2-}$ and the sequence of addition of MS_4^{2-} followed by $MO_2S_2^{2-}$ can be employed to synthesize the octanuclear clusters **1** and **2** .

The Eicosanuclear Clusters are the largest of M-Cu-S compounds. Another Eicosanuclear Cluster $[Bu_4N]_4[Cu_{12}Mo_8S_{32}]$ was obtained via solid-state synthesis at low-heating temperature (about 95°).

The configuration of $[Mo_8Cu_{12}S_{28}O_4]^{4-}$ is shown in Figure 7. The structure consists of Mo atoms in an approximate cubane array and Cu atoms nearly lying along the edges of the cubane between the Mo atoms. As shown in Figure 7, each face of this cubane is composed of a Mo_4Cu_4 octanuclear moiety, similar to the configuration of the square-type cluster $[Et_4N]_4[Mo_4Cu_4S_{12}O_4]$. The X-ray single-crystal analysis shows that sulfur atoms in the anion are in three different environments, namely, μ_2-S, μ_3-S, μ_4-S atoms.

The electronic spectra of Cs_2MoOS_3, $(NH_4)_2MoS_4$, $[Et_4N]_4[Mo_4Cu_4S_{12}O_4]$, **1**; $[Mo_4Cu_4S_{12}O_4(CuTMEN)_4]$, **3**; and $[Et_4N]_4[Mo_8Cu_{12}S_{28}O_4]\cdot$DMF, **5**; are listed in Table II. Except for the absorptions in the range 334 - 420 nm, these spectra exhibit similar absorption bands to those of $[MoOS_3]^{2-}$ in the same regions; additionally, complex **5** also displays similar bands to those of $[MoS_4]^{2-}$. Thus, the principal feature of complexes **1** and **3** can be attributed to the charge-transfer transitions within the $[MoOS_3]^{2-}$ moiety, and that of complex **5** to the charge-transfer transitions within $[MoOS_3]^{2-}$ and $[MoS_4]^{2-}$ moieties. As Cu^{1+} cations are bound to $[Mo_4Cu_4S_{12}O_4]^{4-}$ units, the absorptions at 450 nm are scarcely shifted, but the absorption band at 420 nm in **1** disappears in the spectra of complexes **3** and **5**.

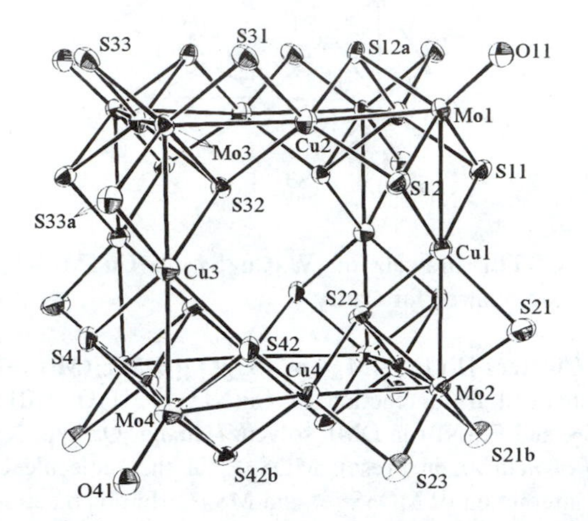

Figure 7. Configuration of $[Mo_8Cu_{12}S_{28}O_4]^{4-}$. (Reproduced with permission from *Polyhedron*, volume 30, number 19, page 3405. Copyright 1996 Elsevier Science.)

Table II. Electronic Spectral Data in DMF

Compd	λ nm ($10^{-3}\varepsilon/dm^3mol^{-1}cm^{-1}$)
$(NH_4)_2MoS_4$[a]	483, 413, 315, 214
Cs_2MoOS_3[a]	441, 385, 320, 255, 215
$[Et_4N]_4[Mo_4Cu_4S_{12}O_4]$	450(11), 420(10), 312(27), 268(23)
$[Mo_4Cu_4S_{12}O_4(CuTMEN)_4]$	450(11), 320(28), 265(27)
$[Et_4N]_4[Mo_8Cu_{12}S_{28}O_4]\cdot DMF$	500(17), 450(24), 330(65), 290(94), 270(83

[a]Data from ref. 2.

One-Dimensional Heterometallic W-Ag-S Polymeric Cluster Complexes

Low-dimensional transition metal chalcogenides have attracted much attention due to their useful properties that arise directly from their structural peculiarity (*48-50*). Although many discrete M-Ag-S cluster complexes have been obtained, such as: $[(PPh_3)_2AgS_2MS_2Ag(PPh_3)\cdot 0.8CH_2Cl_2]$ (linear) (*51*), $[Ag_6S_6M_2(SCMe_3)_2(O)_2(PPh_3)_4]$ (double-cubane-like) (*52*), $[(PPh_3)_4Ag_4W_2S_8]$ (*53*), $[Et_4N][(PPh_3)_2AgS_3MOCu(CN)]$ (butterfly-type) (*33-34*), only three reports concerned the M-Ag-S polymeric complex. Moreover, they all reported only one polymeric configuration (linear chain $[WS_4Ag]_n$). In this section, two new polymeric configurations will be described.

Synthesis of polymeric complexes involving heterometals is still a challenge. In the exploration of polymeric M-Ag-S complexes, the reaction system of $(NH_4)_2WS_4$ and $AgNO_3$ has been employed in our laboratory. The selection of this system is mainly based on two considerations: (1) Self-assembly reactions of $[WS_4Ag]^{1-}$ units are flexible without addition of other soft bases; appropriate small molecules containing N and O atoms or ions added into the reaction can induce self-assembly of $[WS_4Ag]^{1-}$ units and prompt the formation of new polymeric configurations. (2) Assembly reactions are conducive to the combination of metal-sulfur cluster chemistry and supramolecular chemistry. As complementary small molecules contain N and O atoms, and complementary cations are oxygen- and nitrogen-philic, the synergy between polymeric cluster anion, these complementary small molecules (or cations) and solvent not only may urge the formation of new polymeric configurations, but also makes it possible that hydrogen-bonded supramolecular structures form around the polymeric cluster anion. The selection of proper small molecules is worth careful consideration. So far, we have successfully selected some small molecules and ions, and obtained two new M-Ag-S polymeric configurations with complexes simultaneously involving polymeric cluster anions and cationic hydrogen-bonded supramolecular structures.

The Single-Chain Polymeric Cluster Compounds $[MS_4Ag{\cdot}L\text{-}HHistidine{\cdot}DMF]_n$ (M=Mo, W) with Hydrogen-Bond Chain Supramolecular Structure. The polymeric complexes $[MS_4Ag{\cdot}L\text{-}HHistidine{\cdot}DMF]_n$ (M=Mo, **7**; W, **8**) were obtained by reaction of $[NH_4]_2MS_4$ (M=Mo, W) with $AgNO_3$ and L-Histidine in DMF/H_2O solvent (Wang, Q.-M.; Wu, X.-T.; Huang, Q.; Sheng, T.-L. unpublished results). Both compounds crystallize in the triclinic space group *P*1. The structure of a portion of compound **8** is shown in Figure 8. The anion $[WS_4Ag]^-$ extends to form a

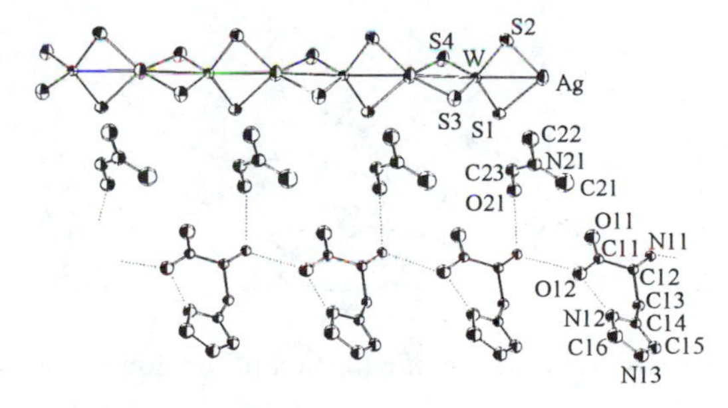

Figure 8. ORTEP drawing of a portion of $[WS_4Ag{\cdot}L\text{-}HHistidine{\cdot}DMF]_n$

polymeric linear chain. Such linear chain structure in $(PPh_4)Ag(MoS_4)$ was first identified by A. Müller from its resonance-Raman spectra in 1985 (*54-55*), and in 1993 the crystal structure of this linear chain anion in complexes $[AgMoS_4{\cdot}\alpha\text{-}PyH]_n$,

$[AgMoS_4 \cdot \gamma\text{-MePyH}]_n$ and $[AgWS_4 \cdot \gamma\text{-MePyH}]_n$ was reported (*56*), however, their cations are discrete.

As shown in Figure 8, $[\text{L-HHis}]^+$ cations (formed by the protonation of L-His) interact with each other as well as DMF molecules through hydrogen bonds to form an infinite one-dimensional chain supramolecular structure. These hydrogen-bond supramolecular cation chains run parallel to the linear $[MS_4Ag]^{1-}$ anion chain. To some extent, the formation of these hydrogen-bond supramolecular cation chains is based on the template function of the linear anion chains; of course, the hydrogen-bond cation chains are also conducive to the stability of the complex.

The Single-Chain and Double-Chain Polymeric Cluster Complexes $[WS_4Ag \cdot NH_3C(CH_2OH)_3 \cdot 2DMF]_n$ (Single-Chain) and $[WS_4Ag \cdot NH_3C(CH_2OH)_3 \cdot H_2O]_n$ (Double-Chain). The single-chain complex $[WS_4Ag \cdot NH_3C(CH_2OH)_3 \cdot 2DMF]_n$, **9**, was prepared from reaction of ammonium tetrathiotungstate, silver nitrate and tris(hydroxy-methyl)-amino-methane in 1:2:1 molar ratio in DMF solution. However, the double-chain polymeric complex $[WS_4Ag \cdot NH_3C(CH_2OH)_3 \cdot H_2O]_n$, **10**, was obtained from addition of a little amount of water into the recrystallization solution of the single-chain complex in ethanol (*57*). It is guessed that water plays a crucial role in the transformation of the single-chain polymeric complex into the double-chain polymeric complex; in the presence of water, the single-chain complex in ethanol solution is probably decomposed into $[S_2WS_2Ag]$ fragments, which self-assemble into the double-chain polymeric complex.

The structure determination shows that the single-chain complex **9** is composed of a polymeric single-chain of $[S_2WS_2Ag]^{1-}$ anions, similar to the polymeric configuration of the anion of **8**.

As in the case of **8** and **9**, the $[S_2WS_2Ag]^{1-}$ anions also act as cyclic units of the double-chain complex **10**, but their linking is different from that in complexes **8** and **9**. The anion structure of **10** can be viewed as two zigzag -SWSAg- chains linked by μ_2-S atoms. The ORTEP diagram of a portion of the polymeric complex **10** is shown in Figure 9.

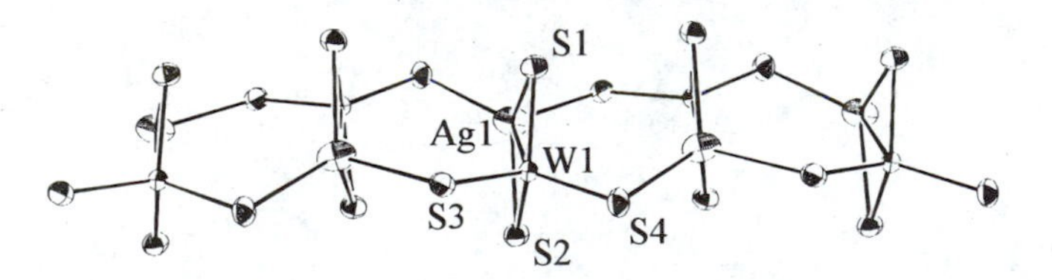

Figure 9. ORTEP drawing of a portion of the double-chain anion.

A supramolecular hydrogen-bond network structure exits in the complex. Consistent with the synthetic requirement, water molecules play a crucial role in the network hydrogen-bond structure. Through hydrogen bonds, H_2O groups bridge three neighboring $NH_3C(CH_2OH)_3^+$ cations. Similarly, a $NH_3C(CH_2OH)_3^+$ cation, as a μ_2 bridge, bridges two neighboring $NH_3C(CH_2OH)_3^+$ cations. This kind of hydrogen bond structure is conducive to the stability of the complex **10**, and may be one of the main factors in transformation of complex **9** into complex **10**. The polymeric double-chain cluster anions are arrayed between two cation hydrogen-bond "walls".

One-Dimensional Polymeric Clusters $[W_4Ag_5S_{16}{\cdot}M(DMF)_8]_n$ (M = Nd and La) with Nonanuclear Fragments as Cyclic Units. These complexes were produced from reaction of ammonium tetrathiotungstate, silver nitrate, lanthanoid nitrate, and p-nitro-aniline (2:4:1:2) in DMF and CH_3CN (*58*). In the reaction, trivalent cations Nd(III) and La(III) acted induce the formation of $[(W_4Ag_5S_{16})^{3-}]_n$ with trivalent anion cluster fragments as cyclic units.

The configuration of a portion of the anion of $[W_4Ag_5S_{16}{\cdot}Nd(DMF)_8]_n$, **11**, is shown in Figure 10. This one-dimensional polymeric anion is propagated by cell translation along the crystallographic C axis, and can be viewed as octanuclear square-type cyclic cluster fragments $[W_4Ag_4S_{16}]^{4-}$ linked through Ag^+ cations. A similar $[W_4Cu_4S_{12}O_4]^{4-}$ square-type configuration has been shown in the cluster $[Et_4N]_4[W_4Cu_4S_{12}O_4]$ (*47*). In a $[W_4Ag_4S_{16}]^{4-}$ fragment, W atoms are arrayed in an approximate square; and the Ag atoms nearly lie along the edges of the square between the W atoms. There exists three kinds of sulfur atoms in the anion, namely, terminal sulfur atoms, μ_2-S atoms, and μ_3-S atoms.

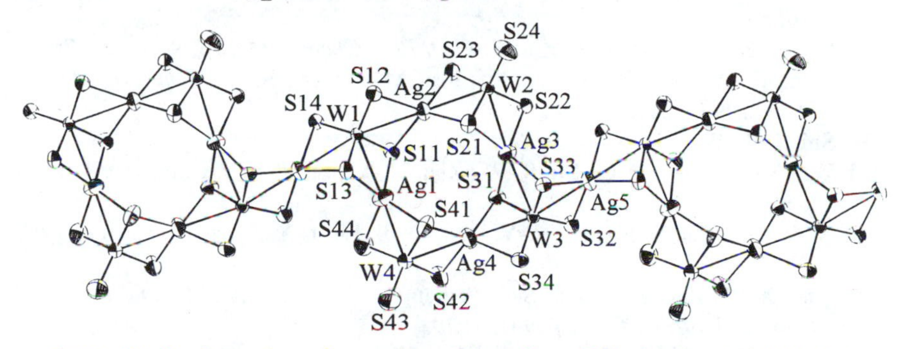

Figure 10. Configuration of a portion of the anion of $[W_4Ag_5S_{16}{\cdot}Nd(DMF)_8]_n$ (Reproduced with permission from reference 58. Copyright 1996 VCH.)

Conclusion

We and others have recognized the importance of the sulfido ligand with lone-pair electrons and unsaturated coordination numbers in the synthesis of transition metal bridging-sulfido clusters. Novel clusters may be obtained by using appropriate active units and appropriate metal complexes. Therefore, the unit construction method can be used to design the synthesis of new clusters. This method has already been successful in some designed rational syntheses of transition metal clusters with bridging sulfido ligands and should be an important method for future use.

There are many main group elements that can act as bridging ligands in transition metal clusters, including Se, Te, P, As, C, N, O, F, Cl etc. Because of the similarity of Se, Te, and S, the discussion about S-bridging clusters should be extendible to the rational syntheses of transition metal Se- or Te-bridging clusters, but the synthetic rules for transition metal clusters containing bridging P, As, C, N, O, F, Cl atom have not yet been defined. It is important to identify the relationship between coordination configuration and electronic factors of these bridging atoms and then to recognize the relationship between properties of the bridging atom and the constructive reactivity.

Acknowledgment. This research was supported by grants from the State Key Laboratory of Structural Chemistry, Fujian Institute of Research on Structure of Matter, Chinese Academy of Sciences, the National Science Foundations of China, and the Provincial Science Foundations of Fujian.

Literature Cited

(1) Hollway, C. E.; Melnik, M. *Reviews in Inorg. Chem.* **1993**, *Vol. 13 (No. 4)*, 233.
(2) Müller, A.; Diemann, E.; Jösters, R.; Bogge, H. *Angew. Chem. Int. Ed. Engl.* **1981**, *20*, 934.
(3) Sarkar, S.; Mishre, S. B. S. *Coord. Chem. Rev.* **1984**, *59*, 239.
(4) Du, S.-W.; Zhu, N.-Y.; Chen, P.-C.; Wu, X.-T. *Angew. Chem. Int. Ed. Engl.* **1992**, *31(No.8)*, 1085.
(5) Shi, S.; Ji, W.; Tang, S.-H.; Lang, J.-P.; Xin, X.-Q. *J. Am. Chem. Soc.* **1994**, *116*, 3615.
(6) Shi, S.; Ji, W.; Lang, J.-P.; Xin, X.-Q. *J. Phy. Chem.* **1994**, *98*(No.*14*), 3570.
(7) Hou, H.-W.; Xin, X.-Q.; Liu, J.; Chen, M.-Q.; Shi, S. *J. Chem. Soc., Dalton Trans.* **1994**, 3211.
(8) Shi, S.; Ji, W.; Xin, X.-Q. *J. Phys. Chem.* **1995**, *99*, 894.
(9) Wu, X.-T.; Chen, P.-C.; Du, S.-W.; Zhu, N.-Y.; Lu, J.-X. *J. Clust. Sci.* **1994**, *5*, 265.
(10) Wu, X.-T.; Lu, S.-F.; Zhu, N.-Y.; Wu, Q.-J.; Lu, J.-X. *Inorg. Chim. Acta* **1987**, *133*, 43.
(11) Lin, X.-T.; Zheng, A.; Lin, S.-H.; Huang, J.-L.; Lu, J.-X. *JIEGOU HUAXUE* (*Chinese J. Struct. Chem.*) **1982**, *1*, 79.
(12) Wei, C. H.; Dahl, L. F. *Inorg. Chem.* **1965**, *4*, 1
(13) Lin, X-T.; Lin, Y.-H.; Huang, J.-L.; Huang, J.-Q. *KEXUE TONGBAO* **1986**, 509.
(14) Zhan, H.-Q.; Zheng, Y.-F.; Wu, X.-T.; Lu, J.-X. *J. Mol. Struct.* **1989**, *196*, 241.
(15) Frisch, P. D.; Dahl, L. F. *J. Am. Chem. Soc.* **1972**, *94*, 5082.
(16) Lu, S.-F.; Zhu, N.-Y.; Wu, X.-T.; Wu, Q.-J.; Lu, J.-X. *J. Mol. Struct.* **1989**, *197*,15.
(17) Saito, T.; Yamamoto, W.; Yamagata, T.; Tmoto, H. *J. Am. Chem. Soc.* **1988**, *110*, 1646.
(18) Adams, R. D.; Babin, J. E.; Natarejan, K.; Tasi, M.; Wang, T. G. *Inorg. Chem.* **1987**, *26*, 3708.
(19) Wu, X.-T.; Wang, B.; Zhang, Y.-F.; Lu, J.-X. *JIEGOU HUAXUE* (*Chinese J. Struct. Chem.*) **1988**, *7*, 47.
(20) Müller, A.; Jorgensen, C. K.; Diemann, E. Z. *Anorg. Allg. Chem.* **1972**, *38*, 391.

(21) Chatt, J.; Mingos, D. M. P. *J. Chem. Soc. A.* **1970**, 1243.
(22) Huang, L.-R.; Wang, B.; Wu, X.-T. *HUAXUE TONGBAO* **1984**, *7*, 15.
(23) Marko, L.; Bor, G.; Almasy, G. *Chem. Ber.* **1961**, *94*, 847.
(24) Coleman, J. M.; Wojcicki, A.; Pollick, P. J.; Dahl, L. F. *Inorg. Chem.* **1967**, *6*, 1236.
(25) Zhan, H.-Q.; Zheng, Y.-F.; Wu, X.-T.; Lu, J.-X. *Inorg. Chim. Acta* **1989**, *156*, 272.
(26) Zheng, Y.-F.; Zhan, H.-Q.; Wu, X.-T.; Lu, J.-X. *Trans. Met. Chem.* **1989**, *14*, 161.
(27) Wu, X.-T.; Zhan, H.-Q.; Zheng, Y.-F.; Lu, J.-X. *J. Mol. Struct.* **1989**, *197*, 33.
(28) Zhu, N.-Y.; Zheng, Y.-F.; Wu, X.-T. *Inorg. Chem.* **1990**, *29*, 2705.
(29) Zhu, N.-Y.; Zheng, Y.-F.; Wu, X.-T. *J. Chem. Soc. Commun.* **1990**, 780.
(30) Du, S.-W.; Zhu, N.-Y.; Chen, P.-C.; Wu, X.-T.; Lu, J.-X. *Polyhedron* **1992**, *11*, 109.
(31) Sheng, T.-L.; Du, S.-W.; Wu, X.-T. *Polyhedron* **1993**, *12*, 111.
(32) Sheng, T.-L.; Du, S.-W.; Wu, X.-T. *J. Clust. Sci.* **1993**, *4*, 89.
(33) Zhu, N.-Y.; Du, S.-W.; Chen, P.-C.; Wu, X.-T.; Lu, J.-X. *J. Coord. Chem.* **1992**, *26*, 35.
(34) Du, S.-W.; Zhu, N.-Y.; Chen, P.-C.; Wu, X.-T.; Lu, J.-X. *J. Chem. Soc. Dalton Trans.* **1992**, 339.
(35) Du, S.-W.; Zhu, N.-Y.; Chen, P.-C.; Wu, X.-T. *Angew. Chem. Int. Ed. Engl.* **1992**, *31*, 1085.
(36) Zhu, N.-Y.; Du, S.-W.; Chen, P.-C.; Wu, X.-T. *J. Clust. Sci.* **1992**, *3*, 201.
(37) Müller, A.; Bögge, H.; Tölle, H. G.; Jostes, R.; Schimanski, U.; Dartmann, M. *Angew. Chem. Int. Ed. Engl.* **1980**, *19*, 654.
(38) Müller, A.; Bögge, H.; Schimanski, U. *J. Chem. Soc., Chem. Commun.* **1980**, 91.
(39) Doherty, R.; Hubbard, C. R.; Mighell, A. D.; Siedle, A. R.; Stewart, J. *Inorg. Chem.* **1979**, *8*, 2991.
(40) Manoli, J. M.; Potvin, C.; Secheresse, F.; Marzak, S. *J. Chem. Soc., Chem. Commun.* **1986**, 1557.
(41) Nicholson, J. R.; Flood, A. C.; Garner, C. D.; Clegg, W. *J. Chem. Soc., Chem. Commun.* **1983**, 1179.
(42) Shibahara, T.; Akashi, H.; Kuroya, H. *J. Am. Chem. Soc.* **1988**, *110*, 3313.
(43) Secheresse, F.; Manoli, J. M.; Potvin, C.; Marzak, S. *J. Chem. Soc., Dalton Trans.* **1988**, 3055.
(44) Potvin, C.; Manoli, J. M.; Salis, M.; Secheresse, F. *Inorg. Chim. Acta* **1984**, *83*, L19.
(45) Lei, X.-J.; Huang, Z.-Y.; Liu, Q.-T.; Hong, M.-C.; Liu, H.-Q. *Inorg. Chem.* **1989**, *28*, 4302.
(46) Li, J.-G.; Xin, X.-Q.; Zhou, Z.-Y.; Yu, K.-B. *J. Chem. Soc., Chem. Commun.* **1991**, 249.
(47) Huang, Q.; Wu, X.-T.; Wang, Q.-M.; Sheng, T.-L.; Lu, J.-X. *Inorg. Chem.* **1996**, *35*, 893.
(48) Murphy, D. W.; Christian, P. A. *Science* **1979**, *205*, 651.

(49) Whittingham, M. S. *Prog. Solid State Chem.* **1978**, *12*, 41.
(50) Chianelli, R. R.; Pecoraro, T. A.; Halbert, T. R.; Pan, H. W.; Stiefel, E. I. *J. Catal.* **1984**, *86*, 226.
(51) Müller, A.; Bögge, H.; Koniger-Ahlborn, E. *Z. Naturforsh* **1979**, *34b*, 1698.
(52) Du, S.-W.; Wu, X.-T. *J. Coord. Chem.* **1993**, *Vol. 30*, 183.
(53) Müller, A.; Bögge, H.; Koniger-Ahlborn, E. *J. Chem. Soc., Chem. Commun.* **1978**, 739.
(54) Müller, A.; Jaegermann, W.; Hellmann, W. *J. Mol. Struct.* **1983**, *100*, 559.
(55) Müller, A.; Hellmann, W. *Spectrochim. Acta* **1985**, *41A*, 359.
(56) Lang, J.-P.; Li, J.-G.; Bao, S.; Xin, X.-Q. *Polyhedron* **1993**, *Vol. 12, No. 7*, 801.
(57) Huang, Q.; Wu, X.-T.; Sheng, T.-L.; Wang, Q.-M. *Inorg. Chem.* **1995**, *34*, 4931.
(58) Huang, Q.; Wu, X.-T.; Wang, Q.-M.; Sheng, T.-L.; Lu, J.-X. *Angew. Chem.* **1996**, *108*, 985; *Angew. Chem. Int. Ed. Engl.* **1996**, *35*, 868.

Chapter 18

Sulfur-Containing Macrocyclic Ligands as Reagents for Metal-Ion Discriminaton

A. M. Groth[1], L. F. Lindoy[1,3], G. V. Meehan[1], G. F. Swiegers[2], and S. B. Wild[2]

[1]School of Molecular Sciences, James Cook University, Townsville, 4811 Queensland, Australia
[2]Research School of Chemistry, Australian National University, Canberra, 2601 Australian Central Territories, Australia

As part of an overall investigation of heavy metal-ion discrimination, the interaction of a selection of transition and post transition metal ions with mixed donor macrocyclic ligands incorporating thioether sulfur donors has been investigated; the ligands incorporate various macrocyclic ring sizes and donor atom combinations as well as exhibiting both single ring and linked ring topologies. In these studies emphasis has been placed on investigating the factors influencing metal ion discrimination, with particular attention being given to achieving selectivity for silver(I) over lead(II).

It is now around a century since Alfred Werner laid the foundations for our modern understanding of metal coordination chemistry. Despite this, it is still often difficult to predict the metal ion preferences of a given ligand - especially when the ligand contains a mixed donor atom set. This is perhaps surprising when it is considered that metal ion recognition is a fundamental attribute of many processes in nature.

For several years the macrocyclic group at James Cook has been involved in an investigation of metal ion recognition involving an extended range of mixed donor macrocyclic ligands (*1-2*). An aim of these studies has been to achieve metal ion recognition for particular heavy metal ions and to understand the reasons for such recognition when it is observed. The above ligand category was chosen for two reasons. First, macrocyclic ligands tend to yield metal complexation behavior in solution that is less complicated than that of their open chain analogs. In part, this is a reflection of the usual restricted flexibility inherent in cyclic systems, often leading to a reduction in the number of coordination modes that can be adopted (*3*). As a result, there is a tendency to yield 1:1 metal:ligand complexes having coordination geometries that are often more readily defined. Secondly, macrocyclic ligands provide a central cavity, which can sometimes be "tuned" for a given metal ion (*3*). This provides an additional ligand-design parameter relative to open-chain systems that is available for use in achieving metal ion discrimination.

[3]Corresponding author

0097–6156/96/0653–0297$15.00/0

Strategy for Achieving Metal Ion Discrimination

In prior studies (*1*), we have employed a straight forward strategy for investigating metal ion recognition (or discrimination if two or more metals are present) by mixed donor macrocycles. In the initial step, a macrocyclic ring system is designed that might be expected to exhibit some degree of metal-ion recognition of the type desired - the design is usually based on analogy with the known behavior of related ligand systems although a measure of intuition is also often involved. Ligand synthesis and characterization is then undertaken. Typically this is followed by an investigation of complexation behavior in solution towards the metals of interest (in particular cases, this has included complementary kinetic and thermodynamic studies). Often, characterization of key complexes in the solid state has also been undertaken at this stage, using both spectroscopic and X-ray diffraction techniques.

A common feature of the solution studies has been the determination of the respective stabilities (log *K* values) of the individual metal complexes using potentiometric means. Where a degree of metal ion discrimination is observed then the factors that appear to promote this discrimination are assessed. The results of this assessment are then fed back to the synthetic program so that a modified ligand structure is produced that might be expected to enhance the desired discrimination behaviour. The entire sequence is then repeated until an optimum level of discrimination is achieved.

As part of the above strategy, we have sometimes employed molecular mechanics to help evaluate the effects of the proposed ligand modifications on the complexation behavior; more recently, we have also used semi-empirical and *ab initio* (DFT) molecular orbital methods predictively for this purpose [Adam, K. R.; Atkinson, I. M.; Lindoy, L. F. unpublished work]. In this manner it has sometimes been possible to obtain a guide to the likely usefulness of a proposed ligand derivative *before* embarking on its synthesis. Such studies thus serve as an additional 'control' for the synthetic program.

Overall, it has been our experience that the iterative procedure just described gives rise to a more complete understanding of the origins of any discrimination observed than when a less systematic, more piecemeal, approach is employed. Further, the collective results frequently provide a background against which the design of new metal-ion reagents can be undertaken.

What are the ligand parameters available for variation in a strategy of the type described above? There are three: (i) donor set (the mix of donor type and their respective numbers), (ii) macrocycle ring size, and (iii) the type and degree of ligand backbone substitution present. In this chapter, modifications of the first type in which thioether sulfur donors feature in the donor set will be given emphasis. However, before looking at representative studies of this type, it is instructive to look more widely at the large effect that donor-set variation within a given ligand framework may have on thermodynamic stability.

Donor Set Variation and Copper(II) Complex Stabilities

The effect of donor set variation on copper(II) complex stability is dramatically illustrated for the complexes of ligands of type (1) in Figure 1 (*4*). Along this series a stability difference spanning ten orders of magnitude is achieved solely by variation of the donor atom type at three of the five donor sites.

All the members of this ligand series were synthesized via sodium borohyride reduction of their corresponding diimine precursors. The latter were obtained by Schiff base condensation between the appropriate dialdehyde and diamine precursors (*5-7*) and the reduction step was carried out *in situ*. A variety of purification procedures, frequently involving an acid extraction step, were used to obtain the individual macrocycles (as solids).

Sulfur-Containing Macrocyclic Systems for Silver(I)/Lead(II) Discrimination

Although our studies have ranged more widely (*1,8-11*), the remainder of this chapter will focus on the development and use of sulfur-containing macrocyclic ligands as reagents for achieving discrimination between silver(I) and lead(II) - two metals which occur together in nature. Although isolated examples of macrocyclic ligands displaying significant discrimination for silver(I) over lead(II) have been reported (*12-14*), in general, less attention has been given so far to the factors influencing such discrimination.

Case Study Involving a Pentadentate Macrocycle Series. The interaction of silver(I) and lead(II) with the ligand series given by (**1**) provides an illustration of the use of the previously mentioned strategy to yield a system showing high discrimination for the first of these ions (*15,16*).

The initial investigation involved the interaction of the parent O_2N_3-donor macrocycle (**1**; X = O, Y = NH) with each of the above ions. In this case discrimination was minimal, with the silver complex being less than an order of magnitude more stable than that of lead. On moving to the analogous 17-membered ring incorporating a N_6-donor set (**1**; X = NH, Y = NH), both metals were observed to form more stable complexes but the discrimination is not enhanced - the stability difference between the respective complexes was again less than an order of magnitude. The third macrocycle to be produced contained oxygen donors in positions X and Y, that is (**1**; X = O, Y = O). This modification led to a drop in stability for each metal complex, reflecting the expected poorer donor properties of ether oxygen relative to secondary nitrogen. Nevertheless, the discrimination for silver(I) was somewhat enhanced compared to the previous systems. Clearly, lead(II) is less tolerant of the replacement of NH in position Y of (1) than is silver(I). At this stage, the effect of introducing thioether sulfur into the donor set was investigated. It was anticipated that, since silver(I) is 'b' class and much 'softer' than lead(II) (*17,18*), enhanced discrimination for the former ion should occur. In fact, the literature contains many reports of thioether-containing macrocycles showing significant affinity for silver(I) (*19-35*). The expected enhanced discrimination was observed - a difference in log K of 4.1 between the silver and lead complexes of (**1**; X = O, Y = S) was observed.

The introduction of sulfur donors in the X positions to yield (**1**; X = S, Y = NH) resulted in the expected increase in the absolute value of the stability of the silver(I) complex. However, in this case the restoration on the N_3-donor string results in a substantial increase in the stability of the lead(II) complex relative to that of (**1**; X = O, Y = S) such that the resulting discrimination of (**1**; X = S, Y = NH) for silver(I) now shows no significant enhancement over the previous system.

Finally, the remaining two ligands in the series have sulfur donors in the X positions and either an oxygen (**1**; X = S, Y = O) or a sulfur (**1**; X = S, Y = S) in the Y position. In accordance with the trends observed so far, both ligands show a substantial enhancement of their affinity for silver(I) relative to lead(II), with the N_2S_3-donor macrocycle giving an approximate selectively of 10^9 for silver over lead!

While caution needs to be exercised in extrapolating solution results to the solid state, it was nevertheless of interest to compare the X-ray structures of particular silver and lead complexes in the above series. By this means it has proved possible to probe the structure/function relationships that appear to underlie the above observed discrimination behaviour.

The X-ray diffraction structures of the 1:1 silver complexes of (**1**; X = S, Y = O) (*16*), (**1**; X = S, Y = NH) (*36*) and (**1**; X = S, Y = S) (*37*), have been reported (Figures 2a - c). In each structure, the silver ion is surrounded by the macrocycle in an irregular 5-coordinate geometry with all the macrocyclic donor atoms coordinated. In each case, the macrocycle 'wraps' around the metal such that a 'tight' complex is formed; there are no long metal-donor bonds present. This result is in keeping with the presence of a strong affinity between metal and ligand in each of this these complexes. In contrast, the X-ray structure of the lead(II) complex of (**1**; X = S, Y = NH) shows that the metal is 7-coordinate with the macrocycle thus providing only part of the coordination shell of the lead(II) (*15*). The coordination sphere is composed of the five macrocyclic donors and two oxygens from perchlorate anions, with the latter occupying axial sites (Figure 3). In this complex, the macrocycle adopts a flatter, more open, configuration with the metal situated below the macrocyclic cavity. The adoption of the 7-coordinate geometry together with evidence for specific metal to donor distances being long (in particular, the Pb-S distances of 3.14 and 3.18 Å are very long) - both suggest that the binding of (**1**; X = S, Y = NH) by lead(II) is reduced relative to that for silver(I).

In further experiments, an investigation of the role of macrocyclic ring size on the extraordinary discrimination shown by the 17-membered system (**1**; X = S, Y = S) for silver(I) over lead(II) was undertaken (*16*). The aim was to investigate how the variation of macrocyclic ring size might affect the discrimination observed for the 17-membered ring. Accordingly, the corresponding 16-membered ring (**2**) and 18-membered ring (**3**) were synthesised and the necessary stability determinations undertaken. For (**2**), log *K* values for the silver(I) and lead(II) complexes of 11.0 and ~3.1 (Δ log *K* = ~7.9), respectively, were obtained while for (**3**) the corresponding values were 10.9 and ~3.5 (Δ log *K* = ~7.2). Hence, while both rings yield a large degree of discrimination for silver, it nevertheless falls short in each case of that exhibited by the corresponding 17-membered ring (**1**; X = S, Y = S).

Calorimetric and NMR Studies. The interaction of silver nitrate with a range of the above 17-membered, macrocyclic ligands has been investigated in acetonitrile (*38*). The nature of the equilibrium present and the enthalpies of formation for the respective complexes in this solvent were determined using complimentary calorimetric and NMR studies. For all systems, the formation of a 1:1 species was observed, with a 1:2 species also occurring in the presence of excess ligand in most cases. The strong affinity of silver(I) for thioether sulfur is clearly evident from the enthalpic data (Table I). The ΔH values for the 1:1 complexes of ligands with X equal to thioether sulfur in

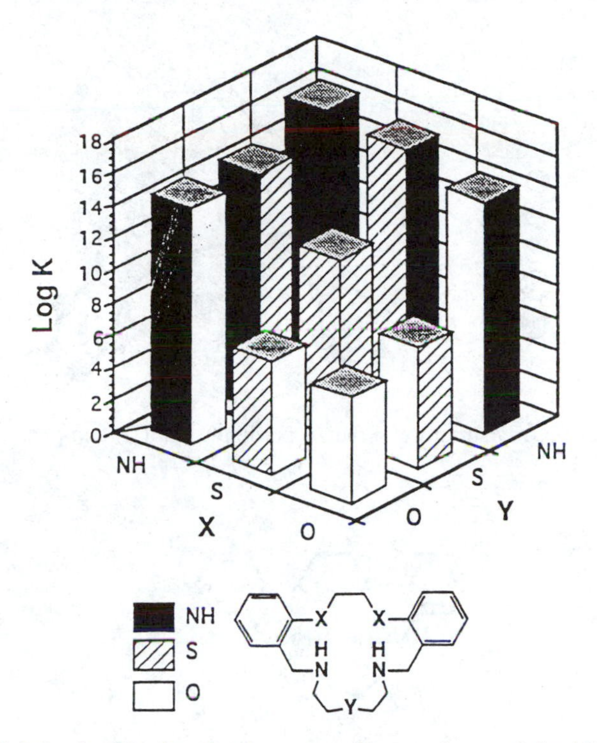

Figure 1. Relative log *K* values for the copper(II) complexes of the 17-membered macrocycles of type (**1**) incoporating different donor atom sets.

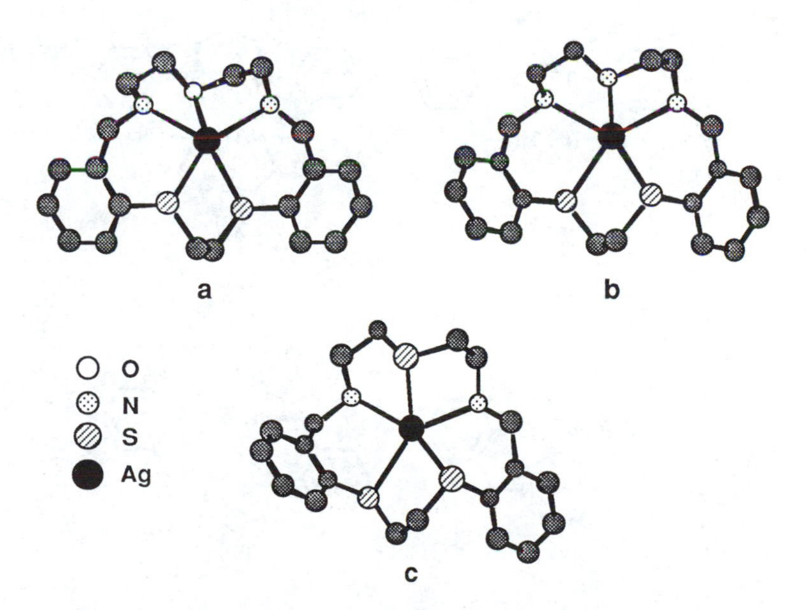

Figure 2. X-ray structures of the silver(I) complexes of (**1**; X = X = S, Y = O) (*16*), (**1**; X = S = S, Y = NH) (*36*) and (**1**; X = S, Y = S) (*37*).

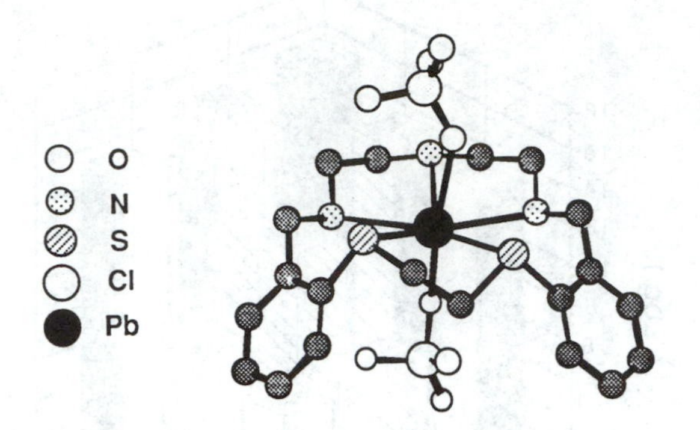

Figure 3. The X-ray structure of the 7-coordinate lead(II) complex of (**1**; X = S, Y = NH) (*15*).

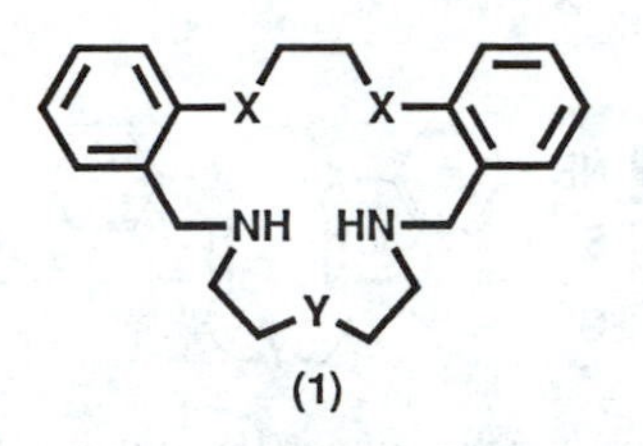

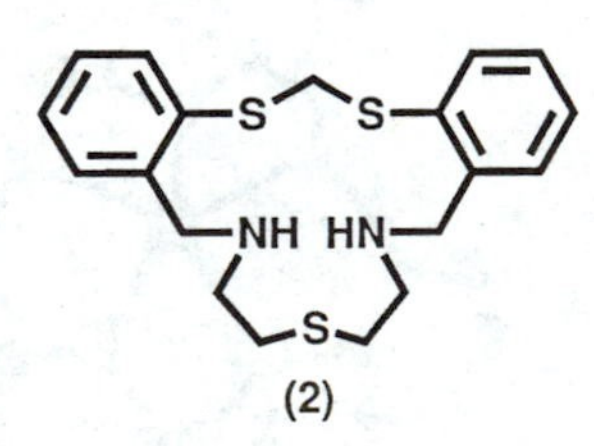

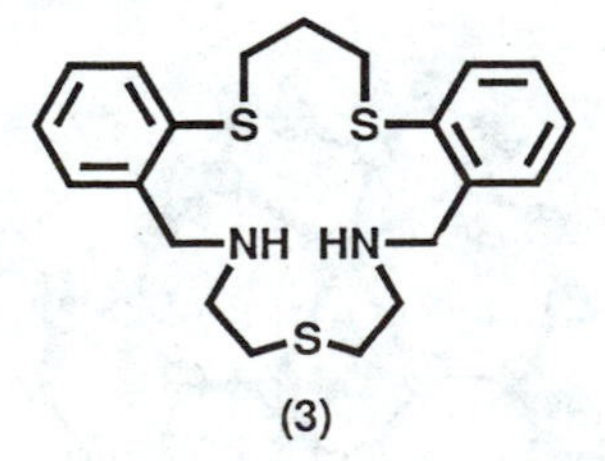

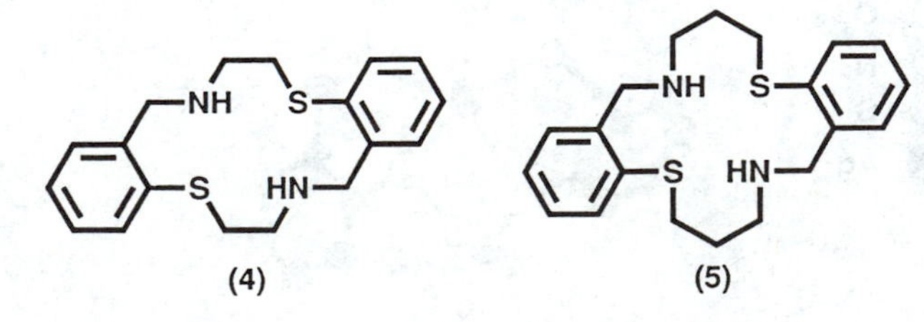

Table I. Calorimetric data for the interaction of silver nitrate with selected ligands of type 1 in acetonitrile[a]

Ligand	$-\Delta_f H^o{}_1$ [a]
1; X = O, Y = N	49
1; X = O, Y = S	51
1; X = S, Y = N	69
1; X = S, Y = S	72

[a] $-\Delta_f H^o{}_1$ is for the reaction Ag + L ⇌ $[AgL]^+$ (kJ mol^{-1}, Ag^+).

the N-X-N sequence are in each case comparable to those for the analogous complexes incorporating secondary amines in this position; that is, for the present systems, both donors appear to show comparable affinity for silver(I).

Smaller Ring, Mixed-Donor Systems. When the coordination chemistry of silver(I) is considered, then a further characteristic seems amenable to exploitation in cyclic ligand design. This is the tendency of silver(I) to assume linear diammine coordination, as occurs in $[Ag(NH_3)_2]^+$. In view of this, the coordination behavior of the 14- and 16-membered rings (**4**) and (**5**) towards silver(I) and lead(II) were investigated. In each of these rings, two amine donors are located in 'opposing' (trans) positions, with sulfur donors positioned between them so that two N-S-N donor strings of the type found favorable for silver(I) coordination [but less satisfactory for lead(II)] are present.

The procedure used to obtain (**4**) and (**5**) is interesting since it involves the initial manipulation of equilibria involving the corresponding small-ring monoimines, the (required) trans-diimine (*39*) and, in the case of the larger ring system, also the corresponding tri-(24-membered)imine and tetra-(32-membered)imine species (*40*). Reduction of the respective 14-membered and 16-membered diimine precursors with lithium aluminium hydride in tetrahydrofuran then yielded (**4**) and (**5**), respectively.

Stability data for the 1:1 complexes of silver(I) and lead(II) are summarized in Table II (*40*). Both ligands yield very respectable discrimination (of approximately 10^5) for silver under the conditions of measurement. Clearly, ring size has little influence on the degree of discrimination exhibited by these systems; however, the absolute log *K* values are marginally higher for the complexes of the 16-membered ring (**5**).

Table II. Stability constants for the ML^{n+} complexes of Silver(I) and Lead(II) with the macrocycles (4) and (5)[a]

Ligand	*Ring size*	*Donor set*	*Ag(I)*	*Pb(II)*
(**4**)	14	*trans*-S_2N_2	8.6	~3.5
(**5**)	16	*trans*-S_2N_2	9.3	4.5

[a] In 95% methanol; I = 0.1, $Et_4N(ClO_4)$, at 25 °C.

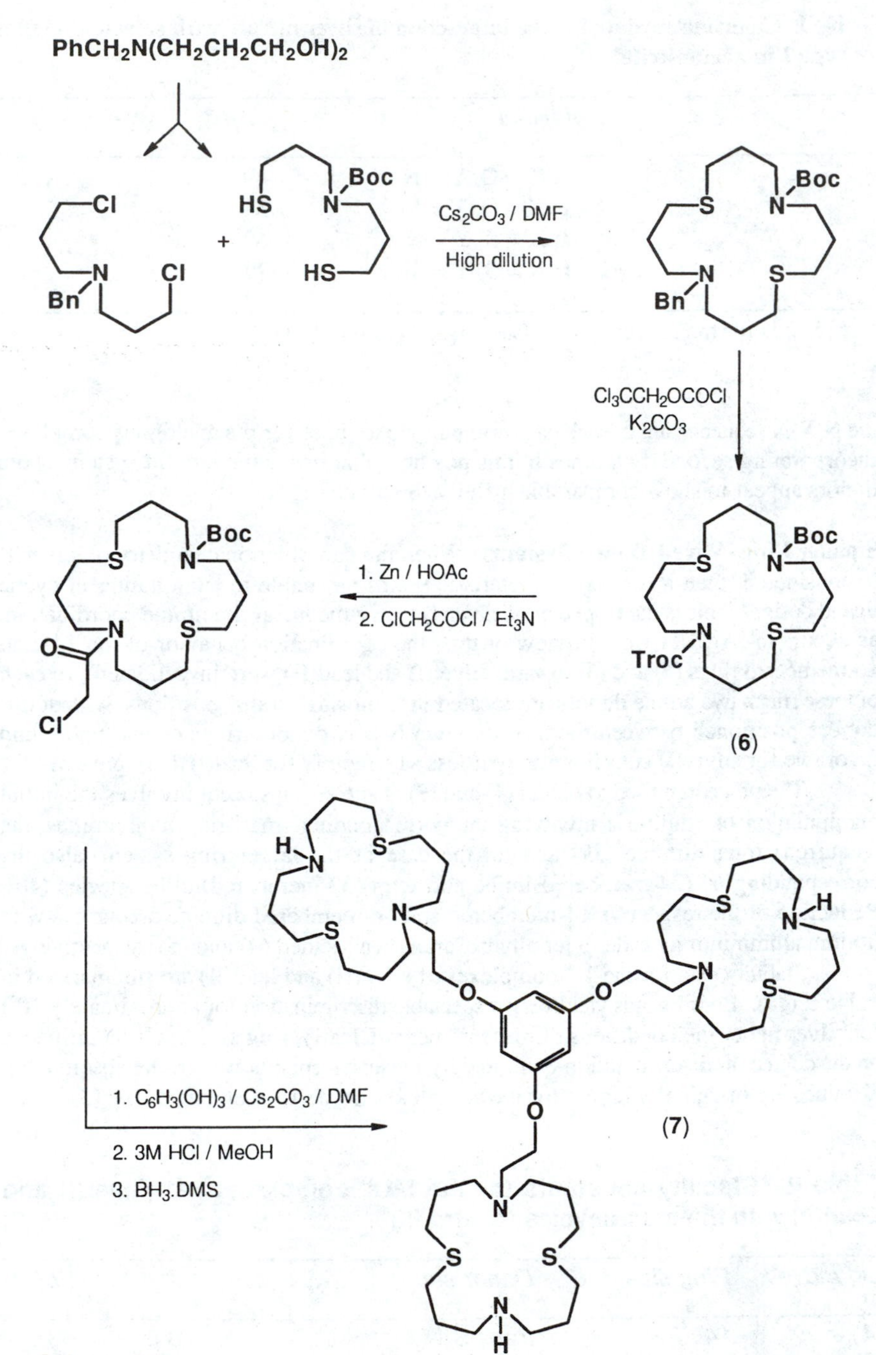
PhCH2N(CH2CH2CH2OH)2
Cl
N
Bn
Cl
+
HS
N
Boc
HS
Cs2CO3 / DMF
High dilution
S
N
Boc
N
S
Bn
Cl3CCH2OCOCl
K2CO3
S
N
Boc
N
S
Troc
(6)
1. Zn / HOAc
2. ClCH2COCl / Et3N
S
N
Boc
N
S
O
Cl
1. C6H3(OH)3 / Cs2CO3 / DMF
2. 3M HCl / MeOH
3. BH3.DMS
H
N
S
N
S
O
O
S
N
H
N
S
O
N
S
S
N
H
(7)

Scheme 1

A Tri-Linked N_2S_2-Donor Macrocyclic Ring System

The linking of mixed-donor macrocyclic rings to form di-linked (*3,41-47*) and, much less commonly, tri-linked mixed-donor (*46,48-50*) systems has been a trend in recent research in macrocyclic chemistry; a large majority of these systems incorporate rings containing combinations of nitrogen and oxygen heteroatoms.

The novel protecting group strategy shown in Scheme 1 has enabled the synthesis of the new tri-linked N_2S_2-donor macrocycle (**7**) containing three 16-membered rings, each of which is attached by spacer groups to a central phloroglucinol core (Groth, A. M.; Lindoy, L. F.; Meehan, G. V., unpublished work).

The procedure involves the initial construction of the required single 16-membered ring incorporating complementary protecting groups on the two nitrogen heteroatoms. This can then be manipulated to provide appropriate mono-N-protected macrocycles suitable for incorporation into more structurally developed systems. In the present case, the N-tert-butoxycarbonyl-N-2,2,2-trichloroethoxy macrocycle (**6**) is an important intermediate in this process.

Scheme **1** outlines a procedure for obtaining *one* category of linked macrocycles. However, the general protecting group approach described should be readily adapted to the synthesis of a variety of other linked systems, using a range of synthetic protocols; studies of this type are at present in progress.

Initial metal-ion binding studies involving the tri-linked species (**7**) confirm its expected affinity for 'soft' silver(I). In an NMR titration experiment, (**7**) was demonstrated to undergo stepwise complexation of this ion in DMSO-d_6/$CDCl_3$ to yield a final product with a silver(I):ligand stoichiometry of 3:1.(Groth, A. M, Lindoy, L. F., Meehan, G. V. unpublished work.)

Finally, solvent extraction (water/chloroform) and related bulk membrane transport (water/chloroform/water) experiments have been performed in which the respective aqueous source phases, buffered at pH 5, contained an equimolar mixture of cobalt(II), nickel(II), copper(II), zinc(II), cadmium(II), silver(I) and lead(II) nitrates, each at 10^{-2} M (initial) concentration. The chloroform phase in both experiments contained (**7**) at 3.3 x 10^{-4}M. For membrane transport, the 'concentric cylinder' cell employed has been described elsewhere (*51*); the aqueous receiving phase was buffered at pH 3 and the experiment was terminated after 24 hours. With all seven metal ions present, both the solvent extraction and the membrane transport experiments yielded results indicating sole discrimination for silver(I) in each case.

Acknowledgments. We thank the Australian Research Council and the Australian Institute of Nuclear Science and Engineering for support.

Literature Cited

(1) Lindoy, L. F. Chapter 2 in *Synthesis of Macrocycles: The Design of Selective Complexing Agents*, Progress in Macrocyclic Chemistry Vol. 3, Eds., Izatt, R. M.; Christensen, J. J., John Wiley, N.Y., 1987, pp. 53-92.

(2) Lindoy, L. F. *Pure and Appl. Chem.* **1989**, *61*, 1575.

(3) Lindoy, L. F. "The Chemistry of Macrocyclic Ligand Complexes", Cambridge University Press, 1989, Cambridge, U.K.

(4) Adam, K. R.; Baldwin, D.; Duckworth, P. A.; Leong, A. J.; Lindoy, L. F.; McPartlin, M.; Tasker, P. A. *J. Chem. Soc., Chem. Commun.* **1987**, 1124.
(5) Adam, K. R.; Lindoy, L. F.; Lip, H. C.; Rea, J. H.; Skelton, B. W.; White, A. H. *J. Chem. Soc., Dalton Trans.* **1981**, 74.
(6) Baldwin, D. S.; Duckworth, P. A.; Erickson, G. R. A.; Lindoy, L. F.; McPartlin, M.; Mockler, G. M; Moody, W. E.; Tasker, P. A. *Aust. J. Chem.* **1987**, *40*, 1861.
(7) Duckworth, P. A.; Lindoy, L. F.; McPartlin, M.; Tasker, P. A. *Aust. J. Chem.* **1993**, *46*, 1787.
(8) Adam, K. R.; Dancey, K. P.; Leong, A. J.; Lindoy, L. F.; McCool, B. J.; McPartlin, M.; Tasker, P. A. *J. Am. Chem. Soc.* **1988**, *110*, 8471.
(9) Adam, K. R.; Antolovich, M.; Baldwin, D. S.; Brigden, L. G.; Duckworth, P. A.; Lindoy, L. F.; Bashall, A.; McPartlin, M.; Tasker, P. A. *J. Chem. Soc., Dalton Trans.* **1992**, 1869.
(10) Adam, K. R.; Antolovich, M.; Baldwin, D. S.; Duckworth, P. A.; Leong, A. J.; Lindoy, L. F.; McPartlin, M.; Tasker, P. A. *J. Chem. Soc., Dalton Trans.* **1993**, 1013.
(11) Adam, K. R.; Arshad, S. P. H.; Baldwin, D. S.; Duckworth, P. A.; Leong A. J.; Lindoy, L. F.; McCool, B. J.; McPartlin, M.; Tailor, B. A.; Tasker, P. A. *Inorg. Chem.* **1994**, *33*, 1194.
(12) Izatt, R. M.; Bradshaw, J. S.; Nielsen, S. A.; Lamb, J. D.; Christensen, J. J.; Sen, D. *Chem. Rev.* **1985**, *85*, 271.
(13) Izatt, R. M.; Pawlak, K.; Bradshaw, J. S.; Bruening, R. L. *Chem. Rev.* **1991**, *91*, 1721.
(14) Lindoy, L. F. in *Cation Binding by Macrocycles* Inoue, Y.; Gokel, G. W., Eds.; Marcel Dekker, New York, 1990; pp. 599-629.
(15) K. R. Adam, K. R.; Baldwin, D. S.; Bashall, A.; Lindoy, L. F.; McPartlin, M.; Powell, H. R. *J. Chem. Soc., Dalton Trans.* **1994**, 237.
(16) Adam, K. R.; Baldwin, D. S.; Duckworth, P. A.; Lindoy, L. F.; McPartlin, M.; Bashall, A.; Powell, H. R.; Tasker, P. A. *J. Chem. Soc., Dalton Trans.* **1995**, 1126.
(17) Ahrland, S.; Chatt, J.; Davies, N. R. *Quart. Rev.* **1958**, *12*, 265.
(18) Pearson, R. G. *J. Am. Chem. Soc.* **1963**, *85*, 3533.
(19) Blake, A. J.; Schroder, M. *Advan. Inorg. Chem.* **1990**, *35*, 1.
(20) Cooper, S. R.; Rawle, S. C. *Struct. and Bonding* **1990**, *72*, 1.
(21) Sevdic, D.; Meider, H. *J. Inorg. Nucl. Chem.* **1977**, *39*, 1403.
(22) Arnaud-Neu, F.; Schwing-Weill, M. J.; Louis, R.; Weiss, R. *Inorg. Chem.* **1979**, *18*, 2956.
(23) Izatt, R. M.; Dearden, D. V.; Brown, P. R.; Bradshaw, J. S.; Lamb, J. D.; Christensen, J. J. *J. Am. Chem. Soc.* **1983**, *105*, 1785.
(24) Tsukube, H.; Takagi, K.; Higashiyama, T.; Iwachido, T.; Hayama, N. *Tetrahedron Lett.* **1985**, *26*, 881.
(25) Poddubnykh, L. P.; Dmitrienko, S. G.; Kuz'min, N. M.; Formanovski, A. A.; Zolotov, Y. A. *Russ. J. Inorg. Chem.* **1986**, *31*, 1040.
(26) Oue, M.; Kimura, K.; Shono, T. *Anal. Chim. Acta* **1987**, *194*, 293.
(27) Blower, P. J.; Clarkson, J. A.; Rawle, S. C.; Hartman, J. A. R.; Wolf, R. E.; Yagbasan, R.; Bott, S. G.; Cooper, S. R. *Inorg. Chem.* **1989**, *28*, 4040.

(28) Craig, A. S.; Kataky, R.; Parker, D.; Adams, H.; Bailey, N.; Schneider, H. *J. Chem. Soc. Chem. Commun.* **1989**, 1870.

(29) Craig, A. S.; Kataky, R.; Matthews, R. C.; Parker, D.; Ferguson, G.; Lough, A.; Adams, H.; Bailey, N.; Schneider, H. *J. Chem. Soc. Perkin Trans. 2*, **1990**, 1523.

(30) Blake, A. J.; Gould, R. O.; Holder, A. J.; Hyde, T. I.; Schroder, M. *Polyhedron* **1989**, *8*, 513.

(31) Wu, G.; Jian, W.; Lamb, J. D.; Bradshaw, J. S.; Izatt, R. M. *J. Am. Chem. Soc.* **1991**, *113*, 6538.

(32) Nabeshima, T.; Nishijima, K.; Tsukada, N.; Furusawa, H.; Hosoya, T.; Yano, Y. *J. Chem. Soc., Chem. Commun.* **1992**, 1092.

(33) Blake, A. J.; Reid, G.; Schroder, M. *J. Chem. Soc., Chem. Commun.* **1992**, 1074 and references therein.

(34) Lockhart, J. C.; Mousley, D. P.; Hill, M. N. S.;Tomkinson, N. P.; Teixidor, F.; Almajano, M. P.; Escriche, L.; Casabo, J. F.; Sillanpää, R.; Kivekas, R. *J. Chem. Soc. Dalton Trans.* **1992**, 2889.

(35) Brzozka, Z.; Cobben, P. L. H. M.; Reinhoudt, D. N.; Edema, J. J. H.; Buter, J.; Kellogg, R. M. *Anal. Chim. Acta* **1993**, *173*, 139.

(36) Kallert U.; Mattes, R. *Inorg. Chim. Acta* **1991**, *180*, 263.

(37) Kallert, U; Mattes, R. *Polyhedron*, **1992**, *11*, 617.

(38) Baldwin, D. S.; Lindoy, L. F.; Graddon, D. P. *Aust. J. Chem.* **1988**, *41*, 1347.

(39) Martin, J. W. L.; Organ, G. J.; Wainwright, K. P.; Weerasuria, K. D. V.; Willis, A. C.; Wild, S. B. *Inorg. Chem.* **1987**, *26*, 2963.

(40) Leong, A. J.; Lindoy, L. F.; Hockless, D. C. R.; Swiegers, G. F.; Wild, S. B. *Inorg. Chim. Acta* **1996**, in press.

(41) Bradshaw, J. S.; Krakowiak, K. R.; Izatt, R. M. *Aza-Crown Macrocycles*, Wiley, New York, 1993.

(42) Oue, M.; Ishigaki, A.; Kimura, K.; Matsui, Y.; Shono, T. *J. Polymer Sci.* **1985**, *23*, 2033.

(43) Bernhardt, P. V.; Comba, P.; Gahan, L. R.; Lawrance, G. A. *Aust. J. Chem.* **1990**, *43*, 2035.

(44) Wallon, A.; Werner, U.; Müller, W. M.; Nieger, M.; Vögtle, F. *Chem. Ber.* **1990**, *123*, 859.

(45) Holý, P.; Koudelka, J.; Bělohradský, M.; Stibor, I.; Závada, J. *Collect. Czech. Chem. Commun.* **1991**, *56*, 1482.

(46) Wang, K.; Han, X.; Gross, R. W. P.; Gokel, G. W. *J. Chem. Soc., Chem. Commun.* **1995**, 641.

(47) Kim, J-H.; Lindoy, L. F.; Matthews, O. A.; Meehan, G. V.; Nachbaur, J.; Saini, V. *Aust. J. Chem.* **1995**, accepted.

(48) Vögtle, F.; Weber, E. *J. Incl. Phen. Mol. Recogn.* **1992**, *12*, 75.

(49) Weber, E. *J. Org. Chem.* **1982**, *47*, 3478.

(50) Vögtle, F.; Wallon, A.; Müller, W. M.; Werner, U.; Nieger, M. *J. Chem. Soc., Chem. Commun.* **1990**, 158.

(51) Chia, P. S. K.; Lindoy, L. F.; Walker, G. W. *Pure & Appl. Chem.*, **1993**, *65*, 521.

Small-Molecule Activation

Chapter 19

Toward Novel Organic Synthesis on Multimetallic Centers: Synthesis and Reactivities of Polynuclear Transition-Metal–Sulfur Complexes

Masanobu Hidai and Yasushi Mizobe

Department of Chemistry and Biotechnology, Graduate School of Engineering, University of Tokyo, Hongo, Bunkyo-ku, Tokyo 113, Japan

Thiolato-bridged diruthenium complexes [Cp*Ru(μ-SPr^i)$_2$RuCp*], [Cp*Ru(μ-SPr^i)$_3$RuCp*], and [Cp*RuCl(μ-SPr^i)$_2$RuCp*][OTf] (Cp* = η^5-C_5Me_5, Tf = CF_3SO_2) have been shown to provide unique bimetallic reaction sites for activation and transformation of various substrates. Novel reactions observed for a series of alkynes are particularly noteworthy, in which the products remarkably depend upon the nature of both the alkyne substituents and the diruthenium sites. On the other hand, the mixed-metal cubane-type cluster complex $[PdMo_3S_4(tacn)_3Cl]^{3+}$ (tacn = 1,4,7-triazacyclononane) exhibits a remarkable catalytic activity for the highly regio- and stereoselective addition of alcohols and carboxylic acids to certain alkynes.

The chemistry of multimetallic complexes is currently attracting much attention. The substrate molecules bound to the multimetallic site may be subjected to cooperative activation by two or more adjacent transition metals and become highly reactive. This possibly leads to the participation of the coordinated substrates in the new chemical transformations, which are not attainable on the monometallic center. In this context, reactivities of a wide variety of transition metal clusters have been studied recently. However, in contrast to the significant advances in the reactions associated with carbonyl clusters, the chemistry of transition metal-sulfur clusters in organic synthesis and catalysis is still poorly explored. Owing to the tendency to make a strong bond with metals and block the reactive site, sulfur has long been recognized as a poison for metal catalysts, and this might be responsible for the limited progress in the chemistry of transition metal-sulfur complexes, especially for those containing noble metals. It is noteworthy, however, that the high affinity for transition metals as well as the multi-electron donating ability implicate sulfur as an excellent bridging ligand for aggregating robust cluster cores. The presence of sulfur in the cluster framework therefore has the advantage of avoiding degradative fragmentation of the cluster under forcing reaction conditions, which often emerges as a serious problem in the reactions of carbonyl clusters containing core structures supported primarily by the metal-metal bonds.

0097–6156/96/0653–0310$15.00/0

Our recent studies in this area have been focused on the exploitation of the novel reactivities displayed by the thiolato-bridged diruthenium complexes (*1*), which have demonstrated that a series of terminal alkynes undergo the diversified transformations on the bimetallic centers in the complexes such as $[Cp^*Ru(\mu\text{-}SPr^i)_2RuCp^*]$ (**1a**) (*2*), $[Cp^*Ru(\mu\text{-}SPr^i)_3RuCp^*]$ (**2a**) (*3*), and $[Cp^*RuCl(\mu\text{-}SPr^i)_2RuCp^*][OTf]$ (**3a**) (Cp^* = $\eta^5\text{-}C_5Me_5$, Tf = CF_3SO_2) (*4*). More recently, we have shown that the cubane-type sulfido cluster containing a noble metal $[PdMo_3S_4(tacn)_3Cl]^{3+}$ (tacn = 1,4,7-triazacyclononane) (*5*) catalyzes the regio- and stereoselective addition of alcohols (*5b*) or carboxylic acids (*6*) to certain alkynes with a remarkably high efficiency. It is to be noted that despite extensive studies on the transition metal-sulfur cubane clusters stimulated by their relevance to metalloproteins and heterogeneous hydrodesulfurization catalysts, little has been reported about their catalytic functions. We wish to summarize herein our recent studies aimed at novel organic synthesis using polynuclear noble metal-sulfur complexes. Emphasis is mainly put upon the transformations of alkynes. Other intriguing reactions using alkyl halides (*2a,7*) and hydrazines (*8*) as substrates will be discussed elsewhere.

Reactions of Thiolato-Bridged Diruthenium Complexes.

The discovery of a convenient method for synthesizing the half-sandwich Ru complex $[Cp^*RuCl(\mu\text{-}Cl)_2RuCp^*Cl]$ (*9*) prompted us to investigate its reactions with various thiolate compounds, which has resulted in the isolation of four types of thiolato-bridged diruthenium complexes depending on the nature of the thiolate and the reaction conditions. These include the Ru(II)/Ru(II) complexes $[Cp^*Ru(\mu\text{-}SR)_2RuCp^*]$ (**1**) (*2*), the formal Ru(II)/Ru(III) complexes $[Cp^*Ru(\mu\text{-}SR)_3RuCp^*]$ (**2**) (*3*), and the Ru(III)/Ru(III) complexes $[Cp^*Ru(\mu\text{-}SR)_3RuCp^*]Cl$ (R = aryl) and $[Cp^*RuCl(\mu\text{-}SR)_2RuCp^*Cl]$ (**4**; R = alkyl) (*10*). Another cationic Ru(III)/Ru(III) complex **3a** containing a coordinatively unsaturated Ru center is readily available from **4a** (R = Pr^i) by treatment with AgOTf (*4*). The structures of these new diruthenium complexes are depicted in Figure 1.

Reactions of 1a with Terminal Alkynes. Treatment of **1a** with excess terminal alkynes HC≡CR resulted in the incorporation of two or three alkynes on the diruthenium center. As shown in Scheme 1, the products markedly depend upon the alkyne substituent. Thus the reactions afford either the μ-ruthenacyclopentenyl complexes **5** — **7** (*11*) or a μ,π-alkyne complex **8** (*12*). The five-membered ruthenacycles in **5** and **6** arise from the coupling of two alkynes on the Ru site, while that in **7** results from an ene-yne moiety in the alkyne bound to the Ru atom. No intermediate stages are isolable in these reactions except for that with HC≡CCOOMe, for which a thiaruthenacyclobutene complex **9** was obtained from the reaction of **1a** with the equimolar alkyne and fully characterized. Complex **9** reacted further with HC≡CR (R = COOMe, Tol; Tol = $4\text{-}MeC_6H_4$) or $HC{\equiv}CSiMe_3$ to give ruthenacyclopentenyl complexes or a μ,π-alkyne complex, respectively (Scheme 2). In related work, the reaction of $[Cp^*Ru(\mu\text{-}SBu^t)_2RuCp^*]$ with HC≡CCOOMe has been reported quite recently. This affords a mixture of four diruthenium complexes in which one, two, three, or five alkyne molecules are incorporated on the diruthenium center, respectively, indicating that the reaction course is also affected significantly by the thiolate substituent (*13*).

Reactions of 2a and 3a with Terminal Alkynes. Complex **2a** reacts with terminal alkynes to generate bis(alkynyl) complexes $[Cp^*Ru(C{\equiv}CR)(\mu\text{-}SPr^i)_2RuCp^*(C{\equiv}CR)]$ (**10**) (eq 1) (*3,14*). The X-ray analysis of **10a** has unequivocally demonstrated its unique structure containing two terminal alkynyl ligands bound to each Ru atom in mutually cis configuration.

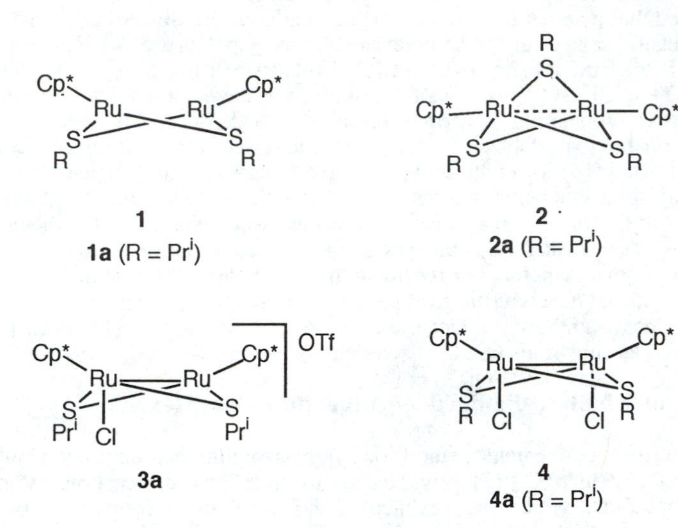

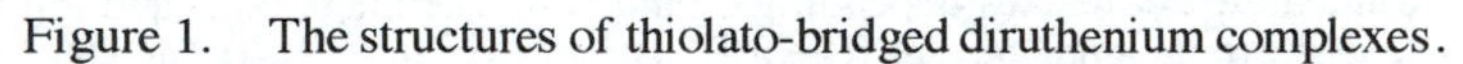
Figure 1. The structures of thiolato-bridged diruthenium complexes.

Scheme 1[a]

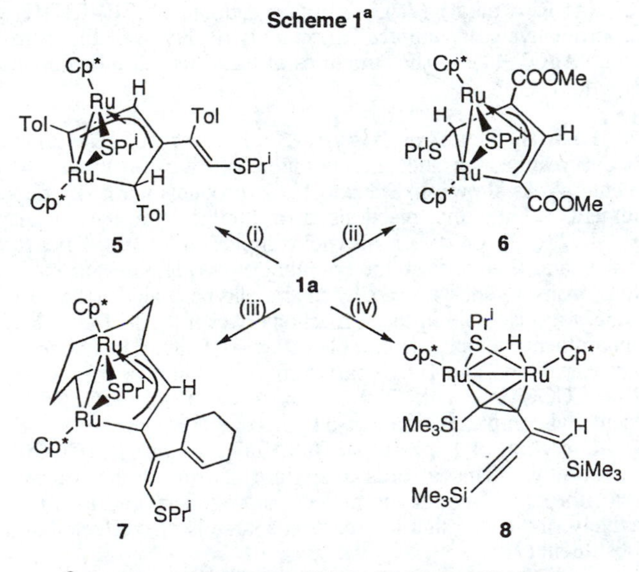

[a] Reagents: (i) HC≡CTol; (ii) HC≡CCOOMe; (iii) HC≡C-C=CH$_2$(CH$_2$)$_3$CH$_2$; (iv) HC≡CSiMe$_3$

When **10a** (R = Tol) and **10b** (R = Ph) were allowed to react with a slight excess of I_2, 1,4-disubstituted-1,3-diynes were obtained in high yields together with $[Cp^*RuI(\mu\text{-}SPr^i)_2RuCp^*I]$ (eq 2). Halogens are well known to add to the β-carbon atom in certain alkynyl ligands to form the corresponding vinylidenes without cleavage of the Ru-C_{sp} bond as demonstrated in the formation of $[CpRu(=C=CXPh)(PPh_3)_2]X_3$ (X = Cl, Br, I) from $[CpRu(C\equiv CPh)(PPh_3)_2]$ and X_2 (*15*). However, the Ru-C_{sp} bonds in diruthenium bis(alkynyl) complexes **10a** and **10b** were readily cleaved to give the coupling products of the two alkynyl ligands. This I_2-induced reaction is quite interesting since it provides a rarc example of the coupling of two organic moieties on two neighboring metal atoms connected by a metal-metal bond. It should be noted that the intramolecular dinuclear elimination from $L_4(R)M\text{-}M(R)L_4$ yielding R-R together with $L_4M=ML_4$ is believed to be a symmetry-forbidden process which requires a large activation energy to follow a C_{2v} concerted least-motion pathway (*16*). The reaction of the dialkyl complex $[Cp^*Ru(CH_2CH_2Ph)(\mu\text{-}SPr^i)_2RuCp^*(CH_2CH_2Ph)]$ with I_2 has also been investigated similarly for comparison. However, $PhCH_2CH_2I$ was the predominant product and the yield of the coupling product $Ph(CH_2)_4Ph$ was much lower than in the alkynyl complexes (*17*).

Quite interestingly, treatment of **10a** and **10b** with HBF_4 led to the unprecedented coupling of these two terminal alkynyl ligands to give the diruthenacyclopentadienoindane complexes **11**. The possible mechanism for the formation of **11** may involve the initial attack of the proton at the β-carbon in one alkynyl ligand to give a vinylidene-alkynyl intermediate, which undergoes the intramolecular vinylidene-alkynyl coupling, forming a dinuclear $\mu\text{-}\eta^1{:}\eta^2$-butenynyl intermediate. Subsequent nucleophilic attack of the aromatic group on the activated C≡C moiety in the butenynyl ligand produces the diruthenacyclopentadienoindene complexes **12**, which are further converted to the final products **11** upon protonolysis (Scheme 3). This mechanism may be supported by the formation of **12a** in 40% yield from the reaction of **10a** with acatalytic amount of HBF_4, which was readily converted into **11a** when treated with HBF_4. Complexes **12**, fully characterized by X-ray crystallography for **12a**, were also available quantitatively upon treatment of **11** with $LiBHEt_3$ or NEt_3 (Scheme 3).

As for the neutral Ru(III)/Ru(III) complex **4a**, reactions with terminal alkynes did not take place. However, the cationic complex **3a** did react with HC≡CR (R = Tol, Ph) to give directly the diruthenacyclopentadienoindane complexes **13**, the OTf analogues of **11**, in high yields (*4*). Extremely high reactivity of **3a** towards terminal alkynes was also manifested by its reaction with $HC\equiv C\overline{C=CH(CH_2)_3C}H_2$, rapidly forming the diruthenacyclopentadiene complex **14**, which was not available from the protonation of the corresponding dialkynyl complex **10c** (R = $\underline{C=CH(CH_2)_3C}H_2$) (Scheme 4). These unique transformations of terminal alkynes may proceed via vinylidene-alkynyl intermediates like **15** generated directly from **3a** and two alkyne molecules.

Importantly, further extention of these reaction systems to that of **3a** and HC≡CFc (Fc = ferrocenyl) has led to the isolation of a dinuclear butenynyl complex $[Cp^*Ru\{\mu\text{-}C(=CHFc)C\equiv CFc\}(\mu\text{-}SPr^i)_2RuCp^*][OTf]$ (**16**) in high yield (*18*) (eq 3). This finding strongly supports the above mechanism, which involves vinylidene-alkynyl coupling to give the butenynyl ligand as a key step. To our knowledge, **16** may represent the first well-defined example of the dinuclear $\mu\text{-}\eta^1{:}\eta^2$-butenynyl complex, although the formation of the η^1 or η^3-butenynyl ligand via the intramolecular migration of the alkynyl ligand to the vinylidene ligand is reported in mononuclear complexes such as $[Ru\{\eta^3\text{-}Me_3SiCCC=CH(SiMe_3)\}\text{-}\{P(CH_2CH_2PPh_2)_3\}]^+$ (*19*) and $[RuCl\{\eta^1\text{-}C(C\equiv CBu^t)=CHBu^t\}(CO)(PPh_3)_2]$ (*20*).

Scheme 2[a]

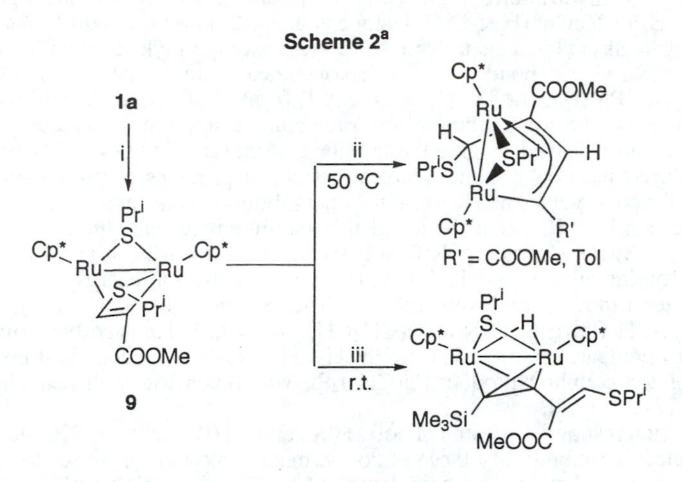

[a] Reagents: (i) HC≡CCOOMe; (ii) HC≡CR'; (iii) HC≡CSiMe₃

Scheme 3

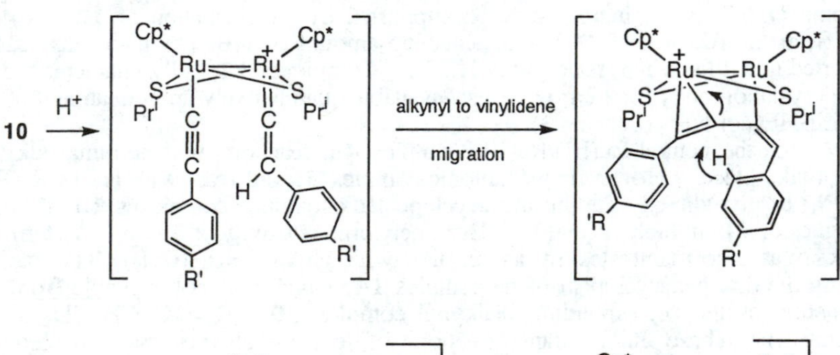

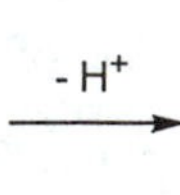

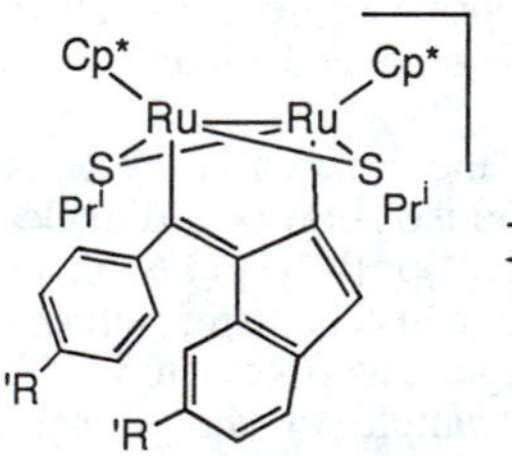

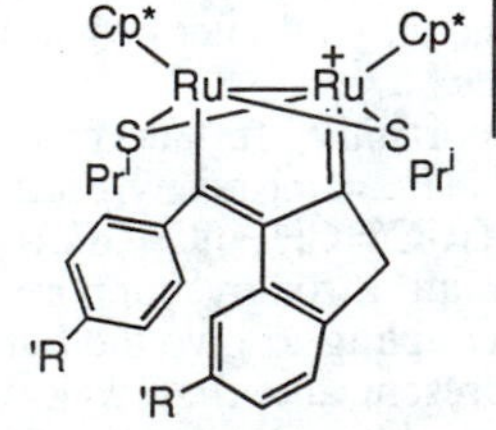

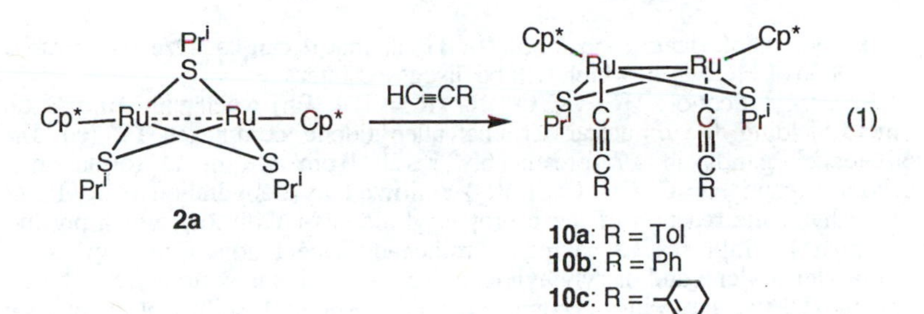
HC≡CR
2a
10a: R = Tol
10b: R = Ph
10c: R =
(1)

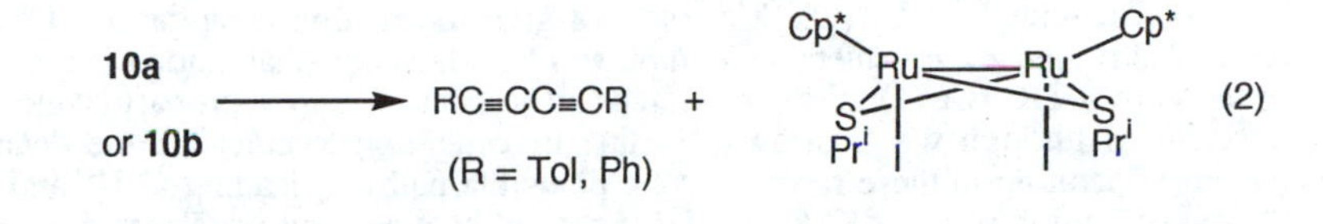
10a
or 10b
RC≡CC≡CR +
(R = Tol, Ph)
(2)

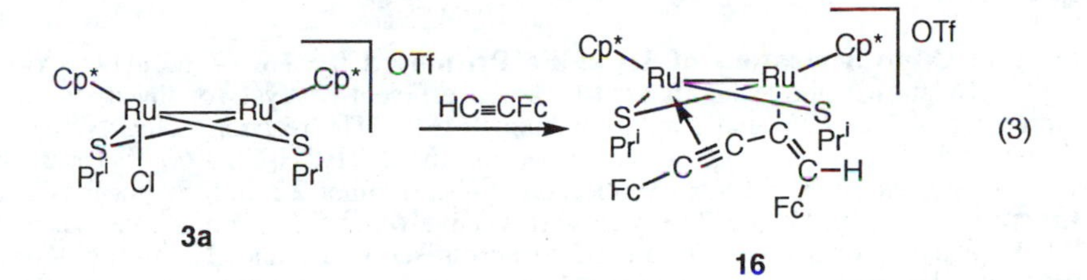
OTf
HC≡CFc
3a
16
(3)

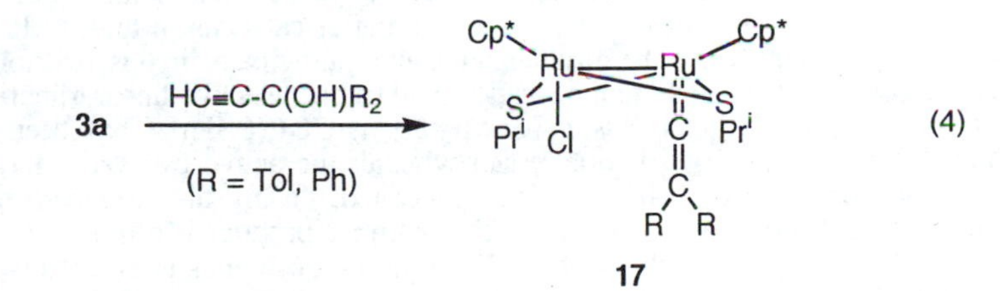
HC≡C-C(OH)R₂
3a
(R = Tol, Ph)
17
(4)

Scheme 4

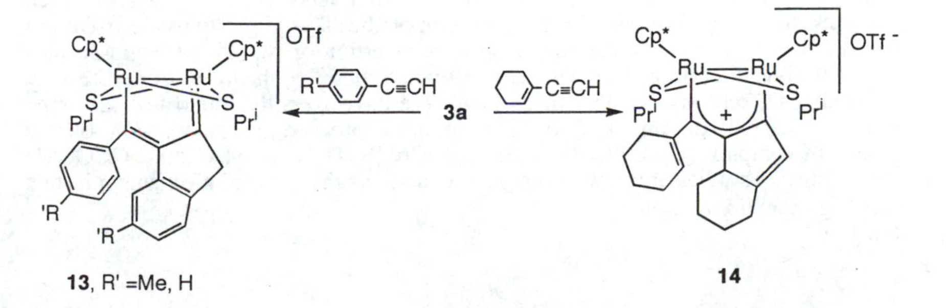
OTf
3a
OTf⁻
13, R' =Me, H
14

Another remarkable feature observed for **16** is that it can catalyze the linear di- and trimerization of HC≡CFe, which will be discussed later.

Propargyl alcohols HC≡CC(OH)R_2 (R = Tol, Ph) react with **3a** in a different manner, yielding the dinuclear terminal allenylidene complexes **17** (eq 4). The allenylidene ligands in **17** presumably result from the initial formation of the vinylidene ligand Ru=C=CH{C(OH)R_2} followed by dehydration (*4*). Particularly noteworthy is the reaction of these propargyl alcohols with **2a**, which produced the quite unusual complexes **18** having a diruthenacyclopentanone framework substituted with diarylmethylene and diarylvinylidene groups. This may arise from the coupling of the two alkyne molecules accompanied by a pseudo-1,3-shift of an oxygen atom (*21*). This reaction sharply contrasts with the formation of bis(alkynyl) complexes in the reactions of **2a** with HC≡CR (R = aryl) shown in eq 1. Furthermore, similar treatment of **2a** with HC≡CC(OH)Me_2 did not give the methyl analogue of **18** but afforded a diruthenacyclopentenone complex **19**, showing that another type of coupling of two HC≡CC(OH)Me_2 molecules proceeded for this propargyl alcohol (Scheme 5). Although we must await further investigation to elucidate the detailed mechanisms operating in these reactions, one plausible pathway leading to **17** and **18** might involve intramolecular C-C bond formation between either allenylidene and alkynyl or vinylvinylidene and alkynyl ligands in intermediates **20** and **21**, respectively.

Catalytic Oligomerization of HC≡CFe Promoted by 16. The μ-butenynyl complex **16** shown above has proved to be an efficient catalyst for linear di- and trimerization of HC≡CFc under mild conditions (*18*). The reaction is highly regio- and stereoselective. In a typical run carried out in $ClCH_2CH_2Cl$ at 60 °C for 30 h using the alkyne/catalyst ratio of 20, the head-to-head dimer **22** and the linear trimer **23** were produced in 62 and 32% yields, respectively (eq 5). Compounds **22** and **23** have unambiguously been characterized spectroscopically and **23** by the X-ray crystallography. The formation of the linear trimer **23** is particularly interesting, since it has never been observed in the oligomerization with mononuclear complexes. For example, the [$Ni(CO)_2(PPh_3)_2$]-catalyzed reaction gives **22**, the acyclic branched trimer FcCH=CHC(C≡CFc)=CHFc, and the cyclic trimer 1,2,4-triferrocenylbenzene (*22*). No byproducts such as cyclic oligomers were detected in the above reaction, and **16** was recovered from the reaction mixture after catalysis in high yield. These results strongly suggest that the diruthenium center participates in this reaction with its bimetallic core intact throughout the catalysis. Highly selective linear trimerization of HC≡CPh using a triangular cluster [$Co_3(\mu_3\text{-}H)(\mu_2\text{-}CO)_3(PMe_3)_6$] has been reported recently (*23*). However, it is not certain whether the active catalyst is the tricobalt species or the monocobalt complex generated from the tricobalt precuror. Furthermore, it should be noted that the linear trimer produced from this Co system, PhC≡CC(Ph)=CHCH=CHPh, has an *E,E* configuration, which is essentially different from **23** with a *Z,Z* configuration.

A possible mechanism for the formation of **22** and **23** is shown in Scheme 6. The butenynyl complex **16** first reacts with two HC≡CFc molecules to form a monoalkynyl-alkyne intermediate **24**, concurrent with liberation of the dimer **22**. The intermediate **24** is then converted into a vinylidene-alkynyl complex which regenerates the starting complex **16** by migration of the alkynyl group to the α-carbon of the vinylidene group. If the trans migratory insertion of two HC≡CFc molecules into the Ru-alkynyl bond in **24** occurs as shown in Scheme 6, the linear trimer **23** may be formed. Of relevance to this mechanism, we have recently found that the trans insertion of an alkyne into the Ru-H bond does proceed in the reaction of the diruthenium complex [Cp*Ru(μ-H)(μ-SPr^i)$_2$RuCp*][OTf] with MeOCOC≡CCOOMe (eq 6). Further studies are now in progress to address the detailed mechanism in this interesting catalytic reaction.

Scheme 5

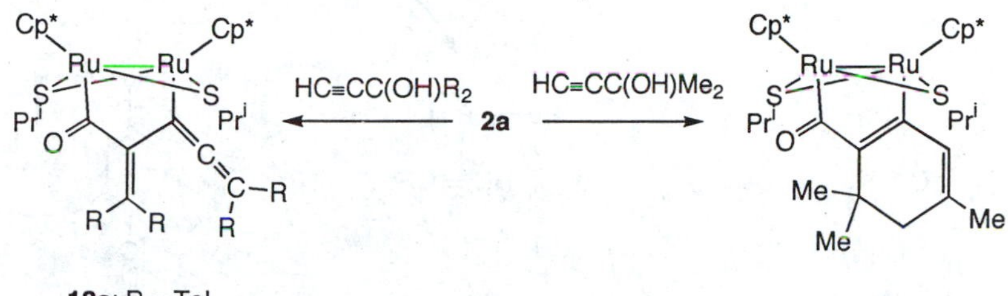
Cp*
Ru
S
Pr^i
O
C
R
HC≡CC(OH)R2
2a
HC≡CC(OH)Me2
Me
18a: R = Tol
18b: R = Ph
19

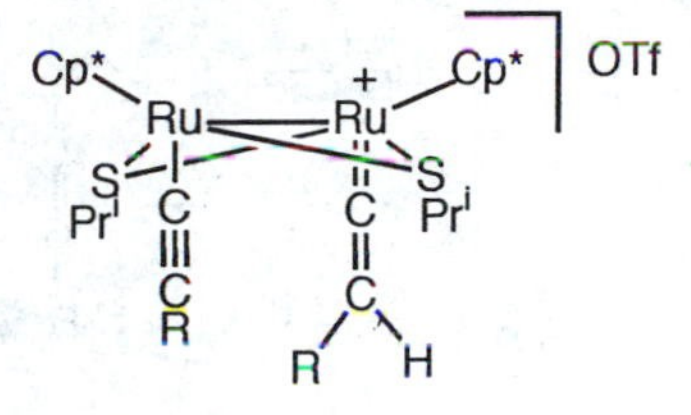
Cp*
+
OTf
Ru
S
Pr^i
C
R
H

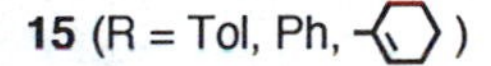
15 (R = Tol, Ph,)

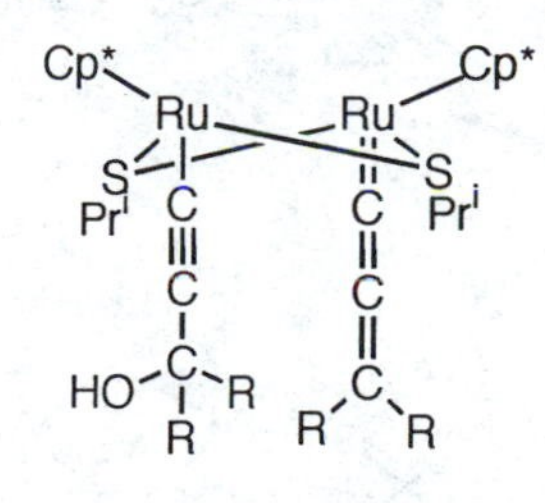
Cp*
Ru
S
Pr^i
C
HO
R
20 (R = Tol, Ph)

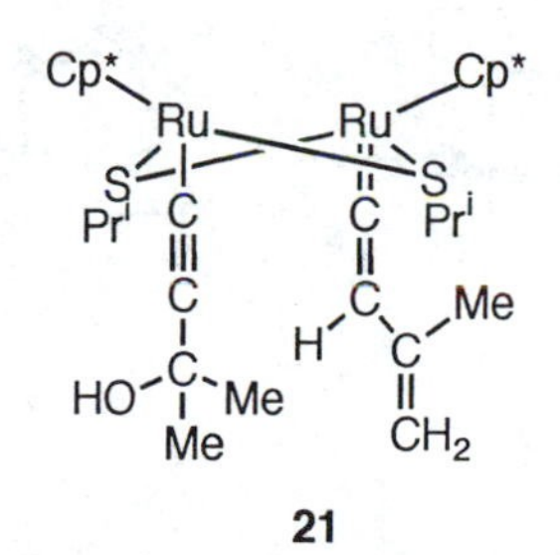
Cp*
Ru
S
Pr^i
C
HO
Me
H
CH2
21

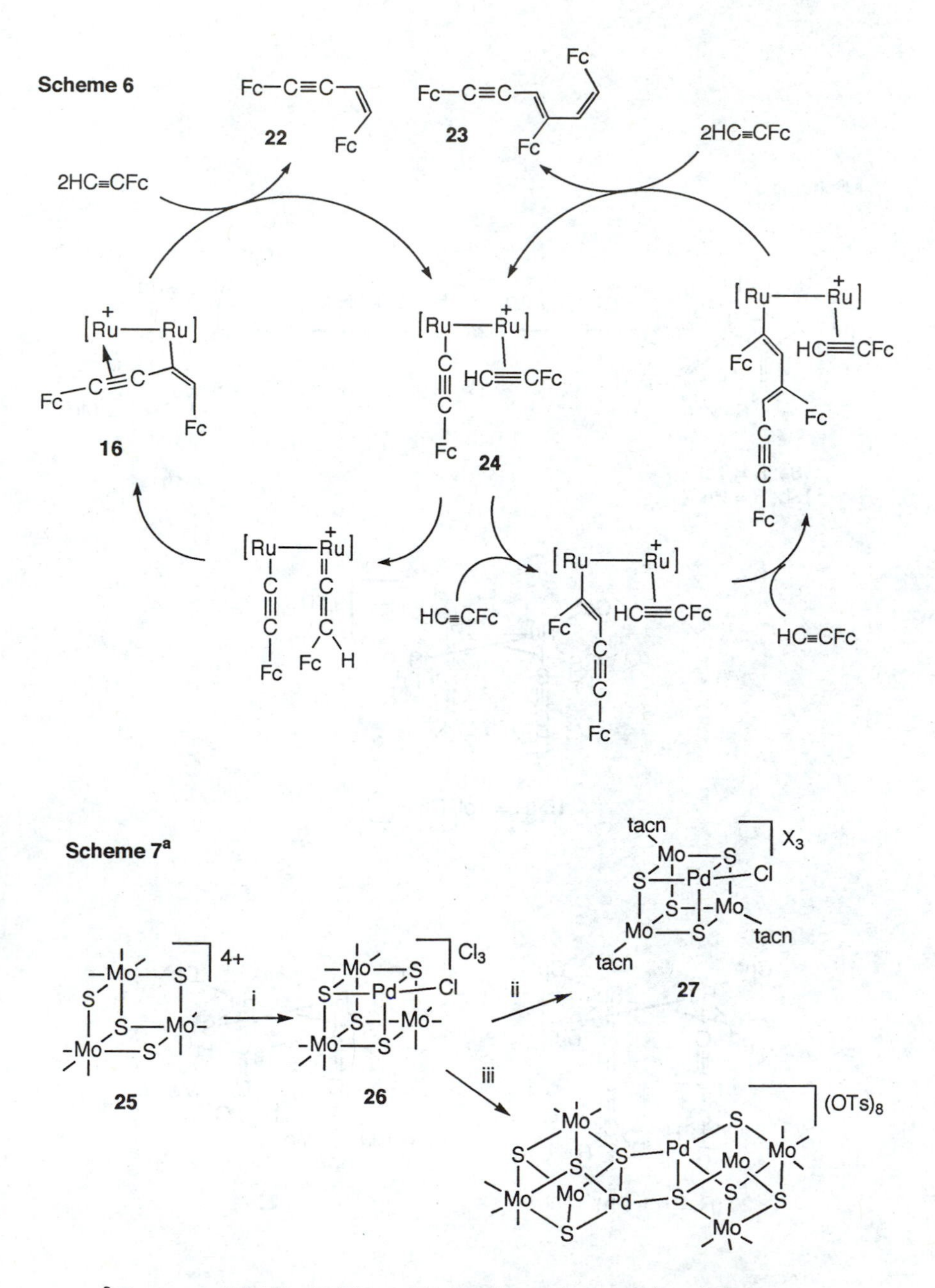

[a] Reagents: (i) Pd black/HCl aq. (ii)1,4,7-triazacyclononane (tacn) then anion metathesis by X^- ($X = ClO_4$, PF_6, BF_4) (iii) TsOH

Heterometallic Cubane Clusters with Bridging Sulfide Ligands.

Transition metal-sulfur cubane clusters, containing the unique M_4S_4 core characterized by two interpenetrating M_4 and S_4 tetrahedra, have been the subject of intensive study, since they may serve as the excellent synthetic models for the protein-bound metal-sulfur clusters occurring in e.g. ferredoxins and nitrogenases and also for the active sites of industrial hydrodesulfurization catalysts. Now a variety of cubane-type homo- and heterometallic clusters are known which contain the metals extended far beyond those present in the biological and industrialsystems. However, the number of the clusters containing noble metals is still limited and clusters that show catalytic activity in a well-defined manner are quite rare, regardless of the metals incorporated.

Synthesis of the $PdMo_3S_4$ Clusters and Reactivity at the Unique Pd Site. Recent studies have shown that the reaction of an incomplete cubane-type aqua cluster $[Mo_3S_4(H_2O)_9]^{4+}$ (**25**) with metals provides a very convenient and versatile route to prepare the mixed-metal MMo_3S_4 cubane clusters (M = Fe, Ni, Co, Cu, Sn, Hg, Sb, etc.) (*24*). We have now found that treatment of **25** with excess Pd black in aqueous HCl affords $[PdMo_3S_4(H_2O)_9Cl]Cl_3$ (**26**) in ca. 90% yield, which demonstrates the first mixed-metal cubane cluster of this type containing a noble metal (*5*). The metrical parameters in the $PdMo_3S_4$ core have been determined for both single and double cubane clusters, $[PdMo_3S_4(tacn)_3Cl]Cl_3$ and $[Pd_2Mo_6S_8(H_2O)_{18}][OTs]_8$ (Ts = 4-$MeC_6H_4SO_2$), derived readily from **26** (Scheme 7). An attractive feature in these clusters is the presence of the tetrahedral Pd site not commonly observed in Pd(II) compounds. Reactivity displayed by this unique Pd site is therefore of particular interest.

Facile coordination of a series of substrates to the $PdMo_3S_4$ core has been demonstrated when the clusters $[PdMo_3S_4(tacn)_3Cl]X_3$ (**27**, X = BF_4, PF_6, ClO_4) were used. Replacement of the Cl anions of **26** in the outer coordination sphere by the poorly coordinating anion X is apparently requisite for isolating the two-electron donor adducts $[PdMo_3S_4(tacn)_3(L)]^{4+}$, where L = CO (**28**), alkenes (**29**), and isocyanides (**30**) (*5*). The X-ray analyses have been undertaken for the clusters $[PdMo_3S_4(tacn)_3(CO)]ClX_3$ (**28a**, X = ClO_4) and **29** (L = *cis*-$HOCH_2CH{=}CHCH_2OH$), which have clearly shown that the coordination of these substrates takes place at the tetrahedral Pd site in the terminal end-on manner for the former and in the side-on fashion for the latter. The IR spectra of **28** show characteristic ν(CO) bands at 2085 for **28a** and 2087 cm^{-1} for **28b** (X = PF_6). These values are significantly lower than those of terminal CO ligands in precedented Pd(II) complexes which commonly appear over 2100 cm^{-1}, but still much higher than those of terminal CO ligands in Pd(0) complexes. This feature can be accounted for by the significant π-donating ability of the Pd atom surrounded by three sulfide ligands. The coordination of isocyanides also occurs at the Pd atom in a terminal end-on manner. The linear Pd-C-N-R linkage in the isocyanide ligand has been confirmed by the preliminary X-ray study of **30** (L = 2,4,6-$Me_3C_6H_2NC$).

Regio- and Stereoselective Addition of Alcohols and Carboxylic Acids to Alkynes Catalyzed by Clusters 27. Interestingly, **27a** (X = PF_6) catalyzes the reaction of alkynic acid esters with alcohols to give the trans addition products under mild conditions with retention of the cluster core (eq 7). The reactions are highly selective, and the cis addition product (*E*)-$R^3OCR^2{=}CHCOOR^1$ and $(R^3O)_2CR^2CH_2COOR^1$ were produced in only negligible amounts, if any (<3% yield) (*5b*). Reactions of the alkynic acid esters of the type $R^2C{\equiv}CCOOR^1$ were

Table I. Conversion of $R^2C{\equiv}CCOOR^1$ into (*Z*)-$(R^3O)CR^2{=}CHCOOR^1$ Catalyzed by 27a[a]

R^1	R^2	R^3	Temp/°C	Time/h	Conv/%	Yield/%[b]
Me	H	Me	40	6	100	82(97)
Me	H	Me	40	24	98	89(95)[c]
Ph	H	Me	40	6	100	94
Me	COOMe	Me	50	24	100	76
Me	Me	Me	30	30	95	82
Me	H	Et	reflux	20	100	83[d,e]
Me	H	$PhCH_2$	reflux	8	96	70[d,f]

[a] Reaction conditions: alkynic acid ester, 3.0 mmol; **27a**, 8.8 μmol (0.3 mol%); MeOH, 2 mL. [b] Isolated yield; the GLC yields in parentheses. [c] **27a**, 0.06 mol%. [d] Acetone (2 mL) was used as solvent. [e] EtOH, 15.0 mmol. [f] $PhCH_2OH$, 9.0 mmol.

considerably slower than those of $HC{\equiv}CCOOR^1$. Reactions of alcohols other than MeOH such as EtOH and $PhCH_2OH$ were performed by adding acetoneas a solvent because of the low solubility of **27a** in these alcohols. Typical results of these reactions are listed in Table I.

Conversion of the alkynic acid esters into the alcohol adducts presumably proceeds at the unique Pd site in the $PdMo_3S_4$ core. This conclusion is supported by the fact that the reaction of $HC{\equiv}CCOOMe$ with MeOH did not take place in the presence of the Mo_3S_4 cluster **25**. Furthermore, it is noteworthy that mononuclear Pd(II) complexes such as $[PdCl_2(PhCN)_2]$, $Na_2[PdCl_4]$, $Pd(OAc)_2$, and $[PdCl_2(PPh_3)_2]$ did not work as effective catalysts.

More recently we have found that **27a** also catalyzes the stereoselective addition of carboxylic acids to alkynic acid esters to give vinyl esters (eq 8) (*6*). In a typical run, when $HC{\equiv}CCOOMe$ dissolved in MeCN was reacted with MeCOOH at 40 °C for 8 h in the presence of **27a** and NEt_3, $MeCOOCH{=}CHCOOMe$ was obtained in 65% yield with a *Z*/*E* isomer ratio of 98/2. Although certain Pd, Ru, Rh, and Ag compounds are known to catalyze the addition of carboxylic acids to alkynes, examples of stereoselective intermolecular additions are limited and no effective catalysts are reported for the reaction of carboxylic acids with terminal electron deficient alkynes to give the *Z* vinyl esters. Selected results obtained in the reactions using **27a** are shown in Table II.

The proposed mechanism for these highly selective trans addition reactions of alcohols and carboxylic acids to alkynic acid esters involves the initial coordination of the alkynic acid esters to the Pd site in the cubane cluster and the successive nucleophilic attack of the alcohols or carboxylate anions on the coordinated alkynes from the outer coordination sphere. Following protonolysis of the Pd-C bond the trans addition products are formed. The mechanism for the reactions of alkynic acid esters with alcohols is shown in Scheme 8. Since the Pd site in **27a** is surrounded by three firmly bonded sulfide ligands and provides only a single vacant site, side reactions such as oligomerization may be suppressed. It should be emphasized that the present reactions offer one of the rare examples in which the well-characterized cluster compounds catalyze organic reactions quite efficiently with retention of the cluster structure.

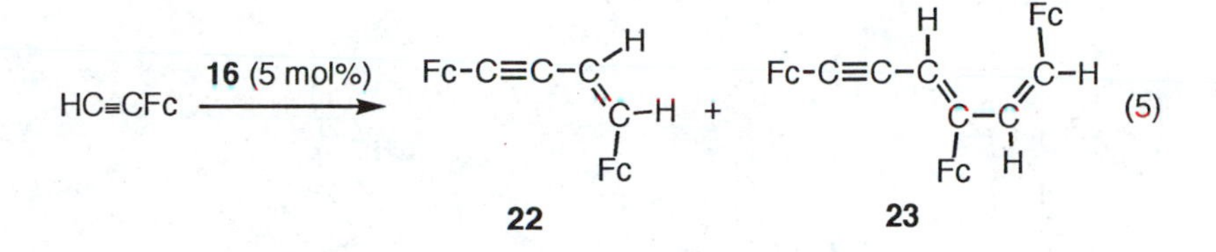

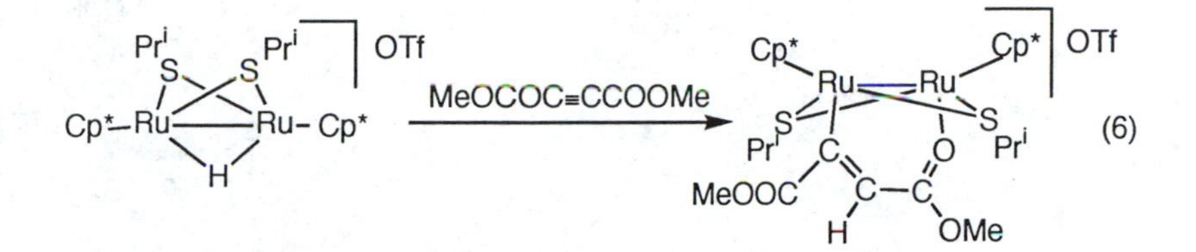

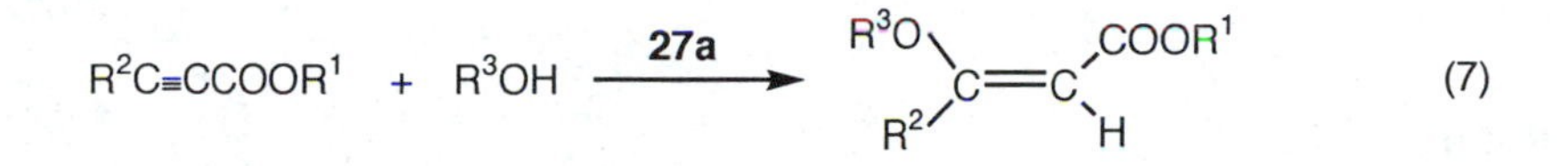

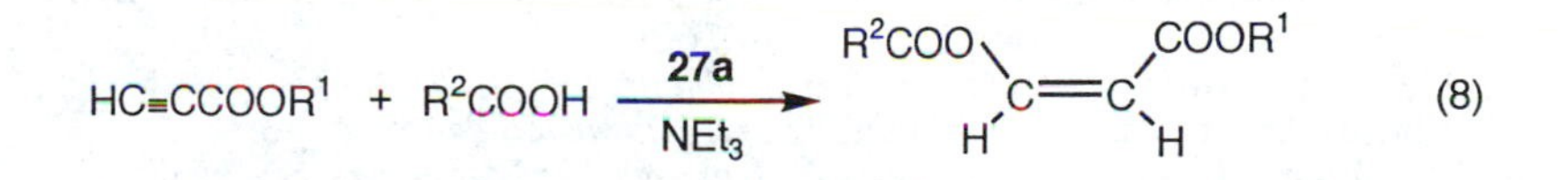

Table II. Conversion of $HC{\equiv}CCOOR^1$ into (*Z*)-$R^2OCOCH{=}CHCOOR^1$ Catalyzed by 27a[a]

R^1	R^2	Time/h	Conv/%	Yield/%[c]
Me	Me	8	90	62
Me	Me	72	97	66[b]
Me	Ph	5	92	76
Et	Me	11	97	58
Bu^t	Me	12	95	71

[a] Reaction conditions: alkynic acid ester, 3.0 mmol; carboxylic acid, 9.0 mmol; NEt_3, 0.15 mmol; MeCN, 2 mL; **27a**, 9.0 μmol; 40 °C. [b] No NEt_3 added.
[c] Isolated yield.

Scheme 8

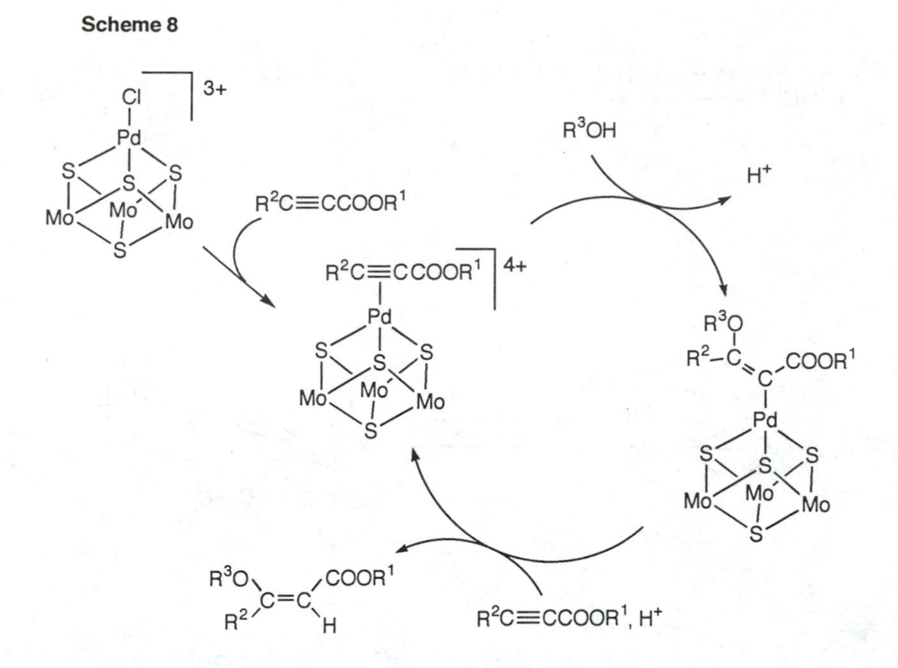

References

(1) Hidai, M.; Mizobe, Y.; Matsuzaka, H. *J. Organomet. Chem.* **1994**, *473*, 1.

(2) (a) Takahashi, A.; Mizobe, Y.; Matsuzaka, H.; Dev, S.; Hidai, M. *J. Organomet. Chem.* **1993**, *456*, 243. (b) Hörnig, A.; Rietman, C.; Englert, U.; Wagner, T.; Koelle, U. *Chem. Ber.* **1993**, *126*, 2609.

(3) Dev, S.; Mizobe, Y.; Hidai, M. *Inorg. Chem.* **1990**, *29*, 4797.

(4) Matsuzaka, H.; Takagi, Y.; Hidai, M. *Organometallics* **1994**, *13*, 13.

(5) (a) Murata, T.: Gao, H.; Mizobe, Y.; Nakano, F.; Motomura, S.; Tanase, T.; Yano, S.; Hidai, M. *J. Am. Chem. Soc.* **1992**, *114*, 8287. (b) Murata, T.; Mizobe, Y.; Gao, H.; Ishii, Y.; Wakabayashi, T.; Nakano, F.; Tanase, T.; Yano, S.; Hidai, M.; Echizen, I.; Nanikawa, H.; Motomura, S. *J. Am. Chem. Soc.* **1994**, *116*, 3389.

(6) Wakabayashi, T.; Ishii, Y.; Murata, T.; Mizobe, Y.; Hidai, M. *Tetrahedron Lett.* **1995**, *36*, 5585.

(7) Takahashi, A.; Mizobe, Y.; Hidai, M. *Chem. Lett.* **1994**, 371.

(8) Kuwata,S.; Mizobe, Y.; Hidai, M. *Inorg. Chem.* **1994**, *33*, 3619.

(9) (a) Tilley, T. D.; Grubbs, R. H.; Bercaw, J. E. *Organometallics* **1984**, *3*, 274. (b) Oshima, N.; Suzuki, H.; Moro-oka, Y. *Chem. Lett.* **1984**, 1161.

(10) (a) Dev, S.; Imagawa, K.; Mizobe, Y.; Cheng, G.; Wakatsuki, Y.; Yamazaki, H.; Hidai, M. *Organometallics* **1989**, *8*, 1232. (b) Hidai, M.; Imagawa, K.; Cheng, G.; Mizobe, Y.; Wakatsuki, Y.; Yamazaki, H. *Chem. Lett.* **1986**, 1299.

(11) (a) Nishio, M.; Matsuzaka, H.; Mizobe, Y.; Hidai, M. *Organometallics* **1996**, *15*, 965. (b) Nishio, M.; Matsuzaka, H.; Mizobe, Y.; Tanase, T.; Hidai, M. *Organometallics* **1994**, *13*, 4214. (b) Nishio, M.; Matsuzaka, H.; Mizobe, Y.; Hidai, M. *J. Chem. Soc., Chem. Commun.* **1993**, 375.

(12) Matsuzaka, H.; Mizobe, Y.; Nishio, M.; Hidai, M. *J. Chem. Soc., Chem. Commun.* **1991**, 1101.

(13) Koelle, U.; Rietmann, C.; Tjoe, J.; Wagner, T.; Englert, U. *Organometallics* **1995**, *14*, 703.
(14) Matsuzaka, H.; Hirayama, Y.; Nishio, M.; Mizobe, Y.; Hidai, M. *Organometallics* **1993**, *12*, 36.
(15) Bruce, M. I.; Koutsantoun, G. A.; Liddell, M. J. *J. Organomet. Chem.* **1987**, *320*, 217.
(16) Trinquier, G.; Hoffmann, R. *Organometallics* **1984**, *3*, 370.
(17) Takahashi, A.; Mizobe, Y.; Tanase, T.; Hidai, M. *J. Organomet. Chem.* **1995**, *496*, 109.
(18) Matsuzaka, H.; Takagi, Y.; Ishii, Y.; Nishio, M.; Hidai, M. *Organometallics* **1995**, *14*, 2153.
(19) Bianchini, C.; Peruzzini, M.; Zanobini, F.; Frediani, P.; Albinati, A. *J. Am. Chem. Soc.* **1991**, *113*, 5453.
(20) Wakatsuki, Y.; Yamazaki, H.; Kumegawa, N.; Satoh, H.; Satoh, J. *J. Am. Chem. Soc.* **1991**, *113*, 9604.
(21) Matsuzaka, H.; Koizumi, H.; Takagi, Y.; Nishio, M.; Hidai, M. *J. Am. Chem. Soc.* **1993**, *115*, 10396.
(22) Pittman, C. U., Jr.; Smith, L. R. *J. Organomet. Chem.* **1975**, *90*, 203.
(23) Klein, H.-F.; Mager, M.; Isringhausen-Bley, S.; Flörke, U.; Haupt, H.-J. *Organometallics* **1992**, *11*, 3174.
(24) Shibahara, T. *Adv. Inorg. Chem.* **1991**, *37*, 143 and references therein.

Chapter 20

Thio-Tungsten Chemistry

Aston A. Eagle, Simon Thomas, and Charles G. Young[1]

School of Chemistry, University of Melbourne, Parkville, Victoria 3052, Australia

The synthesis and characterization of a variety of thio-tungsten complexes containing the hydrotris(3,5-dimethylpyrazol-1-yl)borate ligand (L) are described. High-valent syntheses featuring oxo/thio ligand exchange and atom transfer reactions yield a variety of $LW^{VI}OSX$, $LW^{VI}S_2X$, $LW^{V}SX_2$, and $LW^{IV}SX$ complexes (X = monodentate or ambidentate anion). These and other thio-tungsten complexes may be prepared using a low-valent approach involving successive oxidation of $[LW^0(CO)_3]^-$, $LW^{II}X(CO)_2$ or $LW^{II}X(\eta^2\text{-}MeCN)(CO)$, and $LW^{IV}EX(CO)$ (E = O, S). In this approach, oxygen atom transfer or sulfur atom transfer reagents generate the terminal oxo or thio ligands, respectively. The spectroscopic, structural and chemical properties of these thio-tungsten complexes are presented and discussed.

Important biological and industrial processes are catalysed at metal-sulfur centers (*1*, Chapter 1, this volume). The group 6 elements molybdenum and tungsten, in combination with a variety of sulfur-donor ligands, are vital components of the active sites of nitrogenase (*2*) and the various molybdopterin enzymes (vide infra) (*3*). As well, alumina supported molybdenum- and tungsten-sulfide catalysts are extensively employed in commercial hydrotreating processes such as hydrodesulfurization, hydrogenation, isomerization and hydrocracking (*4*). While the biological importance of molybdenum has been recognized for many decades, the role of tungsten in biological systems has become apparent only in recent years (*3a*). Specifically, a number of thermally stable tungsten oxido-reductases have been isolated from hyperthermophilic organisms (*5*). The extreme oxygen sensitivity of these enzymes, the presence of oxo ligands in the inactive forms of the enzymes, and the enhancement of activity in the presence of added sulfide, provide indirect support for thio-tungsten centers in some enzymes (*3a*). Accordingly, we have been exploring thio-tungsten chemistry with a view to ultimately assessing its relevance to biological systems. As well, we have sought to develop strategies for the selective synthesis and stabilization of thio-tungsten complexes, so that this area of chemistry, in its own right, can be further developed. Over recent years we have designed complementary high-valent and low-valent approaches for the synthesis of mononuclear oxo-thio- and bis(thio)-tungsten(VI) complexes and a variety of related or intermediate thio-tungsten(IV) and thio-tungsten(V) complexes. This paper summarizes the synthesis and characterization

[1]Corresponding author

0097–6156/96/0653–0324$15.00/0

of thio-tungsten complexes containing the hydrotris(3,5-dimethylpyrazol-1-yl)borate ligand (L), designed to maintain the mononuclearity of the metal center and restrict the potential thio-ligand chemistry to three mutually *cis* coordination sites. We begin by describing the synthetic approaches adopted and the spectroscopic and structural attributes of the complexes prepared to date. This is followed by a comparison of the electrochemical properties of the complexes (and their oxo analogues) and a brief survey of their reactivity. The synthesis of an enzyme active site model by a novel reaction generating an ene-1,2-dithiolate is an interesting outcome of the reactivity survey. [Note: Thio is the recommended and preferred name for the formally S^{2-} ligand. In Chemical Abstracts Service index nomenclature, this name is modified to thioxo when a terminal ligand is specified (μ_x-thio specifies bridging ligands). Sulfido is a commonly used alternative name for the S^{2-} ligand. In this article we use the term thio to refer to the ligand in M=S units (*6*)].

Synthetic Approaches and Characterization Data

Mononuclear, high-valent thio-tungsten complexes are more numerous and more stable than related thio-molybdenum species, although they are still much rarer than oxo-tungsten and polyoxotungstate complexes (*7-9*). Thio-tungsten complexes have been prepared by two principal methods: 1) Oxo/thio ligand exchange effected by sulfidation of oxo complexes, e.g., the synthesis of $[WO_{4-n}S_n]^{2-}$ (*10*), $WSE(R_2NO)_2$ (E = O, S throughout) (*11*), Cp^*WOSR (*12*) and $WS_3(CH_2CMe_3)$ (*13*) from their oxo analogues; and 2) Ligand exchange involving thio-halide complexes, e.g., the synthesis of $WS(OR)_4$ (*14*) and $[WS(S_2CNR_2)_3]^+$ (*15*) complexes from WCl_4. Oxygen or sulfur atom transfer, effectively two-electron oxidations, have not been widely applied to the synthesis of such species. Oxo/thio ligand exchange and atom transfer reactions feature in the high-valent and low-valent approaches detailed below.

High-Valent Approach. Reaction of WO_2Cl_2 with KL in *N,N*-dimethylformamide results in the formation of LWO_2Cl, obtained as a pure, white crystalline compound by Soxhlet extraction of the crude precipitated product of the reaction (*16, 17*). Reaction of LWO_2Cl with B_2S_3 (*16*) or P_4S_{10} in refluxing benzene results in the formation of pink LWOSCl and green LWS_2Cl. Significant reduction to W(V) occurs when this reaction is performed 1,2-dichloroethane, and large amounts of blue $LWOCl_2$ and orange $LWSCl_2$ are generated. Methathesis of the LWE_2Cl complexes with various alkali metal salts in toluene or tetrahydrofuran is assisted by the presence of 18-crown-6, and produces LWE_2X (X = OPh, SPh, SePh, mentholate) complexes. Reaction of $LWO_2(S_2PPh_2)$ with B_2S_3 (but not P_4S_{10}) forms $LWS_2(S_2PPh_2)$ [the complex $LWOS(S_2PPh_2)$ can be prepared by a low-valent route (vide infra)]. In contrast to the pale yellow/white dioxo complexes, the thio complexes are intensely colored; the stability and solubilities of these crystalline solids depend on X.

The oxo-thio-W(VI) complexes exhibit strong infrared bands at ca. 940 and 480 cm^{-1}, which are assigned to ν(W=O) and ν(W=S) vibrational modes, respectively, as well as bands due to L. The ^1H NMR spectra are consistent with the presence of molecules with C_1 symmetry. Two diastereomers of LWOS(mentholate) are produced upon methathesis of LWOSCl with sodium (1*R*,2*S*,5*R*)-(-)-mentholate in toluene [these are chiral at tungsten and are assigned (*R*)- and (*S*)- configurations by taking the W··B vector as the imaginary, single binding point of the L ligand, which is assigned lowest priority]. The X-ray crystal structure of (*R,S*)-LWOS(mentholate) has been determined (Figure 1) (*18*); the six-coordinate molecules exhibit distorted octahedral geometries with average W-O(1), W-S(1), and W-O(2) distances of 1.734(9) Å, 2.118(6) Å and 1.850(7) Å, respectively. The O(1)-W-S(1) angles average 102.9(5)°. Only one other octahedral *cis*-oxo-thio-tungsten complex, viz., $[WOS(NCS)_4]^{2-}$, has been structurally characterized (*19*). The incorporation of the chiral ligand was important for the crystallographic distinction of the oxo- and thio- ligands.

The LWS_2X complexes exhibit $\nu_s(WS_2)$ and $\nu_{as}(WS_2)$ infrared bands at ca. 495 and 475 cm^{-1}, respectively, and ^1H NMR spectra consistent with molecular C_s symmetry. The structures of $LWS_2(OPh)$ (*18*) and $LWS_2(S_2PPh_2)$ have been determined by X-ray diffraction. In the structure of $LWS_2(S_2PPh_2)$ (Figure 2), two equivalent W=S bond distances (av. 2.140 Å) are observed. The S(1)-W-S(2) angle (102.3°) and S(1)··S(2) separation (3.33 Å) confirm the presence of a *cis*-bis(thio)-W(VI) moiety. Structurally characterized *cis*-bis(thio)-metal complexes are extremely rare (*11,12,20,21*) and $LWS_2(OPh)$ and $LWS_2(S_2PPh_2)$ are the only two which possess biologically relevant two-electron nitrogen donor atoms and an octahedral geometry.

The Low-Valent Approach. In principle, successive two-electron oxidative atom transfer reactions beginning at the W(II) level should be capable of selectively generating oxo- and thio-W(IV) and oxo-thio- and bis(thio)-W(VI) complexes. Our work in this area was initiated to improve synthetic control, especially in the generation of oxo-thio complexes. As well, we were intrigued by the prospect of generating complexes with unusual combinations of π-acid and π-base ligands, especially (carbonyl)thio- and (carbonyl)oxo-W(IV) complexes. The sequence of reactions begins with yellow $NEt_4[LW(CO)_3]$, first prepared by Trofimenko in 1967 (*22*). Halogen oxidation leads to the formation of carbonyl-halide complexes of the type $LWX(CO)_3$ (*23-25*). The reactions are carried out in acetonitrile using elemental halogen or $PhICl_2$ (in the case of unreported chloro complexes). The tricarbonyl complexes may be thermally decarbonylated to the corresponding dicarbonyl complexes, $LWX(CO)_2$, which are brown, paramagnetic complexes (*23-25*). We have optimized the syntheses for the full series of carbonyl-halides $LWX(CO)_x$ ($x = 2$, 3) and prepared a number of derivatives, including the dithio-acid complexes $LW(S_2PR_2)(CO)_2$ by methathesis reactions. Reaction of these complexes with dioxygen (halides only) or one equivalent of the oxygen atom donor pyridine *N*-oxide results in clean conversion to the (carbonyl)oxo-W(IV) complexes LWOX(CO) (*24,26*). The corresponding reaction with the sulfur atom donor propylene sulfide results in the formation of the dinuclear μ-thio complex $[LW(CO)_2]_2(\mu\text{-}S)$ (*25*). For this complex, the W-S distances of 2.181 Å and W-S-W angle of 171.6° are consistent with considerable π-bonding in the trinuclear core. There is no evidence for the formation of (carbonyl)thio-W(IV) complexes, LWSX(CO), in these reactions. In order to effect the oxidation of the tungsten center by a sulfur atom donor, it is necessary to employ a monocarbonyl complex as precursor. Prolonged thermal decarbonylation of the $LWX(CO)_2$ complexes yields novel η^2-acetonitrile complexes, $LWX(\eta^2MeCN)(CO)$ (*26*). These complexes are ideal starting materials for forming thio-W(IV) complexes and their reaction with propylene sulfide results in (carbonyl)thio-W(IV) complexes, LWSX(CO). The low-valent approach converges with the high-valent approach at the W(IV) level, as W(IV) complexes can also be accessed from the LWE_2X complexes by oxygen or sulfur atom abstraction reactions (vide infra). Further, selective atom transfer reactions involving the W(IV) complexes are capable of generating the LWE_2X complexes previously described. The development of optimal syntheses at each stage of the sequence beginning with $NEt_4[LW(CO)_3]$ was important to the eventual success of this approach. A number of other dicarbonyltungsten complexes undergo similar reactions, e.g., $LW(CO)_2(S_2CNEt_2)$ reacts with pyridine *N*-oxide and sulfur or episulfides to give $LWE(S_2CNEt_2)$, E = O or S, respectively (*27*).

We now provide a description of the spectroscopic and structural properties of the novel η^2-acetonitrile and (carbonyl)thio complexes accessed in the low-valent approach. The nitrile complexes exhibit a strong $\nu(CO)$ band at ca. 1910 cm^{-1}, a weak $\nu(C{\equiv}N)$

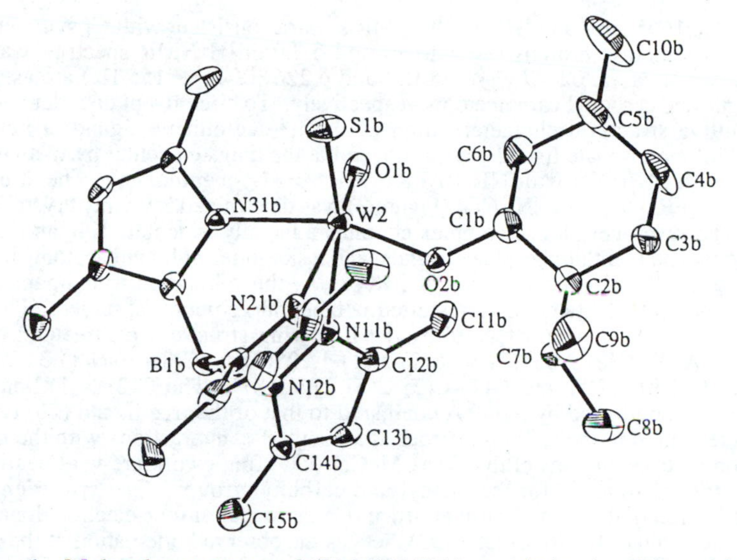

Figure 1. Molecular structure of (R)-LWOS{(-)-mentholate}. Selected bond lengths (Å) and angles (deg) include the following: W(2)-O(1b) 1.71(2), W(2)-S(1b) 2.132(8), W(2)-O(2b) 1.86(1), W(2)-N(11b) 2.26(2), W(2)-N(21b) 2.31(2), W(2)-N(31b) 2.18(1), O(2b)-C(1b) 1.40(2); O(1b)-W-S(1b) 102.0(6).

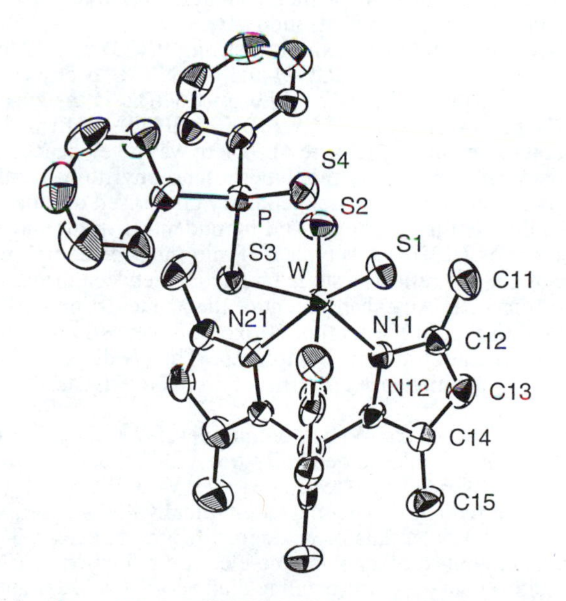

Figure 2. Molecular structure of $LWS_2(S_2PPh_2)$.. Selected bond lengths (Å) and angles (deg) include the following: W-S(1) 2.139(2), W-S(2) 2.138(2), W-S(3) 2.418(2), W-N(11) 2.151(6), W-N(21) 2.301(6), W-N(31) 2.330(7), P-S(3) 2.085(3), P-S(4) 1.954(3); S(1)-W-S(2) 102.37(9).

band at ca. 1695 cm^{-1} and ^1H and ^{13}C NMR spectra consistent with C_1 symmetry. The acetonitrile methyl protons resonate around δ 3.9 in ^1H NMR spectra. Carbon-13 resonances at ca. δ 202 ($^1J_{WC}$ ~ 26 Hz) and δ 226 ($^1J_{WC}$ ~ 155 Hz) are assigned to the nitrile and carbonyl carbon atoms, respectively. To circumvent disorder and permit a definitive structural characterization of the η^2-acetonitrile ligand, a new chiral dialkyldithiophosphate ligand was introduced at the tungsten center by methathesis of LWI(η^2-MeCN)(CO) with $NH_4[S_2PR^*_2]$ [R* = (-)-mentholate]. The structure of (*R*)-LW($S_2PR^*_2$)(η^2-MeCN)(CO) (Figure 3) was determined by X-ray crystallography (*26*). The seven-coordinate complex contains a facially tridentate L ligand, carbonyl and monodentate dithiophosphate ligands, and a side-on *C,N*-bound acetonitrile ligand. Considering the small-bite-angle of 34.9(3) Å, the η^2-acetonitrile ligand may be considered monodentate, and a distorted octahedral geometry assigned. The W(η^2-MeCN) fragment is characterized by the following structural parameters: W-N(1) 2.033(6) Å, W-C(2) 2.051(7) Å, C(2)-N(1) 1.225(9) Å, C(2)-W-N(1) 34.9(3)°, W-C(2)-C(3) 150.0(7)°, and N(1)-C(2)-C(3) 138.1(8)°. The C(2)-N(1) bond in the complex is lengthened by 0.066 Å compared to that of the free ligand (*28*) consistent with a reduction in the C-N bond order as a result of π-interactions with the tungsten. The non-hydrogen atoms of the W(η^2-MeCN)(CO) framework are very nearly planar with a *syn* relationship for the methyl and carbonyl groups. The *syn* orientation is sterically preferred but an electronic structure akin to that of four-electron donor alkyne complexes such as LWI(RC≡CR)(CO) dictates the observed orientation of the carbonyl and acetonitrile ligands. Structurally and spectroscopically, the W(η^2-MeCN) moieties of LWX(MeCN)(CO) are closely related to those of [WCl(η^2-MeCN)(bpy)-$(PMe_3)_2]PF_6$ (*29a*) and WCl_2(η^2-MeCN)$(PMe_3)_3$ (*29b*) the only other known four-electron-donor η^2-acetonitrile complexes. All other η^2-acetonitrile complexes exhibit structures and spectroscopic properties consistent with a two-electron donor description (*30*). The activation of nitriles by metal centers has many potential synthetic applications and attracts considerable attention (*31*).

The (carbonyl)thio complexes exhibit a strong ν(CO) band at ca. 1960 cm^{-1}, a weak ν(W=S) band at ca. 511 cm^{-1} and ^1H and ^{13}C NMR spectra consistent with C_1 symmetry. A ^{13}C resonance at ca. δ 273 ($^1J_{WC}$ ~ 164 Hz) is assigned to the carbonyl carbon atom. The crystal structure of LWS(S_2PPh_2)(CO) confirmed the presence of the (carbonyl)thiotungsten moiety (Figure 4). There was no evidence of disorder in this structure, the first available for any monomeric (carbonyl)thio-metal complex. The molecule exhibits a distorted octahedral geometry comprised of a facially tridentate L ligand, and mutually *cis* thio, carbonyl and monodentate dithiophosphinate ligands. The W-S(1) distance of 2.132(4) Å is typical of thio-tungsten complexes (*32*). The W-C(1)-O(1) angle of 168(1)° reflects a steric interaction between the carbonyl ligand and uncoordinated sulfur S(3). An enhancement of the π-interaction in the (carbonyl)thio complexes relative to that in the (carbonyl)oxo complexes is reflected in the lower energy of the ν(CO) band of the thio complexes. The *cis* disposition of the carbonyl and thio ligands is nicely rationalized in terms of orbital overlap (*33*).

Electrochemistry. The dioxo-W(VI) complexes LWO_2X (X = Cl, NCS, OMe, OPh, SPh, SePh, S_2PPh_2) undergo generally irreversible, one-electron reductions in the range E_{pc} = -1.05 (X = S_2PPh_2) to $E_{1/2}$ = -1.71 V (X = OMe, reversible) vs SCE in acetonitrile. These reduction potentials are typically 450 - 690 mV more negative than the Mo(VI)/Mo(V) potentials of analogous Mo complexes (*34*). The oxo-thio complex LWOSCl exhibits a reversible, one-electron reduction at $E_{1/2}$ = -0.84 V, ca. 400 mV more positive than the corresponding dioxo complex. A slightly more positive (irreversible) reduction potential is observed for LWS_2Cl (E_{pc} = -0.82 V). The complexes LWS_2X (X = OPh, SPh, SePh) exhibit reversible electrochemistry with reduction potentials $E_{1/2}$ in the range -0.99 to -0.86 V. Reduction potentials for the thio-W complexes are in the range of those established for related $[MoO_2]^{2+}$ complexes

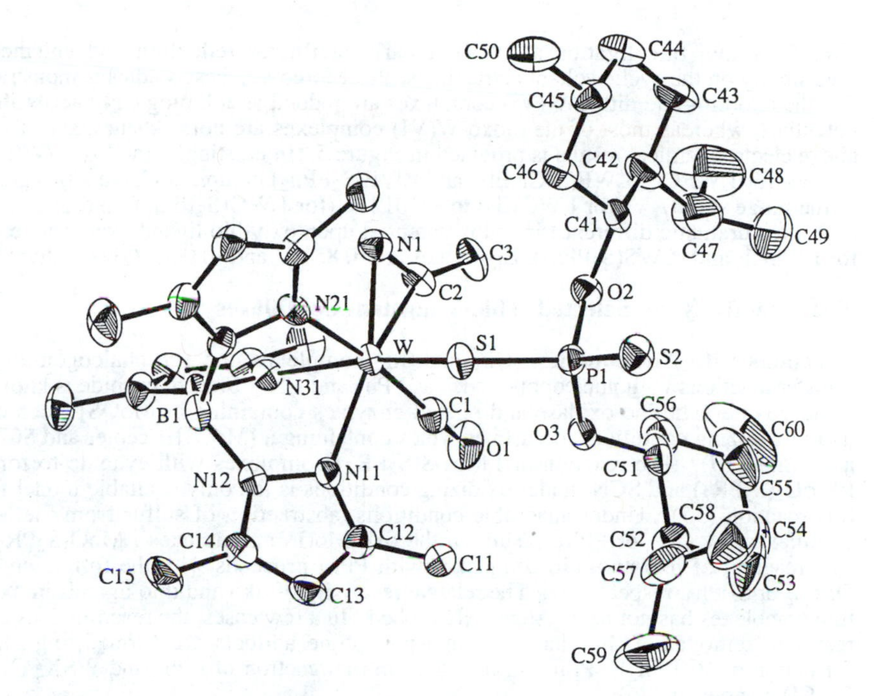

Figure 3. Molecular structure of (*R*)-LW($S_2PR^*_2$)(η^2-MeCN)(CO). Selected bond lengths (Å) include the following: W-C(1) 1.940(8), W-N(1) 2.033(6), W-C(2) 2.051(7), W-S(1) 2.435(2), W-N(11) 2.241(6), W-N(21) 2.262(6), W-N(31) 2.183(6), N(1)-C(2) 1.225(9).

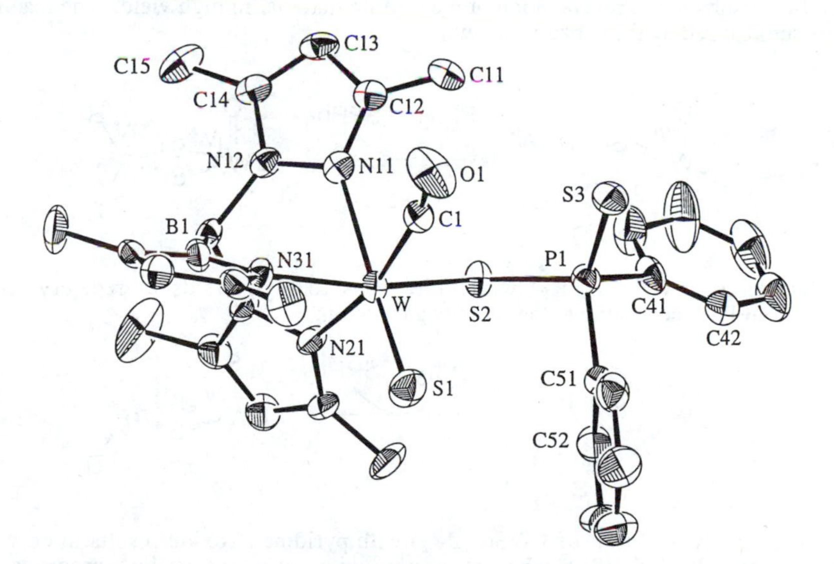

Figure 4. Molecular structure of LWS(S_2PPh_2)(CO). Selected bond lengths (Å) include the following: W-S(1) 2.132(4), W-S(2) 2.446(3), W-C(1) 2.01(1), W-N(11) 2.29(1), W-N(21) 2.18(1), W-N(31) 2.19(1), C(1)-O(1) 1.11(2).

(*34*). Terminal thio ligation at W(VI) clearly facilitates reduction and enhances reversibility on the cyclic voltammetric timescale. Moreover, these studies demonstrate that the oxothio- and dithio-W(VI) complexes are reducible at biologically accessible potentials, whereas most of the dioxo-W(VI) complexes are not. A summary of the above electrochemical results is provided in Figure 5. Interestingly, the W(V)/W(IV) couples for $LWECl_2$, $LWE(S_2CNEt_2)$ and $LWE(S_2PPh_2)$ complexes have $E_{1/2}$ values in the range -0.907 V (for $LWOCl_2$) to +0.012 V (for $LWO(S_2PPh_2)$) and there is a much less dramatic difference in redox potential upon oxo/thio ligand exchange, e.g., for $LWSCl_2$ and $LWS(S_2PPh_2)$, $E_{1/2}$ values are -0.851 V and -0.066 V, respectively.

The Reactivity of Selected Thio-Tungsten Complexes

Reactions with Nucleophiles. Oxo and thio complexes undergo chalcogen atom transfer reactions with nucleophiles such as PPh_3 and CN^-. Indeed, cyanide is known to deactivate xanthine oxidase and related enzymes containing a $[MoOS]^{2+}$ center; under oxidizing conditions, desulfo enzymes containing a $[MoO_2]^{2+}$ center and SCN^- are formed (*3*). The reaction of $LMoOS(S_2PR_2)$ complexes with cyanide to form $LMoO_2(S_2PR_2)$ and SCN^- under oxidizing conditions is the only available model for this reaction (*35*). Under anaerobic conditions, abstraction of sulfur from the Mo complexes, by CN^- or PPh_3, results in the oxo-Mo(IV) complexes $LMoO(S_2PR_2)$. The reaction of oxo and thio complexes with PPh_3 proceeds with the formation of $OPPh_3$ or $SPPh_3$, respectively. The relative reactivity of oxo and thio ligands in oxo-thio complexes has not been extensively probed. In a few cases, the reaction does not result in removal of the chalcogen and phosphine adducts are formed, e.g., the formation of $W(PPh_3)(SPPh_3)X_4$ results from the reaction of PPh_3 and WSX_4 (X = Br, Cl) complexes (*36*).

Using atom transfer reactions, we have been able to interconvert oxo-W(IV), oxo-thio-W(VI), thio-W(IV), and bis(thio)-W(VI) complexes. In a reaction paralleling that observed for the related $LMoOS(S_2PR_2)$ complexes, reaction of PPh_3 with $LWOS(S_2PR_2)$ complexes results in the formation of $LWO(S_2PR_2)$ and $SPPh_3$. The reaction of $LWO(S_2PR_2)$ with elemental sulfur or the sulfur atom donor propylene sulfide results in the regeneration of the starting material in high yield. The reactions are summarized by the equation below:

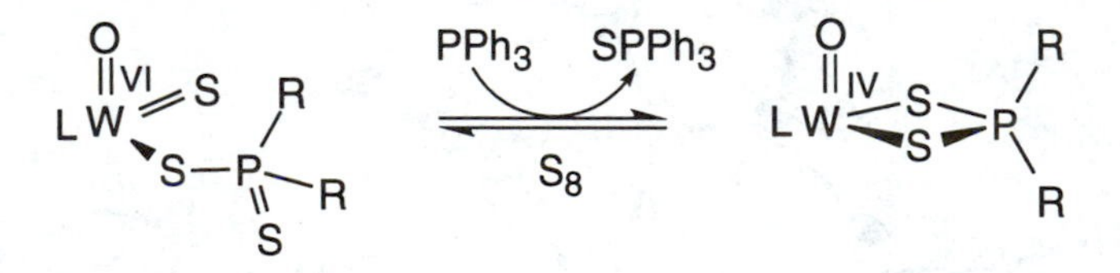

Reaction of $LWS_2(S_2PR_2)$ with PPh_3 leads to high yields of red, crystalline $LWS(S_2PPh_2)$, according to the following equation:

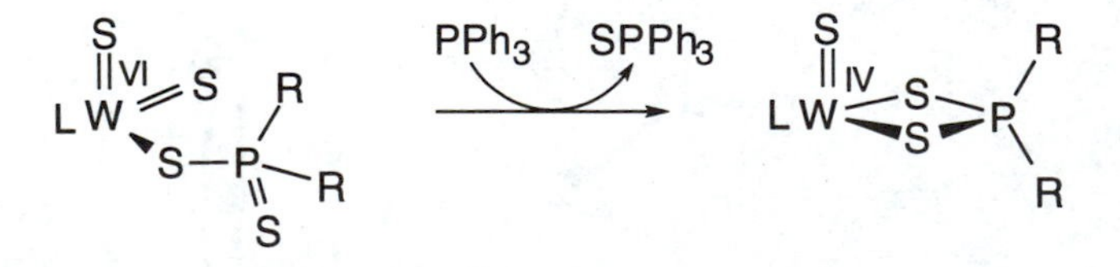

Unexpectedly, reaction of $LWS(S_2PR_2)$ with pyridine *N*-oxide results in only low yields of $LWOS(S_2PR_2)$, whereas reaction with sulfur results in a brown, as yet unidentified complex.

We have already noted that the reaction of cyanide with the $[MoOS]^{2+}$ center of xanthine oxidase and $LMoOS(S_2PR_2)$ results in the abstraction of sulfur from the metal center. In contrast, it has been observed that several tungsten enzymes are not deactivated by cyanide, and this may be interpreted to indicate the absence of thio-tungsten centers in the enzymes. However, we have established that there is no reaction between LWS_2Cl and NEt_4CN in refluxing benzene over 24 hours. The stability of this thio-tungsten complex toward cyanolysis indicates that it is dangerous to dismiss the possible presence of thio-tungsten centers in enzymes on the basis of their failure to react with cyanide.

Reactions Leading to Ene-1,2-dithiolate Complexes. EXAFS of the selenocysteine-containing formate dehydrogenase from *Clostridium thermoaceticum* (CT-FDH) is consistent with an active site devoid of terminal chalcogen ligands, but containing molybdopterin and selenocysteine moieties (I) (*37*). We have sought to model this center by incorporating an L ligand in place of the three facial O- or N-donor ligands, and have sought to develop strategies for the generation of the ene-1,2-dithiolate ligand in particular. We were intrigued by the possibility of converting a *cis*-bis(thio) moiety into an ene-1,2-dithiolate and investigated the reactions of bis(thio) complexes with alkynes.

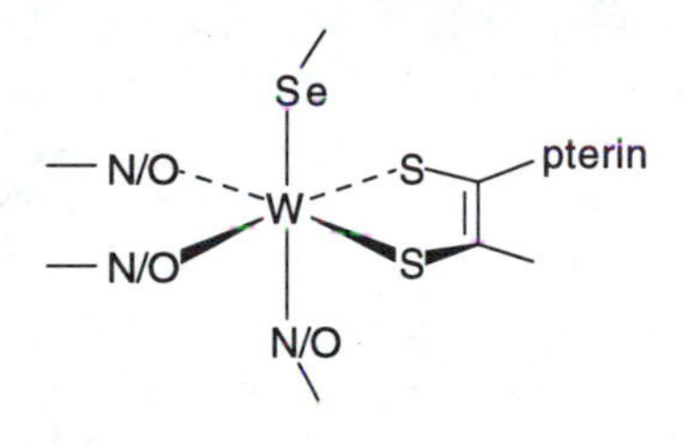

(I)

Indeed, solutions of LWS_2X react with alkynes to form the desired (ene-1,2-dithiolato)tungsten(IV) complexes, e.g., $LWCl(S_2C_2H_2)$, $LWX\{S_2C_2(CO_2Me)_2\}$ (X = Cl, OPh), $LWX\{S_2C_2(Ph)(2\text{-quinoxalinyl})\}$ (X = SPh, SePh), in a new strategy for the synthesis of ene-1,2-dithiolate complexes, viz:

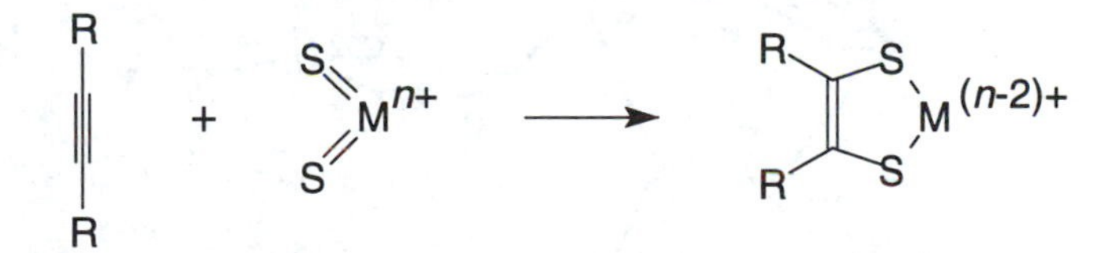

These remarkably facile reactions are very rapid at room temperature even with unactivated alkynes. The structure of orange $LW(OPh)\{S_2C_2(CO_2Me)_2\}$ (Figure 6) confirms the formation of the ene-1,2-dithiolate ligand. The W-S(1) and W-S(2) bond lengths of 2.267(4) and 2.279(4) Å, respectively, are considerably shorter (by 0.07 Å) than any previously reported for a tungsten enedithiolate; the average W-S distance for such compounds is 2.40 Å. The $W\text{-}S_{av}$ distance for CT-FDH is virtually identical to this average at 2.39 Å (*37*). The short W-S distances in $LW(OPh)\{S_2C_2(CO_2Me)_2\}$ are indicative of considerable $d\pi\text{-}p\pi$ bonding between the W and S atoms. The ene-1,2-dithiolate S_2C_4-framework is planar with a mean atom deviation of 0.035 Å and a

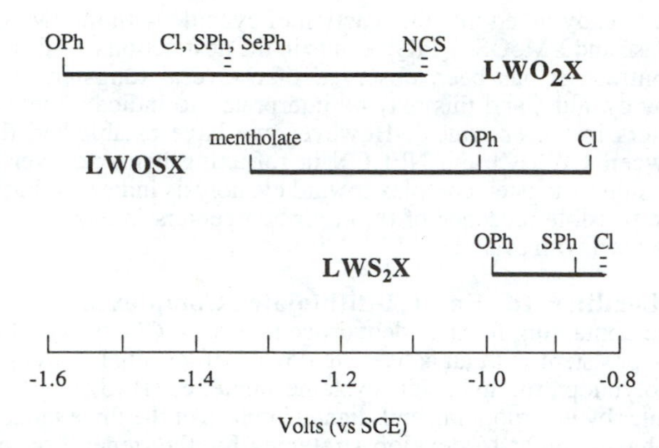

Figure 5. Selected electrochemical data for LWO_2X, LWOSX and LWS_2X complexes. The vertical markers indicate the potential of the W(VI)/W(V) couple for specific complexes (within each series, X is indicated above the markers). Reversible couples are indicated by solid markers and irreversible couples are indicated by dashed markers.

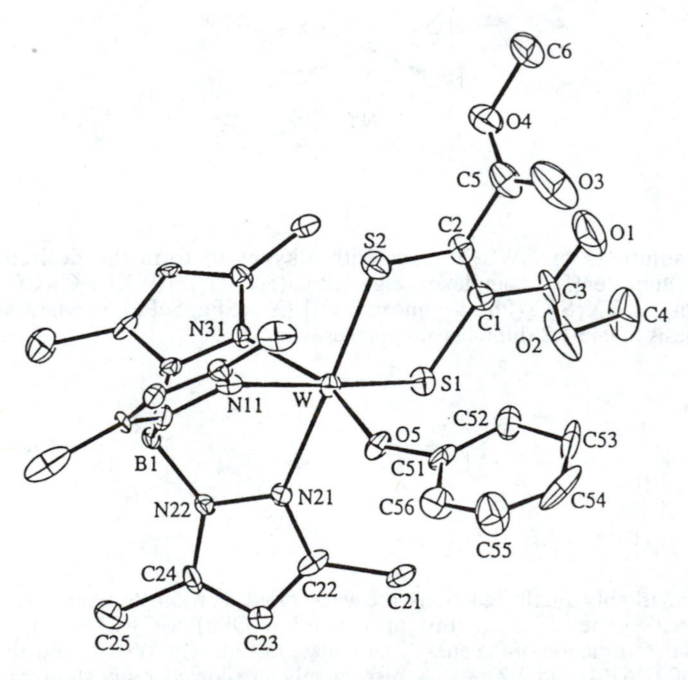

Figure 6. Molecular structure of $LW(OPh)\{S_2C_2(CO_2Me)_2\}$. Selected bond lengths (Å) include the following: W-S(1) 2.267 (4), W-S(2) 2.279 (4), W-O(5) 1.850 (8), W-N(11) 2.19 (1), W-N(21) 2.18 (1), W-N(31) 2.19 (1), S(1)-C(1) 1.77 (2), S(2)-C(2) 1.74 (2), C(1)-C(2) 1.33 (2), O(5)-C(51) 1.34 (2).

C(1)-C(2) distance consistent with a double bond. The dihedral angle between this plane and the W-S(1)-S(2) plane is 8.7°. The W-O(5) distance of 1.850(8) Å and W-O(5)-C(51) angle of 146(1)° are consistent with dπ-pπ bonding between the W and O(5) atoms. Very few mono(enedithiolate) complexes have been reported to date (*21,38*). Similar chemistry was recently reported for the complex $[Cp^*WS_3]^-$ which reacts with alkynes to give $[Cp^*WS(S_2C_2R_2)]^-$ (*21*). The complex LW(SePh){S_2C_2(Ph)(2-quinoxalinyl)} provides a close model and potential structural and spectroscopic benchmark for the active site of CT-FDH. Interestingly, in methanolic solutions it decomposes rapidly in air to give $LWO_2(OMe)$, diphenyldiselenide and (II) (characterized by spectral comparison with an authentic

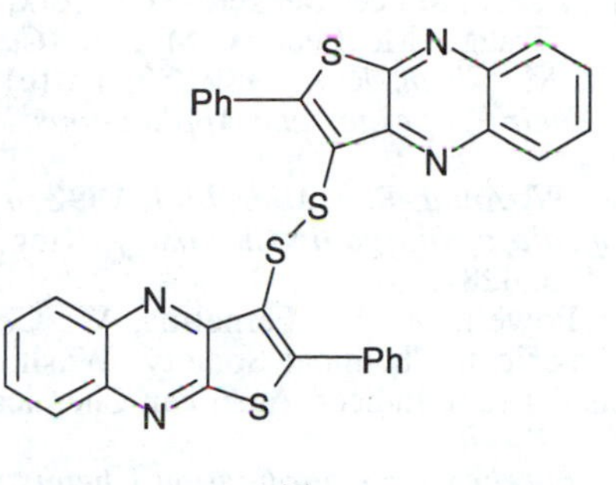

(II)

sample) (*39*). Burgmayer et al. (*39*) reported that (II) is formed in the oxidative decomposition of $[Mo\{S_2C_2(Ph)(2\text{-quinoxalinyl})\}_3]^{2-}$ and related this reaction to the formation of urothione upon degradation of the molybdopterin-containing molybdenum cofactor. On the basis of our isolation of (II) upon oxidation of LW(SePh)-{S_2C_2(Ph)(2-quinoxalinyl)}, the degradation of molybdopterin-containing tungsten enzymes might also be expected to produce urothione.

Conclusion

This paper describes significant developments in thio-tungsten chemistry. High-valent and low-valent strategies exploiting oxo/thio ligand exchange reactions and oxidative and reductive oxygen and sulfur atom transfer reactions have permitted the controlled synthesis of oxo-thio-W(VI), bis(thio)-W(VI), and thio-W(IV) and -W(V) complexes. These have been fully characterized by a plethora of physical and spectroscopic techniques and by X-ray crystallography. The synthesis of derivatives designed especially to avoid lattice disorder has permitted the determination of precise structures. As well, a number of novel tungsten complexes combining electronically disparate carbonyl and chalcogenido ligands, and complexes featuring rare four-electron-donor nitrile ligands have been prepared and characterized. The structures of the complexes are consistent with the mutual electronic requirements of the π-acid and π-base ligands involved in each case. The reactivity of the thio-tungsten complexes has provided further avenues for synthetic manipulations and has provided insights into the behavior of various tungsten enzymes. Our work in this area continues.

Acknowledgments

We wish to gratefully acknowledge the crystallographic expertise of Drs. E. R. T. Tiekink and R. W. Gable, who solved the crystal structures described in this work and most of our cited literature. The financial support of the Australian Research Council is also gratefully acknowledged.

Literature Cited

(1) Müller, A.; Krebs, B. *Sulfur. Its Significance for Chemistry, for the Geo-, Bio- and Cosmosphere and Technology*; Elsevier: Amsterdam, 1984.
(2) (a) Stiefel, E. I.; George, G. N. In *Bioinorganic Chemistry*; Bertini, I., Gray, H. B., Lippard, S. J., Valentine, J. S., Eds.; University Science Books: Mill Valley, CA, 1994; Chap 7, p. 365 and references therein.
(3) (a) Enemark, J. H.; Young, C. G. *Adv. Inorg. Chem.* **1993**, *40*, 1. (b) Young, C. G.; Wedd, A. G. In *Encyclopedia of Inorganic Chemistry*, King, R. B. Ed.; Wiley: New York, 1994, p. 2330. (c) Pilato, R. S.; Stiefel, E. I. In *Inorganic Catalysis*; Reedijk, J., Ed.; Marcel Dekker: New York, 1993, p. 131.
(4) (a) Chianelli, R. R.; Daage, M.; Ledoux, M. *Adv. Catal.* **1994**, *40*, 177. (b) Rakowski DuBois, M. *Chem. Rev.* **1989**, *89*, 1. (c) Weisser, O.; Landa, S *Sulfide Catalysts, Their Properties and Applications*; Pergamon: New York, 1973.
(5) (a) Adams, M. W. W. *Annu. Rev. Microbiol.* **1993**, *47*, 627. (b) Adams, M. W. W. in *Encyclopedia of Inorganic Chemistry*; King, R. B. ed.; Wiley: New York, 1994, Vol. 8, p. 4284.
(6) (a) Block, B. P.; Powell, W. H.; Fernelius, W. C. *Inorganic Chemical Nomenclature*, American Chemical Society, Washington, **1990**, p.41. (b) Chemical Abstracts Guide to Indices, American Chemical Society: Washington, Appendix 4,
(7) Stiefel, E. I. In *Comprehensive Coordination Chemistry*, G. Wilkinson, R. D. Gillard and J. A. McCleverty, Eds.; Pergamon: Oxford, 1987; Chapter 36.5, pp. 1375.
(8) Garner, C. D.; Charnock, J. M. In *Comprehensive Coordination Chemistry*, G. Wilkinson, R. D. Gillard and J. A. McCleverty, Eds.; Pergamon: Oxford, 1987; Chapter 36.4, pp. 1329.
(9) Dori, Z. In *Comprehensive Coordination Chemistry*, G. Wilkinson, R. D. Gillard and J. A. McCleverty, Eds.; Pergamon: Oxford, 1987; Chapter 37, pp. 973.
(10) Müller, A.; Diemann, E.; Jostes, R.; Bögge, H. *Angew. Chem., Int. Ed. Engl.* **1981**, *20*, 934.
(11) (a) Wieghardt, K.; Hahn, M.; Weiss, J.; Swiridoff, W. *Z. Anorg. Allg. Chem.* **1982**, *492*, 164. (b) McDonell, A. C.; Vasudevan, S. G.; O'Connor, M. J.; Wedd, A. G. *Aust. J. Chem.* **1985**, *38*, 1017.
(12) Faller, J. W.; Kucharczyk, R. R.; Ma, Y. *Inorg. Chem.* **1990**, *29*, 1662.
(13) Feinstein-Jaffe, I.; Dewan, J. C.; Schrock, R. R. *Organometallics* **1985**, *4*, 1189.
(14) Chisholm, M. H.; Huffman, J. C.; Pasterczyk, J. W. *Polyhedron* **1987**, *6*, 1551.
(15) Young, C. G.; Kocaba, T. O.; Sadek, M.; Brownlee, R. T. C.; Tiekink, E. R. T. *Aust. J. Chem.* **1994**, *47*, 2075.
(16) Eagle, A. A.; Tiekink, E. R. T.; Young, C. G. *J. Chem. Soc., Chem. Commun.* **1991**, 1746.
(17) Eagle, A. A.; Young, C. G.; Tiekink, E. R. T. *Organometallics* **1992**, *11*, 2934.
(18) Eagle, A. A.; Harben, S. M.; Tiekink, E. R. T.; Young, C. G. *J. Am. Chem. Soc.* **1994**, *116*, 9749.
(19) Potvin, C.; Manoli, J. M.; Marzak, S.; Secheresse, F. *Acta. Crystallogr.* **1988**, C44, 369.
(20) Rabinovich, D.; Parkin, G. *J. Am. Chem. Soc.* **1991**, *113*, 5904.
(21) Tatsumi, *J. Am. Chem. Soc.* **1995**, *117*, 3885.
(22) Trofimenko, S. *J. Am. Chem. Soc.* **1967**, *89*, 6288.

(23) Feng, S. G.; Philipp, C. C.; Gamble, A. S.; White, P. S.; Templeton, J. L. *Organometallics* **1991**, *10*, 3504.
(24) Philipp, C. C.; Young, C. G.; White, P. S.; Templeton, J. L. *Inorg. Chem.* **1993**, *32*, 5437.
(25) Thomas, S.; Tiekink, E. R. T.; Young, C. G. *Inorg. Chem.*, **1994**, *33*, 1416.
(26) Thomas, S.; Tiekink, E. R. T.; Young, C. G. *Organometallics* in press.
(27) Young, C. G.; Bruck, M. A.; Enemark, J. H. *Inorg. Chem.* **1992**, *31*, 593.
(28) Karakida, K.; Fukuyama, T.; Kuchitsu, K. *Bull. Chem. Soc. Jpn.* **1974**, *47*, 299.
(29) (a) Barrera, J.; Sabat, M.; Harman, W. D. *J. Am. Chem. Soc.* **1991**, *113*, 8178. (b) Barrera, J.; Sabat, M.; Harman, W. D. *Organometallics* **1993**, *12*, 4381.
(30) (a) Thomas, J. L. *J. Am. Chem. Soc.* **1975**, *97*, 5943. (b) Bullock, R. M.; Headford, C. E. L.; Kegley, S. E.; Norton, J. R. *J. Am. Chem. Soc.* **1985**, *107*, 727. (c) Wright, T. C.; Wilkinson, G.; Motevalli, M.; Hursthouse, M. B. *J. Chem. Soc., Dalton Trans.* **1986**, 2017. (d) Anderson, S. J.; Wells, F. J.; Wilkinson, G.; Hussain, B.; Hursthouse, M. B. *Polyhedron* **1988**, *7*, 2615. (e) Chetcuti, P. A.; Knobler, C. B.; Hawthorne, M. F. *Organometallics* **1988**, *7*, 650.
(31) (a) Feng, S. G.; Templeton, J. L. *J. Am. Chem. Soc.* **1989**, *111*, 6477. (b) Feng, S. G.; Templeton, J. L. *Organometallics* **1992**, *11*, 1295. (c) Feng, S. G.; White, P. S.; Templeton, J. L. *Organometallics* **1993**, *12*, 1765. (d) Young, C. G.; Philipp, C. C.; White. P. S.; Templeton, J. L. *Inorg. Chem.* **1995**, *34*, 6412.
(32) Orpen, A. G.; Brammer, L.; Allen, F. H.; Kennard, O.; Watson, D. G.; Taylor, R. *J. Chem. Soc., Dalton Trans.* **1989**, S1.
(33) Su, F.-M.; Bryan, J. C.; Jang, S.; Mayer, J. M. *Polyhedron* **1989**, *8*, 1261.
(34) Roberts, S. A.; Young, C. G.; Kipke, C. A.; Cleland, W. E., Jr.; Yamanouchi, K.; Carducci, M. D.; Enemark, J. H. *Inorg. Chem.* **1990**, *29*, 3650.
(35) Eagle, A. A.; Laughlin, L. J.; Young, C. G.; Tiekink, E. R. T. *J. Am. Chem. Soc.* **1992**, *114*, 9195.
(36) Britnell, D.; Fowles, G. W. A.; Rice, D. A. J. *Chem. Soc. Dalt. Trans.* **1975**, 213.
(37) Cramer, S. P.; Liu, C.-L.; Mortenson, L. E.; Spence, J. T.; Liu, S.-M.; Yamamoto, I.; Ljungdahl, L. G. *J. Inorg. Biochem.* **1985**, *23*, 119.
(38) Debaerdemaeker, T.; Kutoglu, A. *Acta. Crystallogr.* **1973**, *B29*, 2664.
(39) Soricelli, C. L.; Szalai, V. A.; Burgmayer, S. J. N. *J. Am. Chem. Soc.* **1991**, *113*, 9877.

Chapter 21

Carbon–Sulfur Bond Cleavage of Thiolates on Electron-Deficient Transition Metals

Kazuyuki Tatsumi and Hiroyuki Kawaguchi

Department of Chemistry, Faculty of Science, Nagoya University, Furo-cho, Chikusa-ku, Nagoya 464–01, Japan

Reactions of Cp^*MCl_4 (M=Ta, Mo, W) with aliphatic thiolates were examined, shedding light on C-S bond activation by electron-deficient early transition metals. From the reactions of Cp^*TaCl_4 with varying amounts of LiS^tBu, $Cp^*Ta(S^tBu)_2Cl_2$ (**1**), $Cp^*Ta(S^tBu)_3Cl$ (**2**), and $Cp^*Ta(S)(S^tBu)_2$ (**3**) were isolated, and C-S bond rupture was found to occur specifically in the reaction between **2** and LiS^tBu. Upon warming a toluene solution of $Cp^*Ta(SCH_2CH_2S)_2$ at 100°C, we isolated $Cp^*Ta(SCH_2CH_2S)\{SCH_2CH_2SCH(CH_3)S\}$ (**9**), while similar treatment of $Cp^*Ta\{SC(CH_3)_2CH_2S\}_2$ gave rise to $Cp^*Ta\{SC(CH_3)_2CH_2S\}\{SSC(CH_3)_2CH_2S\}$ (**12**). The reactions of Cp^*MoCl_4 with LiS^tBu, $Li_2(SCH_2CH_2S)$, and $Li_2(SCH_2CH_2CH_2S)$ yielded the Mo(IV) complexes, $Cp^*Mo(S^tBu)_3$ (**13**), $(PPh_4)[Cp^*Mo(SCH_2CH_2S)_2]$ (**14**), and $(PPh_4)[\{Cp^*Mo(SCH_2CH_2CH_2S)_2\}_3Li_3Cl]$ (**15**), respectively. On the other hand, the analogous reactions of Cp^*WCl_4 with LiS^tBu and $Li_2(SCH_2CH_2S)$ lead to W(VI) sulfide complexes, $Cp^*W(S)_2(S^tBu)$ (**19**)and $Cp^*W(S)_3$ (**20**).

Aliphatic thiolate complexes of early transition metals occasionally undergo C-S bond cleavage reactions (*1-3*), resulting in formation of sulfide and sulfide/thiolate mixed ligand complexes. Christou et al. and Coucouvanis et al. have independently reported that the reaction of $NbCl_5$ with NaS^tBu generates a variety of anionic complexes, $[Nb(S)Cl_4]^-$, $[Nb(S)_2(S^tBu)_2]^-$, and $[Nb(S)(S^tBu)_4]^-$ (*4*). On the other hand, the reaction between $Zr(CH_2Ph)_4$ and 4 equiv of HS^tBu was found to produce a trinuclear sulfide/thiolate cluster, $Zr_3(S)(S^tBu)_{10}$ (*5*). In these intriguing reactions, C-S bond activation occurs concurrently with substitution of chloride or the benzyl group with t-butyl thiolate. For these d^0 transition metal systems, the C-S bond cannot be activated via reductive fission processes. Questions regarding mechanism revolve about which stage of these multi-step reactions the C-S bond is actually broken. Insight into the factors that influence the activation of C-S bonds is of fundamental importance in

0097–6156/96/0653–0336$15.00/0

sulfur-based coordination chemistry and it is germane also to understanding the mechanism of the industrially important hydrodesulfurization processes (*6*). In this context, it is necessary to find systems with which one can trace individual steps of thiolate substitution and C-S bond cleavage. This paper summarizes our recent studies on C-S bond cleavage found in the reactions of Cp^*MCl_4 ($Cp^*=\eta^5\text{-}C_5Me_5$; M=Ta, Mo, W) with aliphatic thiolates.

Reactions of Cp^*TaCl_4 with LiS^tBu

We have previously reported syntheses of pentamethylcyclopentadienyl tantalum dithiolate complexes , $Cp^*Ta(SCH_2CH_2S)_2$, $Cp^*Ta(SCH{=}CHS)_2$, and $Cp^*Ta(ndt)_2$ (ndt = norbornane-exo-2,3-dithiolate) from the reactions of Cp^*TaCl_4 with lithium salts of the corresponding dithiolates (*7*). Only bis-(dithiolate) complexes were isolated even from the reactions between Cp^*TaCl_4 and Li_2(dithiolate) in 1:1 molar ratios, and formation of mono-dithiolate complexes, e.g., $Cp^*TaCl_2(SCH_2CH_2S)$, was not discernible (Scheme 1).

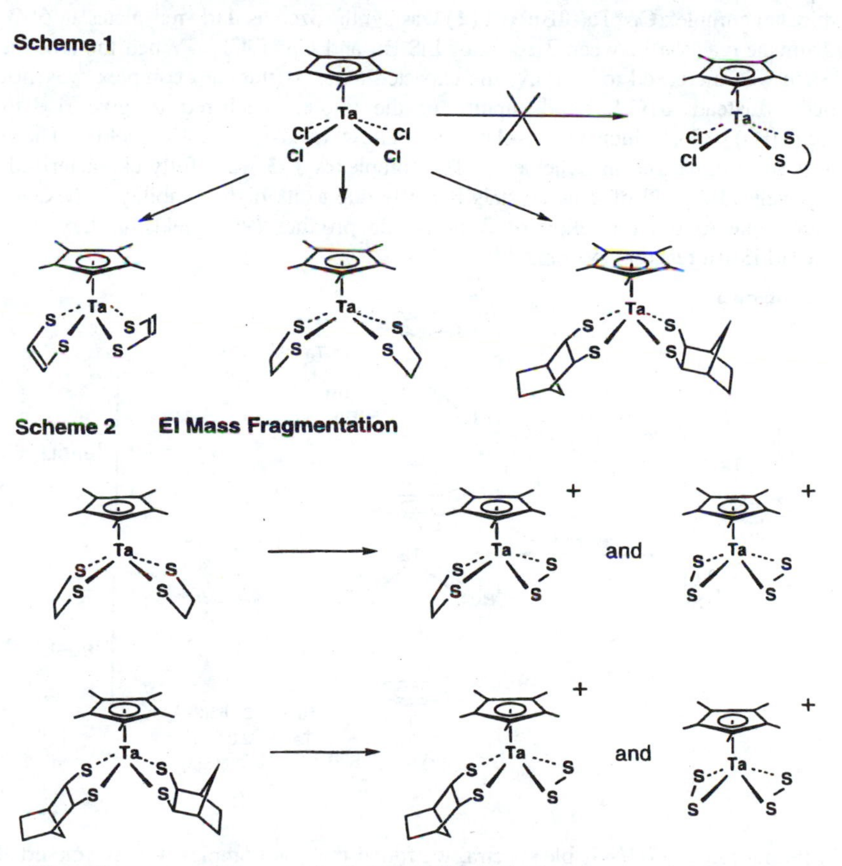

The EI mass spectrum of $Cp^*Ta(SCH_2CH_2S)_2$ shows signals derived from $Cp^*Ta(S_2)(SCH_2CH_2S)^+$ and $Cp^*Ta(S_2)_2^+$ in addition to the parent ion isotopic cluster. No other signals were observed in this mass region. Likewise, the EI fragmentation of $Cp^*Ta(ndt)_2$ consists of $Cp^*Ta(S_2)(ndt)^+$ and $Cp^*Ta(S_2)_2^+$. The stepwise liberation of the alkane portions in the spectra suggest that C-S bonds of these alkanedithiolates are broken cleanly in the EI mass condition (Scheme 2).

In contrast to the synthesis of dithiolate complexes, the reaction of Cp^*TaCl_4 with LiS^tBu resulted in stepwise replacement of chloride in Cp^*TaCl_4 (*8*). When 1 equiv of LiS^tBu in THF was added at 0°C to a THF solution of Cp^*TaCl_4, the color of the solution turned immediately from orange to red-brown and a yellow powder gradually precipitated. The IR spectra of a red-brown powder, which was obtained by quick evaporation of the solvent, indicated the presence of Cp* and the t-butyl group. However, attempts to isolate the mono-thiolate complex $Cp^*Ta(S^tBu)Cl_3$ were unsuccessful, because of instability of the red-brown powder, which decomposes in solution to give an insoluble yellow powder. On the other hand, we were able to isolate the bis-(thiolate) complex, cis-$Cp^*Ta(S^tBu)_2Cl_2$ (**1**) as red needles in 81% yield from a similar treatment of 2 equiv of LiS^tBu with Cp^*TaCl_4 in THF followed by a standard workup and subsequent recrystallization from hexane. Analogously, the tris-(thiolate) complex $Cp^*Ta(S^tBu)_3Cl$ (**2**) was synthesized as dark-red plates in 50% yield from the reaction between 3 equiv of LiS^tBu and Cp^*TaCl_4. When the amount of LiS^tBu was increased to 4 equiv, the expected tetrakis-(thiolate) complex was not obtained. Instead, a C-S bond rupture of the thiolate occurred to give rise to $Cp^*TaS(S^tBu)_2$ (**3**), which was isolated as yellow needles in 54% yield. These reactions are summarized in Scheme 3. The complexes **1**-**3** were fully characterized. The relatively low yield of **2** as crystals is partly due to its high solubility in hexane, and partly due to the formation of **3** as a side product (9% yield) in the (1:3) Cp^*TaCl_4/LiS^tBu reaction system.

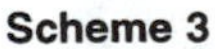

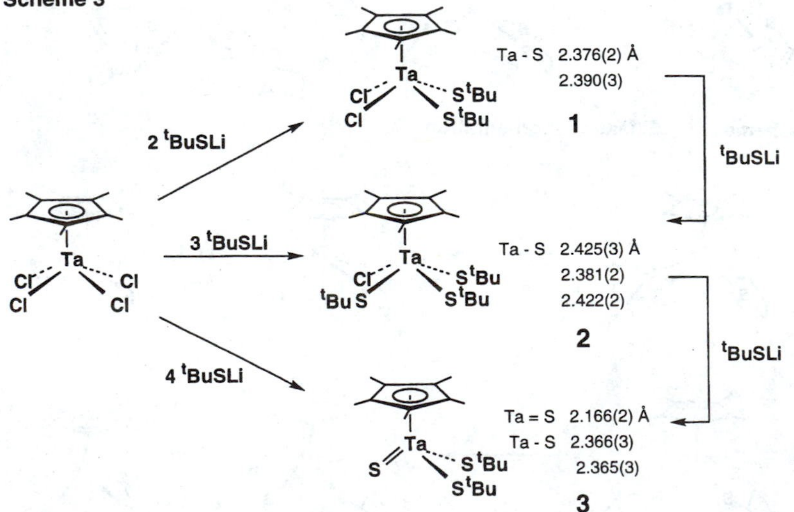

By monitoring the UV-visible spectra, we found that the complex **1** was converted cleanly to **2** when **1** was treated with 1 equiv of LiS^tBu. Transformation of **2** into **3** occurred, again cleanly, upon addition of 1 equiv of LiS^tBu to **2**. Thus, the C-S bond was cleaved specifically in the process where the last chloride of Cp^*TaCl_4 was replaced by t-butylthiolate. Isobutane, $^tBuS^tBu$, and $^tBuSS^tBu$ were detected by a GC-MS study of the byproducts of the latter reaction.

The complex **3** can also be synthesized by treating **1** with Li_2S. On the other hand, the reaction of $Cp^*Ta(S^tBu)_2Cl_2$ (**1**) with 1 equiv of $Li_2(SCH_2CH_2S)$ in THF gave $Cp^*Ta(S^tBu)_2(SCH_2CH_2S)$ (**4**) in 80% yield. In the case of a similar reaction of **1** with 2 equiv of LiSPh, the major product was $Cp^*TaS(StBu)(SPh)$ (**5**), and a small amount of trans-$Cp^*Ta(S^tBu)_2(SPh)_2$ (**6**) was also isolated (Scheme 4). The GC-MS analysis showed that the organic byproducts were ^tBuSPh and $^tBuSS^tBu$. It seems likely that the tetrakis-thiolate complex **6** is the precursor, which, in turn, is converted into **5** and liberates ^tBuSPh.

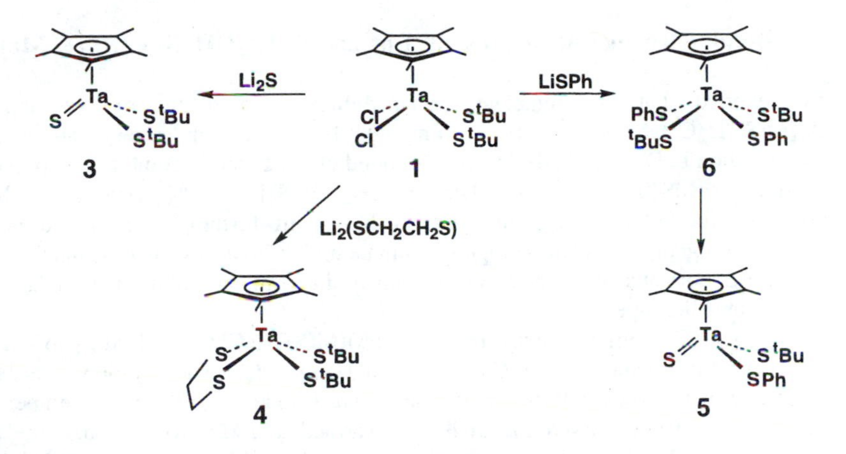

The mechanism of transformation from $Cp^*Ta(S^tBu)_3Cl$ (**2**) to $Cp^*TaS(S^tBu)_2$ (**3**) is not yet certain. Addition of LiStBu to **2** could first lead to $Cp^*Ta(S^tBu)_4$ if the Cp*Ta fragment accommodates four sizable t-butylthiolate ligands, or LiStBu could attack a thiolate ligand of **2** instead of chloride.

Upon warming a toluene solution of **3** at 100℃, the color turned from yellow to dark reddish purple, and a diamagnetic trinuclear cluster $Cp^*_3Ta_3(S_3)(S)_4$ (**7**) was isolated as black crystals in 31% yield. Under these conditions, all the C-S bonds of **3** were broken, and Ta atoms were partially reduced resulting in a formal mixed-valence state of 2Ta(IV) + Ta(V). The two d-electrons reside in the low-lying "a_1" symmetry molecular orbital which is bonding between Ta atoms. The structure of **7** consists of a triangular Ta_3 skeleton with the Ta-Ta distances of 3.092(1)-3.154(1)Å, and a Cp ring caps Ta at each corner. Three sulfur atoms bridge Ta centers below the edges of the Ta_3 triangular plane, and a μ-S and a doubly-bridging S_3 unit coordinate from the other side of the plane (Figure 1).

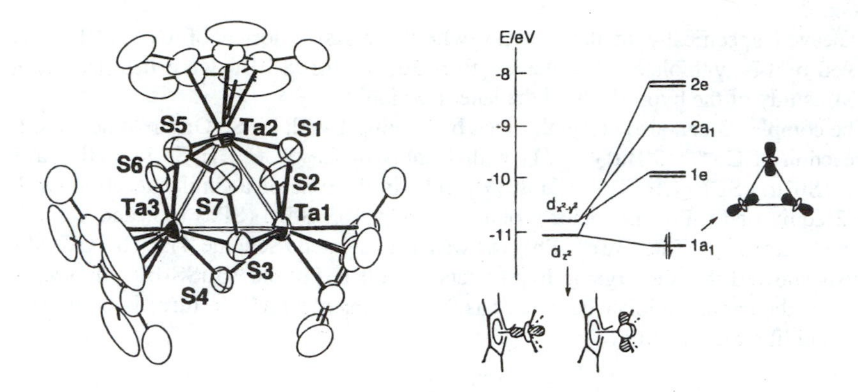

Figure 1. Structure of **7** and the low-lying "a_1" orbital.

C-S Bond Cleavage Reactions of $Cp^*Ta\{SC(R)_2CH_2S\}_2$ (R=H, Me)

We reported that the homoleptic ethanedithiolate complexes of Nb and Ta, $(A)[M(SCH_2CH_2S)_3]$ (A=NEt_4, PPh_4; M=Nb, Ta) underwent an unusual isomerization in DMF at 70-100℃ via C-S bond cleavage and recombination processes to give $(A)[MS(SCH_2CH_2S)(SCH_2CH_2SCH_2CH_2S)]$ in 70% yield (*9*). In its structure, six sulfur atoms surround an Nb atom forming a nearly octahedral coordination geometry. This complex is unique in that it carries three distinctive types of sulfur donors at a single metal center; namely the terminal sulfide, the thiolate type, and the thioether type.

The facile C-S bond cleavage found for $(A)[M(SCH_2CH_2S)_3]$ lead us to examine reactivity of ethanedithiolate in $Cp^*Ta(SCH_2CH_2S)_2$ (**8**). The complex **8** is highly moisture sensitive in a solution, but remains intact under argon at room temperature. However, as a toluene solution of **8** was warmed at 100°C for 6 hours, the color turned from red to brownish red, from which we have isolated $Cp^*Ta(SCH_2CH_2S)\{SCH_2CH_2SCH(CH_3)S\}$ (**9**) as yellow crystals in 60% yield (*10*). The four thiolate type sulfurs are bound to Ta from lateral sites of the Cp*Ta fragment, while the thioether-type sulfur of 1-methyl-3-thiabutane-1,4-dithiolate coordinates from the bottom. Also obtained was an intriguing tetranuclear sulfide/oxide cluster, $Cp^*_4Ta_4(\mu^4\text{-}O)(\mu^3\text{-}S)(\mu\text{-}S)_6$ (**10**), as reddish brown crystals in 5% yield (Scheme 5). The oxygen atom sits at the center of the molecule, and binds four tantalum atoms with a tetrahedral geometry. One sulfur caps a triangle face of the tetrahedral Ta_4 skeleton, so that the molecule has an pseudo-3-fold axis.

The conversion of **8** to **9** obviously involves a C-S bond cleavage of ethanedithiolate and then a recombination process. However, in contrast to the isomerization reaction of $(A)[M(SCH_2CH_2S)_3]$ into $(A)[MS(SCH_2CH_2S)(SCH_2CH_2SCH_2CH_2S)]$, 1,2-hydride shift must take place before the C-S bond recombination step. A plausible mechanism for the transformation of **8** to **9** is shown in Scheme 6. The occurrence of the 1,2-hydride shift suggests formation of a carbocation through heterolytic C-S bond cleavage. The steric congestion caused by the Cp* ligand might prevent an attack of a thiolate sulfur at the cation center, thus

Scheme 5

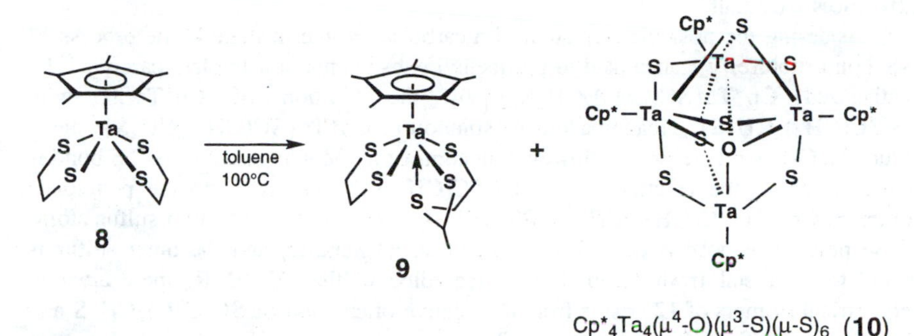

providing a vicinal hydrogen with sufficient time for 1,2-migration. The subsequent disproportion process, leading to **9** and **10**, seems to proceed relatively cleanly, because an NMR experiment on this reaction gives nearly exclusively Cp* proton signals associated with these complexes. Although we used pre-dried solvents for the synthesis, perhaps a trace amount of water trapped a very reactive sulfide intermediate to give the sulfide/oxide cluster **10**.

Scheme 6

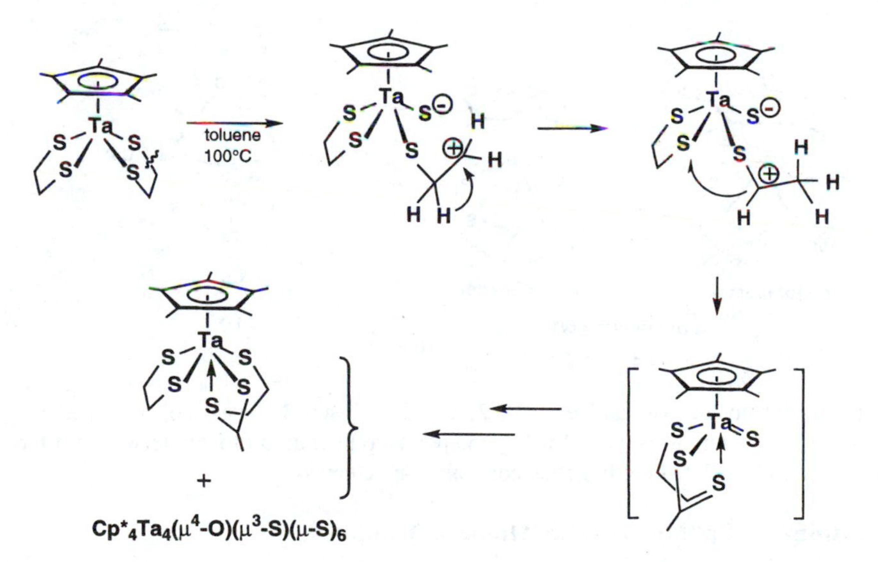

It should be noted here that while $Cp^*Ta(SCH_2CH_2S)_2$ (**8**) is moisture sensitive, and undergoes isomerization to **9** at 100°C in dry toluene, the propanedithiolate analogue $Cp^*Ta(SCH_2CH_2CH_2S)_2$ (**11**) is very stable. The complex **11** survives in wet solvents for days, and does not show a sign of C-S bond rupture in hot toluene. In spite of the clear difference in stability between **8** and **11**, their crystal structures are very much alike. The average Ta-S-C bond angle and the S-Ta-S bite angle are both smaller for **8** (Ta-S-C; 104.5° (**8**), 107.8° (**11**): S-Ta-S; 82.3° (**8**), 85.4° (**11**)), but the differences are small.

Considering the possible formation of a carbocation intermediate in the process of C-S bond rupture, we attempted to put methyl substituents at a thiolate carbon. We synthesized $Cp^*Ta\{SC(CH_3)_2CH_2S\}_2$ by the reaction of Cp^*TaCl_4 with $Li_2(SC(CH_3)_2CH_2S)$. Heating a toluene solution of $Cp^*Ta\{SC(CH_3)_2CH_2S\}_2$ again induced a C-S bond cleavage. However, in contrast to the way in which a C-S bond is activated for **8**, the reaction of $Cp^*Ta\{SC(CH_3)_2CH_2S\}_2$ leads to the perthiolate complex, $Cp^*Ta\{SC(CH_3)_2CH_2S\}\{SSC(CH_3)_2CH_2S\}$ (**12**). The two sulfur atoms of the perthiolate portion coordinate at Ta in an η^2 manner, and the inner sulfur is bound to the metal from the bottom. According to the ^1H NMR, there are two geometrical isomers of **12** arising from the relative orientation of $SC(CH_3)_2CH_2S$ and $SSC(CH_3)_2CH_2S$ as shown in Scheme 7.

Scheme 7

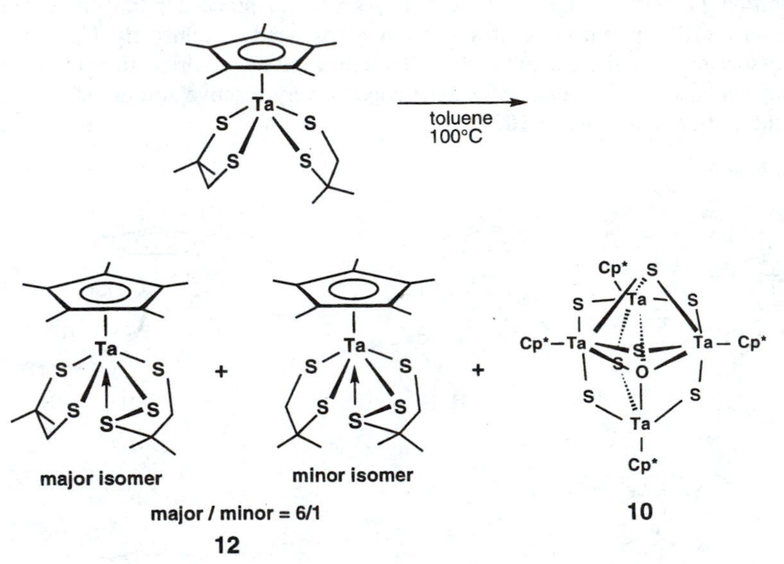

In addition to the two isomers of **12**, the Ta_4 cluster **10** was isolated as a side product. Formation of the persulfide ligand points to liberation of isobutene so that the two C-S bonds of 1,1-dimethyl ethanedithiolate are cleaved.

Reactions of Cp^*MoCl_4 with Aliphatic Thiolates

Addition of a THF solution of 4 equiv of LiS^tBu to a suspension of Cp^*MoCl_4 in THF resulted in a homogeneous red solution. Evaporation of the solvent and recrystallization of the residual solid from hexane gave $Cp^*Mo(S^tBu)_3$ (**13**) as dark red plates in 85% yield. The reaction of Cp^*MoCl_4 with 2.5 equiv of $Li_2(SCH_2CH_2S)$ in THF followed by cation exchange with PPh_4Br in CH_3CN generated (PPh_4) $[Cp^*Mo(SCH_2CH_2S)_2]$ (**14**) as dark red crystals in 66% yield. From the analogous $Cp^*MoCl_4/Li_2(SCH_2CH_2CH_2S)$ reaction system in THF, a dark red crystalline solid formulated as $Li_4(thf)_n\{Cp^*Mo(SCH_2CH_2CH_2S)_2\}_3Cl$ was isolated. Interestingly, only one lithium cation was replaced by cation exchange with PPh_4Br in CH_3CN giving rise to $(PPh_4)[\{Cp^*Mo(SCH_2CH_2CH_2S)_2\}_3Li_3Cl]$ (**15**). The structure of the anion part of **15** consists of three $Cp^*Mo(SCH_2CH_2CH_2S)_2$ units linked by three lithium cations at sulfurs of the thiolates, and the lithium cations are bound to the Cl anion which sits at the center of the cluster molecule (Scheme 8).

Scheme 8

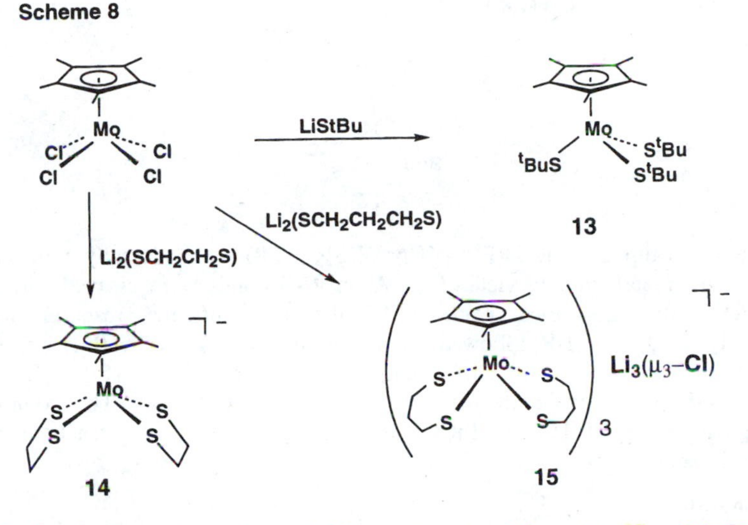

In the above reactions, the molybdenum is reduced from Mo(V) to Mo(IV), and no C-S bond activation was observed. However, $Cp^*Mo(S^tBu)_3$ (**13**) was found to react with H_2 (1 atm) cleanly in hexane generating $\{Cp^*Mo(S^tBu)_2\}(\mu\text{-}S)$ (**16**) in 92% yield (Scheme 9). Under H_2 atmosphere, the oxidation state of molybdenum remains Mo(IV) even though a C-S bond was cleaved. The Mo-S-Mo bond is linear indicating the presence of π interactions between Mo and S. On the other hand, treatment of **13** with $PhHNNH_2$ or $N(CH_3)_2NH_2$ yielded a mononuclear Mo(VI) sulfide complex; $Cp^*Mo(S)_2(S^tBu)$ (**17**) again in high yield. Also, the complex **13** reacts cleanly with O_2 to give $Cp^*Mo(O)_2(S^tBu)$ (**18**).

Reactions of Cp^*WCl_4 with Aliphatic Thiolates

The reactions of Cp^*WCl_4 with LiS^tBu and $Li_2(SCH_2CH_2S)$ are very different from those of Cp^*MoCl_4. Addition of a THF solution of LiStBu (4 equiv) to an orange suspension of Cp^*WCl_4 in THF at room temperature gave a red solution. Removal of the solvent and extraction of the resulting residue by toluene gave $Cp^*W(S)_2(S^tBu)$ (**19**) as red crystals in 68% yield. During this reaction, tungsten was oxidized from W(V) to W(VI), and the oxidation was induced by the C-S bond cleavage of t-butylthiolate. Treatment of Cp^*WCl_4 with 2 equiv of $Li_2(SCH_2CH_2S)$ in THF at 0°C gave at first a red suspension, which dissolved to produce an orange-red homogeneous solution in seconds. The UV-visible spectrum of this solution indicated formation of $[Cp^*W(S)_3]^-$. In fact, subsequent cation exchange with PPh_4Br in CH_3CN followed

Scheme 9

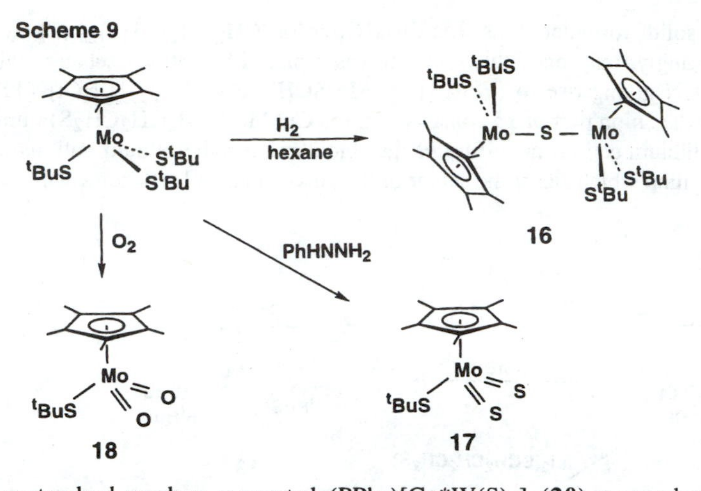

by a standard workup generated $(PPh_4)[Cp^*W(S)_3]$ (**20**) as moderately moisture-sensitive orange crystals in 82% yield (*11*). Again C-S bonds were cleaved and the W(V)-to-W(VI) oxidation took place. In the case of the reaction with $Li_2(SCH_2CH_2CH_2S)$ in THF followed by a cation exchange with Ph_4PBr, the situation is similar to that of Cp^*MoCl_4. The product is $(PPh_4)[Cp^*W(SCH_2CH_2CH_2S)_2]$ (**21**), where the metal center is reduced to W(IV). This complex is the tungsten analog of **15**, but LiCl is not incorporated. These reactions are summarized in Scheme 10.

Scheme 10

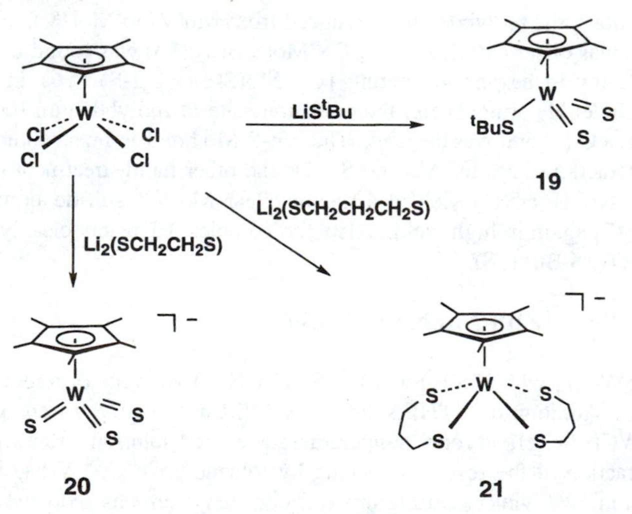

The X-ray structure of the anion of **20** is practically the same as that of $(Et_3NH)[Cp^*W(S)_3]$ reported by G. L. Geoffroy et al (*12*). The latter complex was isolated in low yield from the $Cp^*WCl_4/H_2S/Et_3N$ reaction system, which also produced a mixture of $Cp^*{}_2W_2(S)_2(\mu\text{-}S)_2$, $Cp^*{}_2W_2(S)_2(\mu\text{-}S_2)$, and (Et_3NH) $[Cp^*W(S)_2O]$. The difference in reactivity between Cp^*MoCl_4 and Cp^*WCl_4 may arise from the tendency of tungsten to favor higher oxidation states than molybdenum. The unexpected formation of **20** turned out to be a novel, convenient route to the trithio tungsten complex.

The facile high-yield synthesis of **20** allowed us to examine the reactivity of the terminal sulfides on W(VI), as summarized in Scheme 11. The reactions of **20** with 1 equiv of $PhCH_2Br$, ^tBuBr, or CH_3I in CH_3CN are straightforward, generating the neutral dithio/thiolato W(VI) complexes $Cp^*W(S)_2(SR)$ (R = CH_2Ph (**22**), ^tBu (**19**), CH_3 (**23**)) as red crystals in 91% and 83% yield, respectively. The remaining two W=S bonds do not react with excess alkyl halide. The complex **22** assumes a normal three-legged piano-stool geometry in which the thiolate S-C bond orients parallel to the Cp*(centroid)-W vector, thus allowing the occupied sulfur p orbital (p_y) to interact with the vacant W d_{xy}-p_y hybrid. The W=S bond distances (2.149(3)Å) are shorter by 0.043Å than the mean W=S distance of **20** (2.192Å), which is theoretically understandable. The IR bands of **22** arising from the W=S stretching vibrations appear at 480cm^{-1} and 493cm^{-1} (481cm^{-1} and 494cm^{-1} for **19**), and they are clearly shifted to higher energies relative to the corresponding IR bands of **20** (437cm^{-1} and 466cm^{-1}).

Scheme 11

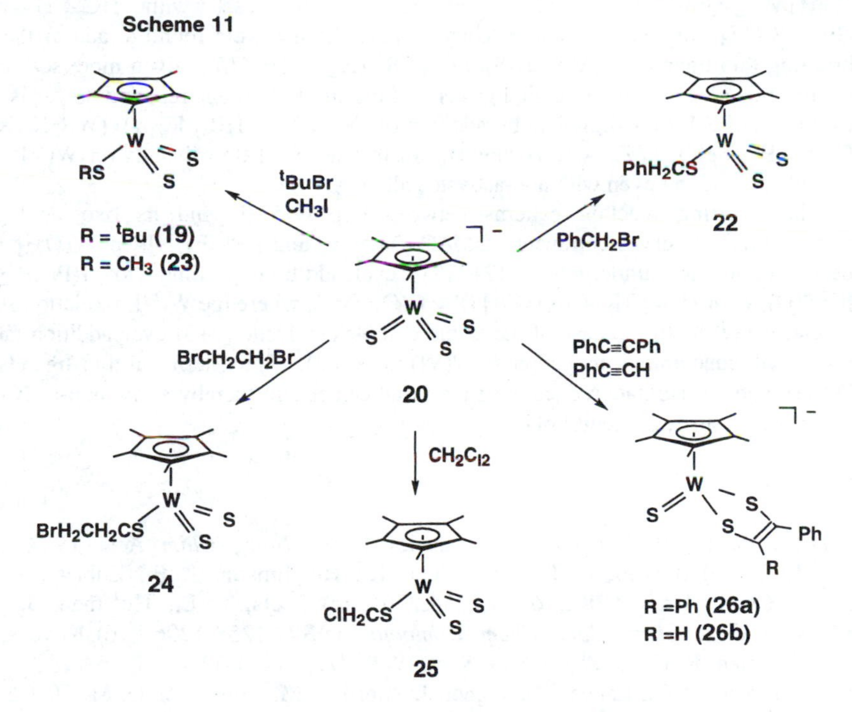

1,2-dibromoethane also reacts cleanly with **20** in CH_3CN to give $Cp^*W(S)_2$-(SCH_2CH_2Br) (**24**). The trithio tungsten complex (**20**) is capable of activating even

CH_2Cl_2. Upon standing, a CH_2Cl_2 solution of **20** slowly formed $Cp^*W(S)_2$-(SCH_2Cl) (**25**). Successful introduction of haloalkyl groups at a terminal sulfide of **20** will allow us to modify thiolate substituents extensively.

The most intriguing reaction that we have found for **20** is formation of thio/1,2-enedithiolate complexes upon treating **20** with 5 equiv of PhC≡CPh or PhC≡CH in CH_3CN. The reaction with PhC≡CPh is complete at room temperature in 1 day, and $(PPh_4)[Cp^*W(S)(S_2C_2Ph_2)]$ (**26a**) was isolated as yellow-green crystals in 76% yield. The reaction with PhC≡CH is faster, and is complete in several hours, giving $(PPh_4)[Cp^*W(S)(S_2C_2PhH)]$ (**26b**) in 80% yield. In the structure of the anion of **26a**, coordination of the 1, 2-enedithiolate ligand occurs with a nearly planar WS_2C_2 framework. The W-S distances of 2.318(2) and 2.334(2)Å fall in the normal range for W(IV)-S bond lengths, and are comparable to the W-S length of **22**. The W=S distance (2.186(2)Å) of **26a**, on the other hand, is longer by 0.037Å than those of **22**, and this lengthening is consistent with the lower W=S stretching frequency (465 cm^{-1}) observed for **26a** relative to **22**. The difference in oxidation state between these thio complexes, W(IV) *vs.* W(VI), is probably one reason behind the trend.

Reaction of alkynes with polysulfides has occasionally been observed, though activated alkynes such as MeO(O)CC≡CC(O)OMe (DMAC) and $CF_3C{\equiv}CCF_3$ are used in most cases (*13*). The direct addition of an alkyne to bridging monothio ligands is more unusual. The μ-S ligands in $(C_5H_5)_2Mo_2(\mu\text{-}S)_2(\mu\text{-}SH)_2$ (*14*) and $(C_5H_4Me)_2V_2(\mu\text{-}S)_2(\mu\text{-}S_2)$ (*15*) were reported to react with HC≡CH and $CF_3C{\equiv}CCF_3$, respectively, and recently various alkynes were found to add to the μ-bridging thio ligands in $[Mo_3(\mu_3\text{-}S)(\mu\text{-}O)(\mu\text{-}S)_2(H_2O)_9]^{4+}$ (*16*). Even more scarce is the reaction of terminal monothio ligands, and the single example reported so far is the formation of a 1,2-enedithiolate by addition of DMAC to $\{HB(Me_2pz)_3\}WS_2X$ (X = OPh, SPh, SePh) (*17*). Our finding shows that terminal thio ligands on W(VI) are capable of reacting even with non-activated alkynes.

The differing reaction patterns between $[Cp^*W(S)_3]^-$ and its oxo analogue $[Cp^*W(O)_3]^-$ deserves comment. DMAC (2 equiv) and $[(Ph_3P)_2N][Cp^*W(O)_3]$ has been found to undergo a [2+2+2] cycloaddition leading to $[(Ph_3P)_2N]$ $[Cp^*(O)_2WOC(R){=}C(R)\text{-}C(R){=}CR]$ (R = C(O)OMe), where the W(VI) oxidation state is retained (*12*). In the case of the trithio complex, a facile [2+3] cycloaddition takes place with concomitant reduction of W(VI) to W(IV). Thus, terminal thio ligands at W(VI) seem to facilitate reduction of the metal center and thereby allow non-activated alkynes to form 1, 2-enedithiolates.

References

(1) (a) Boorman, P.M.; Chiveri, T.; Mahadev, K. N. *Inorg. Chim. Acta* **1976**, *19*, L35-L37. (b) Blower, P. J.; Dilworth, J. R.; Hutchinson, J. P.; Zubieta, J. A. *Inorg. Chim. Acta* **1982**, *65*, L225-L226. (c) Seela, J. L.; Huffman, J. C.; Christou, G. *J. Chem. Soc., Chem. Commun.* **1987**, 1258-1206. (d) Kovacs, J. A.; Bergman, R. G. *J. Am. Chem. Soc.* **1989**, *111*, 1131-1133.

(2) (a) Levert, P. C.; Lance, M.; Vigner, J.; Nierlich, M.; Ephritikhine, M. *J. Chem. Soc., Dalton Trans.* **1995**, 237-244. (b) Skripkin, Y. V.; Eremenko, I. L.; Pasynskii, A. A.; Struchkov, Y.; Shklover, V. E. *J. Organomet. Chem.* **1984**, *267*,

285-292. (c) Piers, W. E.; Koch, L.; Ridge, D. S.; MacGillivray, L. R.; Zaworotko, M. *Organometallics* **1992**, *11*, 3148-3152.
(3) Blower, P. J.; Dilworth, J. R. *Coord Chem. Rev.* **1987**, **76**, 121-185, and references therein.
(4) (a) Seela, J. L.; Huffman, J. C.; Christou, G. *Polyhedron* **1989**, 1797-1799. (b) Coucouvanis, D.; Al-Ahmad, S.; Kim, C. G.; Koo, S.-M. *Inorg. Chem.* **1992**, *31*, 196-2998. (c) Coucouvanis, D.; Chen, S. -J.; Mandimutsira, B. S.; Kim, C. G. *Inorg. Chem.* **1994**, *33*, 4429-4430.
(5) (a) Coucouvanis, D.; Hadjikyriacou, A.; Kanatzidis, M. G. *J. Chem. Soc., Chem. Commun.* **1985**, 1224. (b) Coucouvanis, D.; Hadjikyriacou, A.; Lester, R.; Kanatzidis, M. G. *Inorg. Chem.* **1994**, *33*, 3645-3655.
(6) (a) Angelici, R. J. *Acc. Chem. Res.* **1988**, *21*, 387. (b) Angelici, R. J. *Coord. Chem. Rev.* **1990**, *105*, 61. (c) Rauchfuss, T. B. *Prog. Inorg. Chem.* **1991**, *39*, 259. (d) Wiegand, B. C.; Friend, C. M. *Chem. Rev.* **1992**, *92*, 491. (e) Bianchini, C.; Herrera, V.; Jimenez, M. V.; Laschi, F.; Meli, A.; Sanchez-Delgado, R.; Vizza, F.; Zanello, P. *Organometallics* **1995**, *14*, 4390-4401.
(7) (a) Tatsumi, K.; Takeda, J.; Sekiguchi, Y.; Kohsaka, M.; Nakamura, A. *Angew. Chem., Int. Ed. Engl.* **1985**, *24*, 332-333. (b) Tatsumi, K.; Inoue, Y.; Nakamura, A. *J. Organomet. Chem.* **1993**, *444*, C25-C27.
(8) (a) Tatsumi, K.; Tahara, A.; Nakamura, A. *J. Organomet. Chem.* **1994**, *471*, 111-115. (b) Kawaguchi, H.; Tatsumi, K.; Tahara, A.; Tani, K. *to be published.*
(9) (a) Tatsumi, K.; Sekiguchi, Y.; Nakamura, A.; Cramer, R. E.; Rupp, J. J. *Angew. Chem., Int. Ed. Engl.* **1986**, *25*, 86-86. (b) Tatsumi, K.; Matsubara, I.; Sekiguchi, Y.; Nakamura, A.; Mealli, C. *Inorg. Chem.* **1989**, *28*, 773-780. (c) Tatsumi, K.; Sekiguchi, Y; Nakamura, A; Cramer, R. E.; Rupp, J. J. *J. Am. Chem. Soc.* **1986**, *108,* 1358-1359.
(10) Kawaguchi, H.; Tatsumi, K. *to be published.*
(11) Kawaguchi, H.; Tatsumi, K. *J. Am. Chem. Soc.* **1995**, *117,* 3885-3886.
(12) Rau, M. S.; Kretz, C. M.; Geoffroy, G. L.; Rheingold, A. L. *Organometallics* **1993**, *12*, 3447-3460.
(13) (a) Coucouvanis, D.; Toupadakis, A.; Lane, J. D.; Koo, S. M.; Kim, C. G.; Hadjikyriacou, A. *J. Am. Chem. Soc.* **1991**, *113,* 5271-5282. (b) Pilato, R. S.; Eriksen, K. A.; Greaney, M. A.; Stiefel, E. I.; Goswami, S.; Kilpatrick, L.; Spiro, T. G.; Taylor, E. C.; Rheingold, A. L. *J. Am. Chem. Soc.* **1991**, *113,* 9372-9374, and references therein.
(14) (a) Weberg, R.; Haltiwanger, R.C.; Rakowski DuBois, M. *Organometallics* **1985**, *4*, 1315-1318. (b) Rakowski DuBois, M.; VanDerveer, M. C.; Dubois, D. L.; Haltiwanger, R. C.; Miller, W. K. *J. Am. Chem. Soc.* **1980**, *102,* 7456-7461.
(15) Bolinger, C. M.; Rauchfuss, T. B.; Rheingold, A. L. *J. Am. Chem. Soc.* **1983**, *105,* 6321-6323.
(16) Shibahara, T.; Sakane, G.; Mochida, S. *J. Am. Chem. Soc.* **1993**, *115,* 10408-10409.
(17) Eagle, A. A.; Harben, S. M.; Tiekink, E. R. T.; Young, C. G. *J. Am. Chem. Soc.* **1994**, *116,* 9749-9750.

Author Index

Affiliation Index

Subject Index

F

G

H

R

S